Lecture Notes in Applied and Computational Mechanics

Volume 11

Series Editors

Prof. Dr.-Ing. Friedrich Pfeiffer
Prof. Dr.-Ing. Peter Wriggers

Springer

Berlin
Heidelberg
New York
Hong Kong
London
Milan
Paris
Tokyo

Dynamic Response of Granular and Porous Materials under Large and Catastrophic Deformations

Kolumban Hutter
Nina Kirchner (Eds.)

Springer

Professor Dr. Kolumban Hutter
Dr. Nina Kirchner
Technical University of Darmstadt
Department of Mechanics
Hochschulstr. 1
D-64289 Darmstadt
GERMANY

e-mail: hutter@mechanik.tu-darmstadt.de
e-mail: kirchner@mechanik.tu-darmstadt.de

With 150 Figures

Cataloging-in-Publication Data applied for
Bibliographic information published by Die Deutsche Bibliothek
Die Deutsche Bibliothek lists this publication in the Deutsche Nationalbibliografie; detailed bibliographic data is available in the Internet at <http://dnb.ddb.de>

ISBN 3-540-00849-7 Springer Verlag Berlin Heidelberg New York

Springer-Verlag Berlin Heidelberg New York
a member of BertelsmannSpringer Science+Business Media GmbH

http://www.springer.de

Printed in Germany

Cover design: design & production GmbH, Heidelberg
Typesetting: Digital data supplied by author

Printed on acid-free paper 62/3020Rw-5 4 3 2 1 0

Preface

A "Sonderforschungsbereich" (SFB) is a programme of the "Deutsche Forschungsgemeinschaft" to financially support a concentrated research effort of a number of scientists located principally at one University, Research Laboratory or a number of these situated in close proximity to one another so that active interaction among individual scientists is easily possible.

Such SFB are devoted to a topic, in our case "*Deformation and Failure in Metallic and Granular Materials*", and financing is based on a peer reviewed proposal for three (now four) years with the intention of several prolongations after evaluation of intermediate progress and continuation reports. An SFB is terminated in general by a formal workshop, in which the state of the art of the achieved results is presented in oral or/and poster communications to which also guests are invited with whom the individual project investigators may have collaborated. Moreover, a research report in book form is produced in which a number of articles from these lectures are selected and collected, which present those research results that withstood a rigorous reviewing process (with generally two or three referees).

The theme *deformation and failure of materials* is presented here in two volumes of the *Lecture Notes in Applied and Computational Mechanics* by Springer Verlag, and the present volume is devoted to granular and porous continua. The complementary volume (*Lecture Notes in Applied and Computational Mechanics*, vol. 10, Eds. K. HUTTER & H. BAASER) is dedicated to the *Deformation and Failure in Metallic Materials.*

The SFB "Deformation and Failure in Metallic and Granular Materials" lasted from October 1994 until December 2002, thus a total of slightly more than eight years, and had an interdisciplinary focus: Teachers, researchers, including Ph. D. students from various University Departments were involved, namely Mathematics, Mechanics, Material Sciences, Civil and Mechanical Engineering. Many projects were headed by researchers from two different departments. Each project had one – sometimes two – principal researchers who either as Ph. D. students or Postdoctoral Assistants would perform the actual research under the supervision of the proposers of the individual proposals.

This volume tries to summarise the obtained results not so much in a form as it would appear in specialised peer reviewed periodicals, but such that a broader community is able to follow the red lines of the arguments in each article. This required that authors were asked to include also items that may be obvious to specialists. We hope that this endeavour has been achieved.

This book on the *Dynamic Response of Granular and Porous Materials under Large and Catastrophic Deformations* is divided into three major parts,

(I) debris and mud flows,
(II) porous and granular materials, fundamentals and dynamical processes,
(III) porous and granular materials, sub-scale and micro-mechanical effects.

These three topics define the main classes into which the 14 different articles may fit. They reflect a fairly objective cross section of the research that was conducted during the eight years which this SFB lasted.

Part I. Avalanches, Debris and Mud Flows

This largest part of the book consists of eight articles which all are concerned with the rapid shear flow of a finite mass of granular materials down an inclined plane, a chute or an arbitrary topography. For such flows SAVAGE & HUTTER developed in the late eighties a simple model for the flow in a confined straight and inclined chute, known now as the SAVAGE–HUTTER (SH) theory. This model has attracted considerable interest, and it has obviously also been a centre of activity in this SFB.

- GRAY summarises the extensions the SH-model encountered in two-dimensional confined flows and emphasises in particular the possibility of shock formation and the associated numerical peculiarities that go along with it.
- PUDASAINI, HUTTER & ECKART demonstrate how the original SH-model can be generalised to situations for the flow of a cohesionless granular mass down specially curved and twisted topographies and demonstrate that the emerging model equations derived from a thickness averaging of the balance laws of mass and momentum of an incompressible Coulomb fluid enjoy a very similar hyperbolic structure as did the equations of the original SH-model.
- The article of KOSCHDON & SCHÄFER deals with the numerical integration of such hyperbolic equations using an arbitrary Eulerian–Lagrangian finite-volume method paired with a Riemann integration scheme.
- ISSLER reviews experiments conducted primarily in the field of dry and wet snow avalanches and puts inferences in a balanced context with actual and possible theoretical concepts, pointing also at processes such as snow entrainment much ignored in the available literature. His concluding section presents a long list of pressing questions not sufficiently known to date in avalanche research.
- ZWINGER, KLUWICK & SAMPL are concerned with the more realistic situation; namely, snow avalanches often consist of a dense granular lower layer overlain by a turbulent two-phase flow of snow particles and air. The authors derive for the dense particle layer a SH-type model and connect this at the interface with the powder snow avalanche layer which

is described as a turbulent binary mixture involving $k-\varepsilon$ turbulent closure through suitably formulated transition conditions. The authors test their numerical implementation of the model with data from a real field event.

- The article by ECKART, GRAY & HUTTER touches upon one fundamental assumption of the SH-model. This is the postulate of a dry Mohr–Coulomb friction law that is assumed to hold at the interface between the solid base and moving granular array that rapidly flows over it. Using particle image velocimetry techniques, surface and basal velocities of a granular layer moving down an inclined plane are measured for various different inclination angles of the plane and different lengths of the plane. The experiments seem to suggest that the Mohr–Coulomb basal sliding assumption is not unanimously valid; a viscous sliding component, increasingly important for small inclination angles, appears to be needed to account for the possibility to reach a steady asymptotic velocity far downstream of the inclined plane.

The spirit of emphasis is different in the following two contributions:

- FARWIG is concerned with several models of avalanching flows of granular materials with and without interstitial fluid, including the SH-model. Existence proofs of spatially one-dimensional initial boundary value problems are sketched. Distinguished feature is that the mathematical methods used are elementary and do not go beyond the application of simple analysis; a large number of similarity solutions for the SH-equations are constructed.
- The same solutions (and a few more) are also constructed by CHUGUNOV, GRAY & HUTTER, however using the more sophisticated Lie group theory. Comparison of the simpler analysis method with this more sophisticated group theoretic method is particularly interesting.

Part II. Porous and Granular Materials. Fundamentals and Dynamical Processes

Distinguished features of granular and porous materials are that either the pore volume affects the dynamics of the gross behaviour of such materials and/or that the particulars of the grain interactions directly determine the form of the flux and production terms in the emerging dynamical equations. There are several levels of complexity by which these special features are accounted for.

- In the hypo-plasticity theory the constitutive equation for the stress tensor takes a form in which the pore space is accounted for by making it depending on the void ratio. OSINOV & GUDEHUS in their article of the dynamics of hypo-plastic materials demonstrate that normal stress effects lead to a coupling of shear and compression waves. Of particular interest in the analysis are questions of well-posedness of the dynamic problem, cyclic shearing, the coupling between transverse and longitudinal displacement components, the consequential double-frequency effect and problems of liquefaction in saturated granular materials. The spirit of the analysis is non-linear wave dynamics and its numerical treatment.

- In the article on acoustic waves in porous solid-fluid mixtures by WILMAŃSKI & ALBERS, the role of the pore space is dynamically incorporated in the model by postulating for it a balance law and presenting a full dynamical theory that is based on a careful derivation from thermodynamical principles. Three waves are shown to exist in the bulk – two P-waves and one S-wave. Rayleigh and Love waves are equally analysed as surface waves in contact with a vacuum and a fluid, respectively. Critical issues in the high and low frequency range are equally touched upon.
- CHRZANOWSKA & EHRENTRAUT study the dynamics of a two-dimensional array of hard needles. Molecular dynamic methods are used with proper interaction laws and with these the influence of dissipation on the "cooling" of the needles is studied. By this the authors mean a reduction of fluctuation energy of the needles relative to their mean motion. Orientational clusters can occur both in the isotropic as well as the nematic phase. It is shown that clusters of needles behave differently than genuine liquid crystals, because cooling in clusters of needles does not under all circumstances lead to an increase of the order parameters.
- In their second article, EHRENTRAUT & CHRZANOWSKA study induced anisotropy in rapid flows of non-spherical granular materials. In particular, emphasis is placed upon the role of spin. Thus they scrutinise the SH-theory and assume that the Cauchy stress tensor may be non-symmetric. Their analysis culminates in the statement that the anisotropy of the stress tensor becomes only important when higher order effects are investigated. However, the symmetry condition cannot be an a priori statement. It must be a statement that determines the evolution of the internal structure of the granular medium. The analysis of it is dealt with in the second part of their article.

Part III. Porous and Granular Materials Sub-scale and Micromechanical Effects

The present literature assigns to the pore space, or more precisely the volume fraction of the individual constituents in a mixture, the role of an additional kinematic variable accounting (at least partially) for the sub-scale or micromechanical effects. It is the believe of some of the authors of this part of the book that volume fractions and in particular their gradients in a mixture of particles of different size give rise to the phenomenon of particle size separation. This separation can also be viewed as a localisation phenomenon and then explains the necessity of appearance of volume fraction gradients. Localisation can also arise due to other microscale effects, e.g. abrasion of grain material by friction or fractionation.

- Kirchner & Hutter attempt to model particle size segregation by a granular mixture model of constituents of different size. Their thermodynamic model uses a balance law of the configurational force type and they present a full thermodynamic theory for a mixture of N constituents. Of significance for the modelling of the size separation effect is that the so-called principle of phase separation in the formulation of constitutive equations for the peculiar constituents is abandoned. Stringent application of the second law of thermodynamics seems to be important.
- Fried & Roy employ a different theoretical model derived earlier by Fried, Gurtin & Hutter and apply it to a column bounded below by an impermeable base, above by an evolving free surface and subject to gravity. Existence of solutions of the emerging equations is shown as is uniqueness for the case of two particle sizes only. Analytical and numerical results are constructed for a mixture of particles of three particle sizes. Numerical results are also compared with experiments.

Most of the research reported in this book has wholly or partially been supported by the Deutsche Forschungsgemeinschaft through our SFB. Salaries for Ph. D. students, post-doctoral fellows, experimental equipments, computer soft- and hardware were financed as were subsidies for our travel and guest programmes. Complementary support came also from the Darmstadt University of Technology. All this has helped us in advancing in our scientific work which has in many respects also supported us in improving our teaching. Supporting in the achievement of this improvement were the referees who accompanied us through the eight years. To the Deutsche Forschungsgemeinschaft, to the University, and to them we express our gratitude.

Darmstadt,
November 2002

Kolumban Hutter
Nina Kirchner

Contents

Part III. Porous and Granular Materials. Subscale- and Micromechanical Effects.

Index

Acknowledgements

The articles published in this book have all been peer reviewed by internationally known specialists. The editors wish to acknowledge the help of the following individuals in judging the scientific contents of the articles:

- Hans-Dieter Alber
- Roger P. Denlinger
- Reinhard Farwig
- Ralf Greve
- Morton E. Gurtin
- Hans Herrmann
- Axel Klar
- Pavel Krejci
- Gilberto M. Kremer
- Renato Lancellotta
- Wolfgang Muschik
- Martin Oberlack
- Vladimir Osinov
- Hans Christian Öttinger
- Gerald Ristow
- Yih-Chin Tai
- Cameron Tropea
- Thomas Zwinger

Part I

Avalanches, Debris and Mud Flows

Rapid Granular Avalanches

John Mark Nicholas Timm Gray

Department of Mathematics, University of Manchester,
Oxford Road, Manchester, M13 9PL, U.K., e-mail: ngray@maths.man.ac.uk

received 1 Oct 2001 — accepted 24 Apr 2002

Abstract. Granular avalanches are one of the fundamental grain transport mechanisms in our natural environment and in many industrial grain-processing flows. In recent years significant progress has been made in describing the flow of granular avalanches over complex rigid topography. Often the fluid-like granular avalanches flow over a region of solid-like grains at which there may be erosion or deposition. These phase-transition flows are very poorly understood, yet are of crucial importance in industrial rotating tumblers in which powders and grains are mixed. This paper reviews recent work on granular avalanche flows and presents a generalisation of the Savage–Hutter theory that includes both surface and basal erosion and deposition. Elementary similarity and shock wave solutions are reviewed and a simple mass-coupling to a rigid body of grains is used to investigate the steady mixing of grains of equal size in a rotating tumbler of circular cross-section.

1 Introduction

Rapid granular free surface flows, or granular avalanches, are one of the most important particle transport mechanisms in both industrial processes and our natural environment. In industry they often occur as part of complicated granular phase-transition flows, in which the granular material behaves as a fluid-like avalanche close to the free surface and as a solid in a large region beneath. This is typical of flows in partially-filled rotating drums , which are used to mix or segregate dissimilar grains [11,15,18,29,31,34,36,41,42,49], or whenever particles are poured out of containers or deposited to form heaps [3,11,12,30,37]. The magnitude of industrial applications is huge, Shinbrot & Muzzio [42] estimate that US production of granular pharmaceuticals, foods and bulk chemicals amounts to a trillion kilograms per year alone, and most of these materials will be mixed, segregated, poured or deposited in piles at some stage of the manufacturing process. In nature, snow-slab avalanches and debris flows are prime examples of granular free surface flows. In fact, observations suggest that granular avalanches are scale-independent over a huge range, starting with industrial flows on the scale of a few centimetres in length and going right up to geophysical avalanches with a total volume of up to a million cubic metres.

Over the past decade significant progress has been made in the description of granular avalanches. Savage & Hutter [38] derived a theory for the two-dimensional plane motion of an incompressible Mohr-Coulomb material sliding down a rigid impenetrable surface, at which the body experiences dry Coulomb friction. The shallowness of the flow was exploited to integrate the leading order mass and momentum balances through the avalanche depth to obtain a one-dimensional theory along the flow direction for the avalanche thickness and the downslope velocity. Savage & Hutter [39] introduced a slope fitted curvilinear coordinate system to compute the flow over complicated two-dimensional slopes, from initiation on a steep slope to run-out on a horizontal plane. Laboratory experiments on an exponentially curved chute [19] and on concave and convex chutes [16,22] showed that the model was in very good agreement with experiment. The theory has also been generalised to describe three-dimensional flows on plane chutes [17] and over fully three-dimensional complex topography [14]. These models have been very effective for modelling laboratory flows on inclined planes and on chutes both with and without lateral curvature [27,51]. The assumption that a rapid granular avalanche behaves as a Mohr-Coulomb material may be rather strong, and an alternative hydraulic closure is provided by an inviscid fluid model [10]. Gray et al. [14] and Gray [15] have found that this simpler model also gives very promising results on steep slopes, but is not as good, when the avalanche runs out and stops on a horizontal plane.

Recently normal shock waves, at which there are rapid changes in the avalanche thickness and velocity, have been observed propagating upslope in pattern formation experiments [11]. These are often observed on non-accelerative slopes, where there is no net driving force and the constitutive relation simplifies to a pressure field. In fact, in this situation the Savage–Hutter theory actually reduces to a system that has the same mathematical structure as the shallow water equations of hydrodynamics, and the shock wave can be interpreted as the granular equivalent of a *hydraulic jump* or a *bore* in shallow water theory [13,47]. These granular shocks or bores have sparked the development of shock capturing numerical methods (e.g. [25,33,48,52]) for the Savage–Hutter equations [13,46]. The analogy with shallow water, and indeed with gas dynamics, is a good one, which helps many other frequently observed phenomena to be understood. For instance, *supercritical* (shooting) and *sub-critical* flows can easily be identified, in which the avalanche velocity is above and below the local wave propagation speed, respectively. Furthermore, the avalanche boundary, where the thickness is equal to zero, has the same degenerate mathematical structure as the vacuum boundary in gas dynamics. Particle-free regions are, therefore, termed *granular vacua*. Tai *et al.* [46] have developed highly accurate front tracking algorithms to follow the vacuum boundary in one-dimension and have tested them against parabolic cap similarity [38] and granular bore solutions to the Savage–Hutter equations [13]. Granular vacua are particularly significant, because they can be used to protect buildings and structures at risk from snow slab avalanches or debris flows [45]. The experiments also exhibit a number of features that are not observed in either gas dynamics or shallow water theory. The most striking of these is the formation of a *dead zone* in which absolutely stationary regions form adjacent to regions of rapid flow.

In this article the one-dimensional Savage–Hutter theory is extended to allow mass transfer across the surface and basal interfaces, elementary solutions are reviewed and a theory is developed for granular flow in slowly rotating drums.

2 The Savage–Hutter Avalanche Theory

2.1 Conservation Equations and Boundary Conditions

Granular materials exhibit dilatancy effects during plastic yield [35]. Once failure has occurred and the material is fluidised, however, it is reasonable to assume that it is incompressible with constant uniform density, ρ_0. The conservative form of the mass and momentum balances reduce to

$$\nabla \cdot \boldsymbol{u} = 0, \tag{1}$$

$$\partial_t(\rho_0 \boldsymbol{u}) + \nabla \cdot (\rho_0 \boldsymbol{u} \otimes \boldsymbol{u}) = -\nabla \cdot \boldsymbol{p} + \rho_0 \, \boldsymbol{g}, \tag{2}$$

where ∇ is the gradient operator, $\boldsymbol{u}$ is the velocity, ∂_t indicates differentiation with respect to time, $\otimes$ is the tensor (or dyadic) product, $\boldsymbol{p}$ is the pressure tensor (the negative Cauchy stress) and $\boldsymbol{g}$ is the gravitational acceleration.

The avalanche lies between an upper free surface, $F^s(\boldsymbol{x}, t) = 0$, and a basal interface, $F^b(\boldsymbol{x}, t) = 0$, on which the body slides. At these interfaces the body is subject to kinematic boundary conditions

$$F^s(\boldsymbol{x}, t) = 0 : \ \partial_t F^s + v_n^s \boldsymbol{n}^s \cdot \nabla F^s = 0, \tag{3}$$

$$F^b(\boldsymbol{x}, t) = 0 : \ \partial_t F^b + v_n^b \boldsymbol{n}^b \cdot \nabla F^b = 0, \tag{4}$$

where v_n^s and v_n^b are the normal speeds of the surface and basal interfaces along the outward normals $\boldsymbol{n}^s = \nabla F_s / |\nabla F_s|$ and $\boldsymbol{n}^b = \nabla F_b / |\nabla F_b|$, respectively. The superscripts s and b are used to indicate that a variable is evaluated at the surface or the basal interface. The body is also subject to a traction free boundary condition at the free surface and a Coulomb sliding friction law at the basal interface

$$F^s(\boldsymbol{x}, t) = 0 : \ \boldsymbol{p}^s \boldsymbol{n}^s = \boldsymbol{0}, \tag{5}$$

$$F^b(\boldsymbol{x}, t) = 0 : \ \boldsymbol{p}^b \boldsymbol{n}^b = (\boldsymbol{u}^r / |\boldsymbol{u}^r|) \tan \delta (\boldsymbol{n}^b \cdot \boldsymbol{p}^b \boldsymbol{n}^b) + \boldsymbol{n}^b (\boldsymbol{n}^b \cdot \boldsymbol{p}^b \boldsymbol{n}^b), \tag{6}$$

where δ is the basal angle of friction and $\boldsymbol{u}^r = \boldsymbol{u}^{b+} - \boldsymbol{u}^{b-}$ is the velocity difference between the upper side of the basal interface, $\boldsymbol{u}^{b+}$, and the basal topography, $\boldsymbol{u}^{b-}$, on the lower side of the interface. The factor $(\boldsymbol{u}^r / |\boldsymbol{u}^r|)$ ensures that the Coulomb friction opposes the avalanche motion. This definition differs from that of Savage & Hutter [38] to reflect the fact that the basal topography can have a non-zero velocity component parallel to the interface. Additional effects such as air drag, rate-dependent basal drags and momentum thrust terms due to mass interaction are neglected.

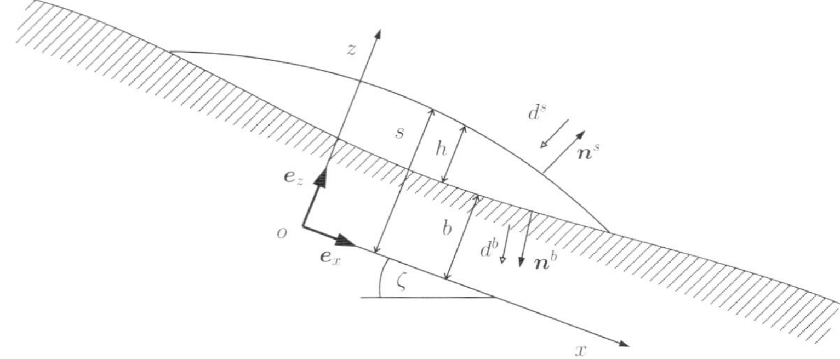

Fig. 1. A sketch of the avalanche and basal topography. A fixed Cartesian coordinate system oxz is defined with the x axis pointing in the downslope direction and inclined at an angle ζ to the horizontal. The avalanche free surface lies at $z = s(x,t)$ and the base at $z = b(x,t)$. The avalanche thickness $h(x,t)$ is measured parallel to the z axis. The accumulation rate d^s and the deposition rate d^b are the equivalent volumes of granular material that are either deposited on the avalanche free surface, or eroded from the avalanche base, respectively, per unit area per unit time

In this article a conservative theory will be derived for the two-dimensional vertical plane motion of an avalanche with mass transfer across the surface and basal boundaries. A fixed Cartesian coordinate system oxz with origin o is defined, so that the x-axis lies approximately parallel to the bed interface, $F^b = 0$, and points in the downslope direction. The x-axis is, therefore, inclined at an angle ζ to the horizontal and the z-axis is chosen along the upward pointing normal. The coordinate system is illustrated in Fig. 1. The velocity $\boldsymbol{u}$ has components u and w in the downslope and normal directions, respectively, and the symmetric pressure tensor has components p_{xx}, p_{xz}, p_{zz}.

In both industrial and geophysical flows the typical avalanche depth, H, is much smaller than its length, R. It is convenient to introduce non-dimensional variables to reflect the *shallowness* of the granular avalanche

$$
\begin{aligned}
(x, z, t) &= (R\tilde{x}, H\tilde{z}, (R/g)^{1/2}\tilde{t}),\\
(u, w, v_n^s, v_n^b) &= (Rg)^{1/2}(\tilde{u}, \varepsilon\tilde{w}, \varepsilon\tilde{v}_n^s, \varepsilon\tilde{v}_n^b),\\
(p_{xx}, p_{zz}) &= \rho_0 g H(\tilde{p}_{xx}, \tilde{p}_{zz}),\\
(p_{xz}) &= \rho_0 g H \mu(\tilde{p}_{xz}),
\end{aligned}
\tag{7}
$$

where the tildes denote non-dimensional variables, g is the constant of gravitational acceleration and $\varepsilon = H/R \ll 1$ is the aspect ratio. Note, that the Coulomb friction law (6) implies that the magnitude of the shear stress is equal to the magnitude of the basal pressure, $\rho_0 g H$, multiplied by the Coulomb friction coefficient $\mu = \tan\delta_0$.

Applying the scalings (7) the non-dimensional mass balance (1) is

$$\partial_x u + \partial_z w = 0, \tag{8}$$

where the tildes are now dropped, and ∂_x and ∂_z denote differentiation with respect to x and z, respectively. The downslope and normal components of the momentum balance (2) are

$$\partial_t u + \partial_x(u^2) + \partial_z(uw) = \sin\zeta - \varepsilon\partial_x(p_{xx}) - \mu\partial_z p_{xz}, \tag{9}$$

$$\varepsilon\left\{\partial_t w + \partial_x(uw) + \partial_z(w^2)\right\} = -\cos\zeta - \varepsilon\mu\partial_x(p_{xz}) - \partial_z p_{zz}. \tag{10}$$

The avalanche free surface, $F^s = z - s(x,t) = 0$, and the basal interface $F^b = b(x,t) - z = 0$, are defined by their heights above the $z = 0$ plane. These definitions ensure that the surface and basal unit normal vectors point outwards from the avalanching material and are given by

$$\Delta^s \boldsymbol{n}^s = -\varepsilon\partial_x s\boldsymbol{e}_x + \boldsymbol{e}_z, \qquad \Delta^b \boldsymbol{n}^b = \varepsilon\partial_x b\boldsymbol{e}_x - \boldsymbol{e}_z, \tag{11}$$

where $\boldsymbol{e}_x$ and $\boldsymbol{e}_z$ are unit vectors in the downslope and normal directions, respectively, and the normalisation factors $\Delta^s = \{1+\varepsilon^2(\partial_x s)^2\}^{1/2}$ and $\Delta^b = \{1+\varepsilon^2(\partial_x b)^2\}^{1/2}$. The surface and basal interfaces are modified by accumulation and deposition. Let the accumulation rate, d^s, be the equivalent volume of granular material deposited on the avalanche free surface per unit area per unit time. Then the normal speed of the interface $v_n^s = \boldsymbol{u}^s \cdot \boldsymbol{n}^s + d^s$. Similarly, if the deposition rate, d^b, is the equivalent volume of granular material deposited at the avalanche base per unit area per unit time, then the normal speed of the basal interface along the outward normal $v_n^b = \boldsymbol{u}^b \cdot \boldsymbol{n}^b - d^b$. It follows that the surface and basal kinematic conditions (3) and (4) are

$$z = s(x,t): \ \partial_t s + u^s\partial_x s - w^s = \Delta^s d^s, \tag{12}$$

$$z = b(x,t): \ \partial_t b + u^b\partial_x b - w^b = \Delta^b d^b. \tag{13}$$

There are many situations in which surface accumulation or basal deposition can occur, e.g. when a heap of granular material is built up on a horizontal plane. The granular material that is poured onto the top of the pile provides a source of surface accumulation to the avalanche and as the granular material comes to rest it is deposited into the solid body region of granular material beneath it. This deposition mechanism can equivalently be viewed as a moving phase boundary between solid and fluid-like regions.

At the free surface the traction free boundary condition (5) has downslope and normal components

$$-\varepsilon p_{xx}^s \partial_x s + \mu p_{xz}^s = 0, \tag{14}$$

$$-\varepsilon\mu p_{zx}^s \partial_x s + p_{zz}^s = 0, \tag{15}$$

and the Coulomb friction law (6) has downslope and normal traction components

$$\varepsilon p^b_{xx}\partial_x b - \mu p^b_{xz} = (\boldsymbol{n}^b \cdot \boldsymbol{p}^b \boldsymbol{n}^b)\left(\Delta^b (u^r/|\boldsymbol{u}^r|)\tan\delta + \varepsilon\partial_x b\right), \tag{16}$$

$$\varepsilon\mu p^b_{zx}\partial_x b - p^b_{zz} = (\boldsymbol{n}^b \cdot \boldsymbol{p}^b \boldsymbol{n}^b)\left(\Delta^b (\varepsilon w^r/|\boldsymbol{u}^r|)\tan\delta - 1\right). \tag{17}$$

Note that as a first approximation the accumulation and deposition are assumed to have no effect on the mechanical boundary conditions at the surface and base of the avalanche. In general more complex boundary conditions may be necessary, which take account of suction effects or the "granular temperature" of the material.

2.2 Integration Through the Avalanche Depth

Following Savage & Hutter [38] the governing equations are now integrated through the avalanche depth to obtain a simplified depth averaged theory in which one spatial dimension is removed from the problem. The avalanche thickness h is equal to the difference between the free surface height, $s(x,t)$, and the height of the basal interface, $b(x,t)$,

$$h = s - b, \tag{18}$$

and the depth averaged value $\overline{f}$ of a function f is

$$\overline{f} = \frac{1}{h}\int_b^s f \,\mathrm{d}z. \tag{19}$$

Leibniz's theorem ([1], 3.3.7) for the differentiation of an integral

$$\frac{\mathrm{d}}{\mathrm{d}c}\int_{b(c)}^{s(c)} f(\xi,c)\,\mathrm{d}\xi = \int_{b(c)}^{s(c)} \frac{\partial}{\partial c} f(\xi,c)\,\mathrm{d}\xi + f(s,c)\frac{\mathrm{d}s}{\mathrm{d}c} - f(b,c)\frac{\mathrm{d}b}{\mathrm{d}c} \tag{20}$$

allows the mass balance (8) to be integrated through the avalanche depth to give

$$\partial_x(h\overline{u}) - [u\partial_x z - w]^s_b = 0, \tag{21}$$

where the square bracket is the difference between the surface and basal values of the enclosed function, i.e. $[f]^s_b = f^s - f^b$. Substituting the kinematic boundary conditions (12) and (13), and using (18), the depth integrated mass balance is

$$\partial_t h + \partial_x(h\overline{u}) = \Delta^s d^s - \Delta^b d^b. \tag{22}$$

Integrating the downslope momentum balance (9) through the avalanche depth using Leibniz's rule implies

$$\begin{aligned}&\partial_t(h\bar{u}) + \partial_x(h\overline{u^2}) - [u\,(\partial_t z + u\partial_x z - w)]_b^s \\ &= h\sin\zeta + \varepsilon\partial_x(h\overline{p_{xx}}) - [\varepsilon p_{xx}\partial_x z - \mu p_{xz}]_b^s\,.\end{aligned} \tag{23}$$

Substituting the kinematic conditions (12) and (13) and the downslope traction conditions (14) and (16) into the square bracketed terms gives the depth integrated downslope momentum balance

$$\begin{aligned}&\partial_t(h\bar{u}) + \partial_x(h\overline{u^2}) - (u^s\Delta^s d^s - u^b\Delta^b d^b) \\ &= h\sin\zeta - \left(\Delta^b(u^r/|\boldsymbol{u}^r|)\tan\delta + \varepsilon\partial_x b\right)(\boldsymbol{n}^b\cdot\boldsymbol{p}^b\boldsymbol{n}^b) - \varepsilon\partial_x(h\overline{p_{xx}}).\end{aligned} \tag{24}$$

2.3 Order of Magnitude Estimates

The shallowness of the granular avalanche is now exploited by making a long-wave approximation based on the small parameter $\varepsilon = H/L$. Since the friction coefficient $\mu = \tan\delta_0$ is less than order unity, but greater than order ε, it is convenient to make the approximation that $\mu = \mathrm{O}(\varepsilon^\gamma)$, for some exponent $\gamma \in (0,1)$. To accurately describe the flow of granular avalanches, terms of order ε must be retained in the theory, but terms of order $\varepsilon^{1+\gamma}$ can be discarded. The effect of this scaling is to retain longitudinal pressure gradients, but to neglect order ε pressure corrections to the Coulomb dry friction law.

Integrating the normal component of the momentum balance (10) with respect to z and applying the free surface boundary condition (15), it follows that

$$p_{zz} = (s - z)\cos\zeta + \mathrm{O}(\varepsilon), \tag{25}$$

and at the base

$$p_{zz}^b = h\cos\zeta + \mathrm{O}(\varepsilon). \tag{26}$$

Hence, the basal normal pressure $\boldsymbol{n}^b\cdot\boldsymbol{p}^b\boldsymbol{n}^b = h\cos\zeta + \mathrm{O}(\varepsilon)$ in the Coulomb dry friction law. It follows that to order $\varepsilon^{1+\gamma}$ the depth integrated mass and downslope momentum balances reduce to

$$\partial_t h + \partial_x(h\bar{u}) = d^s - d^b, \tag{27}$$

$$\partial_t(h\bar{u}) + \partial_x(h\overline{u^2}) - (u^s d^s - u^b d^b) = hD - \varepsilon\partial_x(h\overline{p_{xx}}) - \varepsilon h\cos\zeta\partial_x b, \tag{28}$$

where the net driving force

$$D = \cos\zeta(\tan\zeta - (u^r/|u^r|)\tan\delta), \tag{29}$$

is the difference between the gravitational acceleration and the Coulomb basal friction.

2.4 Constitutive Properties

To complete the theory, a constitutive relation is required for the depth-averaged longitudinal stress, $\overline{p_{xx}}$, in the fluid-like granular flow regime. This is still a subject of current research, and two relations will be considered.

Mohr-Coulomb Savage & Hutter [38] assumed that the avalanche behaved as a Mohr-Coulomb material at yield. This implies on each plane element the normal pressure p and the tangential shear stress $\boldsymbol{\tau}$ are related by a Mohr-Coulomb yield criterion

$$|\boldsymbol{\tau}| = p \tan\phi, \tag{30}$$

where ϕ is the internal angle of friction. This stress state can be conveniently visualised on a normal to shear stress (p, τ)-diagram or Mohr circle diagram. All the allowable stress states lie on the circle

$$(p-a)^2 + \tau^2 = r^2, \tag{31}$$

with radius r and centre

$$a = (p_{xx} + p_{zz})/2. \tag{32}$$

The principal stresses, p_x and p_z, lie on the p axis as illustrated in Fig. 2. When the material is at *yield* the Mohr circle of stress is tangent to the Mohr-Coulomb lines $\tau = \pm p \tan\phi$ and by elementary trigonometry it follows that

$$r = a \sin\phi. \tag{33}$$

The position of the centre of the Mohr circle is obtained by substituting (33) into (31) and solving the quadratic equation for a. This implies

$$a = \sec^2\phi \left(p \pm \sqrt{p^2 \sin^2\phi - \tau^2 \cos^2\phi} \right). \tag{34}$$

The *earth pressure coefficient* K_x relates the *limiting* normal stress in the x and z directions, and was defined by Savage & Hutter [38] as

$$K_x = p_{xx}/p_{zz}. \tag{35}$$

An expression for the earth pressure coefficient is obtained by substituting (32) into (35) to eliminate p_{xx}, i.e. $K_x = 2a/p_{zz} - 1$, and then substituting for the centre of the yield circle, $a = a(p_{zz}, \tau_{xz})$, from (34). It follows that the earth pressure coefficient is given by

$$K_x = 2\sec^2\phi \left\{ 1 \pm \sqrt{\sin^2\phi - \frac{\tau_{xz}^2}{p_{zz}^2} \cos^2\phi} \right\} - 1. \tag{36}$$

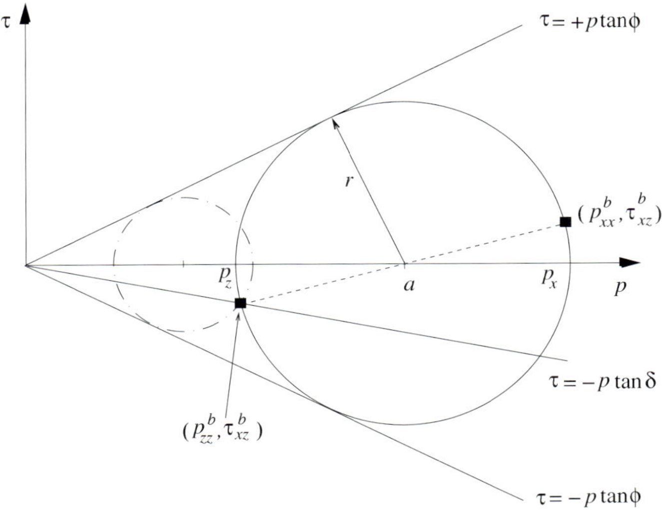

Fig. 2. The stress state within the avalanche is represented on a Mohr circle diagram. The yield criterion corresponds to the two straight lines inclined at angles $\pm\phi$ to the horizontal and the Coulomb basal dry friction is indicated by the line at an angle $-\delta$ to the horizontal. The "passive" basal stress state is indicated by the solid circle of radius r and centred at $p = a$. This circle is both tangent to the yield curves and passes through the point $(p_{zz}, -p_{zz}\tan\delta)$. A second "active" stress state that also satisfies these constraints exists and is indicated by the dot-dash circle

Savage & Hutter [38] argued that at the base of the avalanche the basal shear stress and the normal pressure satisfied a Coulomb dry friction law

$$\tau^b_{xz} = \pm p^b_{zz} \tan\delta, \tag{37}$$

with basal angle of friction δ. On substituting this into (36) the earth pressure coefficient

$$K^b_{x_{act/pas}} = 2\sec^2\phi \left(1 \mp \{1 - \cos^2\phi \sec^2\delta\}^{1/2}\right) - 1, \tag{38}$$

is obtained, which is real valued provided $\delta \leq \phi$.

There are precisely two Mohr stress circles that are both tangent to the Mohr-Coulomb yield lines and that pass through the basal stress point (p^b, τ^b). This results in two possible limiting stress states and, therefore, two possible earth pressure coefficients. An ad-hoc definition associates "active" states with extensive motions and "passive" states with compressive motions

$$K_x = \begin{cases} K_{x_{act}} & \partial_x u \geq 0, \\ K_{x_{pas}} & \partial_x u < 0. \end{cases} \tag{39}$$

Savage & Hutter [38] implicitly assumed that for suitably thin avalanches, equations (35) and (38) hold through the avalanche depth to order ε^γ. Substituting (25) into (35) and integrating over the avalanche depth implies that

$$h\overline{p_{xx}} = K_x \cos\zeta h^2/2 + \mathrm{O}(\varepsilon), \tag{40}$$

which provides an expression for the mean longitudinal pressure. The transition between stress states at $\partial_x u = 0$ is still poorly understood, see however [43] and [44].

Inviscid fluid The assumption that a Mohr-Coulomb yield criterion is appropriate for rapidly flowing granular material is rather strong, as the material may be well beyond the point of yield. An inviscid fluid model proposed by Eglit [10], with an isotropic pressure distribution

$$p_{xx} = p_{zz}, \tag{41}$$

is therefore also considered as an alternative method of closure. Integrating over the avalanche depth, implies that the longitudinal pressure satisfies the simpler relation

$$h\overline{p_{xx}} = \cos\zeta h^2/2 + \mathrm{O}(\varepsilon). \tag{42}$$

A comparison with (40) shows that this is equivalent to assuming that the earth pressure coefficient $K_x = 1$, and that there is no jump in the longitudinal stress. For extensive flow it may be hard to differentiate between the two models, as $K_{x_{act}}$ is close to unity for typical values of ϕ and δ.

2.5 Conservative Formulation

Many observations [8,26,32,40] suggest that the downslope velocity profile with depth is rather blunt, and shear is confined to a very thin layer. To order $\varepsilon^{1+\gamma}$ the downslope velocity is assumed to be independent of depth, implying

$$u^b = \overline{u} + \mathrm{O}(\varepsilon^{1+\gamma}), \qquad u^s = \overline{u} + \mathrm{O}(\varepsilon^{1+\gamma}), \qquad \overline{u^2} = \overline{u}\,\overline{u} + \mathrm{O}(\varepsilon^{1+\gamma}). \tag{43}$$

Using these relations together with the expressions for the longitudinal pressure (40), or (42), the final system of conservation laws for the avalanche motion with accumulation and deposition are

$$\partial_t h + \partial_x(hu) = d^s - d^b, \tag{44}$$

$$\partial_t(hu) + \partial_x(hu^2) + \partial_x(\beta h^2/2) = hD - \varepsilon h\cos\zeta\partial_x b + u(d^s - d^b), \tag{45}$$

where

$$\beta = \varepsilon K_x \cos\zeta \tag{46}$$

and for simplicity the averaging bar is now dropped on the downslope velocity component. The driving force D is given by (29) and the earth pressure coefficient K_x is either defined by (38), or is equal to 1 in the inviscid case. Given a slope inclination angle ζ, basal friction angle δ, internal friction angle ϕ, the basal topography b, the surface accumulation rate d^s and the basal deposition rate d^b this forms a closed system of equations for the avalanche depth h and the depth averaged downslope velocity u.

2.6 Non-conservative Formulation

Expanding the transport terms in (45) and substituting for the depth integrated mass balance (44) the momentum balance can be expressed in a non-conservative reduced form

$$h\partial_t u + hu\partial_x u + \partial_x(\beta h^2/2) = hD - \varepsilon h \cos\zeta \partial_x b, \tag{47}$$

The original Savage & Hutter [38] theory assumed that the spatial gradient of the earth pressure coefficient could be neglected, i.e. $\partial_x K_x$ is of order ε^γ or smaller. This allows the pressure term in (47) to be expanded and the momentum balance can then be divided through by the avalanche thickness, provided $h \neq 0$. With these assumptions the conservative system of equations (44)–(45) reduces to

$$\partial_t h + \partial_x(hu) = d^s - d^b, \tag{48}$$

$$\partial_t u + u\partial_x u + \beta\partial_x h = D - \varepsilon\cos\zeta\partial_x b, \tag{49}$$

Assuming that there is no accumulation or deposition, $d^s = d^b = 0$ this system reduces to the Savage & Hutter [38] equations, with a correction for an error in their basal topography gradient term.

2.7 Classification

The Savage–Hutter equations form a degenerate system of hyperbolic-parabolic equations. The parabolic behaviour arises from the velocity gradient dependence of the earth pressure coefficient $K_x(\partial_x u)$ at the transition between limiting stress states at $\partial_x u = 0$. Most of the time, however, the earth pressure coefficient is constant and the equations degenerate into a first order quasi-linear system.

The degenerate equations can be classified by looking to see if the equations can be transformed into characteristic form (e.g. [6,50]). Gathering the dependent variables into the vector $\boldsymbol{s} = (h, u)^T$, the conservation laws (48)–(49) can be written in vector form as

$$\boldsymbol{a}\partial_t \boldsymbol{s} + \boldsymbol{b}\partial_x \boldsymbol{s} = \boldsymbol{r}, \tag{50}$$

where the coefficient matrices

$$\boldsymbol{a} = \begin{pmatrix} 1 & 0 \\ 0 & 1 \end{pmatrix}, \qquad \boldsymbol{b} = \begin{pmatrix} u & h \\ \beta & u \end{pmatrix} \tag{51}$$

and $\boldsymbol{r} = (d^s - d^b, D - \varepsilon \cos\zeta \partial_x b)^T$ is independent of the derivatives of $\boldsymbol{s}$. A necessary and sufficient condition for a non-trivial transformation of the equations into characteristic form is that the determinant

$$|\boldsymbol{b} - \lambda \boldsymbol{a}| = 0. \tag{52}$$

Each real solution λ determines a characteristic curve of the system. If all k roots are real and distinct then all the characteristics are defined and the system is called *totally hyperbolic.*

For the degenerate form of the Savage–Hutter theory the eigenvalue problem (52) is

$$\begin{vmatrix} u-\lambda & h \\ \beta & u-\lambda \end{vmatrix} = (u-\lambda)^2 - \beta h = 0. \tag{53}$$

The characteristic directions of the system are given by the roots

$$\lambda = u \pm \sqrt{\beta h}. \tag{54}$$

For $h > 0$ the wave speeds are real and distinct and the system is strictly or totally hyperbolic. When $h = 0$ the wave speeds are the same and the system is non-strictly hyperbolic.

The Savage–Hutter avalanche theory has many similarities with the shallow water equations and the equation of gas dynamics. The main mathematical differences are the velocity gradient dependence of the earth pressure coefficient, the scalings that are used and the more complicated source terms. These analogies suggest that it is useful to define the Froude number as the ratio of the velocity to the local wave-speed of the flow

$$\mathrm{Fr} = \frac{u}{\sqrt{\beta h}}, \tag{55}$$

If the Froude number

$$\begin{aligned} &\mathrm{Fr} > 1 \text{ the flow is supercritical,} \\ &\mathrm{Fr} = 1 \text{ the flow is critical,} \\ &\mathrm{Fr} < 1 \text{ the flow is subcritical.} \end{aligned} \tag{56}$$

Recalling that $\beta = \varepsilon K_x \cos\zeta$ it follows that $\mathrm{Fr} \sim \varepsilon^{-1/2} \gg 1$ and therefore many granular flows are expected to be supercritical.

3 Elementary Solutions

Similarity solutions exist for the spreading of a pile of granular material as it accelerates down an inclined plane [20,21,23,24,38]. To derive these solutions it it necessary to introduce a fixed domain mapping, which is achieved by firstly switching to a frame moving with the velocity of the centre of mass and then mapping the front and rear of the avalanche onto a fixed domain.

3.1 Fixed Domain Mapping

An approximation for the velocity of the centre of the avalanche can be obtained by neglecting the thickness gradient and integrating (49) with respect to time

$$u_m(t) = u_{m0} + \int_0^t D \, \mathrm{d}t' . \tag{57}$$

The centre of the avalanche therefore translates a distance

$$x_m(t) = x_{m0} + \int_0^t u_m(t') \, \mathrm{d}t' \tag{58}$$

downslope in time t. It is convenient to define a moving coordinate system (ξ, t), which translates with the centre of mass velocity, and to define the relative velocity $\hat{u}$ in this coordinate system

$$\xi = x - x_m(t), \qquad \hat{u} = u - u_m(t). \tag{59}$$

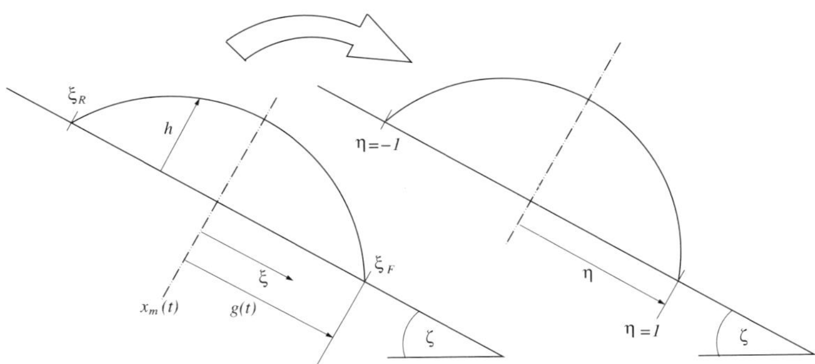

Fig. 3. Mapping form the moving coordinates (ξ, t) to the fixed domain (η, τ)

Solutions are sought that have a symmetric thickness distribution and a skew symmetric velocity distribution about $\xi = 0$, i.e.

$$h(\xi, t) = h(-\xi, t), \qquad \hat{u}(\xi, t) = -\hat{u}(-\xi, t). \tag{60}$$

With these restrictions the front and rear of the avalanche always lie at $\xi_F = g(t)$ and $\xi_R = -g(t)$, respectively. A fixed domain mapping is then used to map the end points onto the interval $[-1, 1]$ by defining new coordinates

$$\eta = \xi / g(t) = [x - x_m(t)]/g(t), \qquad \tau = t, \tag{61}$$

see Fig. 3. With this change of variables the derivatives are

$$\partial_t = \partial_\tau - \left(\eta \frac{g'}{g} + \frac{u_m}{g}\right)\partial_\eta, \qquad \partial_x = \frac{1}{g}\partial_\eta \tag{62}$$

where $g' = dg/d\tau$. Assuming the bed is flat and that there is no surface or basal mass transfer, the non-conservative system of equations (48)–(49) transform to

$$\partial_\tau h + \frac{1}{g}(\hat{u} - \eta g')\partial_\eta h + \frac{h}{g}\,\partial_\eta \hat{u} = 0, \tag{63}$$

$$\partial_\tau \hat{u} + \frac{1}{g}(\hat{u} - \eta g')\partial_\eta \hat{u} + \frac{\beta}{g}\,\partial_\eta h = 0. \tag{64}$$

3.2 Parabolic Cap Similarity Solution

Following Savage–Hutter [38] the transformed system (63)–(64) can be simplified considerably by assuming that the velocity profile has a solution of the form

$$\hat{u} = \eta g', \tag{65}$$

which is consistent with the symmetry requirements (61). Using (65), the momentum balance (64) reduces to

$$\partial_\eta h = -\frac{g g''}{\beta}\eta. \tag{66}$$

Integrating this equation with respect to η from $\eta = -1$ to $\eta = 1$, subject to the boundary condition that $h(\eta = \pm 1) = 0$, implies that the avalanche thickness has a parabolic profile

$$h = \frac{g g''}{2\beta}(1 - \eta^2). \tag{67}$$

Substituting the assumed velocity profile (65) and the computed thickness profile (67) into the mass balance (63) yields an ordinary differential equation for the spreading of the avalanche

$$(g g'')' + g' g'' = 0, \qquad \text{or} \qquad 2g'g'' + g g''' = 0. \tag{68}$$

Multiplying both sides of $(68)_2$ by $g \neq 0$, it is seen that the equation can be rewritten as

$$(g^2 g'')' = 0, \tag{69}$$

which integrates to give

$$g^2 g'' = A, \tag{70}$$

where A is a constant. As the pile is assumed to be of finite size and there is no mass transfer, the total volume, V, must be conserved

$$V = \int_{\xi_R}^{\xi_F} h(\xi, t)\, d\xi = \int_{-1}^{+1} h(\eta, \tau)\, g(\tau)\, d\eta. \tag{71}$$

Substituting for the thickness (67) and integrating the emerging expression determines the unknown constant A in (70)

$$g^2 g'' = \frac{3}{2}\beta V = A. \tag{72}$$

Initially the front and rear of the avalanche are assumed to lie at $\xi = \pm 1$ and the difference velocity is assumed to be zero throughout the flow, which is equivalent to the conditions

$$g(0) = 1, \qquad g'(0) = 0. \tag{73}$$

Changing the independent variable from τ to g and making the substitution $p = g'$ equation (72) becomes

$$p\frac{dp}{dg} = \frac{A}{g^2}. \tag{74}$$

This can be integrated subject to the conditions (73) to give

$$p^2 = 2A(1 - 1/g) \tag{75}$$

and, since $p = dg/d\tau$, this is equivalent to the separable equation

$$\frac{g^{1/2}}{(g-1)^{1/2}} \frac{dg}{d\tau} = \sqrt{2A}. \tag{76}$$

Making the substitution $g = y^2$ transforms the terms on the left-hand side into standard integrable form ([9], integral no: 262.01). The final solution is

$$(g(g-1))^{1/2} + ln|g^{1/2} + (g-1)^{1/2}| = \sqrt{2A}\,\tau \tag{77}$$

which defines τ as a function of g. As $g \to \infty$ this relationship implies

$$g \sim \sqrt{2A}\,\tau. \tag{78}$$

The thickness and velocity solutions can be written in parametric form, with the help of (72) and (76), as

$$\hat{u} = \left[2A(1-1/g)\right]^{1/2}\eta, \qquad h = \frac{A}{2\beta g}(1-\eta^2). \tag{79}$$

This solution is a good test problem as it is typical of the way a granular avalanche spreads as it flows downslope.

3.3 M-wave Solution

The transformed system (63)–(64) can also be solved by seeking a separable variables solution of the form

$$h = l(\tau)H(\eta), \qquad \hat{u} = k(\tau)F(\eta). \tag{80}$$

Substituting these relations yields

$$H - \eta\frac{g'l}{gl'}H' + \frac{kl}{gl'}(HF)' = 0, \tag{81}$$

$$F - \eta\frac{g'k}{gk'}F' + \frac{k^2}{gk'}FF' + \frac{\beta l}{gk'}H' = 0. \tag{82}$$

Primes denote univariate differentiations. (81) and (82) are two ordinary differential equations for H and F provided

$$\frac{g'l}{gl'}, \qquad \frac{kl}{gl'}, \qquad \frac{g'k}{gk'}, \qquad \frac{k^2}{gk'}, \qquad \frac{l}{gk'} \tag{83}$$

are constants. Seeking power solutions of the form

$$g = \tau^{\alpha}, \qquad l = \tau^{\gamma}, \qquad k = \tau^{\delta} \tag{84}$$

the coefficients in (83) are all time independent provided

$$\gamma = 2\delta, \qquad \text{and} \qquad \delta = \alpha - 1. \tag{85}$$

A third relationship follows from the conservation of total volume

$$\int_{-1}^{+1} h(\eta,\tau)\,g(\tau)\,\mathrm{d}\eta = \tau^{\alpha+\gamma}\int_{-1}^{+1} H(\eta)\,\mathrm{d}\eta, \tag{86}$$

which is also time independent, implying

$$\alpha + \gamma = 0. \tag{87}$$

The three linear equations (85) and (87) are easily solved to give

$$\alpha = 2/3, \qquad \gamma = -2/3 \qquad \delta = -1/3. \tag{88}$$

Using (88) and substituting the power solutions (84) into the governing equations (81)–(82) two ordinary differential equations are obtained

$$F'H + (F - 2\eta/3)H' = 2H/3, \tag{89}$$

$$(F - 2\eta/3)F' + \beta H' = F/3. \tag{90}$$

The first of these may be integrated directly to give

$$(F - 2\eta/3)H = C, \tag{91}$$

where C is a constant of integration. Substituting (91) into (90) a single ordinary differential equation is obtain for the thickness profile

$$H' = \frac{H^2}{9}\left(\frac{2\eta H - 3C}{\beta H^3 - C^2}\right). \tag{92}$$

The symmetry requirement in the fixed domain mapping (60) requires that $H(\eta) = H(-\eta)$ and $H'(\eta) = -H'(-\eta)$. Applying these conditions to (92) implies that the constant

$$C = 0, \tag{93}$$

and the solution is simply

$$F = \frac{2}{3}\eta, \qquad H = \frac{1}{9\beta}\left[h_m + \eta^2\right], \tag{94}$$

where h_m is assumed to be greater than zero. It follows that the thickness and velocity profiles are

$$h = \frac{1}{9\beta}t^{-2/3}\left[h_m + \eta^2\right], \qquad \hat{u} = \frac{2}{3}t^{-1/3}\eta. \tag{95}$$

Its shape has prompted it to be called the M-wave solution. The parabolic cap and M-Wave solutions are due to Savage & Hutter [38].

3.4 Shock Wave Propagation on Non-accelerative Slopes

The conservative form of the depth integrated mass balance equation (44) implicitly assumes that h and u have continuous derivatives. To derive conditions for what happens when these derivatives are not defined we must return to an integral form of (44)

$$\frac{\mathrm{d}}{\mathrm{d}t}\int h \,\mathrm{d}x + [hu] = \int (d^s - d^b)\,\mathrm{d}x. \tag{96}$$

Let h and u be discontinuous at $x = X(t)$ and suppose that that discontinuity is contained in the region $x_1 < X < x_2$. Then the integral of the mass balance (44) with respect to x over this region is

$$\frac{\mathrm{d}}{\mathrm{d}t}\left\{\int_{x_1}^{X^-} h\,\mathrm{d}x + \int_{X^+}^{x_2} h\,\mathrm{d}x\right\} + [hu]_{x_1}^{x_2} = \int_{x_1}^{x_2} (d^s - d^b)\,\mathrm{d}x, \tag{97}$$

where the superscripts $\pm$ denote evaluation as $x \to X^{\pm}$. Using (20) to change the order of integration, this can be expressed as

$$\int_{x_1}^{x_2} \partial_t h\,\mathrm{d}x - [\![h]\!]\frac{dX}{dt} + [hu]_{x_1}^{x_2} = \int_{x_1}^{x_2} (d^s - d^b)\,\mathrm{d}x, \tag{98}$$

where the jump bracket $[\![f]\!] = f^+ - f^-$. The mass jump condition across the discontinuity is obtained by letting $x_1 \to x_2$ in (98) to give

$$[\![h(u - v_n)]\!] = 0, \tag{99}$$

where the shock speed $dX/dt = v_n$. A similar argument for the conservative form of the momentum balance (45) yields

$$[\![hu(u - v_n)]\!] = -[\![\beta h^2/2]\!]. \tag{100}$$

If the earth pressure coefficient jumps between the limiting stress states, then a corresponding jump in both the velocity and thickness should be observable in experiments. This is not the case, suggesting that there is at least a smooth transition between the limiting states in the Savage–Hutter [38] theory.

If there is no basal or surface deposition, $d^b = d^s = 0$, there are no basal topography gradients, $\partial_x b = 0$, and the slope is non-accelerative $D = 0$ then the conservation equations (44)–(45) admit constant states,

$$h(x,t) = h^+, \quad u(x,t) = u^+, \quad \text{and} \quad h(x,t) = h^-, \quad u(x,t) = u^-. \tag{101}$$

Such states are commonly observed in experiments, see Fig. 4, and can be coupled together by the jump relations (99) and (100). Thus, given the upslope velocity, u^-, the upslope thickness, h^-, and the downslope velocity u^+ the shock speed v_n and the downslope thickness h^+ may be computed. The shock speed is given by

$$v_n = [\![hu]\!]/[\![h]\!]. \tag{102}$$

Substituting this into the momentum jump (100) it is easily shown that

$$h^+h^-[\![u]\!]^2 = [\![\beta h^2/2]\!][\![h]\!], \tag{103}$$

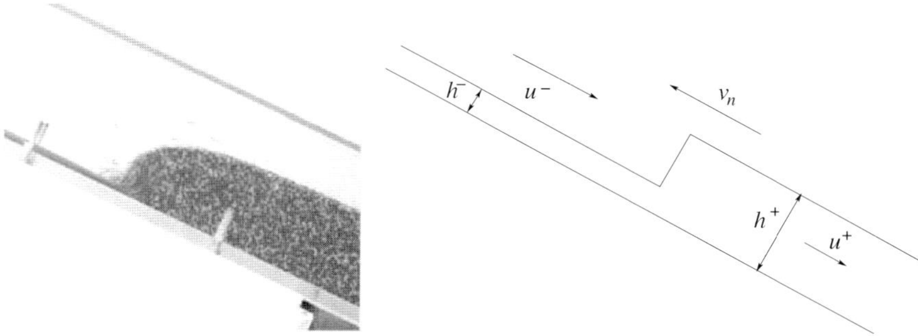

Fig. 4. Upslope propagation of a granular bore or shock. A thin rapidly moving dark layer of granular material enters from the top left in an experiment (left) and the channel is blocked at the bottom right. A shock wave forms and propagates upslope, which brings the grains to rest and increases the avalanche thickness by a factor of ten. A schematic diagram of the shock solution (right) with constant states on either side of the shock. An animation of the experiment can be seen at **www.ma.man.ac.uk/~ngray**

providing $[\![h]\!] \neq 0$. Assuming $[\![K_x]\!] = 0$ it follows that equation (103) yields a cubic equation for the relative thickness on the downslope side of the shock

$$H^3 - H^2 - (1 + 2\mathrm{Fr}_e^2)H + 1 = 0. \tag{104}$$

where the relative thickness H and effective Froude number Fr_e are defined as

$$H = h^+/h^-, \qquad \mathrm{Fr}_e^2 = [\![u]\!]^2/(\beta h^-). \tag{105}$$

Making the substitution $H = \bar{H} + 1/3$ this may be expressed in the reduced form

$$\bar{H}^3 + 3p\bar{H} + 2q = 0, \tag{106}$$

where $p = -2(2+3\mathrm{Fr}_e^2)/9$ and $q = (8-9\mathrm{Fr}_e^2)/27$. The number of real solutions is dependent on the sign of the discriminant $G = p^3 + q^2$ (e.g. [4,2]). In this case the discriminant

$$G = -\mathrm{Fr}^2(8\mathrm{Fr}_e^4 + 13\mathrm{Fr}_e^2 + 16)/27, \tag{107}$$

which is strictly negative for $\mathrm{Fr}_e^2 \neq 0$ implying three real roots. When $\mathrm{Fr}_e^2 = 0$ then $[\![u]\!] = 0$ and the trivial repeated solution is $[\![h]\!] = 0$. The roots can be

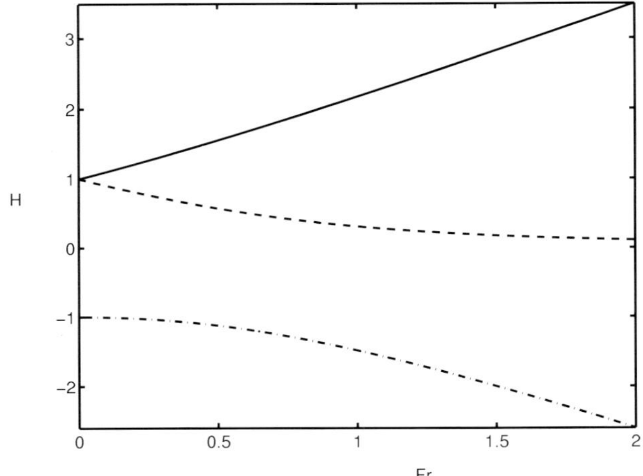

Fig. 5. The three roots of the relative thickness H on the downslope side of the jump are plotted as a function of the effective Froude number Fr_e. The solid line corresponds to root H_1, the dashed line to H_2 and the dot-dash line to H_3

represented in goniometric form

$$H_1 = 1/3 + 2m\cos(\theta/3), \tag{108}$$

$$H_2 = 1/3 - 2m\cos(\theta/3 + \pi/3), \tag{109}$$

$$H_3 = 1/3 - 2m\cos(\theta/3 - \pi/3), \tag{110}$$

where $m = |p|^{1/2}$ and $\cos\theta = -qm^{-3}$. The solutions are plotted in Fig. 5. The third solution H_3 is strictly negative, which is unphysical as it implies negative thickness. The other solutions are strictly positive with $H_1 > 1$ and $H_2 < 1$, implying either an increase, or a decrease, in the thickness on the downslope side of the jump, respectively. Imposing an energy-loss condition requires that the material must enter from the faster and shallow side of the shock, (that is, $u^- > u^+$ and $h^- < h^+$) so H_1 is the solution we are seeking. In general entropy conditions (e.g. [28,7]) are used to pick out the physically relevant solution.

4 Granular Flow in Rotating Drums

4.1 Phase Transitions

The biggest problem with granular materials is that they undergo strong phase transitions between solid, liquid and gaseous states. Solving practical problems, therefore, often involves not only determining the flow in each of the regions, but also the evolution of an unknown interface between two different states. Progress in this area is limited and the phase transitions are still poorly understood. In this section the steady flow in a partially filled rotating drum is used to investigate the coupling between a solid rotating region and a fluid-like avalanche close to the free surface. The key assumption in the theoretical treatment presented here is that the fluid-like avalanche and the solid-like region can be treated as separate bodies with a non-material singular surface between them, at which the field variables are discontinuous. Although this is a mathematical idealisation it appears to be a very good approximation [15].

4.2 Coordinate System

It is convenient to use two Cartesian coordinate systems oxz and OXZ to reflect the different geometries in the avalanching and solid body regions. The avalanche coordinate system oxz was defined in Sect. 2. The OXZ coordinate system is defined so that its axes are parallel to those in oxz, but the origin O is shifted so that it lies on the axis of revolution, as illustrated in Fig. 6. The two coordinate systems are therefore related by

$$Z = l + z, \qquad X = x, \tag{111}$$

where the constant l is defined as the *fill level* of the drum. It follows from (111) that the free surface height, $S(X,t)$, as well as the singular surface height, $B(X,t)$, in the OXZ system are related to their counterparts, $s(x,t)$ and $b(x,t)$ in the oxz system by

$$S = l + s, \qquad B = l + b. \tag{112}$$

The rotating drum is taken to have a radius R. Typical length scales in the solid region will therefore be of order R, in both the X and Z directions. The avalanche is also of length R in the x direction, but because the avalanche is shallow, its thickness is only of order H. As $H \ll R$ the *shallowness approximation* holds for avalanches in rotating drums and the theory of Sect. 2 can be directly applied.

4.3 Governing Equations in the Solid Rotating Region

Velocity and density fields within the solid material will be denoted with the superscript $-$, to avoid confusion with these fields in the avalanche. It is assumed that the solid granular material is a rigid body, of constant uniform density ρ_0^-, which rotates with angular velocity $\Omega(t)$ about the axis of revolution O. The velocity field is therefore simply

$$\boldsymbol{u}^- = \Omega r \hat{\boldsymbol{\theta}}, \tag{113}$$

where r is the distance from the axis of revolution O and $\hat{\boldsymbol{\theta}}$ is the azimuthal unit vector. Positive Ω is counted clockwise. It follows that the downslope and normal velocity components in the solid are

$$u^- = -\Omega Z, \qquad w^- = \Omega X. \tag{114}$$

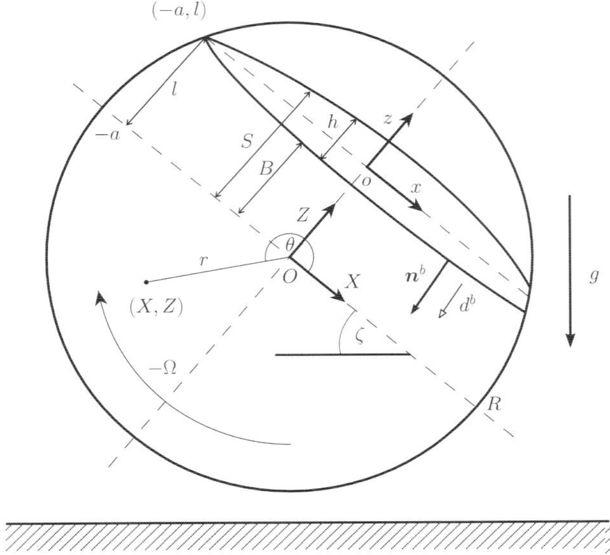

Fig. 6. A sketch of a rotating drum of radius R partially filled with granular material. It is convenient to use a coordinate system $oxyz$ in the avalanche and a system $OXYZ$ in the solid body. The angular velocity Ω is positive for rotation in the counter clockwise sense. The above configuration is for clockwise rotation and the angular velocity is denoted by $-\Omega$

At the singular interface, $F^b = 0$, with the avalanching granular material, the body is subject to a kinematic boundary condition

$$F^b(\boldsymbol{x}, t) = 0 : \qquad \partial_t F^b + v_n^{b-} \boldsymbol{n}^b \cdot \nabla F^b = 0 \tag{115}$$

where v_n^{b-} is the normal speed of the interface in the direction of the normal $\boldsymbol{n}^b$, which points outward from the avalanche. The superscript notation $b+$ and $b-$ is introduced to differentiate between variables that are evaluated on the avalanching side (upper side) and solid body side (lower side) of the singular interface, F^b, respectively.

4.4 Interfacial Conditions and Scalings

At the interface between the avalanche and the rotating granular material there are discontinuities in the velocity and density fields. At such a singular surface the mass jump condition (e.g. [5]) is

$$[\![\rho(\boldsymbol{u} \cdot \boldsymbol{n}^b - v_n^b)]\!] = 0. \tag{116}$$

This provides a coupling condition between the velocities in the avalanche and the solid. If the typical angular velocity magnitude of the rotating granular material is, Ω^*, then the velocities in the solid body are of magnitude $\Omega^* R$. Assuming that density changes are relatively small, a balance between the normal velocity components in (116) implies that $\Omega^* R = \varepsilon (gR)^{1/2}$.

The scalings (7) also hold true for the avalanche in the rotating drum. New scalings are now introduced for the fill height, the free and singular surface variations, and the variables in the solid rotating body

$$\begin{aligned} (X, Z, l, B, S) &= R(\tilde{X}, \tilde{Z}, \tilde{l}, \tilde{B}, \tilde{S}), \\ (\Omega) &= \varepsilon (g/R)^{1/2} (\tilde{\Omega}), \\ (u^-, w^-, v_n^{b-}) &= \varepsilon (Rg)^{1/2} (\tilde{u}^-, \tilde{w}^-, \tilde{v}_n^{b-}), \end{aligned} \tag{117}$$

where the tildes are again used to indicate non-dimensional variables. In the avalanche the downslope and normal length scales are scaled differently to reflect the shallowness of the geometry, whilst in the solid body the coordinates are scaled using the same length scales. Using (111) these differential scalings imply that the non-dimensional coordinates are related by

$$Z = l + \varepsilon z, \qquad X = x, \tag{118}$$

where for simplicity the tildes are dropped. Henceforth all variables are non-dimensional unless stated otherwise. It follows that the free surface height and the interface height are

$$S = l + \varepsilon s, \qquad B = l + \varepsilon b. \tag{119}$$

The velocity components in the solid body are

$$u^- = -\Omega Z, \qquad w^- = \Omega X, \tag{120}$$

and, in particular, the velocity components on the lower side of the singular surface are

$$u^{b-} = -\Omega(l + \varepsilon b), \qquad w^{b-} = \Omega x, \tag{121}$$

in coordinates oxz.

Let d^{b+}, and d^{b-}, be the equivalent volumes of avalanching, and solid, granular material deposited from the avalanche to the solid per unit area per unit time, respectively. It follows that the normal speed v_n^b in the direction of the normal $\boldsymbol{n}^b$ is equal to

$$v_n^{b+} = \boldsymbol{u}^{b+} \cdot \boldsymbol{n}^b - d^{b+}, \tag{122}$$

and equal to

$$v_n^{b-} = \boldsymbol{u}^{b-} \cdot \boldsymbol{n}^b - d^{b-}, \tag{123}$$

where $v_n^b = v_n^{b+} = v_n^{b-}$ is required to conserve mass and prevent void space opening up between the solid and avalanching layers. Assuming that the avalanche density on the upper side of the singular surface is ρ^+ the mass jump condition (116) implies that

$$d^{b+} = (\rho^- / \rho^+)\, d^{b-}, \tag{124}$$

where the ratio of the densities is close to unity. Noting that the velocities on the lower (solid body) side of the singular surface are prescribed by the nature of the rigid rotation (121) the deposition rate on the lower side of the singular surface is

$$d^{b-} = -\varepsilon(\Omega/\Delta^b)(l + \varepsilon b)\partial_x b - (\Omega/\Delta^b)x - v_n^b. \tag{125}$$

The normal velocity $v_n^b = -\partial_t b/\Delta^b$ by (115) and it follows that

$$d^{b-} = -\varepsilon\Omega l \partial_x b - \Omega x + \partial_t b + \mathrm{O}(\varepsilon^2), \tag{126}$$

which together with (124) provides an approximation for the deposition rate d^{b+} to order $\varepsilon^{1+\gamma}$.

4.5 Governing Equations in the Avalanche Region

The fluid-like region in the rotating drum can be modelled as a granular avalanche with erosion and deposition at its basal interface. The theory of Sect. 2 is therefore appropriate and the coordinate system oxz and all the variables, scalings and results are adopted here. In addition the avalanche is assumed to have constant density $\rho^+ = \rho_0$ and the surface accumulation, d^s, defined in Sect.2 is taken equal to zero. The conservation laws therefore reduce to

$$\partial_t h + \partial_x(hu) = -d^{b+}, \tag{127}$$

$$\partial_t(hu) + \partial_x(hu^2) + \varepsilon\partial_x(K_x \cos\zeta h^2/2) = hD - \varepsilon h \cos\zeta \partial_x b - ud^{b+}, \tag{128}$$

to $\mathrm{O}(\varepsilon^{1+\gamma})$, where the deposition rate d^{b+} is given by (124) and (126), and the earth pressure coefficient is defined in (38). In the rotating drum experiments there is a downslope velocity component in the solid region. The scalings (7) and (117) imply that the relative slip velocity $u^r = u - \varepsilon\Omega l + \mathrm{O}(\varepsilon^2)$. This is incorporated into the source term (29) to give

$$D = \cos\zeta(\tan\zeta - \mathrm{sgn}(u - \varepsilon\Omega l)\tan\delta). \tag{129}$$

At slow rotation rates there are essentially two flow regimes; intermittent flow and continuous flow. Intermittent flow occurs at rotation periods above 100 seconds per revolution. In intermittent flow the rotation of the drum increases the inclination of the free surface until the whole or part of it reaches the maximum angle of repose (Hungr & Morgenstern 1984a,b). Failure then occurs along an internal slip line and and an avalanche is released, which flows rapidly downslope. When the avalanche front reaches the drum wall a shock

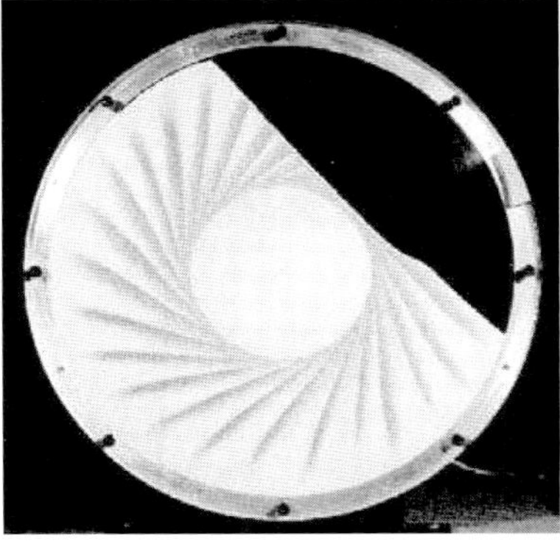
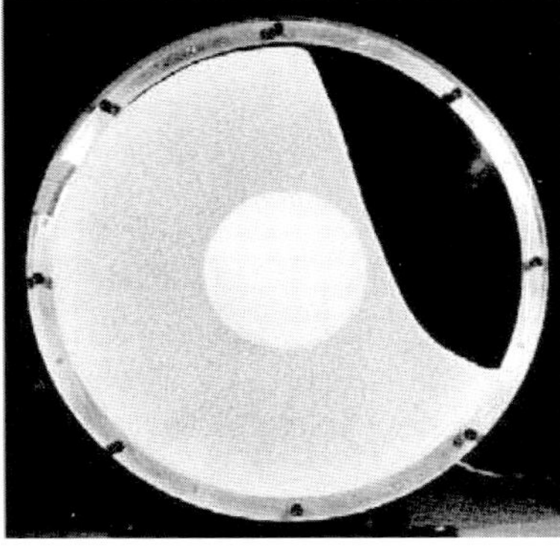

Fig. 7. Two photographs illustrating the differences between the intermittent. (left) and continuous (right) flow regimes. The drum contains small dark particles and large white particles, which segregate in the avalanche so that the large particles over-lie the small ones. The patterns that are formed strongly reflect the differences between the bulk flow regimes changing from time dependent to steady state

wave (Gray & Hutter, 1997) is generated, that propagates upslope bringing the avalanche to rest. The free surface inclination angle now lies below the maximum angle of repose and must wait until the drums rotates it up to the maximum angle of repose, before failure occurs again. The intermittent flow regime is therefore characterised by discrete avalanche events and shock wave propagation.

Continuous flow occurs in the range between 10-100 seconds per revolution. The avalanche flows continuously downslope and both the basal interface and the free surface are spatially fixed, which implies that there is continuous steady erosion and deposition across the avalanche/solid-body interface. It follows that, material is fed to the top half of the avalanche by slow rotation of the solid body region, the avalanche then transports this material rapidly downslope, where it is re-absorbed into the solid body region and transported to the top again. All processes take place continuously and this flow configuration is stable in a wide range of rotation rates.

The differences between continuous and intermittent flow become immediately apparent when a mixture of small dark particles and large white particles is placed into a thin drum and rotated (Gray & Hutter, 1997, 1998). Fig. 7 shows the patterns that are generated in these two flow regimes. The dynamics of the bulk flow is virtually the same as when a single grain size is present, however, within the avalanche the particles are segregated by size, so that the larger ones over-lie the smaller ones. In the intermittent flow regime avalanches are periodically released and come to rest by upslope shock wave propagation. Each discrete avalanche event leaves behind a stripe at the free surface, which is composed of large white particles overlying the small dark ones, that is subsequently rotated around the axis of revolution by the solid body motion. In contrast, in the steady flow regime there is continuous erosion and deposition at the basal interface, which gives rise to a continuous distribution of particle sizes away from the central core.

4.6 An Exact Solution for Steady Flow

Steady state solutions to the rotating drum theory are now investigated. It is therefore assumed that solid granular material rotates with constant angular velocity, Ω_0, and that all derivatives with respect to time are zero, $\partial_t() = 0$. The basal friction angle, $\delta(x)$ is assumed to be equal to a constant value δ_0. It follows that if the coordinate system oxz is chosen such that its angle of inclination $\zeta = \delta_0$, then the net driving force, D, (defined in (129)) is equal to zero, providing that the magnitude of the avalanche velocity is greater than $|\varepsilon\Omega_0 l|$. For classical smooth solutions the conservative form of the momentum balance (128) can be simplified with the help of the mass balance (127) to yield

$$\partial_x(hu) = (\rho^-/\rho^+)(\varepsilon\Omega_0 l\partial_x b + \Omega_0 x), \tag{130}$$

$$hu\partial_x u + \varepsilon\partial_x(K_x\cos\zeta h^2/2) = -\varepsilon h\cos\zeta\partial_x b. \tag{131}$$

The constitutive parameter ρ^-/ρ^+ is included in the theory to account for the fact that the solid material must dilate for the grains to move past each other in the fluid regime. Although this effect does not greatly effect the dynamics of the avalanche, it may be important for future theories which seek to model inter-particle percolation during the size segregation process. Typically the amount of dilatation is of the order of 5 to 10%.

Usually the avalanche equations (130)–(131) are solved for the avalanche thickness, h, and velocity, u, given suitable initial and boundary conditions and basal topography, b. Here a special class of solutions is considered in which the downslope avalanche velocity is assumed to be constant,

$$u = u_0, \tag{132}$$

and equations (130)–(131) are solved for the avalanche thickness, h, and the basal topography, b, given suitable boundary conditions. Since $\partial_x u = 0$ the earth pressure coefficient is equal to a constant value K_0 throughout the avalanche. For the inviscid model $K_0 = 1$, in the Savage & Hutter [38] theory, $K_0 = K_{act}$, and in the regularised model of Tai & Gray [43], $K_0 = 2\sec^2\phi - 1$. For constant $K = K_0$ the governing equations reduce to

$$-\lambda\partial_x h = \varepsilon l\partial_x b + x, \tag{133}$$

$$K_0\partial_x h = -\partial_x b, \tag{134}$$

where λ is the order unity constant

$$\lambda = -\frac{\rho^+ u_0}{\rho^-\Omega_0}. \tag{135}$$

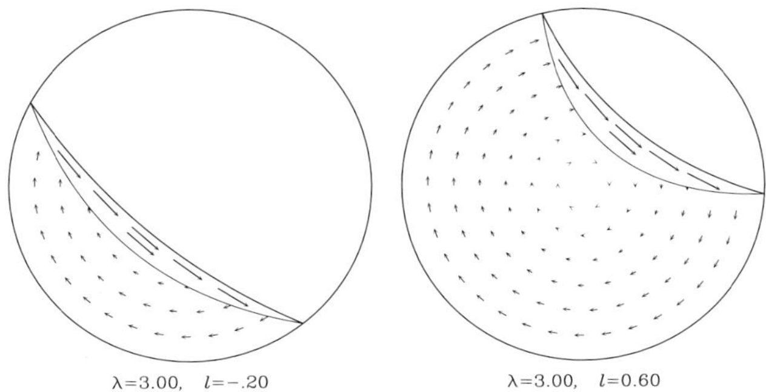

Fig. 8. The steady state solution for the flow of granular material in a partially filled rotating drum is illustrated for two differing fill heights. In both cases $R = 1$, $\rho^-/\rho^+ = 1.05$, $K_0 = 2.0$ and $\delta_0 = 40°$. The position of the avalanche free surface and the interface between the avalanching and solid regions are illustrated, as well as velocity vectors to show the relative speed and direction of the flow

For solutions with positive avalanche thicknesses we shall show that λ is positive. It may therefore be assumed, without loss of generality, that the downslope avalanche velocity is positive, $u_0 > 0$, and hence that the angular velocity is negative, $\Omega_0 < 0$.

The uppermost point of intersection between the avalanche free surface and the drum wall defines the top of the avalanche and lies at $(-a, l)$ in the OXZ system, where l is the fill level and $a = (R^2 - l^2)^{1/2}$. Integrating equations (133) and (134) subject to the boundary conditions $h = 0$ and $b = 0$ at $x = -a$ implies

$$-\lambda h = \varepsilon l b - (a^2 - x^2)/2, \tag{136}$$

$$K_0 h = -b. \tag{137}$$

It follows that the avalanche thickness and basal topography are

$$h = \quad h_0(a^2 - x^2)/a^2, \tag{138}$$

$$b = -K_0 h_0 (a^2 - x^2)/a^2, \tag{139}$$

where the constant

$$h_0 = \frac{a^2}{2(\lambda - \varepsilon l K_0)} \tag{140}$$

is the avalanche thickness at $x = 0$. For positive order unity values of λ the constant h_0 and hence the avalanche thickness h are positive, confirming our original assumption. Given free parameters λ, δ_0, l and R, and using one of the constitutive models to determine K_0 the system of equations can be used to predict the avalanche thickness h and the position of the free surface b.

The explicit solution is plotted in Fig. 8, assuming that $K_0 > 1$, to give the avalanche a characteristic *crescent* shape with a concave free surface. In the inviscid model ($K_0 = 1$) the solution is similar except the free surface is straight. If $K_0 < 1$ then the free surface would be convex. In the experiments presented here the free surface is slightly concave.

In the un-stretched coordinate system, when ε is set equal to unity, the non-dimensional parameter λ becomes large. Using the scalings (7) and (117) it is easily shown that $\lambda = (\rho^+ u_0^*)/(-\rho^- \Omega_0^* R)$, where the starred variables u_0^* and Ω_0^* are typical dimensional magnitudes of the downslope and angular velocities, respectively. The downslope avalanche velocity, u_0^*, is therefore much larger than the maximum speed in the rotating granular material, $R\Omega_0^*$. The parameter λ is a measure of the ratio of the maximum avalanche to solid body speeds, and typically lies in the range $10 \leq \lambda \leq 30$. The solutions in Fig. 8 are presented for the case $\lambda = 3$, i.e. downslope avalanche velocities are three times larger than the solid body velocity close to the drum wall. This value of λ allows the velocity vectors to be plotted on the same diagrams and illustrates the effect of the ε terms in (140). For positive fill levels the coupling with the solid body velocity field implies that the avalanche is thicker than for negative values of l. In addition the curvature of the basal topography is larger for increasing fill levels, which has also been observed in the experiments. As λ is increased the avalanche thickness and curvature decrease and the velocity increases.

4.7 Particle Paths

The particle paths within the avalanching flow can be determined by integrating the differential equations

$$\frac{dx}{dt} = u_0, \qquad \frac{dz}{dt} = w, \tag{141}$$

subject to the initial conditions that $x = x_0$ and $z = z_0$ at $t = 0$. The derivative d/dt is the rate of change as observed when moving with a fixed particle. It follows that

$$x = u_0 t + x_0, \tag{142}$$

since the downslope velocity is constant. The normal velocity w can be determined from the incompressibility relation (8) subject to the interfacial condition (116). Since the downslope velocity is constant incompressibility implies that $\partial_z w = 0$. Integrating this with respect to z

$$w = w^{b+}, \tag{143}$$

where the normal velocity at the base of the avalanche

$$w^{b+} = u_0 \partial_x b + (\rho^-/\rho^+)\Omega_0(\varepsilon l \partial_x b + x) + \mathrm{O}(\varepsilon^2). \tag{144}$$

Substituting (143) and (144) into the differential equation for the normal velocity component (141) and using (142) to make the change of variables $t = (x - x_0)/u_0$ implies that

$$\lambda \frac{dz}{dx} = \lambda \partial_x b - (\varepsilon l \partial_x b + x) + \mathrm{O}(\varepsilon^2). \tag{145}$$

Integrating with respect to x implies that the particles' height as a function of position is

$$z = z_0 + b - (\varepsilon l/\lambda) b - x^2/(2\lambda) + \mathrm{O}(\varepsilon^2), \tag{146}$$

where z_0 is the initial height of the particle.

To aid the process of tracking particles it is convenient to introduce a new parameter α to label the particle paths that pass through the avalanching domain. The maximum height of the avalanche is attained at $x = 0$ and therefore all the avalanche particle paths cross the z-axis. Let α be the relative height of the avalanche particle path to the maximum avalanche height h_0 as it crosses the z-axis. It follows that α is linear in z and equal to zero at the base of avalanche and unity at the free surface. The value of α uniquely labels the particle paths that pass through the avalanche.

It shall now be shown that the avalanching particle paths form closed curves that pass through the fluid- and solid-like regions. Using the definition of α above it follows from (146) that the particles height

$$z = b + h_0 \alpha - (\varepsilon l/\lambda)(b - b_0) - x^2/(2\lambda) + \mathrm{O}(\varepsilon^2), \tag{147}$$

where b_0 is defined as position of the solid/avalanche interface at $x = 0$ and is given by $b_0 = -K_0 h_0$. Equation (147) can be simplified by substituting for the basal topography, from (139) and (140), in the third term on the right-hand side to give

$$z = b + h_0 \alpha - h_0 (x/a)^2 + \mathrm{O}(\varepsilon^2). \tag{148}$$

A particle crosses the interface between the avalanche and the solid rotating granular material when its height is equal to the height of the local basal topography, i.e. when $z = b$. In this case (148) becomes a quadratic equation for the intersection positions and has two real roots

$$x_{b_1} = a\sqrt{\alpha} =: x_b, \qquad x_{b_2} = -a\sqrt{\alpha} = -x_b. \tag{149}$$

These points lie an equal downslope distance on either side of the z-axis. This is illustrated in Fig. 9. The associated normal components of the intersection points are

$$z_{b_1} = b(x_{b_1}), \qquad z_{b_2} = b(x_{b_2}), \tag{150}$$

respectively, where b is given by (119) and (139). As the intersection points lie an equal downslope distance on either side of the origin and since the

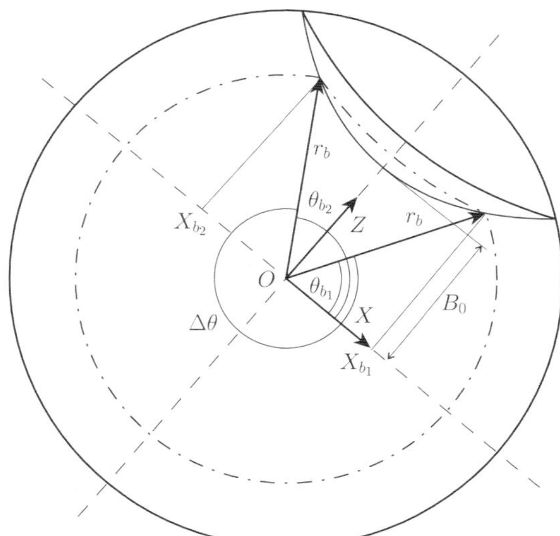

Fig. 9. A sketch showing a closed particle path (dot dash line) for steady granular flow inside a partially filled drum. At position $[r_b, \theta_{b_1}]$ particles leave the avalanching region and rotate rigidly until they reach position $[r_b, \theta_{b_2}]$, where they re-enter the avalanche and are transported to $[r_b, \theta_{b_1}]$ again to complete the circuit. The angle between θ_{b_1} and θ_{b_2} measured through the region of solid rotation is equal to $\Delta\theta$. The base of the avalanche at $X = 0$ lies at B_0. If B_0 is greater than zero a solid core develops in the centre of the drum and the particles in this region never enter into the avalanche

basal interface, b, is an even function, it follows that the intersection points have the same normal components $z_{b_1} = z_{b_2}$ ($= z_b$ say). Assuming that the velocity u_0 is positive, a particle will cross from the solid to the avalanche at $\{x_{b_2}, z_b\}$ and pass back from the avalanche to the solid at $\{x_{b_1}, z_b\}$, where the curly brackets $\{\,,\,\}$ denote components in the oxz system. In the OXZ system these points correspond to (X_{b_2}, Z_b) and (X_{b_1}, Z_b), respectively, where $X_{b_1} = x_{b_1}$, $X_{b_2} = x_{b_2}$ and $Z_b = l + \varepsilon z_b$. Both of these positions lie an equal distance,

$$r_b = \sqrt{X_b^2 + Z_b^2}, \tag{151}$$

from the axis of rotation at the centre of the drum.

The particles in the solid granular material are in rigid rotation about the origin and their paths are given by solving

$$\frac{dX^-}{dt} = -\varepsilon \Omega_0 Z^-, \qquad \frac{dZ^-}{dt} = \varepsilon \Omega_0 X^- \tag{152}$$

subject to the initial conditions that $X^- = X_0^-$ and $Z^- = Z_0^-$ at $t = 0$. It is simple to show that the particles move in circular arcs that are parameterised

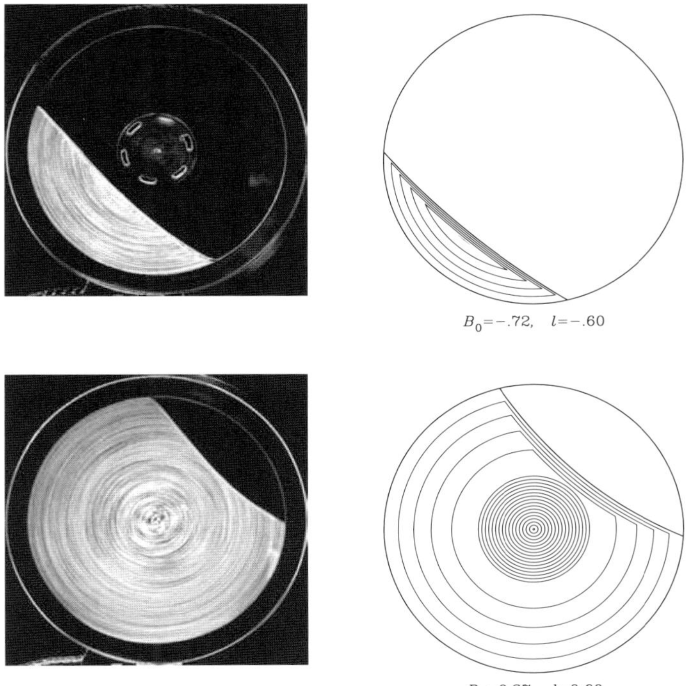

Fig. 10. Long time exposures showing the particle paths and the corresponding steady flow solutions. In the solid rotating region the particles describe circular arcs. For $B_0 \leq 0$ all the particle paths pass through the avalanche. For $B_0 > 0$ some of the particles never intersect with the free surface and perform closed circuits entirely within the solid body region

by

$$X^- = r\cos\theta, \qquad Z^- = r\sin\theta, \tag{153}$$

where the azimuthal angle

$$\theta = \varepsilon\Omega_0 t + \theta_0 \tag{154}$$

and $\theta_0 = \tan(Z_0^-/X_0^-)$.

An avalanche particle path intersects with the basal interface at positions (X_{b_1}, Z_b) and at (X_{b_2}, Z_b). In polar coordinates these points have positions $[r_b, \theta_{b_1}]$ and $[r_b, \theta_{b_2}]$, respectively, where the angles

$$\theta_{b_1} = \arccos(X_{b_1}/r_b), \tag{155}$$

$$\theta_{b_2} = \arccos(X_{b_2}/r_b). \tag{156}$$

It follows that after a particle has crossed from the avalanche to the solid at $[r_b, \theta_{b_1}]$ it then describes a circular arc until it reaches $[r_b, \theta_{b_2}]$, where it re-enters the avalanche. As both the starting and re-entry positions lie on the same circular arc it follows that the particle paths in the rotating drum form closed curves that extend through both the avalanching and solid body regions. Each of these closed curves is identified by the relative height α as it crosses the x-axis in the avalanche. The solutions are illustrated in Fig. 10 for two different fill heights.

A particularly interesting feature of the flow is that if the height of the basal topography lies above the axis of rotation at $x = 0$, a solid central core develops in which the particles never enter into the avalanche. That is, if the height of the basal interface at $x = 0$

$$\left.\begin{array}{l} B_0 > 0 : \text{a solid core develops,} \\ B_0 \leq 0 : \text{no solid core exists,} \end{array}\right\} \tag{157}$$

where $B_0 = l + \varepsilon b_0$. For positive B_0 the radius of the central core is $B_0 = l - \varepsilon K_0 h_0$.

4.8 Circuit Times

The time a particle takes to complete a circuit around the drum is extremely sensitive to the fill level of the drum, l, and the particle path, α. This has very important consequences for the mixing of granular material, which shall be shown experimentally in the next section. A particle entering the avalanche at position $\{-x_b, z_b\}$ takes a time

$$t_a = 2x_b/u_0, \tag{158}$$

to reach the other end of the avalanche at $\{x_b, z_b\}$. Once the particle leaves the avalanche it travels along a circular arc in the solid region from $[r_b, \theta_{b_1}]$ to $[r_b, \theta_{b_2}]$ taking a time

$$t_s = \Delta\theta/(-\varepsilon\Omega_0) \tag{159}$$

where

$$\Delta\theta = \text{ang}(\theta_{b1}, \theta_{b2}) \tag{160}$$

is the angle between the intersection points measured through the solid region. The angle lies in the range $0 \leq \Delta\theta \leq 2\pi$. The factor ε^{-1} in (159) implies that the time spent in the solid region is large compared to the time spent in the avalanche, i.e. $t_s \gg t_a$. Note, that the solutions in this and subsequent sections are all plotted with the aspect ratio 1:1, which is achieved by setting $\varepsilon = 1$ in the avalanche flow solutions to un-stretch the coordinates. For simplicity, the angular velocity is chosen so that the drum performs one

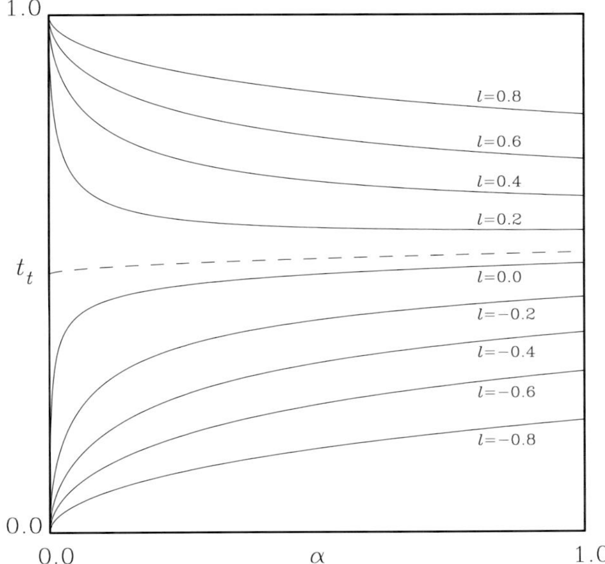

Fig. 11. The total time for a particle to complete one circuit around the drum, t_t, is plotted as a function of the particle path α for a series of fill levels, l. In each case $R = 1$, $\lambda = 15$, $\rho^-/\rho^+ = 1.05$, $K_0 = 2.0$, $\delta_0 = 40°$ and $\Omega_0 = -2\pi$. Recall, that when $\alpha = 1$ the particles move around paths adjacent to the drum wall and the free surface of the avalanche. Whilst when $\alpha = 0$ the particles move around the circumference of the solid core or lie at the singular point $(0, B_0)$. The dashed line marks the transition between flows with a central solid core above the line and those without below the line

complete revolution in a single non-dimensional time unit, i.e. $\Omega_0 = -2\pi$. It follows that the particles travel through the avalanche in a time, $t_a \ll 1$.

The total time t_t taken for a particle to perform one complete circuit of the drum is the sum of the time spent in the avalanche t_a and the time spent in solid rotation t_s,

$$t_t = t_a + t_s. \tag{161}$$

This is plotted in Fig. 11 as a function of the particle path α for a series of fill levels, l. In general, when a solid core develops ($B_0 > 0$) grains that start off on higher particle paths, α, perform complete circuits of the drum faster than those on lower ones. Close to the interface $B_0 = 0$ interface curvature effects can cause the minimum circuit time to occur internally, but this is a weak effect. At the point where the base of the avalanche $(0, B_0)$ coincides with the axis of revolution, i.e. $B_0 = 0$, there is a very sudden transition, which is indicated by the dashed line in Fig. 11. For $B_0 \leq 0$ the solid core disappears and grains that start off on higher particle paths, α, take longer to perform a circuit than those on lower paths. At the point $(0, B_0)$ there is singular behaviour since it takes zero time to perform a circuit, which leads to very intense mixing. These features shall be explained with the help of the experiments of the next section.

4.9 Steady Mixing of Granular Material

The difference between the total time, t_t, to complete a circuit on different particle paths, α, leads to an apparent shearing or mixing of the granular material. Metcalfe et al. [31] designed a simple experiment to make this deformation and mixing visible. The granular material is simply divided into two differently coloured regions of equal mass, as shown in Fig. 12, and then the drum is rotated. In Metcalfe et al.'s [31] experiments the drum was rotated so slowly that they were in the intermittent flow regime, in which discrete avalanches were observed that stopped completely before the next one began. In contrast, the experiments presented here are in the steady continuous flow regime [15].

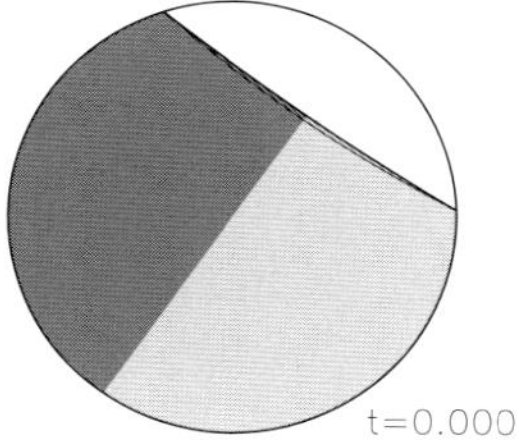

Fig. 12. The granular material is initially divided into two regions of equal mass and the drum is rotated with constant angular velocity

The steady flow solution (138)–(140) together with the particle tracking and circuit time information can easily be used to model continuous mixing in a partially filled rotating drum. A series of marker points are used to track the position of the interface between the light and dark sand. In the theory there is no transport or diffusion across this interface and the marker points simply move around the particle paths. If a marker point is placed on a known particle path α it stays on this path for all time.

Given the initial position of a marker point the problem is to find its position at any given time T. Each circuit around path α takes a known time t_t, which is defined in (158)–(161). This makes tracking the particle easy as the total number of complete rotations around the circuit α is simply

$$n = \mathrm{int}(\mathrm{T}/\mathrm{t_t}), \tag{162}$$

where 'int' is the integer part. The problem therefore reduces to finding the position of a particle in the finite time interval $0 \le \tau < t_t(\alpha)$, where τ is the

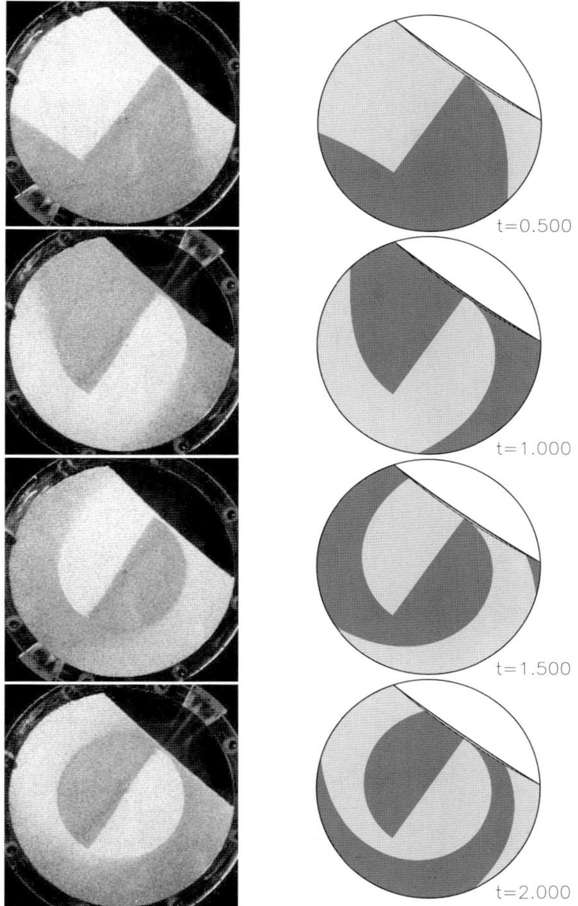

Fig. 13. Photographs (left) and exact solutions (right) for the mixing of mono-disperse granular material during steady continuous flow for a fill level above the axis of revolution. One revolution of the drum equates to one non-dimensional time unit. An animation of the exact solution can be seen at **www.ma.man.ac.uk/˜ngray**

truncated time

$$\tau = T - nt_t. \tag{163}$$

Since the path and the velocity along the path are known it is a simple matter to compute the position of the particle at the truncated time τ.

For $B_0 > 0$, Fig. 13, the central solid core is immediately apparent as no deformation, or mixing occurs, in this region and the interface remains straight. Outside the central core grains on higher particle paths, α, perform

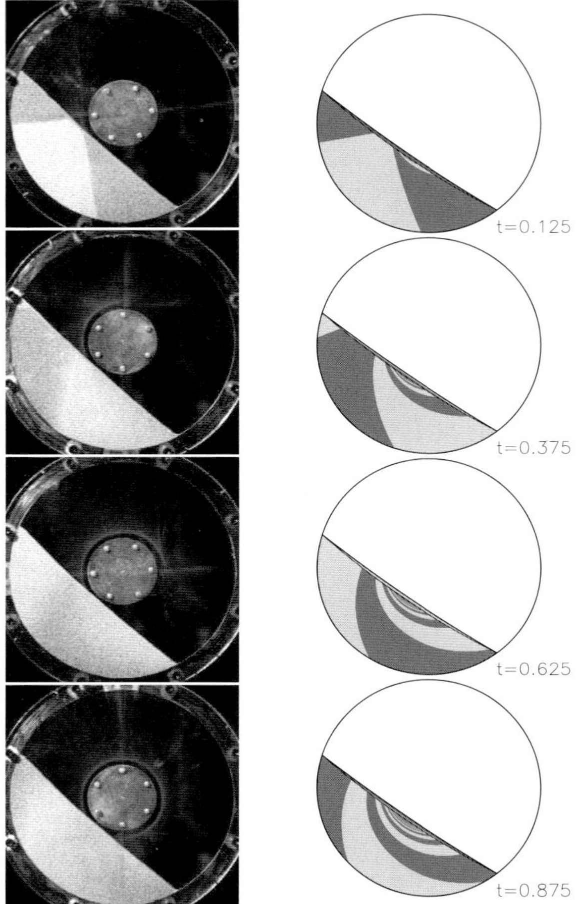

Fig. 14. Photographs (left) and exact solutions (right) for the mixing of monodisperse granular material during steady continuous flow for a fill level below the axis of revolution. One revolution of the drum equates to one non-dimensional time unit

circuits faster than those on lower paths. The reason for this is not because the particles themselves travel any faster, but simply because of where they cross the singular surface, b. The angle between the entry and exit points measured through the solid body, $\Delta\theta$, which was defined in (160), is smaller for higher particle paths, α, than for lower ones. It follows that the grain simply spends less time in slow solid rotation, and gets re-absorbed into the solid body ahead of particles of lower particle paths, α. When the fill level lies below the centre of the drum the character of the explicit solutions and the experimental results change dramatically, see Fig. 14. In direct contrast to the case, $B_0 > 0$, grains on higher particle paths, α, move round the drum slower than those on lower ones. Again the reason for this is simply the position at which they cross the singular surface, b. The angle between the entry and exit points measured through the solid body, $\Delta\theta$, is larger for higher particle paths, α, than for lower ones, when the interface lies below the axis of revolution. These particles must therefore spend more time in slow solid body rotation.

After about half a drum revolution the interface has been deformed so much that the explicit solution develops extremely fine scale features close to $(0, B_0)$, which lie well below the grain scale. The cause of this is that in the limit as $\alpha \to 0$, the time t_t for particles to perform circuits tends to zero, as shown by the graph in Fig. 11. The interface therefore wraps itself around the singular point very quickly, passing through the avalanche many times and undergoing a large amount of diffusion.

Conclusions

This article has reviewed the Savage-Hutter [38] theory and generalised it to include the effects of surface and basal accretion and deposition. An inviscid fluid model, first proposed by Eglit [10], has also been considered. The Savage-Hutter theory is shown to be a degenerate hyperbolic-parabolic system of equations, which reduces to a non-strictly hyperbolic system, that is similar to the inviscid fluid model when the internal angle of friction is equal to the basal angle of friction. The parabolic cap and M-wave solutions [38] have been reviewed and a new shock solution has been derived, which is equivalent to a bore in hydraulic flows.

A simplified theory for granular flow in rotating drums has also been presented, which treats the avalanching and solid rotating regions as two separate bodies that interact across a non-material singular surface. This model captures the essential features of the mixing in the case of steady flow, and there is excellent agreement between theory and experiment. Whilst the extended Savage–Hutter and inviscid theories provide a full mechanical description for the fluid-like avalanching phase of the motion, there is still much work to be done. A complete description requires the stress field in the solid phase to be determined in order to solve for the failure line between the solid- and fluid-like regions, which is at present prescribed.

References

1. Abramowitz, M., Stegun, I. (1970) Handbook of mathematical functions. 9th edition, Dover Publ. Inc. New York
2. Bartsch, H.J. (1994) Taschenbuch mathematischer Formeln. Fachbuchverlag Leipzig–Köln

3. Baxter, J. Tüzün, U., Heyes, D., Hayati, I., Fredlund, P. (1998) Stratification in poured granular heaps. Nature **391,** 136
4. Bronstein, I.N., Semendjajew, K.A.(1975) Taschenbuch der Mathematik. Teubner, Leipzig
5. Chadwick, P. (1976) An introduction to continuum mechanics. George, Allen & Unwin
6. Courant, R., Hilbert D. (1962) Methods of mathematical physics. Part II. Partial differential equations. Interscience Publishers, John Wiley and Sons, New York, London
7. Dafermos, C.M. (2000) Hyperbolic conservation laws in continuum physics. A series of comprehensive mathematical studies, Springer **325**
8. Dent, J.D., Burrel, K.J., Schmidt, D.S., Louge, M.Y., Adams, E.E., Jazbutis, T.G. (1998) Density, velocity and friction measurements in a dry-snow avalanche. Annal. Glac. **26,** 247–252
9. Dwight, H.B. (1968) Tables of integrals and other mathematical data. Macmillan
10. Eglit M.E. (1983) Some mathematical models of snow avalanches. In: Shahinpoor, M. (Ed.) Advances in mechanics and the flow of granular materials, Clausthal-Zellerfeld and Gulf Publishing Company **2,** 577–588
11. Gray, J.M.N.T., Hutter K. (1997) Pattern formation in granular avalanches. Contin. Mech. & Thermodyn. **9**(6), 341–345
12. Gray, J.M.N.T., Hutter, K. (1998) Physik granularer Lawinen. Physikalische Blätter **54**(1), 37–43
13. Gray, J.M.N.T., Tai Y.C. (1998) Particle size segregation, granular shocks and stratification patterns. In: Herrmann et al. (Eds.) NATO ASI series, Physics of dry granular media, Kluwer Academic, 697–702
14. Gray, J. M. N. T., Wieland, M., Hutter K. (1999) Free surface flow of cohesionless granular avalanches over complex basal topography. Proc. Roy. Soc. London A **455,** 1841–1874
15. Gray, J.M.N.T. (2001) Granular flow in partially filled slowly rotating drums. J. Fluid. Mech. **441**, 1–29
16. Greve, R., Hutter, K. (1993) Motion of a granular avalanche in a convex and concave curved chute: experiments and theoretical predictions. Phil. Trans. Roy. Soc. A **342,** 573–600
17. Greve, R., Koch, T., Hutter, K. (1994) Unconfined flow of granular avalanches along a partly curved surface. Part I: Theory. Proc. R. Soc. London, A **445,** 399–413
18. Hill, K.M., Kharkar, D.V., Gilchrist, J.F., McCarthy, J.J., Ottino, J.M. (1999) Segregation-driven organization in chaotic granular flows. Proceedings of the National Academy of Sciences of the United States of America, **96** (21), 11689–12210
19. Hutter, K., Koch, T. (1991) Motion of a granular avalanche in an exponentially curved chute: experiments and theoretical predictions. Phil. Trans. Roy. Soc. A **334,** 93–138
20. Hutter, K., Greve, R. (1993) Two-dimensional similarity solutions for finite mass granular avalanches with Coulomb and viscous-type frictional resistance. J. Glac. **39**, 357–372
21. Hutter, K., Siegel, M., Savage, S.B., Nohguchi, Y. (1993) Two-dimensional spreading of a granular avalanche down an inclined plane. Part I Theory. Acta Mech. **100**, 37–68

22. Hutter, K., Koch, T., Plüss, C., Savage, S. B. (1995) The dynamics of avalanches of granular materials from initiation to runout. Acta Mech. **109**, 127–165
23. Hutter, K., Nohguchi, Y. (1990) Similarity solutions for a Voellmy model of snow avalanches with finite mass. Acta Mech. **82**, 99–127
24. Hutter, K. (1996) Avalanche Dynamics In: Singh, V.P. (Ed.) Hydrology of disasters, Water Sciences and Technology, Kluwer Academic **24**, 317–394
25. Jiang, G., Tadmor, E. (1997) Non-oscillatory Central Schemes for Multidimensional Hyperbolic Conservation Laws. SIAM J. Sc. Comp. **19**(6), 1892–1917
26. Keller, S., Ito, Y., Nishimura K. (1998) Measurements of the vertical velocity distribution in ping pong ball avalanches. Annal. Glac. **26**, 247–252
27. Koch, T., Greve, R., Hutter, K. (1994) Unconfined flow of granular avalanches along a partly curved surface, Part II: Experiments and numerical computations. Proc. R. Soc. London, A **445**, 415–435
28. Le Veque, R.J. (1990) Numerical methods for conservation laws. Birkhäuser Verlag, Basel, Boston, Berlin
29. McCarthy, J.J, Wolf, J.E., Shinbrot, T., Metcalfe, G. (1996) Mixing of granular materials in slowly rotated containers. AIChE **42**, 3351–3363
30. Makse, H.A., Havlin, S., King, P.R., Stanley, H.E. (1997) Spontaneous stratification in granular mixtures. Nature **386,** 379–382
31. Metcalfe, G, Shinbrot, T, McCarthy, J.J., Ottino, J.M. (1995) Avalanche mixing of granular solids. Nature **374**, 39–41
32. Melosh, J. (1986) The physics of very large landslides. Acta Mech. **64**, 89–99
33. Nessyahu, H., Tadmor, E. (1990) Non-oscillatory Central Differencing for Hyperbolic Conservation Laws. J. Comput. Phys. **87,** 408–463
34. Rajchenbach, J. (1990) Flow in powders: From discrete avalanches to continuous regime. Phys. Rev. Lett. **65,** 2221
35. Reynolds O. (1885) Phil. Mag. **20,** 469
36. Ristow, G.H. (1996) Dynamics of granular materials in a rotating drum. Europhys. Lett. **34,** 263
37. Savage, S.B., Lun, C.K.K. (1988) Particle size segregation in inclined chute flow of dry cohesionless granular solids. J. Fluid. Mech. **189**, 311–335
38. Savage, S. B., Hutter, K. (1989) The motion of a finite mass of granular material down a rough incline. J. Fluid Mech. **199,** 177–215
39. Savage, S.B., Hutter, K. (1991) The dynamics of avalanches of granular materials from initiation to runout. Part I: Analysis. Acta Mech. **86,** 201–223
40. Savage, S.B. (1993) Mechanics of granular flows. In: Hutter, K. (Ed.) Continuum mechanics in environmental sciences and geophysics, CISM No. 337 Springer, Wien-New York, 467–522
41. Shinbrot, T., Alexander, A., Muzzio, F.J. (1999) Spontaneous chaotic granular mixing. Nature **397,** 675
42. Shinbrot, T., Muzzio F.J. (2000) Nonequilibrium patterns in granular mixing and segregation. Physics Today, March 2000 , 25–30
43. Tai, Y.C., Gray, J.M.N.T. (1998) Limiting stress states in granular avalanches. Annal. Glac. **26,** 272–276
44. Tai, Y.-C., Hutter, K.,Gray, J.M.N.T. (2000) Steady motion of a finite granular mass in a rotating drum. The Chinese Journal of Mechanics, **16**(2), 67–72
45. Tai Y.C., Gray J.M.N.T., Hutter K., Noelle S. (2001) Flow of dense avalanches past obstructions. Annal. Glac., **32**, 281–284
46. Tai, Y.C., Noelle, S., Gray J.M.N.T., Hutter K. (2002) Shock capturing and front tracking methods for granular avalanches. J. Comput. Phys. **175**, 269–301

47. Tan, W. (1992) Shallow water hydrodynamics. Elsevier Oceanography series, 55
48. Tóth, G., Odstrčil, D. (1996) Comparison of some Flux Corrected Transport and Total Variation Diminishing Numerical Schemes for Hydrodynamic and Magnetohydrodynamic Problems. J. Comput. Phys. **128(1),** 82–100
49. Ullrich, M. (1969) Entmischungserscheinungen in Kugelschüttungen. Chemie-Ing.-Techn. **41**(16) 903–907
50. Whitham, G.B. (1985) Linear and non-linear waves. John Wiley & Sons
51. Wieland, M., Gray, J.M.N.T., Hutter K. (1999) Channelised free surface flow of cohesionless granular avalanches in a chute with shallow lateral curvature. J. Fluid. Mech. **392**, 73–100
52. Yee, H.C. (1989) A class of high-resolution explicit and implicit shock-capturing methods. NASA TM-101088

Gravity-Driven Rapid Shear Flows of Dry Granular Masses in Topographies with Orthogonal and Non-Orthogonal Metrics

Shiva P. Pudasaini, Kolumban Hutter, and Winfried Eckart

Institute of Mechanics III, Darmstadt University of Technology,
Hochschulstrasse 1, 64289 Darmstadt, Germany
e-mail: {pudasain, hutter, eckart}@mechanik.tu-darmstadt.de

received 11 Jul 2002 — accepted 15 Nov 2002

Abstract. In this paper we present a comparison of *i*) a two-dimensional depth-integrated theory for the gravity-driven free surface flow of a granular avalanche over a complex basal topography with *ii*) a helicoidal topography and *iii*) an arbitrary channel with non-uniform curvature and torsion. All these models are important extensions of the original SAVAGE & HUTTER theory. In contrast to other previous extensions, the local coordinate systems for the helicoidal topography as well as for arbitrary channels are based on generating curves with curvature and torsion. Their derivations were necessary because real avalanches are often guided by rather strongly curved and twisted corries. The motions of the avalanches follow the talwegs in all cases. We have shown that such motions are mainly characterised by the underlined metrics and the topographies themselves. These theories allow relatively easy access to comparisons with laboratory experiments and field events in nature. The emerging theories are believed to be capable of predicting the flow of dense granular materials over moderately curved and twisted channels. Physical significance of all three theories is discussed in detail. Each of them has different values depending on the topography and the focus of investigation. The model equations are tested by constructing analytical similarity solutions. The qualitative behaviour of both the (traditional) orthogonal theory for complex basal topography and its non-orthogonal as well as general extensions for a helicoidal and an arbitrary channel agrees well through these similarity solutions.

1 Motivation

Avalanches, debris and mud-flows but equally also landslides are natural phenomena that occur in mountainous regions on our Globe on a regular basis. They are as such therefore a common phenomenon to inhabitants of mountainous regions such as the Alps or the Himalaya, who have learned to accept their occasional occurrence and to avoid the damages that are accompanied with them. Nevertheless, accidents involving damage of property and life and devastating singular incidences have regularly occurred in the past. These are the major reasons why, in such regions, the study of avalanches is a topic of public concern that is of permanent significance. The *physics of the formation* of the rapid motion of a large mass of soil, gravel or snow and the *dynamics of the motion* must be understood, if the danger of the release of a certain mass of gravel or snow should be avoided, or the impact of a moving mass on the avalanche track or on obstructing buildings be estimated. Understanding this physical basis will automatically enable us to take the appropriate measures under certain prescribed conditions.

The last few years have witnessed increased efforts devoted to the physical understanding of the avalanche formation and motion in complex topography. More specially, whilst any forecast of avalanche occurrence and estimation of size is still largely a question of experience, the motion of a given loose mass of gravel or snow is better amenable to analysis. This is so, because the physics of the motion of a finite mass of soil or snow is less difficult to understand than the physics of the mass released from a soil or snow slope at rest.

2 Granular Avalanches and Their Dynamics

By the term *avalanche* we mean a rather large mass of snow, ice or rock that slides rapidly down the side of a mountain. There are different types of avalanches: *rock avalanches*, which can bring down millions of tons of granite or other mountain-forming materials, which may change the land forever; *soil avalanches* and *mud-slides* which are caused when water-saturated soils break loose and flow down a mountainside; also *ice avalanches*, which always contain some snow but are primarily made up of glacier ice. More generally and broadly speaking, a granular avalanche represents the gravity-driven free surface flow of a continuum granulate medium down a steep slope, often initiated by an instability of a granular layer. These phenomena, collectively referred to as avalanches, can be physically characterised as multiphase gravity flows, which consist of randomly dispersed, interacting phases, whose properties change with respect to both time and space [30]. In this sense, an exact analysis of the motion of an avalanche is perhaps an unattainable goal. Thus, we may define an avalanche as follows:

An avalanche can be described as a transient, three-dimensional gravity-driven free surface motion of a mass system made up of an assemblage of granular fragments initiated by an instability of a granular layer and flowing down to the run-out zone on an arbitrarily steep topography with any surface resistance.

The science of avalanche dynamics was not well advanced till the middle of the 20th century. Perhaps, the main reason could be a lack of measured data for avalanche velocity and *impact pressures* and the complicated geometric features over which the flow takes place. Methods to predict the avalanche velocity, its run-out zones and associated impact forces were first developed in Switzerland in the 1950s due to the availability of historical and initial experimental data.

3 Survey on Avalanche Models

This section is devoted to a brief discussion of existing, "classical" avalanche models, and it is then explained why a new model, e.g. the SAVAGE & HUTTER–model, is needed.

3.1 A Brief Discussion of Some Classical Avalanche Models

Statistical Model In mountainous regions *mapping models* are used to determine *avalanche zoning* for the land use and planning. It normally demands either accurate knowledge of past avalanche spreads or methods for computing boundaries of the avalanches. There are several statistical methods for that. Two widely used stochastic models are due to LIED & BAKKEHØI, and MCCLUNG & LIED [32,34]; they establish correlations between the run-out distances and the underlying topographic parameters. These parameters include the location of the initiation point, an intermediate point somewhere in the transition zone (to the run-out) and the position of the stopping point. A simple continuous curve (e.g. a parabola) is used to fit the natural path of the avalanche in the down-hill direction by assuming that the longitudinal profile of the avalanche path governs its dynamics. The *average inclination angle* to the horizontal of the avalanche path is determined by a straight line joining the initial point and the intermediate point. The position of the stopping point of the avalanche motion is described by using a *stopping angle*, that is the angle of a straight line joining the starting and stopping point to the horizontal. This angle is called *Pauschalgefälle*. Using regression methods, this angle can be expressed as a function of the average inclination angle, thus providing the one-dimensional extent and therefore identifying the boundaries of the avalanches. For more detail, see [1,32].

In the last two decades, several extentions of this method have been made. They consist in variations of the parameters and a fit of the model with a particular topography. Although such simple statistical models have been extensively used in practice and give fairly reliable and objective results for fixed sites, many shortcomings are attached with these approaches. This method needs a long-return period, typically 100 years, of avalanches for a given avalanche track. The dynamics of avalanches is governed not only by topographic features of its paths but also depends on many other rheological and mechanical properties of the material such as *basal* and *internal angles of friction* of the base and the material, respectively. This model is limited to one-dimensional path profiles and thus cannot predict the spread of the avalanche which, among others, is one of the most important features of avalanche mapping. The topographic parameters cannot be measured in the laboratory or in the field, and the results rely on past events.

Mass Point Model Till 1989, the most widely used and applied avalanche models utilised a centre of mass approach and were based on the ideas first suggested by VOELLMY [48], who related the shear traction at the base of the flow to the square of the velocity and postulated an additional COULOMB friction contribution to it. On the one hand, VOELLMY assumed uniform and steady conditions, whilst on the other hand, in this model a number of subjective parameters must be predetermined in order to obtain results which match observed data. The simplicity of the model constitutes its power, because, depending upon the parameter choice, it may be applicable to flow as well as powder snow avalanches, but this flexibility makes it also difficult to handle. Many successful attempts were undertaken to improve VOELLMY's model, e.g., by SALM, GUBLER and PERLA et al. [1,35,40]. Unfortunately, none of these extensions could be advanced beyond the centre of mass approach. They are not able to provide information as to the spatial and temporal properties of an avalanche such as the velocity distribution and the evolution of the avalanche height and spread. These are certainly not constant throughout the dimensions of the flowing mass and the time [6,7]. The height of the flow may merely be included as a parameter value, but is not calculated as a function of space and time.

Hydraulic Model Some other, hydraulic, models attempt to idealise complicated materials as linear NEWTONIAN fluids, as discussed by BRUGNOT, DENT & LANG, leading to the NAVIER-STOKES equations which may be solved numerically [4,6]. Although it is perhaps not feasible to assume that this type of constitutive relation adequately describes the media, some success has been achieved in modelling certain aspects, e.g. the geometric properties of the motion of the avalanche, with this approach. Several hypotheses were proposed to explain the mechanisms of fluidisation which occurs in a thin layer close to the basal surface [8,16,18,28,44]. For fluidised granular material, apparent viscosities may be measured but the range of conditions which sustain fluidisation vary greatly among solid-fluid gravity current systems [31].

Kinetic Theory and Molecular Dynamics Model Such models were developed by HAFF, JENKINS & SAVAGE, JENKINS & RICHMAN, LUN et al., HWANG & HUTTER, and many others, to describe the rheology of the granular materials [17,25–27,33]. However, these theories are difficult to apply to avalanche flows as shown by GUBLER, HUTTER, SZIDAROVSZKY & YAKOWITZ, SALM & GUBLER [16,19,20,41]. A kinetic theory would involve the solution of an additional energy equation for the *granular temperature*, velocity and density variations. These solutions would involve the use of very complex boundary conditions, e.g., for granular temperatures, velocities and stresses. It has been demonstrated in [19,20] that the construction of solutions to the related problem, even of chute flows, is very complicated. There is also another problem with the kinetic granular theory in the limit of low particle volume fraction, [17]. This approach can not deal with free surfaces and is not suitable to account for transition of particles on the free surface into suspension. This means that it is not capable of dealing with powder snow avalanche formation. Setting these difficulties aside, these formulations, however, attempt at a correct approach, namely to derive theoretical models from first physical principles. It is likely that with growing computer capacities and increasing computational speed in the next computer generations, these models may well return as competitive alternatives to today's models.

Necessity of a New Model The simple two parameter mass point models, e.g., of VOELLMY [48], SALM [40] and PERLA et al. [35], have only been tested against experimental data as far as runout distances are concerned, and this comparison shows considerable scatter. This means that the field data against which theoretical models could be tested were too scarce to calibrate the existing models of that time with sufficient certainty, or that the classical theoretical formulations [40,48] are oversimplified. Indeed, the latter is true. Obviously, because of the mass point assumption, the temporal evolution of the geometry of the moving avalanche can not be calculated in these models. Runout distances and deposition areas cannot be predicted by these models with significant accuracy partly due to the difficulties of the parameter identification, but more likely due to an inadequate description of (i) the physical (rheological) properties, (ii) the sliding conditions and (iii) the geometries of the moving avalanches [23]. Furthermore, *these models do not allow the determination of the spreading of the avalanching mass and thus cannot give information on the mass distribution in the runout zone* [24]. There are other models describing the avalanches as linear NEWTONIAN fluids [4,6] leading to the NAVIER-STOKES equations. These are in principle able to somehow describe the geometrical properties of the motion. However, *real avalanches are governed by nonlinear constitutive relations*, so that the above ansatz can only serve as a very rough approximation. Alternatively, statistical models are limited to one-dimensional situations and depend on both the topographic feature and long-run period. They cannot incorporate rheological and mechanical behaviour of the material. Moreover, kinetic theory and molecular dynamics models are complicated to handle even for the simplest geometries, [16,19,20,41].

The SAVAGE-HUTTER theory [42] (that will be discussed in detail in the following sections) and their various extensions and generalisations incorporate a great number of important features of granular avalanches. It is a *complete* theory in the sense that it provides a clear formulation of the problem in physical-mathematical terms and leads to well defined initial boundary value problems in problems of practical relevances. Thus, advanced numerical techniques can be developed and successfully implemented. Consequently, theoretical predictions can be and have been validated by many different laboratory experiments. Hence, it may, in many situations, provide a complete knowledge of the avalanche motion from initiation on a steep slope to run-out on a shallow slope.

3.2 A Continuum Mechanical Theory

It is probably fair to state that Savage & Hutter [42], in 1989, developed the first *continuum mechanical theory*, abbreviated by "SH–theory", capable of describing the evolving geometry of a finite mass of a granular material and the associated velocity distribution as an avalanche slides down inclined surfaces. Several simplifying, but nevertheless realistic, assumptions were made that streamlined the mathematical formulation. They are as follows:

- The moving mass is supposed to be *volume preserving*. This assumption is based on laboratory observations: that possible volume expansion and compactions arise at the initiation and stillstand, whilst, during its motion, the moving mass is nearly preserving its volume. Since the dynamics define the motion, supposing volume preserving is an adequate approximation.
- The moving and deforming granular mass is *cohesionless* and obeys a Mohr–Coulomb *yield criterion* both inside the deforming mass as well as at the sliding basal surface, but with different internal, ϕ, and bed, δ, friction angles. This assumption is based on the experimental fact that on any plane, at which shear and normal tractions may act, their ratio is constant and equal to the tangent of ϕ or δ, respectively. This classical criterion is quite appropriate for materials with *rate independent* constitutive properties.
- The *geometries* of the avalanching masses are *shallow* in the sense that typical avalanche thicknesses are small in comparison to the extent parallel to the sliding surface. This shallowness assumption allows the introduction of a shallowness parameter and a simplification of formulas in terms of it.
- The avalanching motion consists of shearing within the deforming mass and sliding along the basal surface. However, on the basis of observations the shearing deformation commonly takes place within a very small basal boundary layer so that it is justified to let this boundary layer collapse to zero thickness and to combine the sliding and shearing velocity to a single sliding law with somewhat larger modelled *sliding velocity*. This then effectively means that variations of the material velocities across the thickness may be ignored and thickness-averaged equations may be employed. This is effectively a method introduced by von Kármán [49] and later refined by Pohlhausen [37] (in which the equations are averaged over depth and the velocity profile is assumed).
- The avalanching body is assumed to be *stress-symmetric* and the shear stresses lateral to the main flow direction can be neglected.

Scaling analysis isolates the physically significant terms in the governing equations and identifies those terms that can be neglected. In order to obtain a spatially one-dimensional theory for the flow down a slope of constant inclination angle, the leading order two-dimensional equations are integrated through the avalanche depth. The simple spatially one-dimensional model of Savage & Hutter [42], applicable along a straight sliding surface, has been generalised in various different ways. In fact, one of the strengths of the procedure is that the model can be generalised to higher dimensions and more complex geometries. There are two main streams of development of this theory depending on the coordinate system and the topography, see [12,14,15,23,42,43,53]. Here we will only focus on three-dimensional extensions of the SH–model and their comparison.

4 Three-Dimensional Granular Avalanche Models

In this section three variants of extensions of the SH–theory will be sketched which are applicable to confined and curved avalanche paths with increasing complexity. We present all three because it turns out that they are mathematically very similarly structured and give rise to unified analytical and numerical solution procedures. This recognition is of much help later on when explicit solutions are constructed. The ultimate goal is to present a critical comparison among them and outline their physical meanings and applicabilities in different configurations.

(1) Orthogonal Complex System Gray et al. [12] extended the SH–theory to model the flow of avalanches over shallow three-dimensional channelised topography whose channel axis is a curve in a vertical plane, see Fig. 1. This led to the first description of the flow of a finite mass of granular material down a valley or corrie. A reference surface that follows the mean down-slope bed topography is used to define a plane orthogonal curvilinear coordinate system, xyz, see Fig. 1. The z-axis is normal to the reference surface and the x- and y- coordinates are tangential to it, with the x-axis oriented in the down-slope direction. The down-slope inclination angle ζ is used to define the reference surface as a function of the down-slope coordinate x. The reference surface does not vary as a function of the cross-slope coordinate y. The chute geometry is superposed by defining its height $z = b\,(x, y, t)$ above the reference surface, $z = 0$, as illustrated in Fig. 1. Even though the local down-slope direction of the basal topography may not coincide with the direction of the x-coordinate, for notational simplicity, the components in the x-direction are referred to as down-slope and components in the y-direction as cross-slope components. The *talweg* in this case is a curve in a vertical plane, but Gray et al. [12] generalised this situation to incorporate talweg curvature and twist. However, this twist was not implemented in the curvilinear coordinate system, because it *only* arose within the steep flat part of the slope and could then be interpreted as a curvature. The situation nevertheless points at an important extension of the one-dimensional SH–theory. The simple curvilinear coordinate system, which is fitted to the mean down-slope chute topography, defines a quasi two-dimensional reference surface. On top of this surface, a shallow three-dimensional basal topography is superposed,

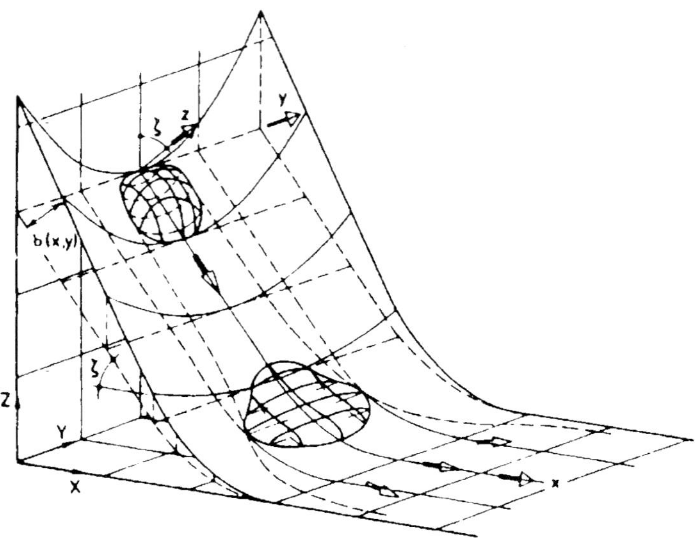

Fig. 1. The rectangular Cartesian coordinate system XYZ is aligned so that the Z-axis is parallel but opposite in direction to the gravity acceleration vector, and the Y-axis is parallel to the cross-slope reference surface coordinate y. The basal topography (solid lines), on which the avalanche slides, $F^b(\boldsymbol{x}) = 0$, is defined by its height, $b = b(x, y)$, above the curvilinear reference surface (x, y) (dashed lines). The shallow complex three-dimensional geometry is therefore superposed on the two-dimensional reference surface

see Fig. 1. With this coordinate system all of the previous theories are reproduced as special cases of this model.

(2) Non-Orthogonal Helical System Pudasaini, Eckart & Hutter [38] recently extended the *SH*–theory to rapid *shear flows* of dry granular masses in a *curved* and *twisted* channel having both *curvature* and *torsion*, see Fig. 3(**a**) in Sect. 7. In particular, they dealt with a helicoidal surface geometry. This theory aims at providing evidence that the *SH*–theory works well not only for topographies having curvature in one direction, but also for rather strongly curved chutes having curvature as well as torsion. Different from the original *SH*–theory [42,43], Pudasaini et al. [38] chose helicoidal coordinates and used them to define a curvilinear coordinate system. They formulated the model equations in terms of these coordinates in the spirit of Gray et al. [12].

(3) Orthogonal General System In a recent contribution Pudasaini & Hutter [39] extended the *SH*–theory to rapid shear flows of dry granular masses in a *non-uniformly curved and twisted channel* having both *curvature and torsion.* In the study of the flow of an avalanche in a channel whose *axial line* (henceforth called the "master curve") is a *generic spatial line*, it is important to choose an appropriate system of coordinates. The curvature and torsion of the master curve, $\kappa = \kappa(s)$, $\tau = \tau(s)$, are assumed to be known as functions of the arc length s. Then, an orthogonal coordinate system along the generic master curve has been introduced (see, e.g., GERMANO and ZABIELSKI & MESTEL [10,11,51,52]), and the SH–equations for a non-steady incompressible dry and cohesionless granular avalanche have been explicitly derived in this frame of reference. PUDASAINI & HUTTER [39] have, thus, studied the effect of torsion on the flow avalanche in arbitrary channels which could not be investigated before.

This theory is designed to model the flow behaviour of the avalanches over curved chutes having general curvature and torsion. This makes the present model amenable to realistic avalanche motions down arbitrary guiding topographies such as valleys and channelised corries. In fact, *Geographic Information Systems* (GIS) applied to mountainous avalanche prone regions can be applied to this model. This theory provides the geometrical basis for an application close to realistic situations and tuned to practical use, and thus lays the theoretical foundation towards this end. Different from the original SH–theory [42,43] and all their previous extensions [12,38,50], an arbitrary space curve is chosen and is used to define an orthogonal-curvilinear coordinate system. The final governing balance laws of mass and momentum are written in these coordinates as in the previous theories.

5 Field Equations

The avalanche is assumed to be an incompressible material with constant density ϱ_0 throughout the entire body. Then the mass and momentum conservation laws reduce to

$$\nabla \cdot \boldsymbol{u} = 0, \tag{1}$$

$$\varrho_0 \left\{ \frac{\partial \boldsymbol{u}}{\partial t} + \nabla \cdot (\boldsymbol{u} \otimes \boldsymbol{u}) \right\} = -\nabla \cdot \boldsymbol{p} + \varrho_0 \boldsymbol{g}, \tag{2}$$

where ∇ is the gradient operator, $\boldsymbol{u}$ is the velocity, $\partial/\partial t$ indicates the differentiation in time, $\otimes$ is the tensor product, $\boldsymbol{p}$ is the pressure tensor (the negative CAUCHY stress tensor) and $\boldsymbol{g}$ is the gravitational acceleration. The granular avalanche is assumed to satisfy a MOHR–COULOMB *yield criterion* in which the internal shear stress $\boldsymbol{S}$ and the normal pressure N on any plane element are related by

$$|\boldsymbol{S}| = N \tan \phi, \tag{3}$$

where ϕ is the *internal angle of friction.* The conservation laws (1) and (2) are complemented by *kinematic boundary conditions* at the free surface,

$F^s(\boldsymbol{x},t)=0$, and at the base, $F^b(\boldsymbol{x},t)=0$, of the avalanche

$$F^s(\boldsymbol{x},t)=0, \quad \frac{\partial F^s}{\partial t}+\boldsymbol{u}^s\cdot\nabla F^s=0, \tag{4}$$

$$F^b(\boldsymbol{x},t)=0, \quad \frac{\partial F^b}{\partial t}+\boldsymbol{u}^b\cdot\nabla F^b=0, \tag{5}$$

where the superscripts 's' and 'b' indicate that a variable is evaluated at the free surface and the base, respectively. The free surface of the avalanche, $F^s=0$, and the basal topography over which the avalanche is assumed to slide, $F^b=0$, are defined by their respective heights above the curvilinear reference

$$F^s\equiv z-s(x,y,t)=0, \quad F^b\equiv -z+b(x,y,t)=0. \tag{6}$$

These definitions guarantee that the free- and basal surface normals point outwards from the avalanching body. There are also *dynamical boundary conditions* that must be satisfied. The *free surface* of the avalanche is *traction free* while the base satisfies a COULOMB *dry-friction sliding law.* That is,

$$F^s(\boldsymbol{x},t)=0, \quad \boldsymbol{p}^s\boldsymbol{n}^s=\boldsymbol{0}, \tag{7}$$

$$F^b(\boldsymbol{x},t)=0, \quad \boldsymbol{p}^b\boldsymbol{n}^b-\boldsymbol{n}^b\left(\boldsymbol{n}^b\cdot\boldsymbol{p}^b\boldsymbol{n}^b\right)=\left(\boldsymbol{u}^b/|\boldsymbol{u}^b|\right)\left(\boldsymbol{n}^b\cdot\boldsymbol{p}^b\boldsymbol{n}^b\right)\tan\delta, \tag{8}$$

where the factor $\boldsymbol{u}^b/|\boldsymbol{u}^b|$ ensures that the COULOMB friction opposes the avalanche motion, and the outward surface and basal normals, respectively, are $\boldsymbol{n}^s=\nabla F^s/|\nabla F^s|$ and $\boldsymbol{n}^b=\nabla F^b/|\nabla F^b|$.

Notice that $\boldsymbol{pn}$ is the negative traction vector, $\boldsymbol{n}\cdot\boldsymbol{pn}$ is the normal pressure and $\boldsymbol{pn}-\boldsymbol{n}\left(\boldsymbol{n}\cdot\boldsymbol{pn}\right)$ is the negative shear traction. It follows that the COULOMB dry-friction law, (8), expresses the fact that the magnitude of the basal shear stress equals the normal basal pressure multiplied by a coefficient of friction, $\tan\delta$. The parameter δ is termed the *basal angle of friction.* Also, $|\cdot|$ stands for the EUCLIDEAN norm of a vector, or a tensor quantity[1].

6 Orthogonal Complex System

We will now introduce an orthogonal coordinate system and proceed with the formulation of the model equations derived for this specific coordinate system.

[1] Notice that the shear traction is assumed to point in the opposite direction to the basal velocity $\boldsymbol{u}^b$ in (8). This implicitly assumes that $\boldsymbol{u}^b\cdot\boldsymbol{n}^b=0$ by (5) and that the basal surface somehow is considered to be fixed. This implies that the basal velocity $\boldsymbol{u}^b$ is tangential to the basal surface. Nevertheless, in the ensuing developments we will retain all those terms which involve the time derivative of the basal surface for theoretical reasons.

6.1 Curvilinear Coordinate System in a Vertical Plane

As introduced by GRAY, WIELAND & HUTTER [12], an orthogonal curvilinear coordinate system, xyz, is defined by a *reference surface*, depicted by dashed lines in Fig 1. In these coordinates, the position vector of a point $\boldsymbol{r}$ is given by $\boldsymbol{r} = \boldsymbol{r}^r(x,y) + z\boldsymbol{n}^r$, where $\boldsymbol{r}^r$ is the position vector of the reference surface and $\boldsymbol{n}^r$ is normal to the reference surface. In Cartesian coordinates, $\boldsymbol{n}^r$ is given by $\boldsymbol{n}^r = \sin\zeta\boldsymbol{i} + \cos\zeta\boldsymbol{k}$, where ζ is the inclination angle of the normal associated to the Z-axis. For ease of notation the identification $(x,y,z) = (x^1,x^2,x^3)$ is made. The associated covariant basis vectors (see, e.g., KLINGBEIL [29]), $\boldsymbol{g}_i$, are given by $\boldsymbol{g}_i = \partial\boldsymbol{r}/\partial x^i = \boldsymbol{r}_{,i}$. The tangent vectors in the reference surface in the x^1- and x^2-directions, respectively, are $\partial\boldsymbol{r}^r/\partial x^1$ and $\partial\boldsymbol{r}^r/\partial x^2$. The orthogonal tangent vectors in the xy-plane are chosen to be $\partial\boldsymbol{r}^r/\partial x^1 = \cos\zeta\boldsymbol{i} - \sin\zeta\boldsymbol{k}$ and $\partial\boldsymbol{r}^r/\partial x^2 = \boldsymbol{j}$. It then follows that

$$\boldsymbol{g}_1 = (1-\kappa z)(\cos\zeta\boldsymbol{i} - \sin\zeta\boldsymbol{k}), \quad \boldsymbol{g}_2 = \boldsymbol{j}, \quad \boldsymbol{g}_3 = \sin\zeta\boldsymbol{i} + \cos\zeta\boldsymbol{k}, \tag{9}$$

where $\kappa = -\partial\zeta/\partial x$ is the local curvature of the reference surface. The directions defined by the tangent vectors $\boldsymbol{g}_1$, $\boldsymbol{g}_2$ and $\boldsymbol{g}_3$ are called the down-slope, cross-slope and normal directions, respectively. The *covariant metric coefficients* are computed from (9) as

$$(g_{ij}) = \boldsymbol{g}_i \cdot \boldsymbol{g}_j = \begin{pmatrix} (1-\kappa z)^2 & 0 & 0 \\ 0 & 1 & 0 \\ 0 & 0 & 1 \end{pmatrix}. \tag{10}$$

The region above the reference surface $z = 0$ can be described by the coordinates xyz that are based on the metric with the square arc length

$$ds^2 = (1-\kappa z)^2\, dx^2 + dy^2 + dz^2. \tag{11}$$

The metric is uniquely defined as long as the z-coordinate is locally smaller[2] than $1/\kappa$. In the theory it is assumed to be satisfied automatically. It is clear that (11) defines an orthogonal metric.

6.2 The Model Equations

Conservative Form The coordinate invariant governing equations of Sect. 5 in a first step are expressed in the curvilinear coordinate system of Sect. 6.1 as shown in Fig. 1. This is done by simultaneously non-dimensionalising the equations via a scaling that introduces an aspect ratio ε = *typical height/typical length* and will be used to simplify the equations. In a second step, the

[2] Physically these points correspond to the positions at which consecutive z-axes, which vary locally, intersect with one another. Therefore, the superposed topography b, should also be a shallow one. Provided the avalanche does not pass through one of these points during the course of its motion the curvilinear coordinates (10) represent a valid coordinate system.

mass and momentum balance equations are integrated through the avalanche depth along the normal of the reference geometry. In this process terms of order higher than $O(\varepsilon)$ are neglected. For an incompressible cohesionless material the continuity equation yields together with the kinematic boundary conditions at the free surface and the base of the avalanche,

$$\frac{\partial h}{\partial t}+\frac{\partial}{\partial x}(hu)+\frac{\partial}{\partial y}(hv)=0, \tag{12}$$

where h represents the evolving geometry of the avalanche and $\boldsymbol{u}=(u,v)$ is the depth-integrated surface-parallel velocity with components u and v in the down-slope and cross-slope directions, respectively. Similarly, the momentum balance equations in the down- and cross-slope directions reduce to

$$\frac{\partial}{\partial t}(hu)+\frac{\partial}{\partial x}\left(hu^2\right)+\frac{\partial}{\partial y}(huv)=hs_x-\frac{\partial}{\partial x}\left(\frac{\beta_x h^2}{2}\right), \tag{13}$$

$$\frac{\partial}{\partial t}(hv)+\frac{\partial}{\partial x}(huv)+\frac{\partial}{\partial y}\left(hv^2\right)=hs_y-\frac{\partial}{\partial y}\left(\frac{\beta_y h^2}{2}\right). \tag{14}$$

Equations (12)–(14), which are in *conservative form*, will henceforth be referred to as an *orthogonal complex system*. The corresponding non-conservative form of model equations were first developed by Gray, Wieland & Hutter [12]. The factors β_x and β_y are defined as

$$\beta_x=\varepsilon\cos\zeta K_x \quad\text{and}\quad \beta_y=\varepsilon\cos\zeta K_y, \tag{15}$$

respectively. The terms s_x and s_y represent the *net driving accelerations* in the down-slope and cross-slope directions, respectively, and are given by

$$s_x=\sin\zeta-\frac{u}{|\boldsymbol{u}|}\tan\delta\left(\cos\zeta+\lambda\kappa u^2\right)-\varepsilon\cos\zeta\frac{\partial b}{\partial x}, \tag{16}$$

$$s_y=\qquad -\frac{v}{|\boldsymbol{u}|}\tan\delta\left(\cos\zeta+\lambda\kappa u^2\right)-\varepsilon\cos\zeta\frac{\partial b}{\partial y}, \tag{17}$$

where δ is the bed friction angle of the granular material with the basal topography, namely b. K_x and K_y are called the *earth pressure coefficients* which are equal to the ratio of the in-plane to vertical pressure in the down- and cross-slope directions, respectively (that is, $K_x=p_{xx}/p_{zz}$ and $K_y=p_{yy}/p_{zz}$). The SH–theory assumes that the down-slope and cross-slope pressures vary linearly with vertical pressure through the depth of the avalanche. The detailed presentation of these statements has been given by Savage & Hutter [42,43] and Gray et al. [12], here we only give a brief account of it. Savage & Hutter [42] made the ad hoc assumption that the down-slope earth pressure K_x is *active* during a down-slope dilatational motion and *passive* during a down-slope compressional motion (for the one-dimensional case, see also [13]):

$$K_x = \begin{cases} K_{x_{act}}, & \partial u/\partial x \geq 0, \\ K_{x_{pas}}, & \partial u/\partial x < 0. \end{cases}$$

Similarly, in the cross-slope direction the four stress ratios were distinguished from one another by considering whether the down-slope and the cross-slope deformations were dilatational or compressional

$$K_y = \begin{cases} K^{x_{act}}_{y_{act}}, & \partial u/\partial x \geq 0,\ \partial v/\partial y \geq 0, \\ K^{x_{act}}_{y_{pas}}, & \partial u/\partial x \geq 0,\ \partial v/\partial y < 0, \\ K^{x_{pas}}_{y_{act}}, & \partial u/\partial x < 0,\ \partial v/\partial y \geq 0, \\ K^{x_{pas}}_{y_{pas}}, & \partial u/\partial x < 0,\ \partial v/\partial y < 0. \end{cases}$$

SAVAGE & HUTTER [42] used elementary geometrical arguments to determine the value of K_x, and HUTTER et al. [23] used the MOHR–circle representations to construct K_y as a function of the internal, ϕ, and basal, δ, angles of friction

$$\begin{aligned} K_{x_{act/pas}} &= 2\sec^2\phi\left(1 \mp \sqrt{1-\cos^2\phi\sec^2\delta}\right) - 1, \\ K^{x}_{y_{act/pas}} &= \frac{1}{2}\left(K_x + 1 \mp \sqrt{(K_x-1)^2 + 4\tan^2\delta}\right). \end{aligned}$$

The conservation laws (12)–(14) are written in non-dimensional form. The non-dimensional variables, $(x, y, h, b, u, v, t, \kappa)$ can be mapped back to their physical counterparts $\left(\hat{x}, \hat{y}, \hat{h}, \hat{b}, \hat{u}, \hat{v}, \hat{t}, \hat{\kappa}\right)$ by applying the scalings

$$(\hat{x}, \hat{y}) = L\,(x, y)\,,\ (\hat{h}, \hat{b}) = H\,(h, b)\,,\ (\hat{u}, \hat{v}) = \sqrt{gL}\,(u, v)\,,\ \hat{t} = \sqrt{L/g}t,\ \hat{\kappa} = \kappa/\mathcal{R},$$

where H is the typical avalanche height, L is the typical length and $\mathcal{R}$ is a typical radius of curvature of the chute in the down-slope direction, while g is the gravitational acceleration. It is, moreover, assumed that both the aspect ratio $\varepsilon = H/L$ and the characteristic curvature of the chute $\lambda = L/\mathcal{R}$, arising in equations (13)–(17), are small.

Let us conclude this section with a few words about the physical meaning of the above model equations. The first term on the right-hand side of (16) is due to the gravitational acceleration and has no contribution in the lateral, y, direction. The second terms of both equations (16) and (17) emerge from the dry COULOMB friction and the third terms are associated with the contribution of the basal topography. Equations (12)–(17) constitute a *two-dimensional conservative system of equations.*

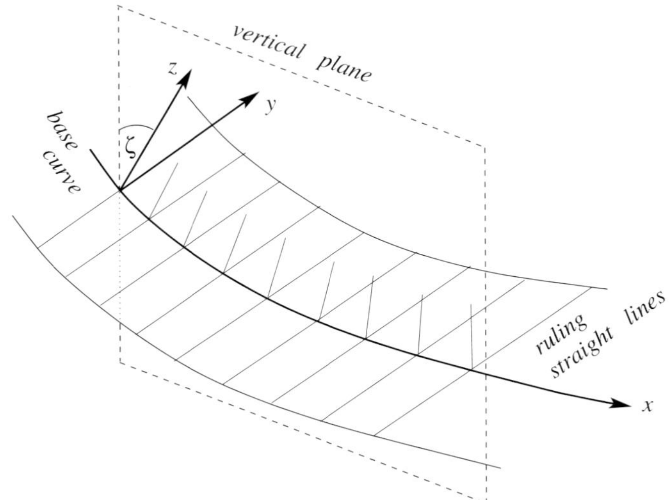

Fig. 2. Ruled surface, constructed from a base curve in a vertical plane and parallel ruling straight lines. The three families of curves define the curvilinear coordinate system x, y, z

7 Flow Through a Helicoidal Channel

The model equations presented in the last section (see, in addition, [12,42,43]) were derived with the intention to be able to describe the motion of a finite mass of granular material down a flat mountain side into a deposition area. The underlined curvilinear coordinate system was based on a so-called *ruled surface*[3] of which the generating curve was in a vertical plane, and the ruling straight lines would be kept parallel to one another, whilst the third coordinate would be perpendicular to these, see Fig. 2. The topographies permissible for this special coordinate system would be small deviations from this reference surface (x, y). Whereas this allows for a large variety of topographies to be studies, see Sects. 5 and 6, the geometries are nevertheless restricted. For instance, the motion in a rather strongly curved channel having both curvature and torsion cannot be analysed with the mentioned curvilinear coordinate system. Such cases do, however, realistically occur. In the transportation of solid materials a finite mass of a dry granular mate-

[3] A ruled surface is a surface which can be swept out by a moving line in space and therefore has a parameterisation (with parameters u and v) of the form

$$R(u, v) = b(u) + vd(u),$$

where b is called the *base curve* (also called the *directrix*) and d is the *director curve*. The straight lines themselves are called *rulings*. The rulings of a ruled surface are *asymptotic curves*.

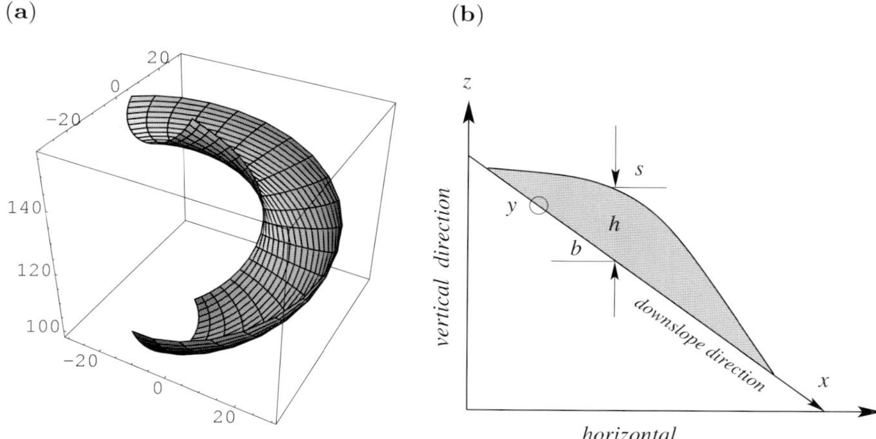

Fig. 3. (a): Helicoidal basal surface. The surface is embedded in a cube of which the edges are parallel to a Cartesian coordinate system. **(b)**: Avalanche depth measured along the direction of gravity

rial may have to be transported through a channel with helicoidal surface, see Fig. 3(**a**). Alternatively, the flow of a snow or debris avalanche down a mountain corrie may be treated as a flow through a channel, of which the talweg is any prescribed three-dimensional curve with curvature and torsion in the physical space. An idealised configuration for this, apt for industrial applications and for laboratory experiments is, e.g., a helix. This geometry also exhibits the advantage of relatively easy experimental study in the laboratory. Both situations give rise to alternative formulations in settings with their own coordinate system. In the ensuing analysis such cases will be considered.

7.1 Non-Orthogonal Helical Coordinate System

Let us consider a curved channel of which the talweg follows a helix with a given pitch. This helicoidal topography is sketched in Fig. 3(**a**). It is therefore natural to consider curvilinear coordinates which follow this helicoidal geometry as closely as possible. For the ensuing analysis, let $\{\boldsymbol{e}_1, \boldsymbol{e}_2, \boldsymbol{e}_3\}$ be a Cartesian basis of the three-dimensional space $\mathbb{R}^3$. Then any point in $\mathbb{R}^3$, referred to this basis, can be represented by its position vector $\boldsymbol{r} = r_1\boldsymbol{e}_1 + r_2\boldsymbol{e}_2 + r_3\boldsymbol{e}_3$, or

$$\begin{pmatrix} r_1 \\ r_2 \\ r_3 \end{pmatrix} = (r_1, r_2, r_3)^T . \tag{18}$$

A curve and a surface in $\mathbb{R}^3$ can be represented by continuous one- and two-parameter functions $\boldsymbol{r} = \hat{\boldsymbol{r}}\left(x^1\right)$ and $\boldsymbol{r} = \hat{\boldsymbol{r}}\left(x^1, x^2\right)$ which will be assumed to be unique mappings from the parameter spaces described by x^1 and $(x^1,\ x^2)$, respectively. For instance,

$$\boldsymbol{r} = \hat{\boldsymbol{r}}\left(x^1\right) := \begin{pmatrix} R\cos x^1 \\ R\sin x^1 \\ a x^1 \end{pmatrix}, \qquad \boldsymbol{r} = \hat{\boldsymbol{r}}\left(x^1,\ x^2\right) := \begin{pmatrix} x^2\cos x^1 \\ x^2\sin x^1 \\ a x^1 \end{pmatrix}, \tag{19}$$

describe a helix with a *pitch* parameter a on a cylinder with radius R, and a screw surface, respectively; x^1 is the polar angle in the plane of the basis vectors $\boldsymbol{e}_1$ and $\boldsymbol{e}_2$, and the cylindrical axis coincides with the direction of the basis vector $\boldsymbol{e}_3$. x^2 is the radial distance from the $\boldsymbol{e}_3$-axis. Coordinate lines on this surface are the helices with radius x^2 ($0 \le x^1 \le 2n\pi$, where $n > 0$ is a real number) and straight lines in the radial direction and parallel to the $(\boldsymbol{e}_1, \boldsymbol{e}_2)$-plane, ($x^1$ fixed; $0 < x^2 < \infty$). It is geometrically obvious that the entire space $\mathbb{R}^3$ is spanned by the three-parameter function

$$\boldsymbol{r} = \hat{\boldsymbol{r}}\left(x^1,\ x^2,\ x^3\right) := \begin{pmatrix} x^2\cos x^1 \\ x^2\sin x^1 \\ x^3 + a x^1 \end{pmatrix}. \tag{20}$$

Adding the coordinate x^3 in the third component of $(19)_2$ generates, for fixed values of x^3, a screw-surface that is translated from the surface $(19)_2$ into the $\boldsymbol{e}_3$-direction by the distance x^3. It is geometrically trivial (it will become clear later) to see that not all coordinate lines of (20) are perpendicular to one another. So the corresponding metric is not orthogonal; this would be only the case for $a = 0$. For ease of notation the identification $(x,\ y,\ z) = (x^1,\ x^2,\ x^3)$ will be made, but the reader is warned that $(x,\ y,\ z)$ are *not* Cartesian components, but rather the coordinates of the helicoidal system. The tangent vectors to the coordinate lines are

$$\boldsymbol{g}_1 = \begin{pmatrix} -y\sin x \\ y\cos x \\ a \end{pmatrix}, \quad \boldsymbol{g}_2 = \begin{pmatrix} \cos x \\ \sin x \\ 0 \end{pmatrix}, \quad \boldsymbol{g}_3 = \begin{pmatrix} 0 \\ 0 \\ 1 \end{pmatrix}. \tag{21}$$

These define the azimuthal, radial and vertical directions, respectively. The *covariant metric coefficients* are given by the matrix

$$(g_{ij}) = \boldsymbol{g_i} \cdot \boldsymbol{g_j} = \begin{pmatrix} y^2 + a^2 & 0 & a \\ 0 & 1 & 0 \\ a & 0 & 1 \end{pmatrix}. \tag{22}$$

The off-diagonal elements arising in the metric tensor, (22), are the manifestation of the non-orthogonality of the coordinates. The metric for this system is given by $ds^2 = d\boldsymbol{r} \cdot d\boldsymbol{r} = g_{ij}\, dx^i\, dx^j$, i.e.,

$$ds^2 = \left(y^2 + a^2\right) dx^2 + dy^2 + dz^2 + 2a\, dx\, dz. \tag{23}$$

The product term $2a\, dx\, dz$ corresponds to the non-orthogonality of the coordinate system under consideration.

7.2 The Model Equations

As in the previous extension of the SH–model by GRAY et al. [12], PUDASAINI, ECKART & HUTTER [38] have recently formulated the balance laws of mass and momentum as well as the boundary conditions of Sect. 5 in terms of the curvilinear helical coordinates of Sect. 7.1, averaged these equations over depth and then non-dimensionalised the averaged equations. There is, however, a slight difference in non-dimensionalising the equations and in the ordering analysis as compared to other previous theories associated to the orthogonal metric of the reference surface. The depth of the avalanche is defined in a new way; *it is taken parallel to the direction of the gravitational acceleration*. This is in contrast to the previous works. Traditionally (i.e., with orthogonal metric) it was taken to be normal to the basal topography. The final governing balance laws of mass and momentum appear to be much less complicated with the averaging operation performed vertically, see Fig. 3(**b**). It is clearly due to the use of a non-orthogonal basis. This direction is also experimentally very convenient. For simplicity, it is also assumed that the basal surface b does not vary with the down-slope coordinate x, i.e. $\partial b/\partial x = 0$.

Conservative Form In the aforementioned circumstances, PUDASAINI et al. [38] developed the following vertically averaged balance laws of mass; and momentum in the down-slope (azimuthal) and cross-slope (radial) directions, respectively,

$$\frac{\partial h}{\partial t} + \frac{\partial}{\partial x}(\psi h u) + \frac{\partial}{\partial y}(h v) + \frac{h}{y} v = 0, \tag{24}$$

$$\begin{aligned} &\frac{\partial}{\partial t}(hu) + \frac{\partial}{\partial x}\left(\psi h u^2\right) + \frac{\partial}{\partial y}(huv) - \psi^2 h y u v + \frac{3h}{y} u v \\ &= -\left(\frac{u}{|\boldsymbol{u}|}\tan\delta + \frac{a}{y}\right) h\chi - \varepsilon\psi \frac{\partial}{\partial x}\left(K_x \nu \frac{h^2}{2}\right), \end{aligned} \tag{25}$$

$$\begin{aligned} &\frac{\partial}{\partial t}(hv) + \frac{\partial}{\partial x}(\psi h u v) + \frac{\partial}{\partial y}\left(h v^2\right) + \frac{h}{y} v^2 - \psi^2 h y u^2 \\ &= -\left(\frac{v}{|\boldsymbol{u}|}\tan\delta + \varepsilon\psi y \frac{\partial b}{\partial y}\right) h\chi - \varepsilon \frac{\partial}{\partial y}\left(K_y \nu \frac{h^2}{2}\right) \\ &\quad -\frac{\varepsilon}{y}\left(K_y \nu \frac{h^2}{2}\right) + \varepsilon\psi^2 y \left(K_x \nu \frac{h^2}{2}\right), \end{aligned} \tag{26}$$

where

$$\psi := \frac{1}{\sqrt{y^2 + a^2}}, \quad \nu := 1 - \frac{2auv}{y\sqrt{y^2 + a^2}}, \quad \chi := \frac{y\sqrt{y^2 + a^2} - 2auv}{y^2 + a^2}. \tag{27}$$

These equations appear here in *conservation form.* Equations (24)–(26) will henceforth be referred to as the *non-orthogonal helical equations.* The non-dimensional variables of system (24)–(26) can be mapped back to their dimensional form via the relations

$$\left(\hat{x}, \hat{y}, \hat{h}, \hat{b}\right) = (x, Ly, Hh, Hb)\,, \;\; (\hat{u}, \hat{v}) = \sqrt{gL}\,(u, v)\,, \;\; \hat{t} = \sqrt{L/g}t, \;\; \hat{a} = \mathcal{R}a. \quad (28)$$

The scalings (28) assume that the avalanche has a typical length-scale L tangential to the reference surface and a typical thickness H normal to it, and $\mathcal{R}$ is the stretching scale of the pitch of the helix. As before, the aspect ratio is assumed small.

Finally, we present meanings of the terms appearing in these model equations from a physical point of view. Let us consider the down-slope (i.e. azimuthal) component of the momentum balance (25). This equation represents the *balance of change of momentum and the stream-wise components of the net driving forces.* That is, the first two terms on the right-hand side are the friction force ($\tan\delta$) and a topographic effect, "a", together with the force associated with the overburden pressure (i.e. χ) at the base. The third term is the (collective) gradient of the earth pressure coefficient, overburden pressure and height of the pile. Similar inferences can be drawn for the cross-slope (radial) component of the momentum balance for which the final two terms are the contributions to the force due to the effects of the curvature and torsion of the topography, geometry of the moving pile and the earth pressure coefficients in two principal directions of the flow.

This system of three equations, (24)–(26), allows three variables h, u and v to be determined, once initial and boundary conditions are prescribed, provided the material parameters δ, ϕ, the pitch parameter a and the basal topography $b(x, y)$ are known.

Remark 1. Before closing this section we ought to mention that the scalings required to arrive at the equations (24)–(27) are partly critically different from those of the original SH-theory, and these differences should be pointed out. The flow depth normal to the bed, $h_\perp$, and that vertically measured, h, are related to one another by the approximate relation $h = \sec\zeta\, h_\perp$ (this relation is based on the slow variation of the depth function with the longitudinal coordinate). Since the length scale remains unchanged, this corresponds to a change in the aspect ratio, ε, for this non-orthogonal case relative to that used for an orthogonal coordinate system, $\varepsilon_\perp$, namely $\varepsilon = \sec\zeta\,\varepsilon_\perp$. As long as $\zeta \in [0, 60°]$ we have $\sec\zeta \in [1, 2]$ which implies that ε and $\varepsilon_\perp$ are both of the same order of magnitude, typically $\varepsilon \in [10^{-3}, 10^{-2}]$. Other problems do not arise since the pitch in equations (24)–(27), $a \approxeq \tan\zeta \le O(1)$ is unproblematic.

More difficult seems the unchanged validity of the earth pressure coefficients as referred to the internal and the bed friction angles, ϕ and δ. A straightforward geometric argument shows that the contravariant stress components p_{xy} may still be assumed to be small as compared to p_{xz} and p_{yz}

so that the element faces with surface normals in the azimuthal direction are still primarily loaded by the pressures p_{yy} normal to these element faces (this is not so for p_{xx}). Provided we consistently work with the contravariant stress components, we then still can apply the earth pressure coefficients (see Sect. 6.2). However, the interpretation is then not exactly according to MOHR-COULOMB, since they applied their yield condition to physical stress components. For small pitch the differences are also small. ◇

8 Orthogonal General System

In this section, we will introduce a general orthogonal coordinate system that is able to describe a finite region of a natural landscape. The model equations are derived in this general coordinate system so as to lay down the mathematical foundation in order to study avalanches in natural terrains.

8.1 Coordinate System

Consider an avalanche prone landscape and a subregion of it where the topography allows the identification of an avalanche track. A single curve, following the landscape topography (e.g. the talweg of the valley) is singled out as a master curve C from which the track topography will be modelled. Let this three-dimensional curve be smooth; in the global coordinate system it may be given by $\boldsymbol{R}(x,y,z)$, where x, y and z are the Cartesian coordinates. A moving coordinate system is constructed, see PUDASAINI & HUTTER [39], by considering this spatial curve described by the position vector $\boldsymbol{R}(s)$, where s is the parameter that measures the arc length from some convenient reference point. At any point of the curve we have the orthonormal triad $\{\boldsymbol{T},\boldsymbol{N},\boldsymbol{B}\}$ which, respectively, comprises of the tangent, principal normal and binormal unit vectors, also expressible as functions of s. The vector pair $\{\boldsymbol{N},\boldsymbol{B}\}$ spans a plane perpendicular to C. Any Cartesian vector $\boldsymbol{X}$ in the three-dimensional space can be expressed as

$$\boldsymbol{X}:=\boldsymbol{X}(s,r,\theta)=\boldsymbol{R}(s)+r\cos(\theta+\varphi(s)+\varphi_0)\boldsymbol{N}(s)+r\sin(\theta+\varphi(s)+\varphi_0)\boldsymbol{B}(s).\tag{29}$$

Here, (r,θ) are polar coordinates spanning the plane normal to the axis of the master curve C in Fig. 4. The origin of the azimuthal angle, θ, in this plane is arbitrary, but measured from the unit vector $\boldsymbol{N}^*$ which is rotated from $\boldsymbol{N}$ by *a phase* $(\varphi(s)+\varphi_0)$. Also φ_0 is an arbitrary constant (in applications often conveniently taken as zero or $\pi/2$) and

$$\varphi(s) = -\int_{s_0}^{s} \tau(s')ds' \tag{30}$$

is the accumulation of the torsion of the curve as it proceeds from the initial point s_0. Hence the torsion, $\tau(s)$, enters into the equation through the auxiliary function $\varphi=\varphi(s)$. From differential geometry (see, KLINGBEIL [29],

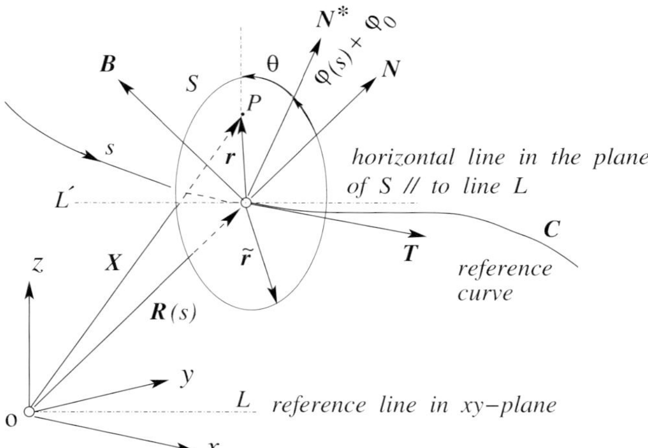

Fig. 4. $\boldsymbol{R}(s)$ describes the reference master curve C embedded in $\mathbb{R}^3$. s is the arc length, $\{\boldsymbol{T}, \boldsymbol{N}, \boldsymbol{B}\}$ is the moving orthonormal unit triad following the curve. (r, θ) are polar coordinates spanning the plane of circle S with radius $\tilde{r}$ normal to the axis of C. The origin of the azimuthal angle, θ, in this plane is arbitrary, but measured from the unit vector $\boldsymbol{N}^*$ which is rotated from $\boldsymbol{N}$ by a phase $(\varphi(s) + \varphi_0)$ for $s \in [s_0, \infty)$, $s_0 \in [0, \infty)$ and $\theta \in (0, 2\pi]$. Also, φ_0 is an arbitrary constant and P is any point in space

Bowen & Wang [2,3]) we recall the following results:

$$\boldsymbol{T}(s) = \frac{d\boldsymbol{R}(s)}{ds}, \quad \boldsymbol{N}(s) = \frac{1}{\kappa}\frac{d\boldsymbol{T}(s)}{ds}, \quad \boldsymbol{B}(s) = \boldsymbol{T}(s) \times \boldsymbol{N}(s), \tag{31}$$

where κ is the curvature of the master curve C and "$\times$" stands for the "cross product" of two vectors. Curvature and torsion can be computed from the function $\boldsymbol{R} = \boldsymbol{R}(x, y, z)$ and are expressible as functions of the arc length s: $\kappa = \kappa(s)$ and $\tau = \tau(s)$. The famous Serret–Frenet formula provides a connection between the curvature and torsion and the changes of $\boldsymbol{T}, \boldsymbol{N}, \boldsymbol{B}$ along s as follows:

$$\frac{d\boldsymbol{T}}{ds} = \kappa\mathbf{N}, \quad \frac{d\boldsymbol{N}}{ds} = \tau\boldsymbol{B} - \kappa\boldsymbol{T}, \quad \frac{d\boldsymbol{B}}{ds} = -\tau\boldsymbol{N}. \tag{32}$$

For ease of notation, the identifications $(x,\, y,\, z) = (x^1,\, x^2,\, x^3) = (s,\, \theta,\, r)$ will be made, *but* the reader is warned that $(x,\, y,\, z)$ are, from now on, *not* Cartesian components. We refer to x, y and z (i.e. the axial, azimuthal & radial) coordinates as down-slope, cross-slope and normal directions, respectively. The tangent vectors to the coordinate lines, i.e., the associated covariant basis vectors, $\boldsymbol{g}_i = \partial\mathbf{X}/\partial x^i$, are given by

$$\boldsymbol{g}_1 = (1 - \kappa z\eta)\,\boldsymbol{T}(x), \quad \boldsymbol{g}_2 = -z\zeta\boldsymbol{N}(x) + z\eta\boldsymbol{B}(x), \quad \boldsymbol{g}_3 = \eta\boldsymbol{N}(x) + \zeta\mathbf{B}(x), \tag{33}$$

where

$$\eta = \cos\left(y + \varphi(x) + \varphi_0\right) \quad \text{and} \quad \zeta = \sin\left(y + \varphi(x) + \varphi_0\right). \tag{34}$$

It is clear that in contrast to the standard Cartesian unit vectors $\boldsymbol{i}, \boldsymbol{j}, \boldsymbol{k}$, the covariant vectors $\boldsymbol{g}_i$ vary as functions of position. The *covariant metric coefficients*, defined as $g_{ij} = \boldsymbol{g}_i \cdot \boldsymbol{g}_j$, are given by the matrix

$$(g_{ij}) = \begin{pmatrix} (1 - \kappa\eta z)^2 & 0 & 0 \\ 0 & z^2 & 0 \\ 0 & 0 & 1 \end{pmatrix}. \tag{35}$$

Therefore, the metric for the chosen coordinate system is given by

$$d\mathbf{X} \cdot d\mathbf{X} = \{1 - \kappa(x) z \eta\}^2 (dx)^2 + (dz)^2 + (z dy)^2 . \tag{36}$$

This corroborates orthogonality of the system (29) and (31). It is also easy to see that when $\tau = 0$, this system of coordinates reduces to the simple toroidal, while it reduces to the cylindrical coordinate system when $\kappa = \tau = 0$.

Because *Geographical Information Systems* (GIS) refer the topography to the global Cartesian system $\{x, y, z\}$, this must in a particular application first be transformed to the orthogonal moving coordinate system. This can be done and is generally done by NURBS (Non-Uniform Rational B-Spline), see PIEGL & TILLER [36]. It will in this paper not be of any concern except to mention the adaptivity of the equations to realistic situations. The most advantageous fact of this moving coordinate system along a curved and twisted line is that we are free to choose the *master curve*. Hence it may have enormous applications in investigations of the flow behaviour of fluid as well as moving granular materials in a curved and twisted channels.

8.2 The Model Equations

As in the previous models of the *SH*–theory, PUDASAINI & HUTTER [39] have recently formulated the balance laws of mass and momentum as well as the boundary conditions of Sect. 5 in terms of the curvilinear coordinates of Sect. 8.1, averaged these equations over depth and then non-dimensionalised the averaged equations. There is, however, a slight difference in non-dimensionalising the equations and in the ordering analysis as compared to other previous theories associated to the orthogonal and non-orthogonal metrics of the reference surfaces. The depth of the avalanche is measured along the normal direction of the reference surface, see Fig. 5. The final governing balance laws of mass and momentum appear to be much less complicated with this averaging operation.

Conservative Form As in Sects. 6 and 7, PUDASAINI & HUTTER [39] developed the following curvilinear non-dimensional depth-averaged balance laws

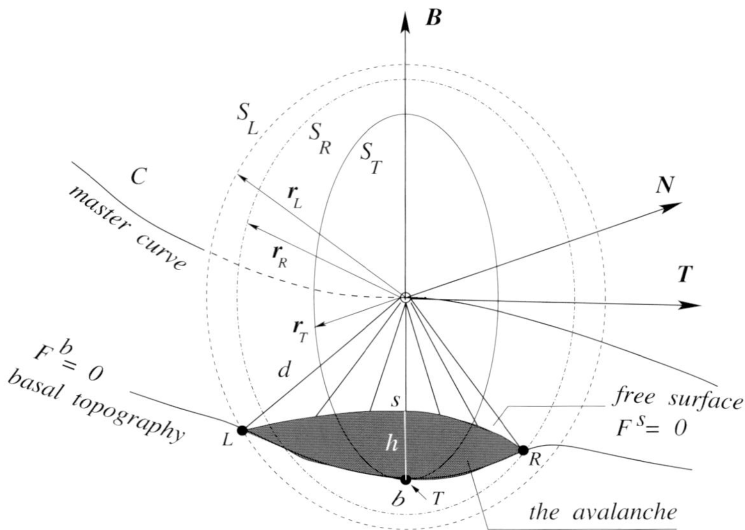

Fig. 5. For a given value of the arc length, the avalanche domain in the lateral direction occupies a region in the plane of the circle $S_T \perp C$ distant from the centre of the moving triad $\{\boldsymbol{T}, \boldsymbol{N}, \boldsymbol{B}\}$. The concentric and coplanar circles, (with the centre at the master curve C and radius $\boldsymbol{r}_T$, $\boldsymbol{r}_R$, and $\boldsymbol{r}_L$), S_T, S_R, and S_L, respectively, pass through the talweg (T) and the left (L) and right (R) marginal points of the avalanche with its basal topography in the lateral direction. The basal topography $F^b = 0$ and the free surface $F^s = 0$ of the avalanche in this plane section are drawn in the figure. The depth of the avalanche in this section is represented by a function h which at different positions is not parallel but radial. Also shown, for instance, is the distance d of the avalanche from the centre line to the circle S_L. Also, the superscripts b and s, respectively, correspond to the base and free surface of the avalanche

of mass and momentum; in the down-slope (axial) and cross-slope (azimuthal) directions, respectively

$$\frac{\partial h}{\partial t} + \frac{\partial}{\partial x}(hu) + \frac{\partial}{\partial y}\left(\frac{1}{b}hv\right) = 0, \tag{37}$$

$$\frac{\partial}{\partial t}(hu) + \frac{\partial}{\partial x}(hu^2) + \frac{\partial}{\partial y}\left(\frac{1}{b}huv\right) = hs_x + \frac{\partial}{\partial x}\left(\frac{\beta_x h^2}{2}\right), \tag{38}$$

$$\frac{\partial}{\partial t}(hv) + \frac{\partial}{\partial x}(huv) + \frac{\partial}{\partial y}\left(\frac{1}{b}hv^2\right) = hs_y + \frac{\partial}{\partial y}\left(\frac{\beta_y h^2}{2}\right), \tag{39}$$

where the factors β_x and β_y are defined, respectively, as

$$\beta_x := \varepsilon g_3 K_x \quad \text{and} \quad \beta_y := g_3 K_y. \tag{40}$$

The terms s_x and s_y represent the *net driving accelerations* in the down-slope and cross-slope directions, respectively, and are given by

$$s_x := g_1 + \frac{u}{|\boldsymbol{u}|} g_3 \tan\delta + \varepsilon g_3 \frac{\partial b}{\partial x}, \tag{41}$$

$$s_y := g_2 + \frac{v}{|\boldsymbol{u}|} g_3 \tan\delta + g_3 \frac{\partial b}{\partial y} + \kappa \zeta u^2, \tag{42}$$

where $|\boldsymbol{u}| = \sqrt{u^2 + \varepsilon^2 v^2}$ is the velocity field tangential to the reference (basal) topography. The non-dimensional variables, $(x, y, h, b, u, v, t, \kappa)$ can be mapped back to their physical counterparts $\left(\hat{x}, \hat{y}, \hat{h}, \hat{b}, \hat{u}, \hat{v}, \hat{t}, \hat{\kappa}\right)$ by applying the scalings

$$(\hat{x}, \hat{y}) = (Lx, y)\,, \quad \left(\hat{h}, \hat{b}\right) = H\,(h, b)\,, \quad (\hat{u}, \hat{v}) = \sqrt{gL}\,(u, \varepsilon v)\,, \quad \hat{t} = \sqrt{L/g}t, \quad \hat{\kappa} = \frac{\kappa}{\mathcal{R}},$$

where H is the typical avalanche height, L is the typical length of the avalanche and $\mathcal{R}$ is a typical radius of curvature of the reference surface in the down-slope direction and g is the gravitational acceleration. It is, moreover, assumed that both the aspect ratio $\varepsilon = H/L$ and the characteristic curvature of the chute $\lambda = L/\mathcal{R}$, arising in equations (40)–(42), are small. Also note that the last term, $\beta_y h^2/2$, on the right-hand side of (39) was initially multiplied by a factor $2/\varepsilon b$, which is assumed to be of order unity. It is so, because due to the geometric construction of the problem under consideration, b is large enough to maintain this ordering and thus is not a constraint in many realistic situations. The system of equations (37)–(39), including (40)–(42), constitutes a *two-dimensional conservative system of equations.* These equations will henceforth be referred to as *general equations* for orthogonal systems.

Before concluding this section, we add some words about the physical meaning of the terms that appear in the present model equations. The first terms on the right-hand side of (41) and (42) are due to the gravitational accelerations in the principal flow directions, i.e., in the down- and cross-slope directions, respectively. The second terms of both equations emerge from the dry COULOMB friction, the third terms are the non-dimensional projections of the topographic variations (with respect to the down- and cross-slope coordinates) along the normal direction and the last term of (42) reflects an additional contribution of the basal topography, namely the curvature and torsion effects. Also, notice that g_1, g_2 and g_3 in these equations are the projections of the gravitational acceleration along the down-slope, cross-slope and normal directions, respectively. Moreover, owing to the sidewise confinement of the terrain the component g_2 is small and thus assumed to be of order ε. Given the guiding (generic) master curve, C (which is used to generate the basal topography), the material parameters δ and ϕ, and the basal topography b; the system of equations (37)–(39) allows three independent variables h, u and v to be computed once appropriate initial and boundary conditions are prescribed.

9 Standard Conservative Form

The three systems of partial differential equations that were derived for the different metrics do all enjoy the same mathematical structure. This is fortunate because it suggests that very similar, if not identical, integration techniques can be used in numerical computations. Here we demonstrate this similarity.

A system of partial differential equations is said be in conservative form *if it can be written as*

$$\frac{\partial \boldsymbol{w}}{\partial t} + \frac{\partial \boldsymbol{f}(\boldsymbol{w})}{\partial x} + \frac{\partial \boldsymbol{g}(\boldsymbol{w})}{\partial y} = \boldsymbol{s}(\boldsymbol{w}), \tag{43}$$

where $\boldsymbol{w}$ denotes the vector of the conservative variables, $\boldsymbol{f}$ and $\boldsymbol{g}$ represent the transport fluxes in the x- and y-directions, respectively, and $\boldsymbol{s}$ is the vector of source terms. Otherwise it is said to be in non-conservative form. For all three systems under consideration, the conservative variables are analogously defined, namely, as h, $m_x = hu$ and $m_y = hv$.

(1) Orthogonal Complex System For the *orthogonal complex system* (12)-(17), $\boldsymbol{w}$, $\boldsymbol{f}$, $\boldsymbol{g}$ and $\boldsymbol{s}$ are

$$\begin{gathered}
\boldsymbol{w} = \begin{pmatrix} h \\ m_x \\ m_y \end{pmatrix}, \qquad \boldsymbol{f} = \begin{pmatrix} m_x \\ (m_x)^2/h + \beta_x h^2/2 \\ m_x m_y/h \end{pmatrix}, \\
\boldsymbol{g} = \begin{pmatrix} m_y \\ m_x m_y/h \\ (m_y)^2/h + \beta_y h^2/2 \end{pmatrix}, \qquad \boldsymbol{s} = \begin{pmatrix} 0 \\ h s_x \\ h s_y \end{pmatrix}.
\end{gathered} \tag{44}$$

(2) Non-Orthogonal Helical System Similarly, the laws (24)–(26) for the *non-orthogonal helical system* take the form (43) with the following vectorial quantities

$$\begin{gathered}
\boldsymbol{w} = \begin{pmatrix} h \\ m_x \\ m_y \end{pmatrix}, \qquad \boldsymbol{f} = \begin{pmatrix} m_x \psi \\ (m_x)^2 \psi/h + \beta_x h^2 \psi/2 \\ m_x m_y \psi/h \end{pmatrix}, \\
\boldsymbol{g} = \begin{pmatrix} m_y \\ m_x m_y/h \\ (m_y)^2/h + \beta_y h^2/2 \end{pmatrix}, \qquad \boldsymbol{s} = \begin{pmatrix} m_y/y \\ s_x \\ s_y \end{pmatrix},
\end{gathered} \tag{45}$$

where

$$\beta_x = \varepsilon K_x \nu, \quad \beta_y = \varepsilon K_y \nu, \quad s_x = s_x^1 + s_x^2, \quad s_y = s_y^1 + s_y^2,$$

and

$$
\begin{aligned}
s_x^1 &= -\left(\frac{u}{|\boldsymbol{u}|}\tan\delta + \frac{a}{y}\right)h\chi, \quad s_x^2 = \frac{m_x m_y}{hy}\left(\psi^2 y^2 - 1\right), \\
s_y^1 &= -\left(\frac{v}{|\boldsymbol{u}|}\tan\delta + \varepsilon\psi y\frac{\partial b}{\partial y}\right)h\chi, \\
s_y^2 &= \left(\frac{m_x^2}{h} + \frac{\beta_x h^2}{2}\right)\psi^2 y - \left(\frac{m_y^2}{h} + \frac{\beta_y h^2}{2}\right)\frac{1}{y}.
\end{aligned}
$$

(3) Orthogonal General System Finally, for (37)–(42), $\boldsymbol{w}$, $\boldsymbol{f}$, $\boldsymbol{g}$ and $\boldsymbol{s}$ are

$$
\begin{aligned}
\boldsymbol{w} &= \begin{pmatrix} h \\ m_x \\ m_y \end{pmatrix}, \qquad \boldsymbol{f} = \begin{pmatrix} m_x \\ (m_x)^2/h - \beta_x h^2/2 \\ m_x m_y/h \end{pmatrix}, \\
\boldsymbol{g} &= \frac{1}{b}\begin{pmatrix} m_y \\ m_x m_y/h \\ (m_y)^2/h - b\beta_y h^2/2 \end{pmatrix}, \qquad \boldsymbol{s} = \begin{pmatrix} 0 \\ h s_x \\ h s_y \end{pmatrix}.
\end{aligned}
\tag{46}
$$

Notice that the explicit representations of the source terms s_x and s_y in all these three systems are different. The same notation is used only for the sake of brevity.

It is evident from these formulas that the flux quantities $\boldsymbol{f}$ and $\boldsymbol{g}$ and the supplies $\boldsymbol{s}$ are differently defined, similarly though but with alternations that are due to the geometric peculiarities of the curve generating the curvilinear coordinate systems. It is also worth noting that system (**3**) includes (**2**) and (**1**), i.e., by accordingly selecting the generating curve in (**3**) the systems (**2**) and (**1**) emerge. For a proof see [39].

10 Characteristic Speeds and Critical Flows

(1) Orthogonal Complex System In order to compute the characteristic speeds of the orthogonal system (43) and (44), we rewrite it as

$$
\frac{\partial \boldsymbol{w}}{\partial t} + \boldsymbol{A}\begin{pmatrix} \dfrac{\partial \boldsymbol{w}}{\partial x} \\[2mm] \dfrac{\partial \boldsymbol{w}}{\partial y} \end{pmatrix} = \boldsymbol{s}, \quad \boldsymbol{A} = \left(\boldsymbol{A}_x,\ \boldsymbol{A}_y\right), \tag{47}
$$

where

$$\boldsymbol{A}_x := \frac{\partial \boldsymbol{f}}{\partial \boldsymbol{w}} = \begin{pmatrix} 0 & 1 & 0 \\ -\left(m_x\right)^2/h^2 + \beta_x h & 2m_x/h & 0 \\ -m_x m_y/h^2 & m_y/h & m_x/h \end{pmatrix},$$
$$\boldsymbol{A}_y := \frac{\partial \boldsymbol{g}}{\partial \boldsymbol{w}} = \begin{pmatrix} 0 & 0 & 1 \\ -m_x m_y/h^2 & m_y/h & m_x/h \\ -\left(m_y\right)^2/h^2 + \beta_y h & 0 & 2m_y/h \end{pmatrix}. \tag{48}$$

The characteristic speeds are only defined in a spatially one dimensional situation. To achieve this at a fixed position $\boldsymbol{x} = (x, y)$ in the avalanche a rotation of the coordinate systems must be performed such that in the rotated coordinate system (identified by the asterisks) equation (47) reduces to

$$\frac{\partial \boldsymbol{w}^*}{\partial t} + \boldsymbol{A}^* \begin{pmatrix} \dfrac{\partial \boldsymbol{w}^*}{\partial x^*} \\ \mathbf{0} \end{pmatrix} = \mathbf{0}, \quad \boldsymbol{A}^* = \left(\boldsymbol{A}_x^*, \ \boldsymbol{A}_y^*\right) \tag{49}$$

implying that the characteristic equation is now given by

$$\det\left(\boldsymbol{A}_x^* - \lambda \boldsymbol{I}_3\right) = 0. \tag{50}$$

Note that the condition $\partial \boldsymbol{w}^*/\partial y^* = 0$ defines the rotation matrix $\mathbf{0}$ of the coordinate system.

We restrict ourselves to the situation for which this rotation does not have to be performed, namely those lines for which either $\partial \boldsymbol{w}/\partial y = 0$ or else $\partial \boldsymbol{w}/\partial x = 0$. Equation (50) then reads

$$\det\left(\boldsymbol{A}_x - \lambda \boldsymbol{I}_3\right) = 0, \quad \text{and} \quad \det\left(\boldsymbol{A}_y - \lambda \boldsymbol{I}_3\right) = 0 \tag{51}$$

with the solutions

$$\lambda_1 = u, \qquad \lambda_{2,3} = \frac{m_x}{h} \pm \sqrt{\beta_x h},$$
$$\lambda_4 = v, \qquad \lambda_{5,6} = \frac{m_y}{h} \pm \sqrt{\beta_y h}. \tag{52}$$

$\lambda_{1,4}$ give as characteristic speed the particle velocity in the x- and y-direction, respectively; alternatively, the other solutions give in each case a subcritical and supercritical speeds in the x- and y-directions, respectively. The general case explained by (50) will also yield these solutions.

(2) Non-Orthogonal Helical System For the non-orthogonal helical system, (43) and (45), the matrix $\boldsymbol{A}_y$ remains unchanged but the matrix $\boldsymbol{A}_x$ must be multiplied by the factor ψ. In a straightforward manner one can compute the following eigenvalues for this system:

$$\begin{aligned} \lambda_1 &= u, \qquad \lambda_{2,3} = m_x/h \pm \sqrt{u^2 - \psi\,(u^2 - \beta_x h)}, \\ \lambda_4 &= v, \qquad \lambda_{5,6} = m_y/h \pm \sqrt{v^2 - \psi\,(v^2 - \beta_x h)}. \end{aligned} \tag{53}$$

(3) Orthogonal General System The matrices $\boldsymbol{A}_x$ and $\boldsymbol{A}_y$ for the orthogonal general system (43) and (46) are

$$\begin{aligned} \mathbf{A}_x &= \frac{\partial \boldsymbol{f}}{\partial \boldsymbol{w}} = \begin{pmatrix} 0 & 1 & 0 \\ -(m_x)^2/h^2 - \beta_x h & 2m_x/h & 0 \\ -m_x m_y/h^2 & m_y/h & m_x/h \end{pmatrix}, \\ \boldsymbol{A}_y &= \frac{\partial \boldsymbol{g}}{\partial \boldsymbol{w}} = \frac{1}{b}\begin{pmatrix} 0 & 0 & 1 \\ -m_x m_y/h^2 & m_y/h & m_x/h \\ -(m_y)^2/h^2 - b\beta_y h & 0 & 2m_y/h \end{pmatrix}. \end{aligned} \tag{54}$$

The eigenvalues analogous to those evaluated in (51) and (52) are here

$$\begin{aligned} \lambda_1 &= u, \qquad \lambda_{2,\,3} = \frac{m_x}{h} \pm \sqrt{-\beta_x h}, \\ \lambda_4 &= \frac{v}{b}, \qquad \lambda_{5,\,6} = \frac{m_y}{bh} \pm \frac{\sqrt{-b\beta_y h}}{b}. \end{aligned} \tag{55}$$

For all these systems written in the principal direction for which $\partial \boldsymbol{w}^*/\partial y^* = 0$ we can conclude the following facts about the characteristics of the flow: When a finite avalanching mass of granular material moves down a steep slope and approaches the runout zones with supercritical speed, a considerable deceleration will suddenly occur and lead to a transition from supercritical to subcritical flow. Any such transition from a supercritical to a subcritical flow state produces a shock. That is accompanied with changes from small heights and larger speeds to large heights and smaller speeds. Tai and Tai et al. [45,46] developed and well implemented the shock capturing numerical schemes for the orthogonal complex system.

11 Some Analytical Solutions

It is possible to construct exact solutions for the *orthogonal complex system*, the *non-orthogonal helical system* and the *orthogonal general system* for a particular geometry for which there is no variation of basal topography in the down-hill direction. Imagine the situation that a finite mass of granular material in a narrow chute has initially parabolic shape and is then released to evolve. The question is, does this mass upon release, perhaps, preserve its shape, thus remain a parabola and only change its aspect ratio? It can be shown that such similarity solutions do indeed exist. The forms of the preserving geometries are in fact parabolas and the solution of their length is described by an ordinary differential equation in time. The main conclusions that are deduceable from these computations are that these results demonstrate useful quantitative physical behaviour but do not indicate whether similarity solutions could be in any way suitable for prognostic purposes [43].

In the following computations we will consider a *confined chute*. Here, we will construct solutions for a helical basal surface. For a confined helicoidal chute we may assume that there is a very small spreading of the pile in the lateral direction and that the whole mass is mainly flowing along the talweg of the basal topography. Also, in real situations the variations of the down-slope earth pressure coefficient with respect to the azimuthal coordinate is small. For diverging motions the avalanche is always extending. So, no shocks will form. Under these circumstances we are allowed to make the following assumptions: $v \sim 0, \;\; \partial/\partial y(\cdot) = 0$. With these assumptions at our disposal, accounting for $\psi = 1/\sqrt{y^2 + a^2}$ and inserting the mass balance in the down-slope momentum balance, equations (24)–(25) reduce to

$$
\begin{aligned}
&\frac{\partial h}{\partial t} + \psi \frac{\partial}{\partial x}(hu) = 0, \\
&\frac{\partial u}{\partial t} + \psi u \frac{\partial u}{\partial x} + \varepsilon \psi K_x \frac{\partial h}{\partial x} = \psi D,
\end{aligned} \tag{56}
$$

where $D = -\left(sgn(u)\, y \tan\delta + a\right)$ is the *net driving force*. Notice that these equations contain y as a *passive parameter*. We set $t := \tilde{t}/\psi$, and $\beta := \varepsilon K_x$. With this transformation, (56) reduces to the standard form

$$
\begin{aligned}
&\frac{\partial h}{\partial \tilde{t}} + \frac{\partial}{\partial x}(hu) = 0, \\
&\frac{\partial u}{\partial \tilde{t}} + u \frac{\partial u}{\partial x} + \beta \frac{\partial h}{\partial x} = D,
\end{aligned} \tag{57}
$$

where $\tilde{t}$ is a transformed time. For notational brevity we will replace $\tilde{t}$ by t in the sequel.

The system of equations (57) exactly reproduces a corresponding system obtained by Savage & Hutter [42,43], in their original paper, for an avalanching motion on a rough inclined chute, which can be obtained from the *orthogonal complex system.* For this purpose we should select a basal topography that satisfies the condition $\psi = 1$. This means, we should consider the flow in a channel whose radius of the cross-section, y, and the helicity, a, constitute the circle $y^2 + a^2 = 1$, but physically admissible values of y and a can only be in the domain $(0, 1)$. Also, note that the values of these variables should be chosen in such a way that the channel can produce the flow. For the orthogonal general system, we can exactly derive system (57) if we consider the corresponding master curve to be almost a plane curve. A detailed derivation of it can be found in [13,38,39]. In this sense the system (57) and its exact solutions (that we will present here) are of fundamental nature in the development of the SH–theory, and here we just write the final solutions. Chugunov et al. and Farwig, both in this volume, equally constructed similarity solutions that go beyond those presented here [5,9].

To find particular solutions to our *moving-boundary-value problem* we apply a *fixed domain mapping* by which the span interval is mapped onto a fixed interval with variable η. Defining a *moving coordinate system* (see [13,42,43]) we may deduce the following exact solution of permanent type

$$h = \frac{E}{2\beta g}\left(1 - \eta^2\right), \quad u = \eta\sqrt{2E\left(1 - \frac{1}{g}\right)}, \tag{58}$$

where $E = \frac{3}{2}\beta V$, V is the total volume of the avalanche body, η has values[4] in the fixed domain $[-1, 1]$, u is the relative velocity from the centre of the granular pile and g is the spreading factor (i.e. half the length) of the support of the avalanche with the base given by the following function against time t

$$\sqrt{g\,(g-1)} + \ln\left|\sqrt{g} + \sqrt{g-1}\right| = \sqrt{2E}t. \tag{59}$$

For large values of g, (59) indicates that g varies linearly with time t, see Fig. 6. From (58), we conclude that the height of the avalanche is *a parabolic profile* and *varies inversely with time.* This is illustrated in Fig. 7.

The technique of *separation of variables* can also be used to solve the system (57); if pursued, the following avalanche thickness and relative velocity profiles emerge:

$$h = \frac{1}{9\beta}t^{-2/3}\left[h_m + \eta^2\right], \quad u = \frac{2}{3}t^{-1/3}\eta. \tag{60}$$

As calculated by Savage & Hutter [42], the value of h_m is given by

$$V = \frac{2}{9\beta}\left(d_m - 1 + \frac{1}{3}\right) = \frac{2}{9\beta}\left(h_m + \frac{1}{3}\right), \tag{61}$$

[4] $\eta = \pm 1$ for the front and rear margin and $\eta = 0$ for the (moving) centre of the avalanche.

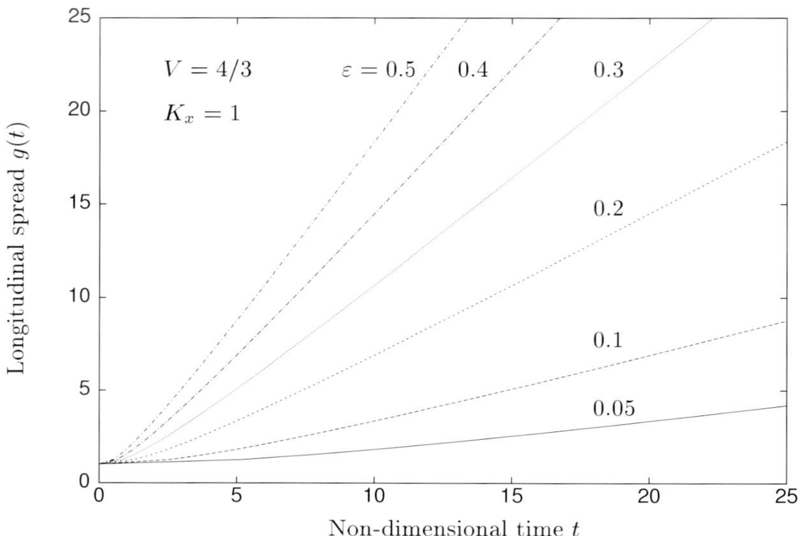

Fig. 6. The spreading factor g for different parameter values of ε

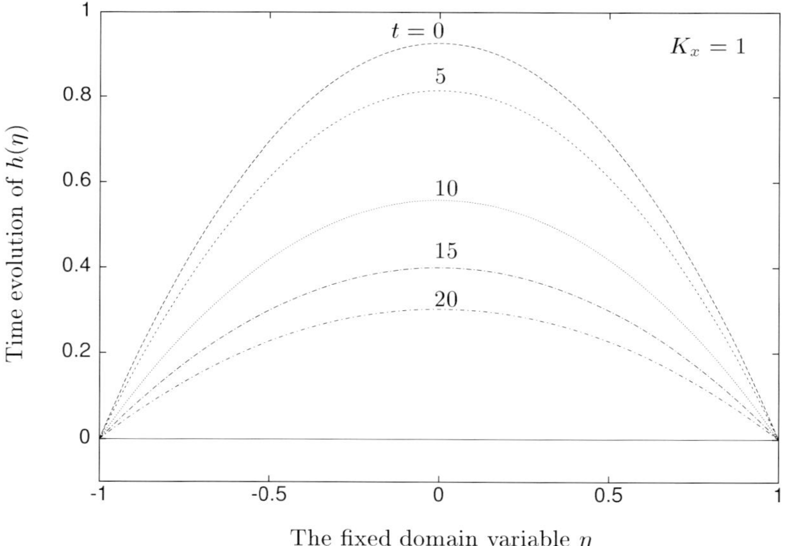

Fig. 7. The time and space evolution of the parabolic cap similarity solution

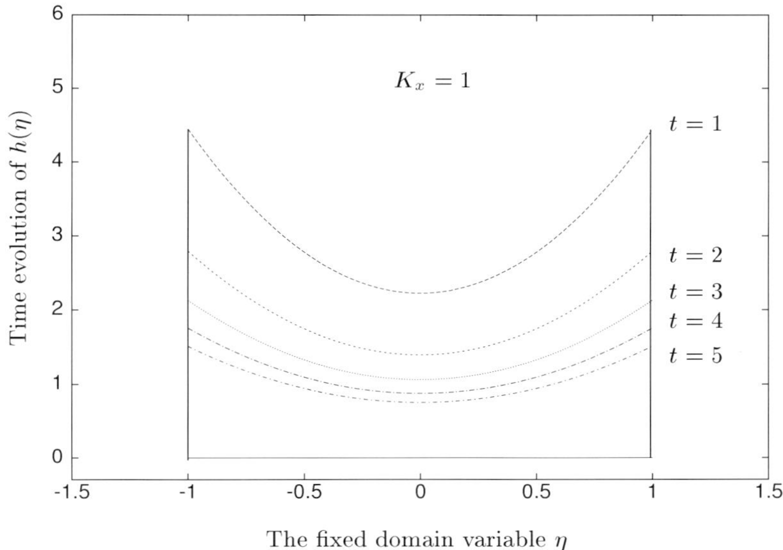

Fig. 8. The time and space evolution of the M-wave similarity solution

where $d_m - 1 = h_m$ is the normalised depth of the avalanche margin and thus assumed to be greater than zero. Equation (61) also implies an interesting result that for $K_x = 1$ we have $\varepsilon > 0.074/V$. This shows that for such kind of similarity solutions, the choice of the parameter ε explicitly depends on the total volume V of the avalanche body. Also the shape of h in (60) prompted SAVAGE & HUTTER to call it the *M-wave solution*. Figures 7 and 8 show the time evolution of the parabolic cap and M-wave similarity solutions for the evolution of the height of the granular avalanches. The reader should keep in mind that these figures are drawn in terms of the fixed domain variables so that one can not directly observe the preservation of the volume. It is seen, however, that the profiles for h are shape preserving and that the avalanche spread and the amplitude of the pile vary with time. The parabolic cap solution is of special importance, particularly in checking the efficiency of numerical codes, as done by TAI and TAI et al. [45–47].

12 Discussion

In this section we will compare the three systems of equations we have considered so far against each other. Practical applicabilities of each model under specific situations are also outlined.

12.1 Orthogonal Complex Versus Non-Orthogonal Helical System

Let us first compare the *non-orthogonal helical system* formulated by PUDASAINI, ECKART & HUTTER [38] against the *orthogonal complex system* derived by GRAY, WIELAND & HUTTER [12] in the SH–spirit for three-dimensional basal topography. At the same time we will state the physical significance of these models.

- **Effects of Curvature and Torsion in Mass Balance** The main idea behind the helical and non-orthogonal coordinate system is to introduce both curvature and torsion in the associated metric. This metric, then, describes the reference topography as well as handles the process of transformation of coordinate free field equations into a non-orthonormal curvilinear coordinate system. The presence of the factor $\psi = 1/\sqrt{y^2 + a^2}$ in the mass balance equation (24) has a strong effect in the mass flux which is not the case in all previous considerations of the SH–theory which were based on the orthogonal metric. For smaller curvature and torsion (i.e. a large), $\psi = 1/\sqrt{y^2 + a^2}$ is also small and the flow is less rapid in the down-slope direction which is seen clearly from the term "$\partial(\psi hu)/\partial x$" in the mass balance. For large value of the pitch parameter a, both ψ and u are small. Consequently, the whole term "$\partial(\psi hu)/\partial x$" is small, indicating lesser mass flux in the down-slope direction.
- **Effects of Curvature and Torsion in the Total Derivative** The operator $d/dt = \partial/\partial t + \psi u \partial/\partial u + v \partial/\partial y$, in the acceleration terms of the left-hand sides of the momentum balances, (25)–(26), contain ψ. This implies that the smaller the curvature and torsion are, the smaller will be the component of acceleration in the down-slope u-direction, which is an intrinsic property of the *non-orthogonal helical system of coordinates*. In the previous theories, e.g., [12], derived by using the *orthogonal system of coordinates*, no such effects of curvature and torsion could be seen in the acceleration operator. Note that, on assuming continuity of the solution, this operator can be obtained by substituting the mass balance, (24), into the momentum balances (25)–(26).
- **Overall Effects in the Model Equations** From the balance equations of mass and momentum for the *non-orthogonal helical system* it is clear that most terms are effected explicitly or implicitly by curvature and torsion of the basal topography and the coordinate system itself. These effects can easily be seen from the terms in the model equations containing either the pitch parameter, "a", or the *stretching factor* of the motion, "$\psi = 1/\sqrt{y^2 + a^2}$", which slows the motion down or boosts it up. Such a direct effect of curvature together with torsion was not possible to investigate before by the use of the *orthogonal metric* [12].
- **Direct Influence in Solutions** Note that for the similarity solution, which is presented here for the non-orthogonal helical system of coordinates, the time t is not the physical time. It is scaled by the stretching

factor ψ, i.e., $t = \psi \times physical\ time$, which was not the case in other models of the SH–equations based on the orthogonal metric. This rescaling of the time made it possible to find semi-analytical solutions. These solutions also take into account the effects of the passive parameter y in the lateral direction. Therefore, for the first time, PUDASAINI, ECKART & HUTTER [38] have incorporated the influence of the curvature and torsion into the test solution of the model equation.

12.2 Orthogonal Complex Versus Orthogonal General System

In situations where the talweg is a plane curve in a vertical plane, it is already proven (for instance, see [12,50]) that the orthogonal complex system can reproduce laboratory experiments to a very good approximation. The major problem was to introduce non-uniform curvature and torsion in the metric that describes the whole flow behaviour. It is made possible by using the orthogonal general system of equations derived by PUDASAINI & HUTTER [39]. In many cases the orthogonal complex system seems to be very useful. In general, the orthogonal general systems may serve as a good theoretical foundation in order to investigate the flow of granular masses in more complicated topographies. Here, we discuss the connection and differences between these two theories.

- **Broad Applicability** The general equations (37)–(39) with the precisions (40)–(42) are formally analogous, almost identical, to those of previous derivations (12)–(14) under (much) simpler situations, see GRAY et al., HUTTER et al. and WIELAND et al. [12,21,50]. However, the applicability of this general system of equations is by far broader than in the previous cases. This has been achieved by use of a different underlying coordinate system than in the previous cases. For this reason, there is no limit of these model equations that would exactly merge into the previous models, only approximately so.
- **Introduction of Non-Uniform Curvature and Torsion** The key idea was the use of an orthogonal curvilinear moving coordinate system that is based on a master line in three dimensions which exhibits non-uniform curvature and torsion. The talweg of a valley or the axis of a three-dimensionally curved and twisted channel or pipe may be the basis for the construction of this master curve. Planes perpendicular to this master curve give rise to the introduction of a polar coordinate system within these planes, of which the origin is the intersection point with the master curve. The topographic profile of the avalanche within these planes can be described in terms of these polar coordinates; the normal (radial) direction determines the direction of the height, the cross-slope (azimuthal) direction embraces its width, see Fig 5. Obviously, for different azimuthal angles the radial directions are not parallel; it is here where this model deviates from previous ones. It implies that the earlier

equations of the SH-model [12,21,50] with torsion free master curves are only exactly reproduced when these master curves are moved infinitely far away such that $\lim z \to \infty$ implies that zdy is incremental. Otherwise, the model determines the rays of a depth function which, at different positions of a cross section, are not parallel but radial. Similarly, whilst the down-slope velocity component is in the entire cross section parallel to the local direction of the master curve, the transverse velocity component follows concentric circles about the centre at the master curve within the cross sectional planes.
- **Flexible and More Realistic** The advantage of this formulation of a depth-integrated avalanche model lies in its flexibility of application. The flow down an inclined plane or within a channel of which the axis lies in a vertical plane but may be curved, and the flow of a granular avalanche in a helicoidal channel of arbitrary cross section can be described, as can the flow down mountain valleys with arbitrarily curved and twisted talwegs. It is this last application which guided PUDASAINI & HUTTER [39] to derive this model, because it is ideally suited to the application in realistic situations in connection with the use of *Geographical Information and Visualisation Systems* (GIVS).

12.3 Non-Orthogonal Helical Versus Orthogonal General System

One may quickly infer that there are some direct connections between these two theories. But, in general this is not the case. There are some connections and some major differences between them.

- **The Underlying Metric** It is clear from the above discussions that the flows in any situation are characterised by the metrics and the topographies in use. Depending on some specific situations of investigations and the interest of the investigators, the non-orthogonal helical system may be more adequate than the orthogonal general system because it is better tuned to the application at hand. However, in general, the latter may be more useful than the former.
- **Deduction of one System from Another** One thing in common may be that both of these systems should be able to produce similar results to some good approximation if we use simply a helix as a master curve in an orthogonal general system. But this may not be true in all cases.
- **Fundamental Differences in the Equations** Different ideas are used to derive these theories. The orthogonal general theory incorporates non-uniform curvature and torsion, whereas the helical system contains only curvature and torsion of a helix. This is one of the major differences between these two models. Another difference concerns the depth integration. For the non-orthogonal helical system, the process of averaging along the direction of gravity is used which ultimately led the theory to a simple state. The model would be much more complicated if we had

not used this particular technique. For the orthogonal general system, the depth integration along the normal of the reference surface is used which makes the equations simpler in this case.

13 Conclusions

This paper presents a comparison between three theories that were formulated in order to extend the original SH–theory [42,43] for granular avalanches. The first one is associated with an orthogonal metric of the reference surface, whereas the second is for a completely new geometry, the so-called helicoidal topography and its associated non-orthogonal metric system. Similarly, the third model is derived for non-uniform curvature and torsion based on a general orthogonal metric of the basal topography. In contrast to other previous extensions (e.g., see Gray et al. [12]), the local coordinate system for the latter two are based on generating curves with curvature and torsion. Their derivations were necessary because real avalanches are often guided by rather strongly curved and twisted corries, and it is useful to corroborate a mathematical model by laboratory experiments as well as field events. In all cases, the motion follows the talweg of the basal topography. The theory that is based on the helicoidal metric is characterised by its non-orthogonality, but it allows relatively easy access to comparison with laboratory experiments. Analogously, the motions for other cases are characterised by their corresponding orthogonal metrics.

All these theories assume a shallow avalanche of a dry cohesionless granular material, incompressible with constant density throughout the motion from initiation to run-out. Balance laws of mass and momentum, kinematic boundary conditions on the basal surface and the free surface, the Coulomb–dry friction law at the base and the tractionless free surface condition constitute the underlying field equations. Depth averaging these equations leads to a set of non-linear partial differential equations for the evolution of the granular pile height and the depth-averaged stream-wise and cross flow velocity distributions of a finite mass of granulates. Coulomb–like constitutive behaviour both for the bed and interior of the granular body is employed. Although some of the details of the flow field are lost, an enormous and significant simplification has been achieved by the depth averaging process. The emerging theories are capable of predicting the flow of dense granular materials over complex basal topography and strongly curved and twisted channels, respectively, depending on the kind of topography and metric system in use.

The model equations are tested by constructing analytical solutions of the simplest kind. These elementary solutions, the so-called parabolic cap and M-wave similarity solutions, were constructed in a fashion analogous to those in the SH–theory [42]. The depth profiles of the moving and deforming avalanches are found to have a parabolic shape and the relative velocity is found to vary linearly with distance from the centre of mass of the moving avalanche. Although it is difficult to validate both similarity solutions by experiments, the similarity solutions are still useful in the sense that they provide insight into the qualitative behaviour which is meaningful as a semi-analytical solution in a comparison against some general numerical code.

There is no direct comparison between the full set of final equations representing the balance laws of all three models. These laws contain the three unknown variables, the avalanche height and the depth-integrated velocity profiles. Nevertheless, the qualitative behaviour of both the traditional theories with orthogonal metric and its other extentions with non-orthogonal as well as orthogonal general metric agree well through the parabolic cap and M-wave similarity solutions. Such solutions for the case of the non-orthogonal helical system also take into account the effect of the passive parameter, the lateral coordinate y, through the definition of the new time scale. For the first time, the effect of curvature and torsion of the basal topography is incorporated into the model equations and their semi-exact solutions through the underlying metric of the curvilinear reference geometry. It is shown that these effects are crucial in understanding the physical behaviour of the flow of dry granular avalanches in curved and twisted channels. This may be a good motivation and a preliminary proof of how the extended theories may work in reality and provide evidence that the SH–theory works well not only for gently curved chutes in one direction, but also for rather strongly curved chutes having both curvature and torsion. This is an important step towards the development of a fully non-orthogonal as well as general orthogonal theory for further complicated topographies that may be encountered in many industrial and geophysical flows. Therefore, these models may find important applications in industrial and geophysical flows that deal with flows in channelised complex basal topographies as well as in curved and twisted chutes. The other aspect of future attention should be directed to extend the non-orthogonal helical model which also incorporates the straight inclined starting gully and horizontal smooth run-out zones.

The proposed model equations for the orthogonal general system are suitable for the prediction of avalanche flows of granular materials down arbitrarily curved and twisted tracks. The equations reduce to earlier models for which the SH–theory has been demonstrated to reproduce results from laboratory experiments well.

The next and immediate goal may be to perform numerical simulations with the intention to provide a general purpose software for practitioners involved with the prediction of avalanche run-out in mountainous regions. The intention is to use *Geographical Information Systems* (GIS) from which digitised realistic topographies in mountainous regions are available. With these GIS, particular avalanche prone subregions can be selected, and for individual sites the master curve and the cross sectional topography can be constructed. From a preselected release of a finite mass of gravel or snow at a breaking zone it is then intended to determine the flow from initiation to run-out. This step requires numerical integration via an avalanche purpose software using total-variation-diminishing and non-oscillating numerical schemes. Its output can, in a final step, be used in visualisation software to identify endangered zones. A multitude of applications will then be at the disposal for practitioners to be investigated with the software. Comparison with observational data in the field taken by photography from helicopters may then become possible.

This optimistic view should not stay isolated without mentioning some caveats which indicate that there is still a lot of work left for avalanche modellers.

First, the numerical implementation ought to be done with shock capturing integration techniques. The unavoidable numerical diffusion can be minimised, if total variation diminishing techniques are implemented. However even with these, unexpected inaccuracies may still arise when these schemes are applied on realistic topographies.

Second, common resolutions of mountain topographies given in digital elevation models are not very good (typically 25×25 m) and may therefore yield smoother surfaces than in reality. Interpolation makes these surfaces even smoother and may then falsify the inferences simply because of the use of an inadequate topography. One of the referees' of this paper expresses the opinion "that introducing an accurate model with the generalised metric probably is an overkill, since the curvature and torsion obtained by interpolation of topographic data most likely will introduce a larger error than the one achieved in the theory." However, this still makes the model suitable at least for laboratory avalanches.

Third, it may be questioned in the realistic situation that the talweg can be defined beforehand. True, and indeed the theory is unable at its present stage to cope with cases where the dense avalanche splits into two different parts, separated by their own talwegs. However, there are still ample situations in which the likely talweg can be a priori guessed and, if needed, corrected. The model has been designed for these situations !

Fourth, the basal friction law that is implemented is restricted to the rate independent MOHR-COULOMB law. There have been numerous suggestions to "amend" this law by incorporating a viscous contribution to improve agreement with observations. Such claims are likely to be correct, but in our opinion do not withstand scrutiny. Careful experimental observation is needed to substantiate these claims. Incorporation into the model is straightforward but may considerably affect the numerics.

Finally, the criticism most justified, is the omission in our model of the entrainment from the snow cover and deposition mechanism of snow to the basal surface. Theoretically, these effects are easily accounted for; an entrainment/deposition function must enter the kinematic boundary condition at the base. However, the phenomenological presentation of this functional relation is difficult – largely unknown. We regard it as *the grand unsolved problem of avalanche dynamics*. It will affect the dynamics of dense flow avalanches dramatically, because situations are known where the deposited mass is up to a factor of five larger than the mass released.

In summary, work will continue in attempts to describe avalanche dynamics for quite some time.

Acknowledgement

We thank the reviewers for their constructive criticisms that improved the paper considerably.

References

1. Ancey, C. (2001) Snow Avalanches. In: Balmforth, N.L., Provenzale, A. (Eds.) Lecture Notes in Physics, LNP Vol. 582: Geomorphological Fluid Mechanics, Springer, 319–338
2. Bowen, R.M., Wang, C.-C. (1976) Introduction to Vectors and Tensors: Volume 1–Linear and Multilinear Algebra. Plenum Press, New York and London
3. Bowen, R.M., Wang, C.-C. (1976) Introduction to Vectors and Tensors: Volume 2–Vector and Tensor Analysis. Plenum Press, New York and London
4. Brugnot, G. (1980) Recent progress and new applications of the dynamics of avalanches. J. Glaciology **26**(94), 515–516
5. Chugunov, V., Gray, J.M.N.T., Hutter, K. (2003) Group Theoretic Methods and Similarity Solutions of the Savage–Hutter Equations. This volume
6. Dent, J.D., Lang, T.E. (1980) Modelling of snow flow. J. Glaciology **26**(94), 131–140
7. Dent, J.D., Lang, T.E. (1982) Experiments on the mechanics of flowing snow. Cold Reg. Sci. Tech. **5**(3), 253–258
8. Erismann, T. (1986) Flowing, rolling, bouncing, sliding: Synopsis of basic mechanisms. Acta Mechanica **64**, 101–110
9. Farwig, R. (2003) Existence of Avalanching Flows. This volume
10. Germano, M. (1982) On the effect of torsion on helical pipe flow. J. Fluid Mech. **125**, 1–8
11. Germano, M. (1989) The Dean equations extended to a helical pipe flow. J. Fluid Mech. **203**, 289–305
12. Gray, J.M.N.T., Wieland, M., Hutter, K. (1999) Gravity-driven free surface flow of granular avalanches over complex basal topography. Proc. R. Soc. London A**455**, 1841–1874
13. Gray, J.M.N.T. (2003) Rapid Granular Avalanches. This volume
14. Greve, R., Hutter, K. (1993) Motion of a granular avalanche in a convex and concave curved chute: experiments and theoretical predictions. Phil. Trans. R. Soc. London A**342**, 573–600
15. Greve, R., Koch, T., Hutter, K. (1994) Unconfined flow of granular avalanches along a partly curved surface. I. Theory. Proc. R. Soc. London A**445**, 399–413
16. H.-U. Gubler (1987) Measurements and modelling of snow avalanche speeds. In: Salm, B., Gubler, H.-U. (Eds.) Avalanche Formation, Movement and Effects IAHS Publ. 162, pp. 405–420

17. Haff, P.K. (1983) Grain flow as a fluid-mechanical phenomenon. J. Fluid Mech. **134**, 401–430
18. Hsü, K. (1975) On sturzstorms - catastrophic debris streams generated by rockfalls. Geol. Soc. Am. Bull. **86**, 129–140
19. Hutter, K., Szidarovszky, F., Yakowitz, S. (1986a) Plane steady shear flow of a cohesionless granular material down an inclined plane: A model for snow avalanches, Part I: Theory. Acta Mech. **63**, 87–112
20. Hutter, K., Szidarovszky, F., Yakowitz, S. (1986b) Plane steady shear flow of a cohesionless granular material down an inclined plane: A model for snow avalanches, Part II: Numerical results. Acta Mech. **65**, 239–261
21. Hutter, K., Koch, T. (1991) Motion of a granular avalanche in an exponentially curved chute: experiments and theoretical predictions. Phil. Trans. R. Soc. London A**334**, 93–138
22. Hutter, K. (1991) Two- and three dimensional evolution of granular avalanche flow – theory and experiments revisited. Acta Mech. **1**[Suppl], 167–181
23. Hutter, K., Siegel, M., Savage, S.B., Nohguchi, Y. (1993) Two-dimensional spreading of a granular avalanche down an inclined plane. I. Theory. Acta Mech. **100**, 37–68
24. Hutter, K., Koch, T., Plüss, C., Savage, S.B. (1995) Dynamics of avalanches of granular materials from initiation to runout. Part II: Experiments. Acta Mech. **109**, 127–165
25. Hwang, H., Hutter, K. (1995) A new kinetic model for rapid granular flow. Continuum Mech. Thermodyn. **7**, 357–384
26. Jenkins, J.T., Richman, M.W. (1985) Grad's 13-moment system for a dense gas of inelastic spheres. Arch. Rat. Mech. Anal. **87**, 355–377
27. Jenkins, J.T., Savage, S.B. (1983) A theory for the rapid flow of identical, smooth, nearly elastic particles. J. Fluid Mech. **130**, 186–202
28. Kent, P.E. (1965) The transport mechanism in catastrophic rockfalls. J. Geol. **74**, 79–83
29. Klingbeil, E. (1966) Tensorrechung für Ingenieure. Bibliographisches Institut, Mannheim: Hochschultaschenbücher
30. Lang, T.E., Dent, J.D. (1982) Review of surface friction, surface resistance and flow of snow. Rev. of Geophys. & Space Phys. **20**(1), 21–37
31. Lang, R.M. (1992) An Experimental and Analytical Study on Gravity-Driven Free Surface Flow of Cohesionless Granular Media. PhD Thesis, Darmstadt University of Technology, Darmstadt, Germany
32. Lied, K., Bakkehøi, S. (1980) Empirical calculations of snow-avalanche run-out distance based on topographic parameters. J. Glaciol. **26**(94), 165–177
33. Lun, C.K.K., Savage, S.B., Jeffrey, D.J., Chepurniy, N. (1984) Kinetic theories for granular flows: inelastic particles in couette flow and slightly inelastic particles in general flow field. J. Fluid Mech. **140**, 223–256
34. McClung, D.M., Lied, K. (1987) Cold Reg. Sci. Technol. **13**(2), 107–119
35. Perla, I.P., Cheng, T.T., McClung, D. (1980) A two-parameter model of snow avalanche motion. J. Glaciology **26**(94), 197–207
36. Piegl, L., Tiller, W. (1997) The NURBS Book. 2nd edition, Springer
37. Pohlhausen, K. (1921) Zur näherungsweisen Integration der Differentialgleichung der laminaren Grenzschicht. ZAMM **1**, 252–268
38. Pudasaini, S.P., Eckart, W., Hutter, K. (2002) Gravity–Driven Rapid Shear Flows of Dry Granular Masses in Helically Curved and Twisted Channels. Mathematical Models and Methods in Applied Sciences (accepted for publication)

39. Pudasaini, S.P., Hutter, K. (2002) Rapid Shear Flows of Dry Granular Masses in Arbitrarily Curved and Twisted Channels Subject to Gravity Forces. J. Fluid Mech. (submitted for publication)
40. Salm, B. (1966) Contribution to avalanche dynamics. IAHS AISH Publ. 69. 199–214
41. Salm, B., Gubler, H. (1985) Measurement and Analysis of dense flow of avalanches. Annals of Glaciol. **6**, 26–34
42. Savage, S.B., Hutter, K. (1989) The motion of a finite mass of granular material down a rough incline. J. Fluid Mech. **199**, 177–215
43. Savage, S.B., Hutter, K. (1991) Dynamics of avalanches of granular materials from initiation to runout. Part I: Analysis. Acta Mech. **86**, 201–223
44. Shreve, R.L. (1966) Sherman landslide, Alaska. Science **154**, 1639–1643
45. Tai, Y.-C. (2000) Dynamics of Granular Avalanches and their Simulations with Shock–Capturing and Front–Tracking Numerical Schemes. Ph.D. thesis, Department of Applied Mechanics, Darmstadt University of Technology, Germany
46. Tai, Y.-C., Gray, J.M.N.T., Hutter, K. (2001) Dense Granular Avalanches: Mathematical Description and Experimental Validation. In: Balmforth, N.L., Provenzale, A. (Eds.) Lecture Notes in Physics, LNP Vol. 582: Geomorphological Fluid Mechanics. Springer, 339–366
47. Tai, Y.-C., Noelle, S., Gray, J.M.N.T., Hutter, K. (2002) Shock-Capturing and Front-Tracking Methods for Granular Avalanches. J. Comp. Phys. **175**(1), 269–301
48. Voellmy, A. (1955) Über die Zerstörungskraft von Lawinen. Schweizeriche Bauzeitung **73**, 159–162, 212–217, 246–249, 280–285
49. von Kármán, Th. (1921) Über laminare und turbulente Reibung. ZAMM **1**, 233–252
50. Wieland, M., Gray, J.M.N.T., Hutter, K. (1999) Channelized free-surface flow of cohessionless granular avalanches in a chute with shallow lateral curvature. J. Fluid Mech. **392**, 73–100
51. Zabielski, L., Mestel, A.J. (1998a) Steady flow in a helically symmetric pipe. J. Fluid Mech. **370**, 297–320.
52. Zabielski, L., Mestel, A. J. (1998b) Unsteady blood flow in a helically symmetric pipe. J. Fluid Mech. **370**, 321–345.
53. Zwinger, T., Kluwick, A., Sampl, P. (2003) Avalanche flow of dry snow over natural terrain. Numerical simulation. This volume

A Lagrangian-Eulerian Finite-Volume Method for Simulating Free Surface Flows of Granular Avalanches

Karl Koschdon and Michael Schäfer

Department of Numerical Methods in Mechanical Engineering,
Darmstadt University of Technology, Petersenstr. 30, 64287 Darmstadt, Germany,
e-mail: {karl, schaefer}@fnb.tu-darmstadt.de

received 31 Jul 2002 — accepted 11 Nov 2002

Abstract. A numerical method for the prediction of gravity-driven, free surface flows of granular materials is presented and investigated. The method is based on an Arbitrary LANGRANGEan–EULERian (ALE) finite-volume method, where unstructured boundary-fitted moving grids are employed to follow the free boundary. The underlying flow solver consists of a GODUNOV-type approach in the space-time domain, and the fluxes are calculated using the solution of RIEMANN problems. Two versions of the scheme on moving grids are developed and the interaction of the capability of shock-capturing and arbitrary grid movement is analysed. The numerical scheme is applied to characteristic test problems and simulations of avalanche flows for which experimental results are also available.

1 Introduction

Snow slab avalanches, landslides, and rock falls are extremely dangerous and destructive natural phenomena. The events endanger human lives and infrastructure. A reliable numerical method for predicting the avalanche paths and maximum run-out distances is therefore of considerable interest to civil engineers responsible for planning in populated mountain regions. However, the development of such a tool turns out to be a rather challenging task, because complex nonlinear source terms, free boundaries, moving grids, shock phenomena, and corresponding interactions have to be treated adequately by the numerical method.

The theory of SAVAGE & HUTTER [10] is an established model to describe an incompressible granular avalanche flowing down an inclined plane. It assumes shallow geometries in the sense that the avalanche thickness is much smaller than its extension parallel to the bed.

GRAY et al. [3] extended the SAVAGE-HUTTER theory for three-dimensional motion of avalanches over shallow topographies. They transformed the governing equations into a conservative form for describing the shock formations. Numerical simulations of stationary shock formations are presented by

IRMER et al. [6], where the case of supercritical flow past a wedge in which oblique shocks have been observed in granular materials is investigated.

TAI [11] studied dynamics of granular avalanches and implemented different shock-capturing methods to describe transient granular shock formations. Besides the formation of shocks in the interior, the margin constitutes a free boundary of the avalanche. A front-tracking, non-oscillatory scheme on a steady grid has been developed by TAI et al. [12].

In this paper a shock-capturing GODUNOV-type method on moving grids is presented. The grid movement does not depend on the local particle velocity but follows the free boundary (margins of the avalanche). In this way the difficulties associated with strong grid distortion when using a pure LAGRANGEan method can be avoided. The main advantage of this approach is that the free boundary is treated in a LAGRANGEan fashion to produce a sharp resolution. In the interior of the computational domain an exact RIEMANN problem solution is used to evaluate the flux function on the moving computational cell boundary. The movement of the numerical grid is defined by the solution of a vacuum RIEMANN problem.

Several characteristic numerical examples are considered in order to illustrate and investigate the properties of the various components of the proposed approach. Here, special emphasis is given to the interaction of shock capturing and grid movement, which is one of the most crucial aspects of the considered type of problems.

2 Governing Equations

The detailed derivation of the SAVAGE-HUTTER theory has been given in [3,10], so only a brief description of the governing equations is presented here. The behaviour of the granular media can be described by the mass and momentum conservation laws which in symbolic coordinate-free notation read

$$\nabla \cdot \boldsymbol{v} = 0\,, \tag{1}$$

$$\rho \left\{ \frac{\partial \boldsymbol{v}}{\partial t} + \nabla \cdot (\boldsymbol{v} \otimes \boldsymbol{v}) \right\} = -\nabla \cdot \boldsymbol{p} + \rho \boldsymbol{g}\,, \tag{2}$$

where the density ρ is assumed to be constant and $\boldsymbol{g}$ is the acceleration due to gravity. The unknowns of the problem are the velocity vector $\boldsymbol{v}$ and the pressure tensor $\boldsymbol{p}$.

The extension of the SAVAGE-HUTTER theory by GRAY et al. [3] describes the motion of the avalanche over a complex three-dimensional topography, which is modelled by defining an orthogonal curvilinear reference surface (see Fig. 1). Curvilinear coordinates x, y, z are defined, where the x and y coordinates lie in the reference surface and the z coordinate is normal to reference surface. The x coordinate is chosen in the main direction of the avalanche motion; it constitutes the arc length along the generic curve in a vertical

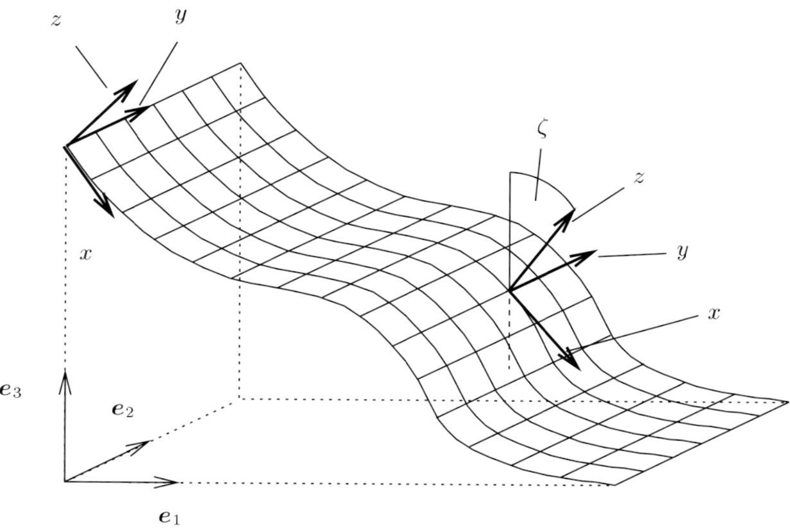

Fig. 1. Curvilinear coordinate system x, y, z and reference surface for the description of the avalanche flow

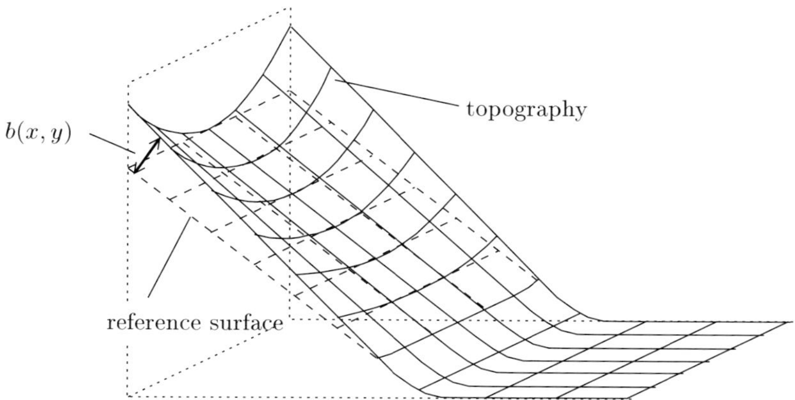

Fig. 2. The basal topography (solid lines) defined by its height $b(x, y)$ above the curvilinear reference surface (dashed lines)

plane, so that the slope angle $\zeta = \zeta(x)$ is a function of x only. This relationship defines the shape of the reference surface completely. The topography is introduced by the definition of the distance between the reference surface and the basal topography $b = b(x, y)$ (see Fig. 2). After depth-integrating the mass and momentum balance equations and introducing the constitutive law by the Mohr-Coulomb closure the final formulation of the system of governing equations takes the form (for a thorough basic derivation of the equations see also GRAY [2] and PUDASAINI et al. [8]).

$$\frac{\partial h}{\partial t} + \frac{\partial}{\partial x}(hv_1) + \frac{\partial}{\partial y}(hv_2) = 0\,, \tag{3}$$

$$\frac{\partial}{\partial t}(hv_1) + \frac{\partial}{\partial x}\left(hv_1^2 + \beta_x \frac{h^2}{2}\right) + \frac{\partial}{\partial y}(hv_1 v_2) = hs_x\,, \tag{4}$$

$$\frac{\partial}{\partial t}(hv_2) + \frac{\partial}{\partial x}(hv_1 v_2) + \frac{\partial}{\partial y}\left(hv_2^2 + \beta_y \frac{h^2}{2}\right) = hs_y\,, \tag{5}$$

where h is the height of the granular material layer, and v_1 and v_2 are the depth-averaged components of the velocity vector $\boldsymbol{v}$. The above equations together with assumption $h \neq 0$ provides a hyperbolic system of conservation laws.

The right hand sides of the momentum equations are defined by

$$s_x = \sin\zeta - \frac{v_1}{|\boldsymbol{v}|}\tan\delta\left(\cos\zeta + \lambda\kappa v_1^2\right) - \epsilon\cos\zeta\frac{\partial b}{\partial x}, \tag{6}$$

$$s_y = - \frac{v_2}{|\boldsymbol{v}|}\tan\delta\left(\cos\zeta + \lambda\kappa v_1^2\right) - \epsilon\cos\zeta\frac{\partial b}{\partial y}. \tag{7}$$

The first term in the x-source term defines the acceleration due to the component of the gravity force parallel to the x-coordinate. The second term corresponds to friction by the normal force with the friction coefficient $\mu = \tan\delta$, where δ is the basal friction angle. The second term on the right constitutes the shear stress due to the normal component of the gravity force, $\cos\zeta$, and the centripetal force, $\lambda\kappa v_1^2$, where $\kappa = \partial\zeta/\partial x$ is the curvature of the reference surface. The friction is pointing in the opposite direction of the velocity vector $\boldsymbol{v}$, which is described by the factors $-v_1/|\boldsymbol{v}|$ and $-v_2/|\boldsymbol{v}|$, respectively. The last term defines the force due the variation of the topography $b(x,y)$. The coefficients ϵ and λ are scale parameters. For our computations in Sect. 4 we always use $\epsilon = \lambda = 1$.

The coefficients

$$\beta_x := \epsilon\cos\zeta K_x \quad \text{and} \quad \beta_y := \epsilon\cos\zeta K_y \tag{8}$$

result from the introduction of the constitutive law. They constitute the net forces in the x- and y-directions due to pressure variations in the downslope and cross slope directions and the earth pressure coefficients K_x,K_y describe the stress ratio on the contact area between the avalanche and the respective surface

$$K_x := p_x/p_z \quad \text{and} \quad K_y := p_y/p_z\,. \tag{9}$$

HUTTER et al. [5] used the MOHR circle representations to determine these values as functions of internal and basal angles of friction (ϕ and δ):

$$K_{x,\mathrm{act/pas}} = \frac{2}{\cos^2\phi}\left(1 \mp \sqrt{1 - \frac{\cos^2\phi}{\cos^2\delta}}\right) - 1, \tag{10}$$

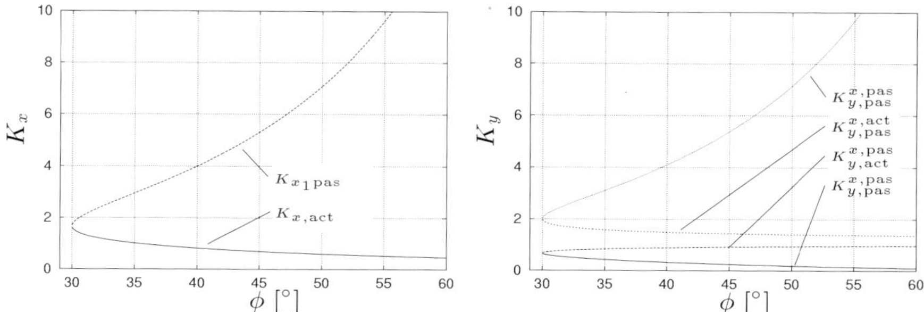

Fig. 3. The pressure coefficients defined by the constitutive law of the Savage-Hutter theory. The values are illustrated for a basal friction angle $\delta = 30°$

$$K_{y,\text{act/pas}} = \frac{1}{2}\left(K_x + 1 \mp \sqrt{(K_x - 1)^2 + 4\tan^2\delta}\right), \tag{11}$$

in which the upper (lower) signs apply for active (passive) conditions. Note that the pressure coefficients are real numbers only for $\delta \le \phi$. Depending on the behaviour of the avalanche, the pressure coefficients can be active or passive. The downslope earth pressure coefficient K_x is active during a downslope dilatational motion and passive during a downslope compressional motion

$$K_x = \begin{cases} K_{x,\text{act}}, & \text{for } \partial v_1/\partial x > 0, \\ K_{x,\text{pas}}, & \text{for } \partial v_1/\partial x < 0. \end{cases} \tag{12}$$

In the cross-slope direction, the four stress ratios are distinguished by considering whether the downslope and the cross slope deformations are dilatational or compressional as follows:

$$K_y = \begin{cases} K^{x,\text{act}}_{y,\text{act}}, & \text{for } \partial v_1/\partial x > 0 \text{ and } \partial v_2/\partial y > 0, \\ K^{x,\text{pas}}_{y,\text{act}}, & \text{for } \partial v_1/\partial x < 0 \text{ and } \partial v_2/\partial y > 0, \\ K^{x,\text{act}}_{y,\text{pas}}, & \text{for } \partial v_1/\partial x > 0 \text{ and } \partial v_2/\partial y < 0, \\ K^{x,\text{pas}}_{y,\text{pas}}, & \text{for } \partial v_1/\partial x < 0 \text{ and } \partial v_2/\partial y < 0. \end{cases} \tag{13}$$

The behaviour of the pressure coefficients as functions of ϕ is illustrated in Fig. 3. The governing equations have a structure similar to the shallow-water equations. The difference consists in the possible jump of the earth pressure coefficients and the source terms hs_x and hs_y on the right-hand side of the momentum conservation law. In particular, this and the moving avalanche boundary makes the development of an appropriate numerical approach to simulate the avalanche flow a rather challenging task.

3 Numerical Method

In vector notation (3)-(5) can be written in the form

$$\frac{\partial \boldsymbol{u}}{\partial t} + \frac{\partial \boldsymbol{f}_1(\boldsymbol{u})}{\partial x} + \frac{\partial \boldsymbol{f}_2(\boldsymbol{u})}{\partial y} = \boldsymbol{b}(\boldsymbol{u}) \,, \tag{14}$$

where $\boldsymbol{u} = \boldsymbol{u}(t, \boldsymbol{x})$ is the vector of conservative variables, $\boldsymbol{f}_1(\boldsymbol{u})$ and $\boldsymbol{f}_2(\boldsymbol{u})$ are the flux vectors and $\boldsymbol{b}(\boldsymbol{u})$ is the source term vector:

$$\boldsymbol{u} = \begin{pmatrix} h \\ hv_1 \\ hv_2 \end{pmatrix} , \qquad \boldsymbol{f}_1(\boldsymbol{u}) = \begin{pmatrix} hv_1 \\ hv_1^2 + \beta_x h^2/2 \\ hv_1 v_2 \end{pmatrix} , \tag{15}$$

$$\boldsymbol{f}_2(\boldsymbol{u}) = \begin{pmatrix} hv_2 \\ hv_1 v_2 \\ hv_2^2 + \beta_y h^2/2 \end{pmatrix} , \qquad \boldsymbol{b}(\boldsymbol{u}) = \begin{pmatrix} 0 \\ hs_x \\ hs_y \end{pmatrix} . \tag{16}$$

In the next subsections we present the various components of the numerical solution method for the above equations. We first consider the corresponding homogeneous system, i.e. (14) with $\boldsymbol{b} = \boldsymbol{0}$, in order to separate the treatment of fluxes and the complex source term. The incorporation of $\boldsymbol{b}$, which will be done via a simple splitting method, is then described in Subsect. 3.7.

3.1 Finite-Volume Formulation

We discretise the spatial domain into N polygonal finite volumes V_i with volume $\Delta V_i = \int_{V_i} 1 \, \mathrm{d}V$ for $i = 1, \ldots, N$. The temporal domain, i.e. the time interval $[0, T]$, is discretised into time steps $t_n = t_{n+1} + \Delta t_n$ for $n = 0, 1, \ldots$ starting with $t_0 = 0$, which, in general, may be of variable size.

Integrating (14) over a cell V_i and a time step Δt_n and applying the divergence theorem yields

$$\int\limits_{t_n}^{t_{n+1}} \int\limits_{V_i(t)} \frac{\partial \boldsymbol{u}}{\partial t} \mathrm{d}V \mathrm{d}t + \int\limits_{t_n}^{t_{n+1}} \int\limits_{S_i(t)} \sum_{k=1}^{2} \boldsymbol{f}_k(\boldsymbol{u}) n_k \mathrm{d}S \mathrm{d}t = \boldsymbol{0} \,, \tag{17}$$

where n_k are the components of the unit outward normal vector to the surface S_i of V_i. Since the volume $V_i(t)$ is time dependent, the order of differentiation and integration is essential. The rate of change of a volume integral over a time dependent domain for an arbitrary quantity φ can be expressed as

$$\frac{\mathrm{d}}{\mathrm{d}t} \int\limits_{V(t)} \varphi \, \mathrm{d}V = \int\limits_{V} \frac{\mathrm{d}\varphi}{\mathrm{d}t} \mathrm{d}V + \int\limits_{V(t)} \varphi \frac{\mathrm{d}(\mathrm{d}V)}{\mathrm{d}t} \,. \tag{18}$$

By taking into account the so-called *geometric conservation law* (THOMAS & LOMBARD [13])

$$\frac{\mathrm{d}}{\mathrm{d}t}\int\limits_V \mathrm{d}V = \int\limits_S \sum_{k=1}^{2} \boldsymbol{v}_k^g n_k \mathrm{d}S\,, \tag{19}$$

where v_1^g and v_2^g are the components of the velocity with which the surface S moves, (18) can be rewritten as

$$\frac{\mathrm{d}}{\mathrm{d}t}\int\limits_{V(t)} \varphi\,\mathrm{d}V = \int\limits_V \frac{\mathrm{d}\varphi}{\mathrm{d}t}\,\mathrm{d}V + \int\limits_{V(t)} \varphi \sum_{k=1}^{2} \frac{\partial v_k^g}{\partial x_k}\,\mathrm{d}V\,. \tag{20}$$

Using the definition of the material time derivative it follows that (20) is equivalent to

$$\frac{\mathrm{d}}{\mathrm{d}t}\int\limits_{V(t)} \varphi\,\mathrm{d}V = \int\limits_V \frac{\partial\varphi}{\partial t}\,\mathrm{d}V + \int\limits_{V(t)} \sum_{k=1}^{2} \frac{\partial}{\partial x_k}\,(\varphi v_k^g)\,\mathrm{d}V\,. \tag{21}$$

Applying this relation to (17) yields

$$\int\limits_{t_n}^{t_{n+1}} \frac{\mathrm{d}}{\mathrm{d}t}\int\limits_{V_i(t)} \boldsymbol{u}\,\mathrm{d}V\mathrm{d}t + \int\limits_{t_n}^{t_{n+1}} \int\limits_{S_i(t)} \left[\sum_{k=1}^{2}(\boldsymbol{f}_k(\boldsymbol{u}) - \boldsymbol{u}v_k^g)n_k\right]\mathrm{d}S\mathrm{d}t = \boldsymbol{0}\,. \tag{22}$$

Equation (22) is a consistently derived arbitrary LAGRANGEan-EULERian (ALE) formulation of the problem in local integral form, which forms the basis for the further numerical treatment.

3.2 GODUNOV's Method

Starting from given general initial data $\boldsymbol{u}(t^n, \boldsymbol{x})$ for (22) at time t_n, in order to evolve the solution to time t_{n+1}, the GODUNOV method first assumes a piecewise constant distribution of the data. Formally, this is realised by defining cell-averages

$$\boldsymbol{u}_i^n = \frac{1}{\Delta V_i}\int\limits_{V_i} \boldsymbol{u}(t^n, \boldsymbol{x})\mathrm{d}V \tag{23}$$

for each cell V_i. Since we consider polygonal control volumes, the boundary of a computational cell V_i consists of M linear edges S_{ij} with lengths ΔS_{ij}. The index j refers to the neighbouring cell V_j (see Fig. 4). Denoting the normal flux component across the edge S_{ij} by

$$\overline{\boldsymbol{f}}_{ij}(\boldsymbol{u}) = \sum_{k=1}^{2}(\boldsymbol{f}_k(\boldsymbol{u}) - \boldsymbol{u}v_k^g)n_k, \tag{24}$$

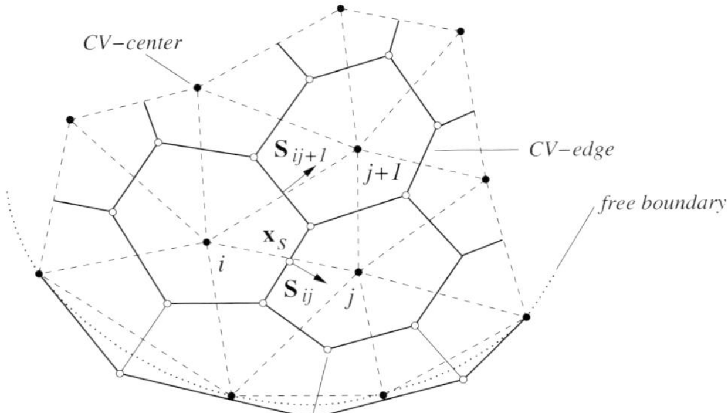

Fig. 4. Control volumes (CV) of the used computational grid. A control volume (—) is obtained by joining the barycentres of the triangles (○). The unknowns $\boldsymbol{u}_i^n$ are stored in the centres (•) of the control volumes

an approximation to (22) can be expressed in the form

$$\boldsymbol{u}_i^{n+1}\Delta V_i^{n+1} = \boldsymbol{u}_i^n \Delta V_i^n - \int\limits_{t_n}^{t_{n+1}} \left\{ \sum_{j=1}^{M} \overline{\boldsymbol{f}}_{ij}(\boldsymbol{u}_s^n)\Delta S_{ij} \right\} \mathrm{d}t \,. \tag{25}$$

Here, the index s denotes the position $\boldsymbol{x}_s$ on the edge, where the flux is defined, i.e. the intersection point of the line connecting the cell centres i and j with the edge (see Fig. 4).

To evaluate the flux, a locally one-dimensional problem normal to the edge at $\boldsymbol{x}_s$ will be considered. When the underlying system is rotationally invariant (as, for instance, is the case for the shallow water equations) one easily obtains a corresponding formulation by the transformation

$$\overline{\boldsymbol{f}} = \boldsymbol{T}^{-1}\boldsymbol{f}_1(\overline{\boldsymbol{u}}) \quad \text{with} \quad \overline{\boldsymbol{u}} = \boldsymbol{T}(\boldsymbol{u}) \tag{26}$$

with the rotation matrix

$$\boldsymbol{T}(\boldsymbol{n}) = \begin{bmatrix} 1 & 0 & 0 \\ 0 & n_1 & n_2 \\ 0 & -n_2 & n_1 \end{bmatrix} . \tag{27}$$

However, the SAVAGE-HUTTER system of equations is not rotationally invariant, because of the anisotropy of the constitutive law. The governing equations (3)-(5) define two different pressure terms $h^2\beta_x/2$ and $h^2\beta_y/2$ in x- and y-directions.

Originally, there is no information about the pressure terms in arbitrary directions. To derive such an expression for this, we consider the *Jacobian matrix* corresponding to the fluxes. The characteristic polynomial of the Jacobian yields for the characteristic celerity in an arbitrary direction $\boldsymbol{n}$

$$a(\boldsymbol{n}) = \sqrt{h(\beta_x n_1^2 + \beta_y n_2^2)}\,, \tag{28}$$

which is related to the corresponding pressure term $p(\boldsymbol{n})$ by

$$a(\boldsymbol{n}) = \sqrt{\frac{\gamma p(\boldsymbol{n})}{h}} \tag{29}$$

with an isentropy coefficient γ. From (28) and (29) we obtain the desired expression for the pressure term:

$$p(\boldsymbol{n}) = \frac{h^2}{\gamma}\left(\beta_x n_1^2 + \beta_y n_2^2\right) . \tag{30}$$

To recover the pressure terms in the x- and y-directions, it is obvious that the isentropy coefficient should be $\gamma = 2$.

With (30) and the definition

$$\overline{\boldsymbol{u}} = \begin{pmatrix} h \\ h\overline{v}_1 \\ h\overline{v}_2 \end{pmatrix} = \begin{pmatrix} h \\ h(v_1 n_1 + v_2 n_2) \\ h(v_2 n_1 - v_1 n_2) \end{pmatrix} , \tag{31}$$

we can formulate the locally one-dimensional flux expression in the form

$$\overline{\boldsymbol{g}}_{ij}(\overline{\boldsymbol{u}}_s^n) = \begin{pmatrix} h(\overline{v}_1 - v_n^g) \\ h(n_1\overline{v}_1 - n_2\overline{v}_2)(\overline{v}_1 - v_n^g) + n_1(\beta_{x_1} n_1^2 + \beta_{x_2} n_2^2)h^2/2 \\ h(n_2\overline{v}_1 - n_1\overline{v}_2)(\overline{v}_1 - v_n^g) + n_2(\beta_{x_1} n_1^2 + \beta_{x_2} n_2^2)h^2/2 \end{pmatrix} , \tag{32}$$

where $v_n^g = n_1 v_1^g + n_2 v_2^g$ denotes the component of the grid velocity perpendicular to the edge adjoining cells i and j.

The next step is to determine $\overline{\boldsymbol{u}}_s^n$ by considering a corresponding RIEMANN problem with the two constant states $\boldsymbol{u}_i = \overline{\boldsymbol{u}}^L$ (left) and $\boldsymbol{u}_j = \overline{\boldsymbol{u}}^R$ (right). The RIEMANN problem is formulated for a locally one-dimensional coordinate system normal to the edge (the coordinate is denoted by $\overline{x}$) with origin in $\boldsymbol{x}_s$

$$\frac{\partial \overline{\boldsymbol{u}}}{\partial t} + \frac{\partial}{\partial \overline{x}} \overline{\boldsymbol{g}}_{ij}(\overline{\boldsymbol{u}}) = \boldsymbol{0} \tag{33}$$

with the initial condition

$$\overline{\boldsymbol{u}}(t^n, \overline{\boldsymbol{x}}) = \begin{cases} \overline{\boldsymbol{u}}_i^n & \text{if } \ \overline{x} < 0 , \\ \overline{\boldsymbol{u}}_j^n & \text{if } \ \overline{x} > 0 . \end{cases} \tag{34}$$

The solution of the RIEMANN problem gives $\overline{\boldsymbol{u}}_s^n$, with which the flux is finally evaluated by (32). To solve the problem an exact RIEMANN solver is considered. This is described in the following section.

3.3 Exact RIEMANN Solver

We describe the solution process for (33) using the primitive variables

$$\overline{\boldsymbol{w}} = \left[h, \overline{v}_1, \overline{v}_2\right]^T , \tag{35}$$

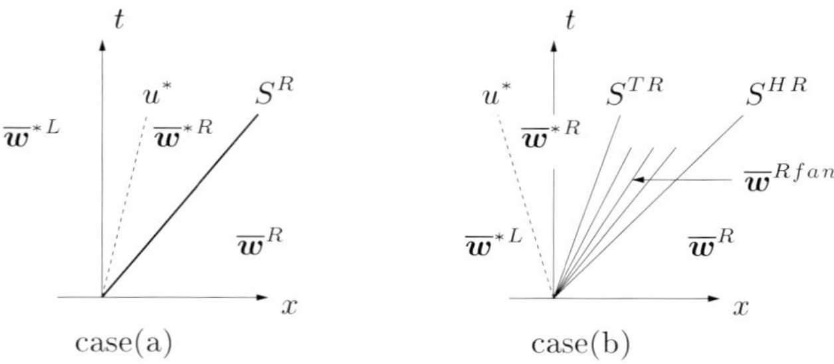

Fig. 5. Two possible wave configurations to the right of the shear wave: (a) right wave is a shock wave and (b) right wave is a rarefaction wave

from which $\overline{\boldsymbol{u}}$ can easily be computed by division of the last two components by h.

Following TORO [15], there are three waves which separate four constant states: $\overline{\boldsymbol{w}}^L$ (left data), $\overline{\boldsymbol{w}}^{*L}$, $\overline{\boldsymbol{w}}^{*R}$, and $\overline{\boldsymbol{w}}^R$ (right data). The states $\overline{\boldsymbol{w}}^{*L}$ and $\overline{\boldsymbol{w}}^{*R}$ comprise the so-called star region defined by a steady frame of reference moving with the velocity of a shock, which occurs at the point $\boldsymbol{x}_s$. These states emerge from the interaction of the states $\overline{\boldsymbol{w}}^L$ and $\overline{\boldsymbol{w}}^R$ (see Fig. 5) and are unknown quantities of the problem. The left and right waves are either shocks or rarefaction waves and the middle wave is always a shear wave. The determination of the type of waves for given initial conditions is part of the solution process. Across the left and right waves both h and $\overline{v}_1$ change, but $\overline{v}_2$ remains constant. Across the middle wave $\overline{v}_2$ changes discontinuously and both h and $\overline{v}_1$ remain constant. We denote the constant values of h and $\overline{v}_1$ in the star region (stationary frame) by h^* and $\overline{v}_1^*$, respectively. The types of non-linear left and right waves are determined by the following conditions:

$$\left.\begin{array}{l} h^* > h^L : \text{left wave is a shock wave,} \\ h^* \leq h^L : \text{left wave is a rarefaction wave,} \end{array}\right\} \tag{36}$$

and

$$\left.\begin{array}{l} h^* > h^R : \text{right wave is a shock wave,} \\ h^* \leq h^R : \text{right wave is a rarefaction wave.} \end{array}\right\} \tag{37}$$

In a first step, we determine the height h^* in the star region from the following single non-linear algebraic equation derived by TORO [15]:

$$\varphi(h^*) \equiv \varphi^L(h^*, h^L) + \varphi^R(h^*, h^R) + (\overline{v}_1^R - \overline{v}_1^L) = 0\,, \tag{38}$$

where the functions φ^L and φ^R are given by

$$\varphi^L = \begin{cases} 2\left(\sqrt{h^*} - \sqrt{h^L}\right), & \text{if } h^* \leq h^L \text{ (rarefaction),} \\ (h^* - h^L)\sqrt{\dfrac{1}{2}\left(\dfrac{h^* + h^L}{h^* h^L}\right)}, & \text{if } h^* > h^L \text{ (shock),} \end{cases} \tag{39}$$

$$\varphi^R = \begin{cases} 2\left(\sqrt{h^*} - \sqrt{h^R}\right), & \text{if } h^* \leq h^R \text{ (rarefaction)}, \\ (h^* - h^R)\sqrt{\frac{1}{2}\left(\frac{h^* + h^R}{h^* h^R}\right)}, & \text{if } h^* > h^R \text{ (shock)}. \end{cases} \tag{40}$$

Since there is no closed-form solution available to (38), we solve it numerically by using the NEWTON-RAPHSON iteration scheme

$$h^{(l+1)} = h^{(l)} - \frac{\varphi(h^{(l)})}{\varphi'(h^{(l)})}, \tag{41}$$

for $l = 0, 1, \ldots$ with the starting value

$$h^{(0)} = \left[\frac{1}{2}(a^L + a^R) - \frac{1}{4}(\overline{v}_1^R - \overline{v}_1^L)\right]^2, \tag{42}$$

where a is the local wave celerity. The iteration (41) is stopped whenever the change in h is smaller than a prescribed tolerance (i.e. $\epsilon = 10^{-7}$).

Knowing h^*, the solution for the velocity $\overline{v}_1^*$ in the star region follows as

$$\overline{v}_1^* = \frac{1}{2}\left(\overline{v}_1^L + \overline{v}_1^R\right) + \frac{1}{2}\left[\varphi^R(h^*, h^R) - \varphi^L(h^*, h^L)\right]. \tag{43}$$

The tangential velocity component, which is dependent on $\overline{v}_1^*$, follows as

$$\overline{v}_2 = \begin{cases} \overline{v}_2^L, & \text{if } \overline{x}/t \leq \overline{v}_1^*, \\ \overline{v}_2^R, & \text{if } \overline{x}/t > \overline{v}_1^*. \end{cases} \tag{44}$$

Next the final solution is determined by a sampling procedure. Given a time t^*, the solution $\overline{\boldsymbol{w}}(t^*, \boldsymbol{x})$ is a function of $\boldsymbol{x}$ only and gives a profile at time t^*. Since the solution is a *similarity solution*, we perform the sampling using the speed $S = x/t^*$. The character of the left and right waves is determined from the ratios h^*/h^L and h^*/h^R respectively; see (36) and (37). The sampling strategy divides the wave pattern into two subregions, namely that to the left of the shear wave $\mathrm{d}\overline{x}/\mathrm{d}t = \overline{v}_1^*(\overline{\boldsymbol{w}}^L \text{ and } \overline{\boldsymbol{w}}^{*L})$ and that to the right of $\mathrm{d}\overline{x}/\mathrm{d}t = \overline{v}_1^*(\overline{\boldsymbol{w}}^R \text{ and } \overline{\boldsymbol{w}}^{*R})$. If the sampling point $(t^*, \boldsymbol{x})$ lies to the right of the shear wave, $S = \overline{x}/t^* > \overline{v}_1^*$, then there are two possibilities illustrated in Fig. 5. If $h^* > h^R$, the right wave is a shock and

$$\overline{\boldsymbol{w}}(t^*, \boldsymbol{x}) = \begin{cases} \overline{\boldsymbol{w}}^{*R} = [h^*, \overline{v}_1^*, \overline{v}_2^R]^T & \text{if } \overline{v}_1^* \leq \overline{x}/t^* \leq S^R, \\ \overline{\boldsymbol{w}}^L = [h^R, \overline{v}_1^R, \overline{v}_2^R]^T & \text{if } S^R \leq \overline{x}/t^*, \end{cases} \tag{45}$$

where S^R is the shock speed (see Fig. 5(a)).

If $h^* \leq h^R$, the right wave is a rarefaction and the solution to the right of the shear wave is

$$\overline{\boldsymbol{w}}(t^*, \boldsymbol{x}) = \begin{cases} \overline{\boldsymbol{w}}^{*R}, & \text{if } \overline{v}_1^* \leq \overline{x}/t^* \leq S^{TR}, \\ \overline{\boldsymbol{w}}^{Rfan}, & \text{if } S^{TR} \leq \overline{x}/t^* \leq S^{HR}, \\ \overline{\boldsymbol{w}}^R, & \text{if } S^{HR} \leq \overline{x}/t^*, \end{cases} \tag{46}$$

where S^{TR} and S^{HR} are the speeds of the *tail* and *head* of the rarefaction, respectively, and $\boldsymbol{w}^{Rfan}$ is the solution inside the rarefaction fan (see Fig. 5(b)):

$$\overline{\boldsymbol{w}}^{Rfan} = \begin{cases} a = \frac{1}{3}\left(-\overline{v}_1^R + 2\overline{a}_1^R + \overline{x}/t^*\right) , \\ u = \frac{1}{3}\left(\overline{v}_1^R - 2\overline{a}_1^R + 2\overline{x}/t^*\right) . \end{cases} \tag{47}$$

If the sampling point $(t^*, \boldsymbol{x})$ lies to the left of the shear wave, a solution can be determined in an analogous way.

3.4 Second-Order Scheme

The scheme presented in the previous sections is only of first order in space and time. To obtain a higher accuracy we also consider a second-order scheme. We employ the MUSCL (*Monotonic Upstream Centred Schemes for Conservation Laws*) by van LEER [16] as a well known and well documented approach (e.g. for the shallow water equations it is described by TORO [15]). In the following description we restrict ourselves to the one-dimensional case (the extension to the two-dimensional case is straightforward).

The distribution of the primitive variable vector $\boldsymbol{w}_i^n$ inside a computational cell V_i at time t_n is defined by

$$\boldsymbol{w}_i(x) = \boldsymbol{w}_i^n + (x - x_i)\frac{\Delta_i}{\Delta x_i} , \tag{48}$$

where x_i is the centre of V_i and Δx_i is its volume. Δ_i denotes the cell mean derivative of the primitive variable vector $\boldsymbol{w}_i^n$ component , which can be approximated by a central difference of the neighbouring states. The MUSCL approach consists of the following three steps.

1. *Data Reconstruction:*
   From the linear distribution of the vector $\boldsymbol{w}$ in cell V_i one obtains the boundary-extrapolated values of the vector $\boldsymbol{w}_i$ at the left, x^L, and right, x^R, boundaries,
   $$\boldsymbol{w}_i^L = \boldsymbol{w}_i^n + (x^L(t) - x_i)\frac{\Delta_i}{\Delta x_i} , \quad \boldsymbol{w}_i^R = \boldsymbol{w}_i^n + (x^R(t) - x_i)\frac{\Delta_i}{\Delta x_i} . \tag{49}$$
   Note that the position of the cell boundaries can change during a time step. In order to avoid the expected spurious oscillations, a TVD constraint is enforced by limiting the slopes Δ_i with the so-called Minmod flux limiter (see e.g. TORO [15]).
2. *Evolution of Extrapolated Values:*
   In each cell the boundary-extrapolated values $\boldsymbol{w}_i^L$ and $\boldsymbol{w}_i^R$ are evolved by a time $\Delta t_n/2$ in terms of the corresponding conserved variables according to the formula
   $$\tilde{\boldsymbol{w}}_i^L = \boldsymbol{w}_i^L + \frac{\Delta t_n}{2\Delta x_i^{n+1/2}}\left[\boldsymbol{f}(\boldsymbol{w}_i^L)^{n+1/2} - \boldsymbol{f}(\boldsymbol{w}_i^R)^{n+1/2}\right] . \tag{50}$$

Note that the flux vector is evaluated at the two boundary extrapolated vector values to provide a flux difference.

3. *The* RIEMANN *Problem:*
At each interface x^R there is a pair of constant states $(\tilde{\boldsymbol{w}}_i^R, \tilde{\boldsymbol{w}}_{i+1}^L)$. One now solves the conventional RIEMANN problem with data $\tilde{\boldsymbol{w}}_i^R, \tilde{\boldsymbol{w}}_{i+1}^L$ to find the similarity solution $\boldsymbol{w}^R(x/t)$. If $\boldsymbol{w}^R(v_n^g)$ denotes the solution along the line $x/t = v_n^g$ the inter-cell flux is

$$\boldsymbol{f}^R = \boldsymbol{f}(\boldsymbol{w}^R(v_n^g)) \,. \tag{51}$$

The scheme defined by the above three steps is second-order accurate in space and time. The third step alone corresponds to the first-order scheme.

3.5 Free Boundary

For the treatment of the free boundary of the granular material by boundary-fitted moving grids, some special considerations are necessary.

We assume that the free boundary coincides with edges of boundary cells. Following MUNZ [7], the velocity of the flow at this margin is equal to the solution of the vacuum RIEMANN problem in normal direction $\overline{x}$ of the corresponding edge. We consider a wave that separates a region of granular material from the region of no material. The corresponding vacuum RIEMANN problem is given by:

$$\overline{\boldsymbol{u}}(t^n, \overline{x}_b) = \begin{cases} \overline{\boldsymbol{u}}^L \neq \overline{\boldsymbol{u}}_0, & \text{if} \quad \overline{x} < \overline{x}_b \,, \\ \overline{\boldsymbol{u}}_0, & \text{if} \quad \overline{x} > \overline{x}_b \,, \end{cases} \tag{52}$$

with

$$\overline{\boldsymbol{u}}^L = \begin{bmatrix} h^L = h_i \\ \overline{v}_1^L = \overline{v}_{1i} \end{bmatrix} \quad \text{and} \quad \overline{\boldsymbol{u}}^R \equiv \overline{\boldsymbol{u}}_0 = \begin{bmatrix} h_0 = 0 \\ \overline{v}_0 = 0 \end{bmatrix} , \tag{53}$$

where $\overline{x}_b$ corresponds to the point on the edge. $\overline{\boldsymbol{u}}^L$ provides the data for the values inside the boundary cell, i.e. h_i and $\overline{v}_{1i}$. Following [7], the wave at the free boundary cannot be a shock wave, since the RANKINE-HUGONIOT *condition* cannot be satisfied in this case. Consequently, it has to be a rarefaction wave. Hence the speed of the contact discontinuity is given by

$$S^{*L} = \overline{v}^L + 2\,\overline{a}^L \,, \tag{54}$$

where $\overline{a}^L$ is the characteristic celerity. The complete solution is

$$\overline{\boldsymbol{w}}^{L0}(t, \boldsymbol{x}) = \begin{cases} \overline{\boldsymbol{w}}^L, \text{ if } \overline{x}/t \leq (\overline{v}_1^L - \overline{a}^L), \\ \overline{\boldsymbol{w}}^{Lfan}, \text{ if } (\overline{v}_1^L - \overline{a}^L) \leq \overline{x}/t \leq S^{*L}, \\ \overline{\boldsymbol{w}}^0, \text{ if } S^{*L} \leq \overline{x}/t \,, \end{cases} \tag{55}$$

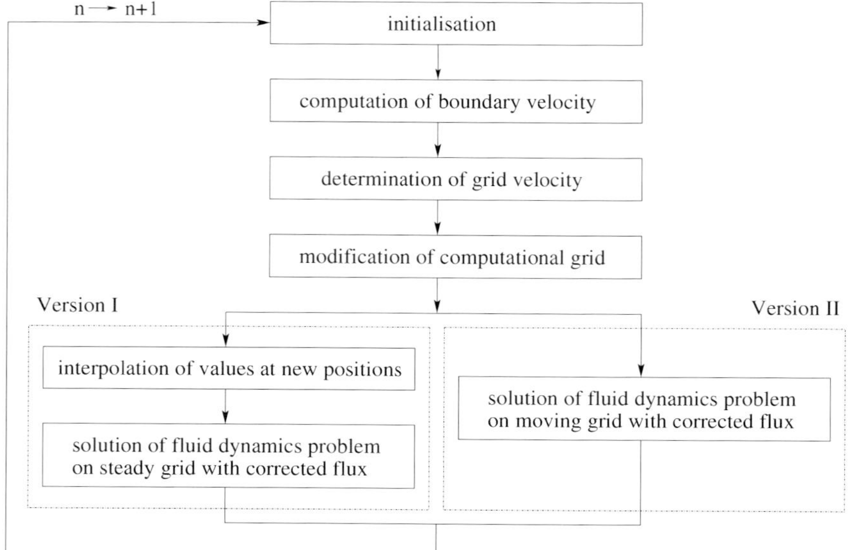

Fig. 6. Two versions of the moving grid algorithm

where $\overline{\boldsymbol{w}}^{Lfan}$ is the solution inside the rarefaction fan.

The movement of the free surface satisfies the kinematic condition

$$\boldsymbol{v} \cdot \boldsymbol{n} = \boldsymbol{v}^b \cdot \boldsymbol{n}\,, \tag{56}$$

which simply means that there is no flux across the free boundary, if the free boundary moves with the fluid velocity. Thus we define the free boundary velocity $\overline{\boldsymbol{v}}^b$ as the speed of the contact discontinuity S^{*L}

$$\boldsymbol{v}^b \equiv \boldsymbol{v}^L + 2\,\overline{a}^L \boldsymbol{n}\,, \tag{57}$$

where $\overline{a}^L$ is the characteristic speed in outward normal direction, see (28). At the beginning of each time step the boundary surface moves with $\boldsymbol{v}^b$.

3.6 Moving Grid Schemes

For the moving grid treatment we consider two different schemes, which are illustrated schematically in Fig. 6. First, they proceed in the same way: the velocity $\boldsymbol{v}_b$ of an exterior edge of each boundary cell is computed by (57). We assume the direction and value of the edge velocity as constant during one time step Δt_n. The displacement of the exterior edge is determined by $\boldsymbol{v}_b/\Delta t_n$, the interior grid velocity $\boldsymbol{v}^g$ of each grid point is found by suitable interpolation. Next, two alternatives are considered:

Version I is based on a splitting according to time and space:

- we approximate the value $\boldsymbol{u}_i(t^n, \boldsymbol{x} + \Delta\boldsymbol{x})$ of each cell after the displacements by a spatial interpolation at the time level n,
- we use the interpolated value $\boldsymbol{u}_i(t^n, \boldsymbol{x} + \Delta\boldsymbol{x})$ to determine the solution $\boldsymbol{u}_i(t^{n+1}, \boldsymbol{x} + \Delta\boldsymbol{x})$ at the end of this time step and solve the governing equations at the position $\boldsymbol{x} + \Delta\boldsymbol{x}$ in the already known way.

The solution of the RIEMANN problem (45), (46) is evaluated for this version at $x/t = 0$. The flux in (32) is determined with the real grid velocity v_n^g. This velocity of each edge is evaluated at the beginning of each time step.

Version II needs no spatial interpolation for solving the governing equations. Only the volume change of the computational cell within the time step is inserted into (22). The RIEMANN problem is determined on the moving edge. The position of the edge needed to define the RIEMANN problem is the one at time $t + \Delta t_n/2$. The solution of the RIEMANN problem determines the flux across the boundary of the computational cell. Because of the grid movement, we need the solution $\boldsymbol{u}_{i,j}(x/t = v_n^g)$. Thus we solve the RIEMANN problem with data $\boldsymbol{u}_L \equiv \boldsymbol{u}_i^n(\text{left})$ and $\boldsymbol{u}_R \equiv \boldsymbol{u}_j^n(\text{right})$ in its own local frame of reference. The sampling procedure identifies the required value along the $(x/t = v_n^g)$ -line. Note that the wave pattern changes with the grid velocity v_n^g (as shown in Fig.5). The flux vector is defined by (32), where $[h, \overline{v}_1, \overline{v}_2]^T$ is defined by the solution $\boldsymbol{u}_{i,j}(v_n^g)$.

3.7 Treatment of Source Terms

So far, we have not considered the complex source term $\boldsymbol{b}(\boldsymbol{u})$ of (14). We do this using a simple splitting scheme suggested by YANENKO [18]. First the homogeneous problem

$$\frac{\partial \boldsymbol{u}}{\partial t} + \frac{\partial \boldsymbol{f}_1(\boldsymbol{u})}{\partial x} + \frac{\partial \boldsymbol{f}_2(\boldsymbol{u})}{\partial y} = \boldsymbol{0} \tag{58}$$

is solved as described in the previous sections yielding an "intermediate" solution $\tilde{\boldsymbol{u}}$. This solution is then used as an initial value for the second step problem

$$\frac{\mathrm{d}\boldsymbol{u}}{\mathrm{dt}} = \boldsymbol{b}(\boldsymbol{u})\,, \tag{59}$$

which is solved by one step of the explicit EULER method:

$$\boldsymbol{u}^{n+1} = \tilde{\boldsymbol{u}} - \Delta t_n\, \boldsymbol{b}(\tilde{\boldsymbol{u}}) \tag{60}$$

It can easily be shown (see YANENKO [18]) that the solution $\boldsymbol{u}^{n+1}$ obtained in this way gives a first order approximation to the solution of the inhomogeneous problem, i.e. (14). It is also possible to define a corresponding second-order splitting scheme, but we will not consider this here.

4 Numerical Results

In this section results of various numerical experiments are presented to illustrate the properties of the numerical schemes. In Subsect. 4.1 we compute the similarity solution for a parabolic avalanche body with a vacuum front at the margins slides down an inclined flat plane to verify the accuracy of the scheme. In Subsect. 4.2 we simulate a flow down a convex and concave curved chute for comparison with experimental data. A travelling shock wave is computed in Subsect. 4.3. Here we inspect the shock capturing capability of our scheme. In Subsect. 4.4 we compute an avalanche, which forms an upward-propagating shock wave when it comes to a halt at a flat runout. In Subsect. 4.5 we show the results of the dam break problem computed by the second order scheme. In the final numerical experiment in Subsect. 4.6 we compute a pile of granular material with an initial spherical cap geometry moving down an inclined plane to illustrate the functionality of the scheme for a real two-dimensional case. It should be remarked that also the other tests, although they are one-dimensional, were carried out with the two-dimensional method (with periodic boundary condition in one direction).

4.1 Parabolic Similarity Solution

We consider the flow of a bulk with an initial velocity v_1^0 on an inclined flat plane with constant inclination angle ζ^0. The analytical solution for this test case is first derived in [10]. This solution was generalised in [11]. A LAGRANGEan moving grid method, which can simulate this problem very well, is introduced in [10]. The initial height distribution is defined by a parabola with a height of one unit at the central point $x = 4$ and a width of 3.2. The initial velocity is chosen as $u^0 = 1.2Fr$, where the FROUDE-number Fr defines the velocity as the ratio of the fluid velocity to the characteristic celerity, i.e. $Fr = u/\sqrt{\beta h}$. The constant inclination angles is chosen to be $40°$ and the internal and basal friction angles are selected to be $30°$. Figure 7 illustrates the simulated results at the dimensionless time units $t = 0, 2, 4, 6, 8$. The avalanche body extends during the motion and keeps the parabolic depth profile. The velocity remains linearly distributed along the avalanche at each point of time.

Version I of the scheme is tested using three spatial discretisations with cell numbers N=67, N=241 and N=914 over the interior of the avalanche. We introduce an error measure for depth h and velocity u by

$$E_h = \frac{\sum_{i=1}^{N} |h_i^{\text{num}} - h_i^{\text{ana}}|}{\sum_{i=1}^{N} h_i^{\text{ana}}} \quad \text{and} \quad E_u = \frac{\sum_{i=1}^{N} |u_i^{\text{num}} - u_i^{\text{ana}}|}{\sum_{i=1}^{N} u_i^{\text{ana}}} . \tag{61}$$

Figure 8 shows the relative error at time $t = 8$ using the three numerical grids. The values, together with the corresponding spatial order, are given in Table 1. Because of the deformation of our computational cells during the

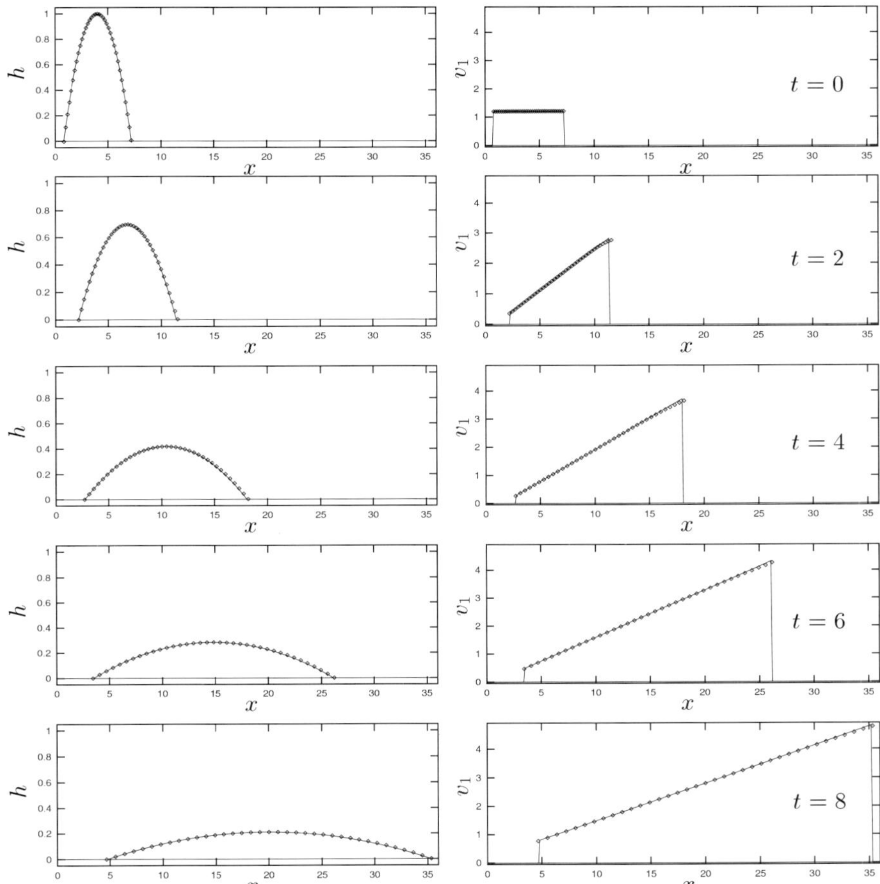

Fig. 7. Depth (left) and the corresponding velocity (right) profiles of the parabolic similarity solution at the dimensionless time units $t = 0, 2, 4, 6, 8$. The solid lines indicate the exact solution introduced in [10]. The points denote the numerical solution of the first version scheme on grid with $N = 241$ cells

avalanche motion the convergence rate slightly deteriorates from the theoretical second order. However, one can conclude that with Version I of our scheme a flow with a vacuum front at the margins can be adequately simulated.

4.2 Convex and Concave Curved Chute

Greve & Hutter [5] conducted experiments with avalanches that slide down a convex and concave curved chute. We take this avalanche motion as a test case for a comparison between experimental and numerical results.

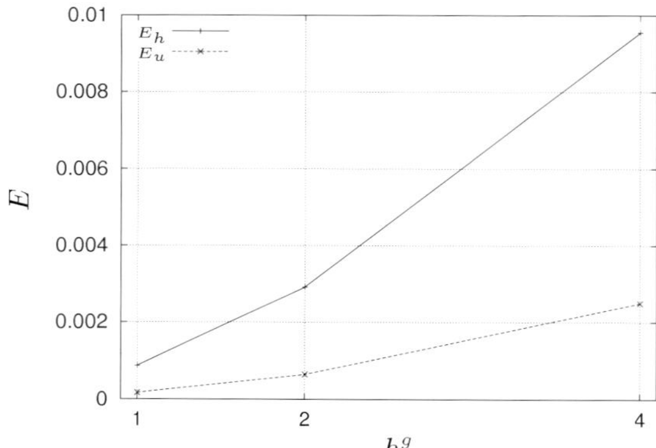

Fig. 8. Relative error of the height and velocity of the parabolic similarity solution at dimensionless time $t = 8$ on different grids with mesh size h^g

The shape of the chute is defined by

$$\zeta(x) = \zeta_0 e^{-0.1x} + \zeta_1 \frac{\xi}{(1+\xi^8)} - \zeta_2 e^{-0.3(x+{}^{10}/_3)^2}, \tag{62}$$

where $\xi = (4/15)(x-9)$ and the constants are given by $\zeta_0 = 60.0°$, $\zeta_1 = 31.4°$, and $\zeta_2 = 37.0°$. The second term in (62) is responsible for the bump around $x = 9$. Due to this bump, an initial single pile of granular avalanche could separate in the course of the motion into two piles which are separately deposited above and below the bump.

The influence of the confining walls of the chute on the bed friction was also determined by replacing the bed friction angle δ by the *effective* bed friction angle δ_{eff}, where the two are related by

$$\delta_{\text{eff}} = \delta_0 + \epsilon k_{\text{wall}} h\,. \tag{63}$$

Here, ϵ is the aspect ratio and k_{wall} is the measured correction factor to account for the side wall effects in the bed friction angle. For the simulations

Table 1. Relative errors of different grid sizes for the parabolic similarity solution

N	E_h	E_u
67	$9.6 \cdot 10^{-3}$	$2.5 \cdot 10^{-3}$
241	$2.9 \cdot 10^{-3}$	$0.6 \cdot 10^{-3}$
913	$0.9 \cdot 10^{-3}$	$0.2 \cdot 10^{-3}$
order	1.70	1.96

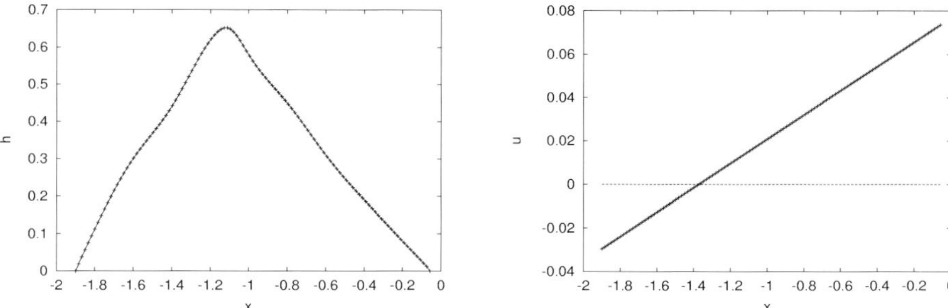

Fig. 9. Interpolation by cubic spline of an initial avalanche profile for simulation of motion in a convex and concave curved chute

with Version I of our scheme, the friction angles are chosen to be $\delta_0 = 26.5°$ and $\phi = 37°$. The wall friction correction in (63) is described by $k_{\text{wall}} = 11°$. We discretise the computational domain (avalanche body) in *vertex* volumes (VORONOI dual mesh). Primarily we have a triangular partition of the domain. The computational cells are associated with vertices of the triangle mesh and are obtained by joining the barycentre of each triangle with its vertices, as shown in Fig. 4. We assume constant values of h in x_2 direction and zero v_2-velocity. The vertical walls are represented by periodic boundary conditions in x_2 direction. The number of cells in x_1 direction is chosen to be $N = 111$. A cubic spline interpolation of experimental data is used by GREVE & HUTTER for the determination of the initial height and velocity distribution (illustrated by Fig. 9). Figure 10 shows the experimental results (solid line) and the results of our numerical simulation (crosses). For better visibility the height of the material layer is scaled by a factor 7. There is a fairly good agreement between the simulated and experimental results. The simulation shows a deposition above the bump and some rest slides over the bump on the flat run-out zone. The predicted deposition above the bump is smaller and, consequently, below the bump it is larger. From the numerical point of view there is no real partition of the grid between the two depositions, but the height of the avalanche in this area is very small. The predicted final position of the avalanche head is very close to the experimental data. A shock formation develops when the avalanche stops on the run-out zone and the material, which overrides the bump still flows. This is also observed in the experiments. Altogether, the results show that the scheme describes the dynamic behaviour of the avalanche satisfactorily.

4.3 Travelling Shock Wave

The travelling shock wave problem serves as a test problem for the numerical scheme with respect to the capability of shock capturing. In this test problem we are concerned with an inclined chute, where the internal and basal friction angles are both presumed to be equal to the inclination angle, $\phi = \delta = \zeta =$

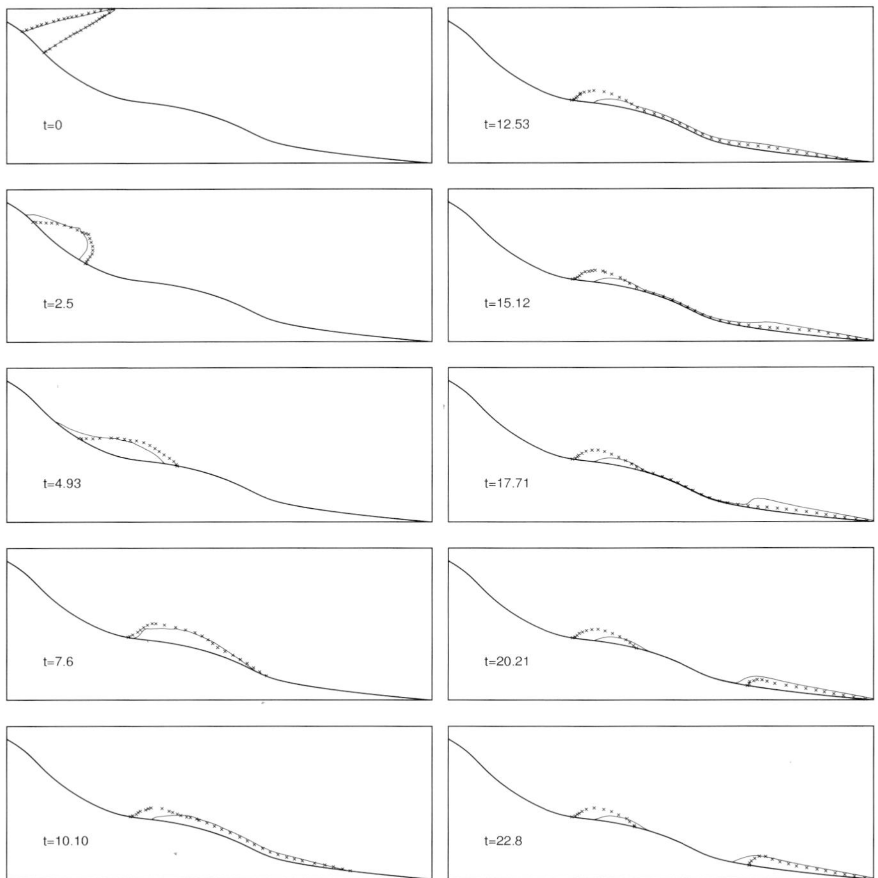

Fig. 10. Height of the granular material at different time-steps for the convex and concave curved chute test case. The results are scaled by a factor 7. The solid lines indicate the experimental data. The points correspond to the numerical solution

40°. That implies that the pressure coefficient $K_x = K_{x,\mathrm{act}} = K_{x,\mathrm{pas}}$ defined by (10) is constant. The initial conditions are defined as

$$h^0(x) = \begin{cases} 0.3, & 0 \leq x \leq 24, \\ 0.9, & 24 \leq x \leq 36, \end{cases} \quad v_1^0(x) = \begin{cases} 1.3148, & 0 \leq x \leq 24, \\ 0.1, & 24 \leq x \leq 36. \end{cases} \tag{64}$$

The velocity of the upslope travelling wave is $V_n = -0.50741585$. We investigate the capability of shock capturing of Version I of our moving grid scheme. The grid movement is defined as follows: the edges of the boundary cell are moving with the fluid velocity at this point, and the velocity of each internal grid line is determined by linear interpolation. Figure 11 shows the analytical

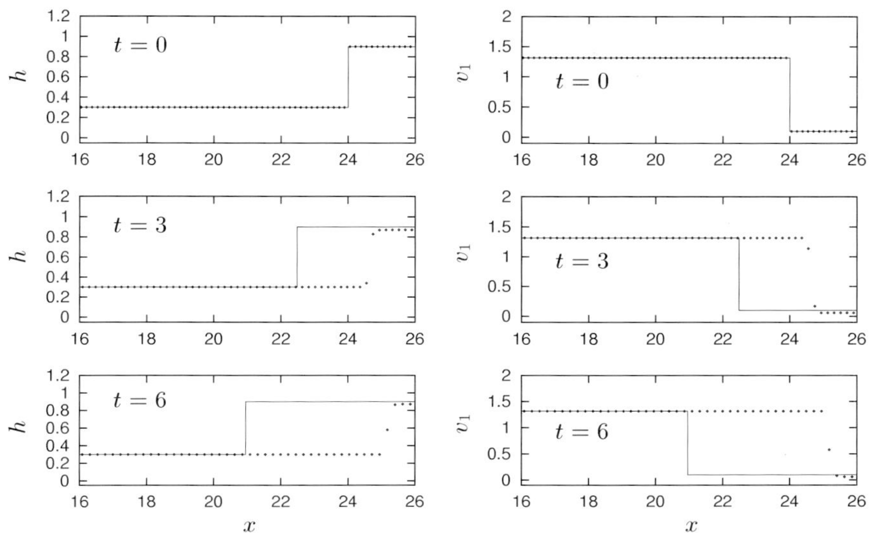

Fig. 11. Height and corresponding velocity profiles of the travelling shock wave simulated with Version I of the scheme

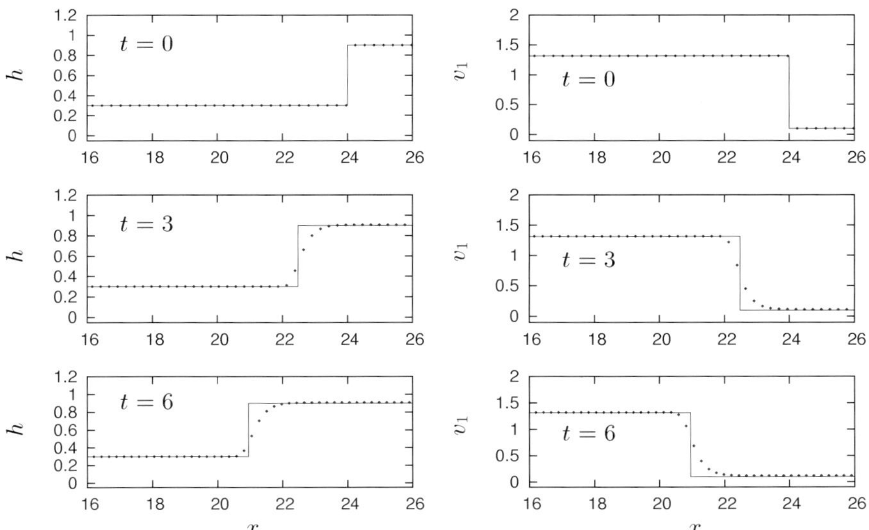

Fig. 12. Height and corresponding velocity profiles of the travelling shock wave simulated with Version II of the scheme

solution (solid line) and the results of the numerical simulation (circles). The simulated shock wave is moving with a wrong velocity in the opposite direction of the analytical solution. This clearly indicates that Version I of the moving grid scheme, which works very well in the absence of shocks, can neither give the right wave transport velocity nor describe the travelling shock wave. The reason for this failure is the spatial and temporal split, which is a basic assumption for this scheme. Thus, the scheme can not handle shocks and grid movement together. For simulations with a locally steady numerical grid the shock capturing capability is still available. In case of the avalanche flow over a convex and concave curved chute in the previous section the scheme computes a correct shock wave motion. Here, the grid motion is bounded by the head and tail of the avalanche. The shock waves develop when the head and tail of the avalanche already stop, such that the local grid velocity in the region of the shock wave is close to zero.

With Version II of the scheme the travelling shock wave can be simulated correctly. The corresponding results can be seen in Fig. 12. The shock is reasonably well resolved by 4 to 5 grid points and moves with the right wave transport velocity. There are no oscillations behind the shock. This indicates that Version II of the scheme can properly handle the combination of shock waves and grid movement. The excessive numerical diffusion inherent in this first-order method is clearly manifested near "corners", where the solution has a discontinuity in its derivative. The corresponding improvements which are possible with a second-order scheme will be demonstrated in Subsect. 4.5

4.4 Upward Moving Shock Wave

As a next test case we consider the granular material released from a parabolic cap sliding down an inclined plane and merging into a horizontal run-out zone. The flat plane is defined by an inclination angle $\zeta^0 = 40°$ and the internal and basal friction angles are $\phi = 38°$ and $\delta = 40°$, respectively. Consequently, we have a jump in the pressure coefficient K_{x_1}, when the flow changes from the expanding to the contracting region. The results are presented in Fig. 13. The avalanche accelerates and extends until the front reaches the horizontal run-out zone. With the approaching mass from the tail, a shock wave propagates backwards (see Fig. 14). At $t = 12$ there is obviously a jump in the velocity taking place at the transition zone, which corresponds to the steep surface gradient. During the whole computation the numerical grid is moving. When the front part stops, the tail is still moving until the end of the avalanche activity. So the scheme is also able to deal correctly with this kind of shock waves in combination with a moving grid.

4.5 Dam-Break Problem

For a comparison between the first-order and second-order schemes (see Subsect. 3.4) we consider the well known dam-break problem. The inclination angle ζ is chosen to be 0. The pressure coefficients are chosen to be one and the right hand side of the governing equations is assumed to vanish (this

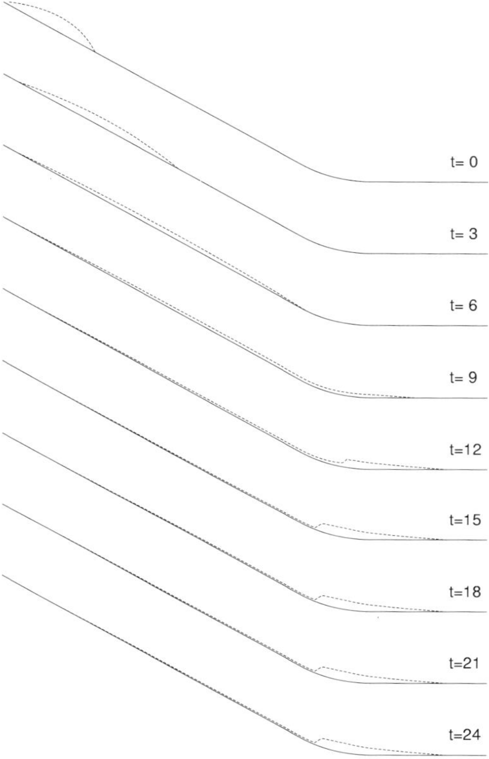

Fig. 13. Avalanche depths at different time steps for upward moving shock problem

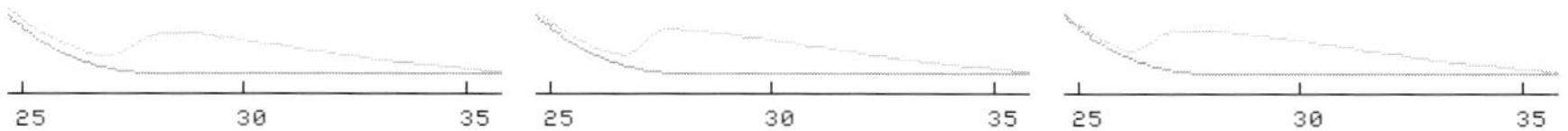

Fig. 14. Height profiles of the upward moving shock at dimensionless time $t = 12.5, 13.5$ and 14.5

situation corresponds to that of the shallow water equations). The initial conditions are defined as

$$h^0 = \begin{cases} 1.0 & \text{for} \quad 0 \leq x \leq 0.5, \\ 0.1 & \text{for} \; 0.5 < x \leq 1.0, \end{cases} \quad \text{and} \quad v_1^0 = 0. \tag{65}$$

We test our moving grid scheme for a grid motion defined as follows: the grid line at the position $x = 0.5$ (jump in the height distribution) moves with the shock wave, the grid velocity at the inflow and outflow boundary is zero, and the velocity of every other grid line is provided by linear interpolation. The solution is computed until time $t = 0.35$ (dimensionless time). Figure 15

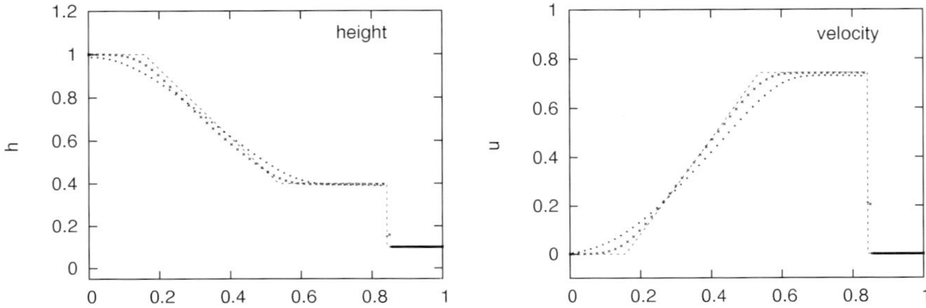

Fig. 15. Depth and velocity profiles for the dam-break problem at dimensionless time $t = 0.35$: analytical solution (—) and numerical solution with first-order (+) and second-order (×) schemes

shows the exact solutions (solid lines) for h and u together with the numerical solutions with the first-order and second-order schemes at $t = 0.35$. The shock is reasonably well resolved by both schemes and no oscillations occur behind the shock. As expected, the results of the second-order scheme are less diffusive, which is clearly observed in the vicinity of corners.

4.6 Spherical Cap

In order to demonstrate the functionality of the proposed method for two-dimensional problems, we consider the flow of a granular material with an initial spherical cap geometry on an inclined flat plane with constant inclination angle. A description of laboratory experiments for this test case can be found in [4]. The initial height distribution is defined by a spherical cap with a total dimensionless volume $V = 13.6$ (cap radius $r = 1.85$). The initial velocity is chosen as $[v_1^0 = 0.5Fr, v_2^0 = 0]^T$. The constant inclination angle is assumed to be 45°. The internal and basal friction angles are defined as $\phi = 39°$ and $\delta = 10°$, respectively. Figure 16 illustrates the simulated results at the dimensionless time units $t = 0, 1, 2$. The results show, that the treatment of the avalanche margin is qualitatively correct. A comparison with the experimental data in [4] shows that the scheme describes the dynamic behaviour of this avalanche, i.e. the development of the circular shape into a tear drop form, satisfactorily.

5 Conclusion

A numerical technique for the solution of the Savage-Hutter equations governing motions of granular avalanches has been presented. The scheme is based on an Arbitrary LagrangeEan-EulerRian (ALE) formulation of the discretised governing equations. The corresponding free boundary problem is solved as a vacuum Riemann problem. The velocity of the free boundary

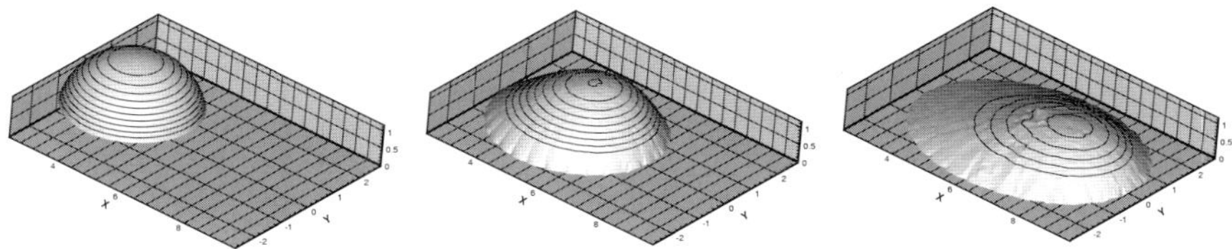

Fig. 16. Calculated time sequence ($t = 0, 1, 2$) for the basal profile of an avalanche on an inclined plane. Positions and times are given in dimensionless representation

determines the movement of the numerical grid in the interior of the computational domain. A splitting method is used for solving the non-homogeneous governing equations. The chosen GODUNOV approach uses the solution of the RIEMANN problem to evaluate the flux on boundaries of the control volumes. Since the edges are not aligned with the Cartesian directions, the definition of the pressure term in arbitrary direction has been necessary.

Special emphasis has been given to the proper treatment of shock phenomena which may occur in granular material flows. For this, two versions of the finite-volume moving grid scheme has been investigated. The first version works very well in the absence of shocks. The scheme is also able to capture the shock wave on steady grids, as shown in the case of the flow down a convex and concave curved chute. However, this version is not able to capture a travelling shock wave on moving a grid accurately. The second version, which works without the spatial interpolation of the variables, simulate the shock waves adequately. This version also produces satisfactory results for the problem of an upward moving shock in opposite direction to the grid movement, a phenomenon, which is also observed in experiments.

Further, an extension of the method to yield a second-order scheme via a MUSCL-approach has been considered and the resulting improvements in accuracy have been illustrated.

In summary, one can conclude that the considered numerical approach appears to be capable of handling the complex interacting phenomena associated with flows of granular avalanches in an adequate way.

Future work will concern the application of the method to more complex two-dimensional problem configurations, which, in particular, makes an improvement of the strategy for the movement of the internal grid necessary.

References

1. Godunov, S.K., Zabrodin, V.A., Ivanov, M.L., Kraiko, A.N. (1976) Numerical Solution of Multidimensional Problems of Gasdynamics. Nauka, Moscow
2. Gray, J. M. N. T. (2003) Rapid Granular Avalanches. This volume
3. Gray, J. M. N. T., Wieland, M., Hutter, K. (1999) Gravity-driven free surface flow of granular avalanches over complex basal topography. Proc. R. Soc. London. A **455**, 1841–1874

4. Greve, R., Koch, T., Hutter, K. (1994) Unconfined flow of granular avalanches along a partly curved surface. I. Theory Proc. R. Soc. London. A **445**, 399–413
5. Hutter, K., Greve, R. (1993) Two-dimensional similarity solutions for finite mass granular avalanches with Coulomb- and viscous-type frictional resistance. J.Glac. **39(132)** 357–372
6. Irmer, A., Schäfer, M., Gray, J.M.N.T., Voinovich, P. (1999) An adaptive unstructured solver for shallow granular flows. 22nd International Symposium on Shock Waves, Imperial College, London, UK, July 18-23
7. Munz, C.-D. (1994) A Tracking Method for Gas flow into Vacuum Based on the Vacuum Riemann Problem. Math. Meth. Appl. Sci. **17** 597–612
8. Pudasaini, S.P., Hutter, K. Eckart, W. (2003) Gravity-Driven Rapid Shear Flows of Dry Granular Masses in Topographies with Orthogonal and Non-orthogonal Metrics. This volume
9. Rodionov, A. V. (1987) Monotonic scheme of the second order of approximation for the continuous calculation of non-equilibrium flows. Zh. vychisl. Mat. mat. Fiz. **27** 4, 585–593
10. Savage, S.B., Hutter, K. (1989) The motion of a finite mass of granular material down a rough incline. J. Fluid Mech. **199** 177–215
11. Tai, Y.-C. (2000) Dynamics of Granular Avalanches and their Simulations with Shock-Capturing and Front-Tracking Numerical Schemes. Ph.D. Thesis, Darmstadt University of Technology
12. Tai, Y.C., Noelle, S., Gray, J.M.N.T., Hutter, K. (2002) Shock-Capturing and Front-Tracking Methods for Granular Avalanches. J. Comput. Phys. **175** 269–301
13. Thomas, P.D., Lombard C.K. (1979) Geometric conservation law and its application to flow computation on moving grids. AIAA J. **17** 1030–1037
14. Toro, E.F. (1990) Riemann problems and the WAF-Method for the Two-Dimensional Shallow Water Equations. College of Aeronautics Report, Cranfield Institute of Technology
15. Toro, E.F. (2001) Shock Capturing Methods for Free-surface Shallow Flows. John Wiley & Sons, Chichester
16. van Leer, B. (1974) Towards the Ultimate Conservative Difference Scheme II. Monotonicity and Conservation Combined in a Second Order Scheme. J. Comput. Phys. **14** 361–370
17. Weiyan, T. (1992) Shallow Water Hydrodynamics. Elsevier, Amsterdam, Elsevier Oceanography Series, 55
18. Yanenko, N.N. (1976) The method of fractional steps for the solution of problems in continuum mechanics. Am. Math. Soc., Translat., II. Ser. **105** 333–346

Experimental Information on the Dynamics of Dry-Snow Avalanches

Dieter Issler

NaDesCoR, Promenade 153, CH–7260 Davos Dorf, Switzerland
e-mail: nadescor@davos-onlinde.ch

received 16 Oct 2002 — accepted 28 Dec 2002

Abstract. This paper is a first step towards a synoptic analysis of the experimental information on snow avalanche flow provided by field observations and dedicated experiments over the past 60 years. Both full-size tests in instrumented avalanche tracks and laboratory experiments with snow or substitute materials are used to extract information on two major questions: (i) Which flow regimes are possible in avalanches and under which conditions do they occur? (ii) By which mechanisms and at which rate do avalanches entrain snow from the snow cover? The major types of sensors used in avalanche experiments are briefly discussed, and it is seen that a large variety of sensors and experimental techniques—including laboratory experiments—have to be combined in order to obtain definitive answers to the open questions.

1 Introduction

Land-use planning based on hazard mapping is the safest and usually also most cost-effective way of responding to the threat that snow avalanches pose to settlements, traffic routes and other infrastructure in mountainous areas winter after winter. Where good records of extreme avalanche events and enough space for development are available, statistical models, e. g. [66], may be sufficient for the task of hazard mapping. Many countries, however, allow construction in areas that are only moderately endangered by avalanches. In these cases, a more detailed mapping procedure recognizing different risk levels and indicating the spatial variation of avalanche pressure is required. Scarcity of areas for construction in mountain resorts puts a high premium on increasing the precision of such mapping. For this purpose, dynamical models of snow avalanche flow are employed. They confine the stochastic element inherent in avalanche phenomena to the dependence of the initial conditions and snow properties on the return period.

The models used in hazard mapping practice today are typically one to three-parameter models of the hydraulic type. Most of them contain a velocity-independent friction term proportional to the flow depth as in granular materials and a drag term proportional to the square of the velocity to keep the latter within realistic bounds. Important effects like snow entrainment are usually neglected. The significant differences in flow behavior between wet and dry snow avalanches, channeled and unconfined flows or between large and small events are only reflected in different choices of the parameters. It is well known that the simulation results are rather sensitive to these parameters [4], so extensive calibration, experience and common sense are required to obtain valid results. Assessment of the same avalanche problem by different groups of experts [33] revealed the amount of subjectivity still involved in hazard mapping.

For about six decades, full-scale and laboratory experiments on avalanches have been conducted in the former Soviet Union, Switzerland, France, Canada, the United States, Norway, Japan, and Italy. In the first stage, the emphasis was on measuring typical impact pressures for dimensioning structures exposed to avalanches. Later on, speeds and runout distances were also measured for calibrating the models then available. The rapid development of measurement techniques and data acquisition systems as well as the emergence of more sophisticated modeling concepts shifted interest more and more towards unraveling the basic physical processes at work in avalanche flow. To this end, experiments were designed to characterize the avalanche structure in terms of vertical profiles of velocity, flow depth, density, and velocity fluctuations. The translated monograph by Bozhinskiy and Losev [9] contains a brief summary of, and references to, Soviet work almost exclusively published in Russian; an overview of the avalanche test sites and some of the relevant laboratory equipment in Western Europe is given in [54]. Work and equipment in Canada and the U.S. may be traced from [16,71,103] while [58,76,80,81,83] contain valuable information on Japanese experiments. The review articles by Hopfinger [44] and Hutter [48] are equally recommended as they tie the observations and measurements to concepts of flow dynamics.

Full-scale avalanche experiments are rather dangerous, unpredictable with respect to timing, difficult and very costly. Not surprisingly, therefore, a wide variety of laboratory experiments have been carried out over the decades. As they often imply thorny similarity issues, they are mostly used for studying basic flow mechanisms in detail and under controlled conditions. At small scale, the majority of experiments use dry granular materials, but fluidized snow has also been used in a coldroom. Turbidity currents (particles suspended in water) have been studied as models of powder-snow avalanches. At an intermediate scale, outdoor chute experiments can be carried out with natural snow; they are particularly suited for investigating impacts on obstacles. In the following sections, laboratory experiments will be discussed together with, and related to, full-scale experiments.

Many of the experiments carried out in the past complement each other with respect to the avalanche types and sizes, measured quantities and sensors. However, to the author's knowledge, few attempts have been made to critically compare the data from these scattered sources and to obtain more definitive answers to the most pressing questions of avalanche dynamics. Only partial descriptions [19,92,113] and analyses [102] of the important experiments carried out at the new and very well equipped Swiss test site Vallée de la Sionne in January and February 1999 have been published to date. Unfortunately, a full review and analysis of avalanche experiments is also beyond the scope of this paper, not the least because of the difficulty of accessing all the sources. Nevertheless, a first step in this direction will be attempted here, in the hope that the importance of this endeavor will become evident and that a larger group of authors with full access to the data will take up this task in the near future. Specifically, four issues will be addressed here:

1. Which flow regimes can be realized in snow avalanches, and under which conditions do they occur?
2. What role does snow entrainment play in avalanche dynamics and what are the dominant mechanisms?
3. What conclusions can be drawn from our present experimental knowledge concerning the suitability of different modeling approaches?
4. Which experiments should be carried out in the future?

The focus of this article is on flow regimes of dry-snow avalanches and mass exchange between the avalanche and the snowcover. For this reason, the significant body of laboratory work on density and turbidity currents as models of powder snow avalanches [6,7,41,45,59,104,105] will be mentioned only in passing as it does not directly pertain to the main questions addressed here. Wet-snow avalanches—even though interesting in their own right—will be mentioned only for contrasting their properties with those of dry-snow avalanches. (Also, wet-snow avalanches have received less attention in research because the extreme runout distances relevant for hazard mapping usually are reached only by dry-snow avalanches.) Slushflows or slush torrents (mixtures of water and snow flowing rapidly on gentle slopes) are an equally fascinating topic that has to be left out in this article; some informative and recent papers from which the older literature may be traced are [10,37,42,43,89,106].

2 Experimental Techniques and Sensor Types

Interpretation of raw experimental data has to take account of the sensor properties and the circumstances of the measurements. For this reason, a brief discussion is given here of a variety of observation techniques and sensor types with regard to their capabilities and limitations. Additional sensor types are described in [68].

2.1 Observations on Avalanche Deposits

The overwhelming majority of avalanches are released with no human observers present, but their deposits can often be investigated for some time after the event. Mapping and describing avalanche events are especially important in areas where hazard mapping and land-use planning are necessary because a cadastre, if elaborated over a sufficiently long period, allows calibration of the model parameters or inverse modeling to infer the relationship between frequency and initial conditions for a particular avalanche path. Moreover, some important scientific information may also be obtained, one particularly successful example being the determination of the mass balance along the avalanche path at the Italian test site at Monte Pizzac [110].

Observations made in the starting zone of an avalanche provide information on the initial conditions (areal extent, volume, mass, type of snow) that is crucial if the event is to be used for model calibration. Along the track, the maximum flow width can usually be determined, sometimes even rough estimates of the maximum flow depth or the flow velocity may be obtained from traces on the vegetation and the radial inclination of the presumed avalanche surface in bends. Usually, these estimates are rather uncertain, but combining them may often give a fairly consistent picture of the event.

Damage done (or not done) to trees and/or man-made structures is often an invaluable indication of the magnitude of an avalanche event. For a classic example of this method, applied to dense-flow avalanches, see [114]; in [24], maps of estimated maximum pressure were derived for several powder snow avalanches from destroyed forest areas and damage to houses. Based on the damage wrought by severe storms, one estimates that stagnation pressures around 1 kPa will break many branches and occasionally fell non-deciduous trees or old fruit trees; except for losing chimneys or roof tiles and old windows, normal buildings will not be affected. At 3 kPa, extensive damage is done in a forest and to old barns; normal houses are still not endangered except for their roofs, doors and windows. Power lines may be torn down in certain cases. At 5–10 kPa, one expects total destruction of large mature forests, and some structual damage may be inflicted on houses not specifically designed for avalanche loads. Note that pressures around 1 kPa can be disastrous to cars and trucks in motion and may derail light-weight narrow-gauge railway coaches.

For a long time, avalanche researchers and practitioners have seen indications that avalanches may entrain substantial amounts of snow along the track; examples of global mass balance estimates may be found, e. g., in [24,56]. In three snow pits from a small and a very large powder snow avalanche [56], the basal old snow layer and the undisturbed fresh snow that fell prior to the avalanche event could be distinguished from the avalanche deposits. The surroundings undisturbed by the avalanche allowed to estimate the total amount of fresh snow deposited by the storm that had caused the avalanches. In this way the mass balance at single locations could be determined. Sovilla et al. [110] extended this method and systematically applied it to four artificially triggered dry-snow and wet-snow avalanches in the instrumented Monte Pizzac avalanche path, digging and analyzing several

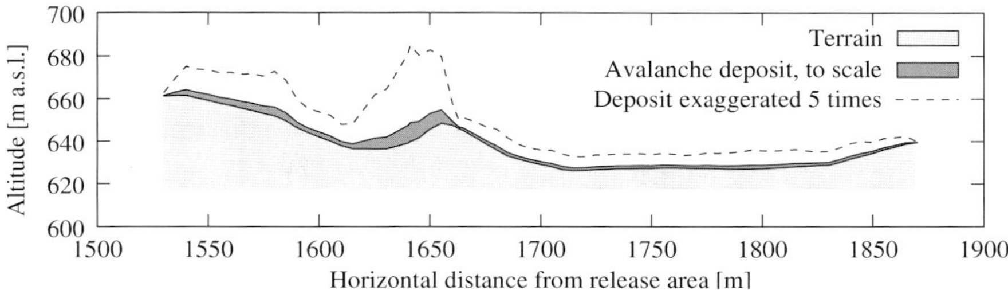

Fig. 1. Example of longitudinal profile of deposit depth at the Norwegian Geotechnical Institute's test site at Ryggfonn, western Norway. From [65], with permission

dozens of cross sections along the entire path. These results are all the more valuable as they concern both wet and dry avalanches and are complemented by measurements of the front velocity, flow height and impact pressures at several locations.

In the runout zone, the runout distance is an obvious parameter to measure, useful for the calibration of all models. The distribution of deposit depths will, however, put models to a much more demanding test as it cannot so easily be adjusted by tuning one or two parameters. Under favorable weather conditions, the spatial distribution of the deposits can be measured to reasonable accuracy, either manually (see Fig. 1) or by using aerial photogrammetry [113]; in the near future, laser-based scanning distometers may provide a more rapid and economic alternative.

In mixed avalanches, one can often distinguish the compact, irregular and slab-like deposits of the dense avalanche core from an extended distal area of fine-grained deposits with dense snowballs distributed over the surface, the size of which diminishes in the distal direction [56]. These observations, especially if combined with measurements during the avalanche flow, hint at a three-layered structure of mixed avalanches and put significant constraints on models.

2.2 Laboratory Experiments at Reduced Scale

Full-scale experiments are dangerous, very expensive and highly dependent on weather conditions; the initial and boundary conditions can be influenced only to a very limited degree. For these reasons, experimental snow chutes have been constructed in Switzerland [95,112], the U.S. [15], Japan [76], France (M. Naaim, pers. comm., May 2001) and possibly elsewhere. On the larger of these chutes, snow volumes on the order of $10\,\mathrm{m}^3$ may be released, attaining speeds in excess of $10\,\mathrm{m\,s}^{-1}$. The same type of instrumentation may be applied as in full-scale experiments, and certain processes may even be studied visually through the sidewalls.

A large number of experiments have also been performed with dry granular materials [29,47,49,50,62,99]. Both two-dimensional and laterally spreading three-dimensional flow configurations on inclined planes and curved surfaces have been studied with high-speed photography and particle-image velocimetry. In this way, the evolution of the edge of the spreading avalanche and the flow depth could be obtained with good precision and resolution. In some cases, velocity profiles near a lateral plexiglass wall were also measured. Recent experiments [111] observed shock-wave formation at obstacles and demonstrated the potential of granular-chute models for efficient design of protection measures against avalanches.

These circumstances make the chutes particularly suited for detailed process studies requiring a large number of experiments with systematic variation of key parameters. It needs to be mentioned, however, that the results may not always be directly transferable to real avalanches because of similarity problems: The stresses scale linearly with the density and with the square of the velocity. If Froude similarity is respected, the square of the velocity scales in the same way as the linear dimensions, which are typically reduced by a factor of 10–100 in the experiments. If snow is used in the chute experiments, its cohesion (typically on the order of 0.5–1 kPa) is similar to that of natural avalanche snow and thus one to two orders of magnitude too large for similarity. Performing the experiments with fine-grained snow at low temperatures should alleviate these problems. Conversely, if dry cohesionless granular materials are used, similarity is violated at least near the avalanche surface where cohesion is not negligible compared to the COULOMBian friction forces. In many situations, the effects of the similarity violations should not be too significant, but careful evaluation is required in each case before model results are applied to the prototype situation (see [9, Sect. 5.3],[48]).

The role that chutes may play in the future development of avalanche dynamics will be discussed in Sect. 6. In the rest of this section, we will summarize measurement techniques developed primarily for full-scale experiments. Most of them are also applicable in chute experiments. On general terms, one may distinguish *invasive techniques* with sensors placed in the avalanche flow and influencing it to some degree, from *non-invasive techniques* where the sensors are placed outside the flow domain. The latter type of sensors probe the flow either actively by emitting electromagnetic or acoustic waves, or passively by measuring signals emitted by the avalanche. Load cells or pressure sensors of various types may be used either way: When placed in the flow they measure impact forces of snow particles or stagnation pressures, but they may also be inserted flush with the ground surface for measuring normal and tangential loads or pore pressure.

2.3 Non-Invasive Techniques

Video recordings. This is the most classical non-invasive measurement method and is quite powerful if the avalanche is released during daylight under good weather conditions. Stereophotography was used early on in the Soviet Union

Fig. 2. DOPPLER radar systems at Vallée de la Sionne. On the left, the pulsed C-band and two Ka-band antennas, in the middle and on the right two narrow-beam and one wide-beam antenna for X-band systems. (Photo courtesy H. SCHAFFHAUSER)

[11] to measure the evolution of the volume and front velocity of avalanches. With high-quality film or video footage, the surface-velocity field of the avalanche may be determined using correlation techniques [26]. For laboratory experiments, particle image velocimetry (PIV) is now a well established, sophisticated analysis technique that is routinely applied, see [21].

Video taping is thus an inexpensive technique yielding fundamental information and should be used in any full-scale experiment (and also in laboratory experiments). Its main limitation is that it does not reveal processes inside the flow; in particular the dense core of the avalanche is usually shrouded in a layer of suspended material.

DOPPLER *radar.* Microwave signals from the C-band (wavelength 5 cm) to the Ka-band (wavelength 0.8 cm) are strongly reflected by typical snow particles in an avalanche yet penetrate sufficiently deeply into the flow to reveal its inner structure. Beyond the Ka-band, atmospheric absorption would be too strong and the necessary electronic components are not readily available. The DOPPLER radar measures the frequency shifts caused by particle motion along the line of sight; it produces spectra representing the distribution of particle velocities inside the probed volume. Interpretation of the spectral amplitudes in terms of density is virtually impossible, however, because the reflectivity depends strongly on the particle size, which varies by orders of magnitude inside an avalanche. Also, the signals from a particle diminish strongly with its distance from the radar. Recently, systems combining C or X-band and Ka-band radar have been used in order to measure the dense-flow and powder-snow parts simultaneously [92], see Fig. 2.

The spatial resolution depends on the wavelength and antenna size and shape. Due to safety and accessibility considerations, the radar will usually "see" the entire width and height of the avalanche. If continuous-wave radar is used, several devices operating at slightly different wavelengths have to be used to measure velocities in specific path segments [36]. More recently, pulsed DOPPLER radar has been used; sampling the echo of an emitted pulse at different intervals allows "range-gating", i. e., a series of spectra from measurement volumes at different distances from the radar are obtained (see [92] and references therein).

Despite the interpretational difficulties mentioned, DOPPLER radar is a reliable and first-rate tool for avalanche experiments. It provides measurements of the following quantities:

- spectra of particle velocities along the line of sight,
- separate velocities of powder-snow layer (small particles) and saltation layer or dense core (large particles) in dual-frewquency systems,
- front velocity in pulsed systems with range gating.

It operates under all weather conditions and can be placed in a relatively safe location. It is able to obtain data from the entire avalanche path and thus ideally complements measurement techniques collecting information at a single location only.

Profiling radars. Pulsed DOPPLER radar cannot be used for measurements with high spatial resolution and at distances below approximately 50 m. One way to obtain spatial information is to emit a signal sweeping rapidly over a frequency interval; the frequency difference between the signals emitted and received simultaneously is then a measure of the distance of the reflector [35]. The frequency spectra obtained in this way represent a reflectivity profile along the line of sight. Placing such a radar in an avalanche track, the following quantities can be measured (see Fig. 3):

- the undisturbed snow cover depth before, during and after an avalanche event, and from this the erosion or deposition rate of the avalanche;
- the flow depth of the avalanche as a function of time;
- the vertical profile of longitudinal velocity from the cross-correlations of two profiling radars placed at a distance of 5–10 m along the flow direction of the avalanche.

The echos from snow particles moving parallel to the ground in the sidelobes of the radar beam exhibit a DOPPLER shift that may significantly smear the distance information. Motion perpendicular to the ground, as expected in fluidized flows, contributes additional DOPPLER broadening of the spectra. If periods without frequency sweep (i. e., essentially the DOPPLER mode) are interlaced with the frequency sweeps, the DOPPLER broadening can be reduced during data analysis. It remains to be seen if information on the vertical fluctuating motion of the snow particles can be extracted from the DOPPLER signals.

Profiling radar thus yields extremely valuable information on the internal dynamics of avalanches at a specific location without influencing the flow. Some problems and disadvantages should also be mentioned, however: First, the equipment is expensive and has to be reliable under harsh conditions because it may not be accessible during the entire winter. It puts high demands

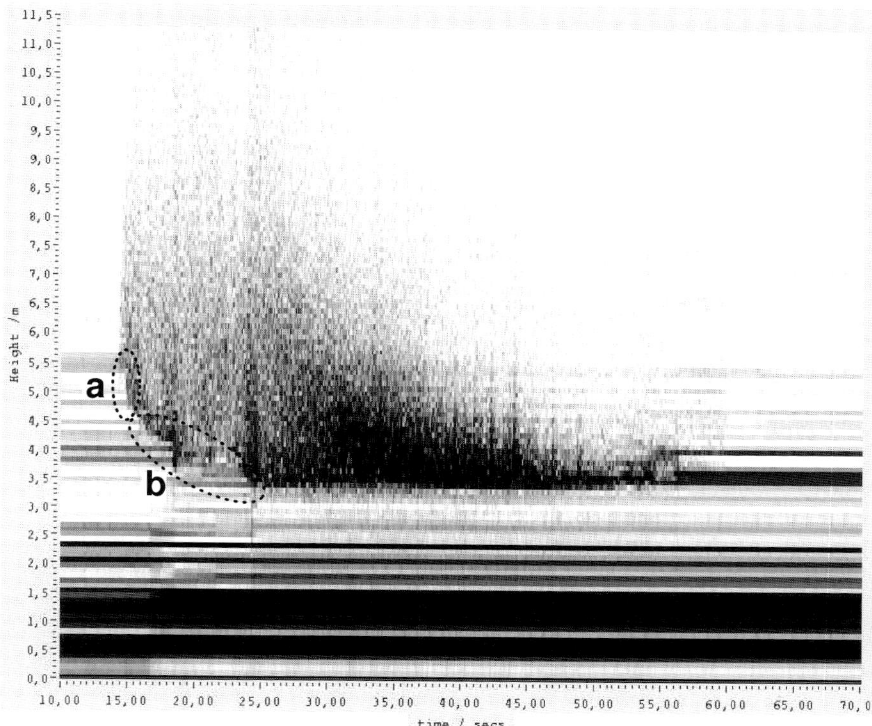

Fig. 3. Echo intensity from frequency-modulated continuous-wave (FMCW) radar as a function of time and electromagnetic distance from the ground. Vallée de la Sionne, 25/02/1999, uppermost radar location. Note the erosion of approx. 1 m of snow within the first second (**a**) and the more gradual removal of 1.3 m (**b**) during the following 9 s until the apparently densest part of the avalanche arrives. From draft of [20]

on the data acquisition and transmission system, and the radar locations have to be chosen carefully. Second, due to the high absorption coefficient of water in the C and X-bands, the radar does not work well under wet-snow conditions. Third, data analysis and interpretation is complicated by the occurrence of multiple reflections and the loss of correlation between neighboring radars due to local topographic disturbances. Despite these difficulties, very important results have been obtained as exemplified in [36, Figs. 10–12] and [19, Fig. 7].

Seismic and acoustic sensors. Seismic recordings of avalanches were carried out at least as early as the 1970s [103], usually to trigger other instruments and to estimate avalanche velocity. More recently, systems have been developed for monitoring avalanching activity on the regional scale by means of a network of seismic stations. Attempts to interpret the signals in terms of the internal dynamics of avalanches were made, e. g., in [79] and [94]. The vibrations produced by the avalanche are transmitted through the snow cover, which strongly damps the high-frequency components, and then through the ground. Attempts are under way to analyze the seismic data with resepct to the temporal evolution of the spectra and to combine them with data from profiling radars (E. Suriñach and B. Sovilla, pers. comm., 2002).

If several geophones are located in or near the avalanche path, the strong rise in amplitude occurring at the passage of the avalanche can be used to determine the mean avalanche velocity from one geophone to the next. Distinct signals are also generated at abrupt slope and direction changes or at the collision with structures. These latter signals are received almost simultaneously at all nearby geophones due to the high propagation velocity in the ground and can also be used to track the progression of the avalanche if the sources of the signals can be identified. It is not yet known, however, whether the different components of an avalanche (dense core, saltation layer, suspension layer) have distinct seismic signatures. This circumstance makes the velocities estimated from geophones rather uncertain.

Low-frequency acoustic signals emitted by avalanches have been recorded with special microphone systems. The tests described in [1] indicate that acoustic goniometry (employing three microphone systems) might be even more effective in monitoring avalanching activity on the regional scale than seismic stations because the propagation velocity of sound in air is relatively low and the spectral signature of avalanches is quite distinct from other phenomena. Obtaining information on the physical processes in avalanches appears to be as difficult as from the seismic signals, however.

Plates for measuring normal and tangential loads. Using strain gauges, load plates can be designed to be sensitive to tangential as well as normal forces, see e. g. [90]. Placing them flush with the ground in an avalanche track, one obtains an effective friction coefficient from the ratio of the tangential to the normal load [16]. Note that this effective friction coefficient is not the same physical quantity as the dry-friction parameter known from several two-parameter dynamical models [85,87,98]: The total friction in those models also contains a contribution quadratic in the velocity which may be thought of as resulting from the inelasticity of snow-particle impacts on the ground (or snow cover surface). It does closely correspond, however, to the bed friction coefficient in the Savage–Hutter model [99].

Great care is required when installing such shear plates and also when interpreting the data they deliver. If the terrain is even slightly concave within a distance of a few times the typical flow height of an avalanche, hard-pressed snow will be deposited in this trough and redistribute the forces over significant distances. The measured loads will be too low in most cases. Even if the terrain is perfectly plane in the vicinity of the load plate, the roughness of the surface should closely match that of the surroundings. If this is not the case, too low tangential loads may be measured if the full shear stress cannot be transmitted from the avalanching snow to the plate surface. It would be desirable to subject load plates to detailed tests under the controlled conditions of chute experiments where the bottom shear stress may be determined independently by measuring the flowing mass and its acceleration.

2.4 Invasive Techniques

Impact pressure measurements with load cells. Before the advent of modern microelectronics, measurements of maximum pressure were carried out in natural avalanches by means of purely mechanical devices based on the so-called indentation test. Though simple and robust, these maximum-pressure cells have low precision and obvious drawbacks when there are multiple avalanche events. Pressure values from those early measurements thus can only be used with utmost caution.

More recent experiments use either electro-mechanical or piezoelectric transducers for measuring impact pressures. If very good time resolution is desired (measurement frequency above 1 kHz), piezoelectric transducers will be chosen. However, at those frequencies, the whole system including the sensors and the instrument support structure has to be designed carefully to suppress low-frequency vibrations that would induce spurious inertial forces. At the test site Vallée de la Sionne of the Swiss Federal Institute for Snow and Avalanche Research (SLF), piezoelectric transducers were mounted in "tandem" configuration at the top of a 5 m high steel wedge and above 5 m at the 20 m high steel mast. The front transducer is exposed to the avalanche forces while the rear one is not, but they are both subjected to the vibrations in the support structure; the avalanche load is given by the difference of the recorded loads. (At high frequencies, the finite speed of sound in steel leads to a phase shift between the two transducers spaced about 20 cm apart; this must be corrected during data analysis [102].)

Analysis and interpretation of the load measurements are significantly more involved and time-consuming than it might appear at first sight. First, the measured pressure depends strongly on the sensor area [71]: Load cells less than about 10 cm in diameter record distinct impacts with very high load peaks (on the order of 1 MPa) while the impacts overlap and average out on large plates with an area of 1 m^2. Second, soon after the first impact, a cone of hard snow may form on the sensor and deflect the approaching snow, biasing the measurements towards too low values. More violent impacts may shatter the snow cone from time to time, however. Third, if the sensor is smaller than the impacting snow clod or the impact is near the sensor edge, only a fraction of the momentum of the clod is transmitted to the sensor. A first attempt at correcting the data for some of these effects and extracting semi-quantitative information on the size and velocity distribution of snow particles was made in [102] with data from the 1999 measurements at Vallé de la Sionne; a much more detailed study including laboratory tests is needed, however, to unlock the full information contained in such data.

This brief discussion of the intricacies of pressure measurements in avalanches shows clearly that comparison of pressure measurements from different sites with different sensor designs is only possible if detailed information on the sensor designs is available.

Pitot tubes. Pitot tubes measure the difference between the stagnation pressure at the tip of the device and the static pressure at its side; if the fluid density is known, the flow velocity can be determined. Application of such devices to the suspension layer of avalanches would be very interesting; it has been quite successful in ping-pong ball avalanches [78] and partly so in the very dilute part of a powder-snow avalanche [91], but the results were more questionable in the denser layers [58,83,84]. The main problems encountered were (i) clogging of the openings in the Pitot tube, (ii) the strong direction sensitivity of the Pitot tubes exposed to a highly irregular, turbulent flow, and (iii) the unknown density of the snow-air mixture.

The first problem might be overcome by closing the openings with suitable membranes or by using tiny load cells. By measuring the pressure simultaneously at four, five or more locations on a sphere instead of a tube promises to solve the second difficulty, although at the expense of a higher price and data volume combined with a more involved analysis procedure. Finally, a vertical array of such sensors measuring the difference to the undisturbed air pressure could be used to determine the snow concentration from the differences in hydrostatic pressure if sufficient precision can be achieved. If further development of such systems proves successful, our knowledge of the suspension layer will advance significantly.

LED–photocell pairs. Variations in the light from light-emitting diodes (LED) reflected by passing snow particles can be detected with photocells. By correlating the signals from two such detectors a short distance apart in the flow direction, the mean particle velocity can be obtained. Arrays of such sensor pairs have been successfully used at Ryggfonn [84] and in relatively small avalanches at Montana State University's Revolving Door site [16]. In the latter experiments, velocity profiles were obtained from arrays of such sensor pairs. Loss of correlation due to vertical motion of the snow particles may be a serious problem, though. The main advantages of this technique are the very low price of the sensors and the relatively simple data-analysis procedure. On the other hand, the technique is invasive and so requires a sturdy and expensive structure for mounting the sensors (which may also be used by other sensors, however). The high data rate from the large number of sensors required for velocity profiles presents no problem in chute experiments but puts high demands on the data acquisition system in a large avalanche site. We may expect to see more widespread use of this sensor type where mounting structures, e. g. for pressure measurements are already available.

Capacitance probes. None of the sensors discussed so far is capable of measuring the density of avalanching snow—a quantity of considerable importance both for understanding the dynamics of avalanches and for calculating impact pressures. Ice being a fairly strong dielectric, capacitance probes can be used for this purpose. While the principle is rather simple, several technical problems involving the geometry of the electrodes, the choice of driving frequency and the electronics had to be resolved to produce fully functional devices [67]. In addition, there are two problems rooted in the physics, namely the sensitivity to free water, which has an even higher electric susceptibility than ice, and the texture dependence of the electric susceptibility of an ice–air mixture. In many cases, the first problem can be handled by calibrating the measurements after the avalanche event against manual density measurements. The texture dependence appears to be tolerable in typical dry-snow avalanches. This technique has been successfully used at Revolving Door [16], and similar, but larger sensors have recently been installed together with LED–photocell pairs on the 19 m high mast at Vallée de la Sionne (B. Sovilla, pers. comm., 2002).

As for the LED–photocell technique, velocity profiles can be obtained from arrays of pairs of capacitance probes. Unfortunately, the high price of the probes are still prohibitive for wide-spread use at present, and the high data rates require an expensive data acquisition system, especially if many probes are installed in an inaccessible avalanche track.

Acoustic techniques. Attempts have repeatedly been made to use the attenuation of sound waves over a fixed distance for measuring the density of snow. The strong damping at high frequencies requires either the use of low-frequency waves with a correspondingly high noise level or very small measurement distances, which lead to difficulties with clogging and very strong signal fluctuations from the passage of single snow particles. For further details see [68].

Ultrasonic anemometers have been used at the Shiai-dani test site in Japan for measuring velocities in powder-snow avalanches [83]. Useful results were obtained in the dilute part of the powder-snow cloud whereas the instruments did not work properly in the saltation layer (see Fig. 6 and Sect. 3.3 for a discussion of these terms). The technique is based on measuring the runtimes of acoustic pulses over a fixed distance in two or three independent directions; if the speed of sound is known, the velocity vector can be determined. The presence of snow particles may, however, reduce the sound velocity by as much as two thirds if the volumetric particle concentration is 1 %. Therefore, the sound velocity should be determined by measuring the signal runtime in both senses for each direction. This makes such an instrument more complicated and expensive, but holds the promise of measuring the density of the snow suspensions at the same time. Some basic investigations of the particle-concentration dependence of the sound velocity at different frequencies and development work to make the instrument robust and to minimize disturbance of the flow patterns will be needed for assessing the potential of this technique.

In watertank experiments on turbidity currents as models of powder snow avalanches, ultrasonic transducers have been very successfully used for measuring both particle velocity and concentration [41,60,105]. However, the concentration measurements are based on experimental calibration curves that are valid only if the particle size distribution does not change appreciably during the flow. With sufficiently high sampling rates, turbulent correlations can be obtained directly [3,8].

3 Main Experimental Results

Over the course of half a century, a variety of different qualitative descriptions of the avalanching process have been put forward and elaborated into mathematical models. While the avalanche was treated as a mass point, i. e., essentially as a solid, in the very first models (reviewed in [63], later work [114] emphasized the fluid-like properties of avalanche flow. Later models were explicitly devised either for dense or suspension flow modes (Voellmy had used essentially the same approach for dense-flow and powder-snow avalanches). From the 1980s, concepts from granular flows began to be applied to dry-snow avalanches [31,32,85,86,88,99,100] while the BINGHAM rheology and extensions thereof were proposed in [15], especially for wet-snow avalanches [69,70,75].

In the following paragraphs, these theoretical concepts are confronted with observations and results from various experiments.

3.1 Inferences from Visual Observations

Video footage of dry-snow avalanches triggered by skiers or explosives shows rapid growth of the destabilized zone in all directions (interpreted as a shear failure along a weak layer in the snow cover), followed by tensile and shear fracture along the crown and flanks. Preliminary analysis of the videos from the 1999 experiments at Vallée de la Sionne [20] revealed fracture propagation velocities around $50\,\mathrm{m\,s^{-1}}$. In medium to large slab avalanches, the released snow layers break up into small snow clods after a relatively short distance (on the order of 50–200 m). It is thus justified to consider a dry-snow avalanche as comminuted from the start.

Significant differences are observed visually in the behavior of wet and dry snow avalanches. Observers agree that wet-snow avalanches are considerably slower than dry ones, but large wet-snow avalanches may nevertheless reach peak velocities around $30–40\,\mathrm{m\,s^{-1}}$. The runout distances are also somewhat shorter than those of dry-snow avalanches in the same path; to the author's knowledge, there is no systematic analysis of this question, which would also have to take into account the flow depth and total volume of the avalanches as they enter the runout zone. In their flow behavior, wet-snow avalanches resemble subaerial debris flows. Considerable normal and shear stresses must be transmitted in the avalanche body: The head is often pushed over large distances by the snow masses from behind. If an obstacle is encountered, the flow direction may change abruptly, sometimes by more than 90°; the small-scale topography determines the flow path to a large degree. Two striking features, especially noticeable in the runout zone of smaller events, are the sharp lateral boundaries of the flowing body and the essentially constant width on open slopes. They indicate that cohesive forces play an essential role at least at low velocities.

In contrast, dry-snow avalanches generally appear to be more "fluid", their lateral boundaries are less sharp, the flow velocities are smaller at the sides than in the center, and there may be considerable lateral spreading. Smaller obstacles are quite easily overflowed, e. g., the bed of a brook is quickly filled with avalanche snow and plays a negligible role in the determination of the flow path. On opposing slopes, dry-snow avalanches may ascend a considerable distance.

Shortly after the release of a dry-snow avalanche, significant mixing with air occurs at its front and along its upper surface; this effect is much less pronounced in wet-snow avalanches. As the mixing layer grows in height, the dense core is lost from sight and its velocity and depth can only be inferred through measurements. If the avalanche path is long and sufficiently steep, the dilution may continue to the point where a large "cloud" of suspended particles is formed. This "cloud" may separate from the dense part of the dry-snow avalanche if the latter is deflected or rapidly stopped. On video tapes taken from the shelter at Vallée de la Sionne, particles of considerable size have been seen in the air at the front of the approaching powder snow cloud as it was ascending the opposing slope; apparently, these particles were ejected violently from within the front.

Such a suspension flow generated by a dry-snow avalanche is often called a powder snow avalanche, but there is considerable confusion about the precise meaning of this term as it is often also used for the entire dry-snow avalanche including a dense core, the cloud of suspended particles and an intermediate layer to be discussed below in connection with other observations and measurements. The terminology used in the present article will be established and explained in Sect. 3.3.

3.2 Information from Deposit Structure

The deposits of different types of avalanches contain a wealth of information that has rarely been fully utilized in the past; remarks on different aspects are scattered in the literature and few papers are available on the subject, among them [56,110]. Characteristic features of wet-snow avalanche deposits are (i) huge numbers of large snow clods, (ii) steep, sharp boundaries, (iii) high densities in the range of 400–600 $\mathrm{kg\,m^{-3}}$, and (iv) mostly in large avalanches

Fig. 4. Snow pit in the runout area of the 1995 powder snow avalanche at Seewis, in an area not reached by the dense flow. Large snowballs are scattered on the surface and fir twigs in the exposed snow prove that the snow was deposited by the avalanche tail after the front had eroded approx. 1 m of fresh snow. From [56]

very smooth, icy shear planes. In dry-snow avalanches, the deposited snow may also be somewhat damp, either because of energy dissipation through friction and collisions or because humid snow was entrained in the lower part of the path. The snow particles are generally smaller than in wet-snow avalanches or they may hardly be discernible. The deposits tend to have less sharp boundaries and to taper off at the sides and at the front. The density may be smaller than in wet-snow avalanches, but it can attain very high values as well if the avalanche was stopped abruptly and compressed thereby. Icy shear planes do not appear to be typical features, however.

There is considerable disagreement, even among avalanche professionals, as to the disctinction between the deposits from the dense-flow or powder-snow parts of a mixed avalanche. However, the three events described in [56] showed certain common features at widely different size scales, combined with fortunate topographical circumstances that facilitate their interpretation and so, in the author's opinion, allow some basic inferences on the flow regime to be made. The 1999 Vallée de la Sionne experiments [20] by and large confirmed the earlier findings, and reports from earlier experiments [58,81,83,103] can also be readily understood in these terms.

A fairly small avalanche (estimated release volume: 20–30 000 m^3) was released near the village of Seewis in north-western Switzerland on January 10 or 11, 1995 towards the end of a strong and rather cold snowstorm. The upper avalanche track progressively contracts into a 10–20 m wide gully. Over a stretch of almost 1 km, typical deposits of a dense-flow avalanche covered the bottom of the gully, with clearly discernible lateral boundaries. No such deposits were seen outside the gully, but damage to the trees proved that the avalanche had exerted considerable pressures there. Snow pits outside the gully (see Fig. 4, to be discussed further in Sect. 5) moreover showed that substantial amounts of snow had first been eroded and later deposited. The deposits consisted of a matrix of very fine-grained and compacted snow, with embedded snowballs from hazelnut to fist size making up a few per cent of the volume. At a sharp turn of the gully, the dense-flow deposits remained clearly confined to the gully but the fluidized and suspended components of the avalanche maintained its direction and climbed the steep lateral slope at the bend onto a gentle open slope. Snowballs with a diameter up to 40 cm were found there on the surface of the deposits, smaller snowballs again being embedded in the fine-grained snow. Note that there was no evidence of a similar layer on top of the dense-flow deposits in the gully.

About two weeks of strong and not so cold westerly winds followed the mentioned snowstorm. During a snowfall of only 30 cm, a large mixed avalanche (release volume approx. 400 000 m^3, crown length 1 km) occurred on the Albristhorn near the resort village of Adelboden, Bernese Oberland. The runout zone on a gently sloping alluvial fan was completely covered with humid avalanche debris, deflected at the embankment of a brook running perpendicular to the avalanche flow direction. On the embankment slope and 100–200 m beyond it, deposits with scattered snowballs very similar to those of the Seewis avalanche were found, with the size of the surface particles diminishing in the distal direction. The deposits were embedded in a layer of fresh snow and could be traced for about 1 km up the gentle slope. At that distance, they contained no longer distinct particles but were clearly distinguishable by their higher density and strength. This deposit was not recognized on top of the debris from the dense-flow part, indicating that the powder snow avalanche moved faster than the dense flow.

The avalanche from Scex Rouge onto the Pillon Pass and the resort village of Les Diablerets in the western Swiss Alps followed two days later and involved initially close to $2 \times 10^6\,m^3$ of snow. After a vertical fall height of 1500 m, the dense-flow part formed an impressive deposit of 10–15 m height with many large shear planes on the valley floor whereas the powder snow part climbed the opposing slope onto a plateau and beyond (altitude difference 100–200 m) and continued for at least 3 km down the valley. The deposits on the pass road reached a depth of 3 m and typical densities were 500–560 $kg\,m^{-3}$. As for the Albristhorn avalanche, the mass of the powder-snow avalanche deposits was about 10 % of the total mass.

From surveys of the deposits of powder-snow avalanches, their mass can be estimated. Pictures of the powder-snow clouds [56] give a rough indication of their volume, leading to estimates of their average density in the range 2–5 $kg\,m^{-3}$ (with large uncertainty).

3.3 Pressure Measurements

As pointed out in [71], the size of the load plate has a strong influence on the results of pressure measurement, especially in the upper parts of the flow. The 10 cm diameter load plates used in the Vallée de la Sionne experiments from 0.9 to 19 m above ground [102] illustrate very clearly that there are strong horizontal and vertical variations of the flow structure, as evidenced by Fig. 5. In the lower part of the flow (up to 1–3 m above the snow surface in large

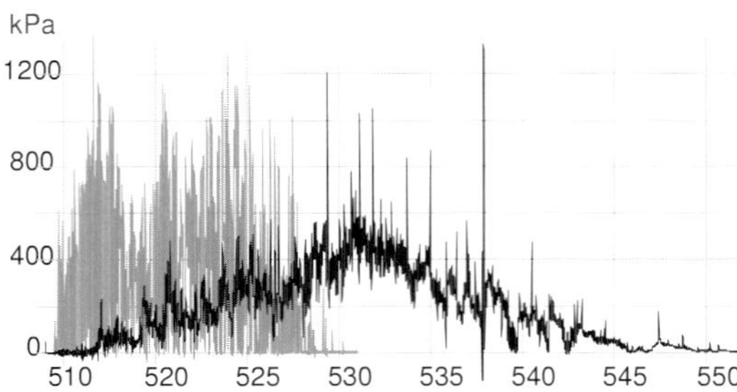

Fig. 5. Vallée de la Sionne test site, avalanche of Febr. 10, 1999: Pressure measurements on 10 cm diameter load cells at 3.0 m (black line) and 3.9 m (gray line) above ground, illustrating the difference between particle impacts in the saltation layer and the dense core. From [102]

avalanches) and possibly after an extended "overture" with isolated impacts, the load is continuous, albeit with significant short-term fluctuations (the black line in Fig. 5). In fast dry-snow avalanches, but not in slow wet-snow flows, there is a 2–5 m thick region above and often also ahead of the layer just identified in which impacts with peak pressures between 300 and 1200 kPa and lasting 10–30 ms are frequent; however, in between impacts, the pressure drops to almost zero, as evidenced by the grey line in Fig. 5. These two measurements were taken simultaneously 0.9 m apart in the vertical direction. At 19 m above ground (and sometimes also at 7 m), pressures are much lower and the peaks do not show the characteristic skew of plastic particle impact but are symmetric.

These measurements confirm the visual observations and the conclusions reached by Schaerer and Salway [103] as well as Grigoryan, Urumbayev and Nekrasov [30] about the layering of dry-snow avalanches with a dense core at the bottom, a regime of "light flow" (also termed the saltation layer) above and possibly also ahead of it, topped by the suspension layer that consists of very small particles and in which the pressure fluctuations are mainly due to turbulence. This conceptual picture of a dry-snow avalanche, schematically depicted in Fig. 6, also immediately explains the observed structure of the deposits described above. Throughout this paper, the term "dry-snow avalanche" will be used for avalanches consisting (mostly) of dry snow; in most cases, they will exhibit the three layers just described, but their fraction of the total mass may vary widely from one avalanche to another and along the path of a single avalanche flow. In the literature, one also finds the terms "mixed avalanche" and (particularly among practitioners) "powder snow avalanche" used more or less synonymously for this entire phenomenon. However, in this article the latter term will be reserved for situations in which the saltation and suspension layer have separated from the dense core. The observations and measurements summarized here and in Sect. 3.2 very strongly suggest that a fully developed suspension layer separated from the dense core is either accompanied by the saltation layer from which it emerged, or generates a saltation layer while eroding the overflowed snowcover.

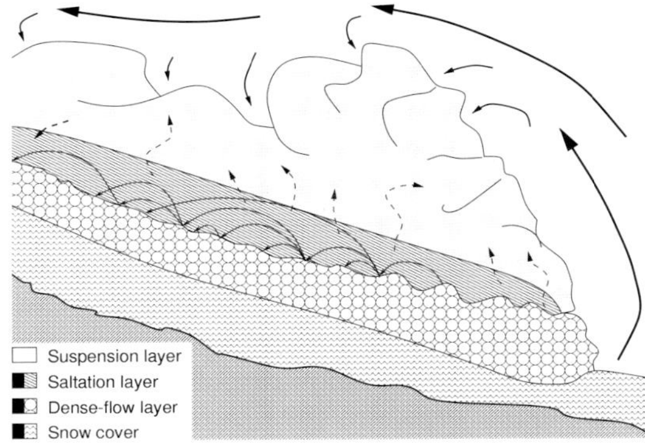

Fig. 6. Schematic representation of the structure of a dry-snow avalanche in the early stage of its formation, i. e., the saltation and suspension layers do not yet move ahead of the dense core. In the dense core, no distinction is made between possible shear and plug zones. From [55]

Assuming that the flow velocity inside the dense core is somewhat smaller than the front velocity of the powder-snow cloud, one concludes that the core density is most likely in the range 250–500 $\mathrm{kg\,m^{-3}}$, i. e., denser than the undisturbed snow cover and slightly less dense than the deposits. The preliminary analysis of the single-particle impacts in [102] yielded densities from 10 to 45 $\mathrm{kg\,m^{-3}}$ in the saltation layer. Pressure measurements at Shiaidani presented in [83] yielded an estimate of 2–4 $\mathrm{kg\,m^{-3}}$ near the front and the bottom of one powder snow avalanche. Only a few significant particle impacts occurred during the measurement, and both the velocity and pressure were low, indicating that no dense core was present and that the saltation layer was only weakly developed in that avalanche.

It should be noted that there is no definitive evidence whether or not the layers described above are separated by fairly sharp interfaces or have smooth transitions. On the basis of the FMCW radar reflection profiles (Fig. 3) and impact measurements (Fig. 5) at Vallée de la Sionne, combined with the evidence of sharply bounded deposits from the channeled dense flow within a much wider powder-snow avalanche at Seewis (Sect. 3.2), the author favors the concept of fairly sharp transitions between layers.

3.4 Velocity Measurements

Front speed measurements are fairly numerous. Examples of peak velocities are 17–20 $\mathrm{m\,s^{-1}}$ for dry snow flows on a chute [18], 6–10 $\mathrm{m\,s^{-1}}$ for small dry-snow avalanches at Revolving Door [16], 18–35 $\mathrm{m\,s^{-1}}$ for dry-snow avalanches at Rogers Pass [103], 33 $\mathrm{m\,s^{-1}}$ in a humid avalanche at Ryggfonn [107], between 23 and 49 $\mathrm{m\,s^{-1}}$ for 12 dry-snow avalanches at Ryggfonn [65], and up to 65 $\mathrm{m\,s^{-1}}$ in the Aulta path in Switzerland [97]. For the huge February 25, 1999 mixed avalanche at Vallée de la Sionne, the velocity of the strongly lobated front of the powder snow cloud fluctuated rapidly by as much as 25 $\mathrm{m\,s^{-1}}$, with peak values approaching 80 $\mathrm{m\,s^{-1}}$ [19]. An intriguing observation is the abrupt reduction of acceleration in the 1984 Aulta avalanche after a horizontal distance of 200 m at a velocity around $45\,\mathrm{m\,s^{-1}}$ [97], interpreted as a flow-regime transition.

In a few experiments, the front and internal velocities could be determined simultaneously; pulsed Doppler radar is particularly suited for such comparisons. In a measurement of a mixed humid avalanche with a powder snow cloud at Ryggfonn in 1997 [107], the front velocity was around $30\,\mathrm{m\,s^{-1}}$ between the tower and concrete structure, coinciding with the velocity at maximum spectral intensity, whereas the fastest components were measured at about $43\,\mathrm{m\,s^{-1}}$, i. e., almost 45 % higher. A similar picture results from the 1999 measurements at Vallée de la Sionne with dry-snow avalanches [20,92]. Figure 7 shows several simultaneous velocity spectra for one of those avalanches; clear differences are seen between different range gates and between the X- and Ka-band. In the February 25 event, the fastest components exceeded $100\,\mathrm{m\,s^{-1}}$. Interestingly, the velocity spectrum in the C-band of the radar (sensitive to large snow particles) is significantly more narrow at the head of the avalanche than in its body. In the Ka-band (more sensitive to small particles) this effect is less pronounced—a fact that may be explained in terms of the vortex structure of the powder cloud with possibly negative velocities at the top. In the break-up phase after release and in the stopping phase, the Doppler spectra tend to be narrow while they are very wide in the middle phase of dry-snow avalanches. The layered structure of dry-snow avalanches cannot be inferred from only the Doppler radar measurements, but the latter are fully compatible with the inferences from the pressure measurements.

Despite the experimental difficulties of localized velocity measurements inside an avalanche, a number of experiments produced velocity profiles $u(z)$ where z is the coordinate perpendicular to the slope and u is the velocity in the main flow direction. Note that methods based on correlation techniques are quite sensitive to the choice of the sampling time interval (see Fig. 6 in [16]). With pairs of FMCW radar devices, velocity profiles could be obtained only in a few cases [34,36] (see right plot in Fig. 8), none are available so far from the 1999 experiments at Vallée de la Sionne. Cross-correlations of pressure signals at Ryggfonn [84] also appear somewhat erratic. More reliable profiles were obtained in chute experiments by means of image analysis [15] and in small avalanches from cross-correlations of photosensor signals [16] (Fig. 8, left plot).

All velocity profiles published so far are from dry-snow avalanches at slope angles of 25–40°, presumably under near-uniform flow conditions. An exception may be Dent's experiments [15] where the profile appears to have been taken after the transition to the horizontal runout section, and preliminary results from the Weissfluhjoch chute [112] where the measurements were performed immediately after an abrupt slope transition from 45° to 32°. Most measurements show a pronounced shear layer at the bottom, extending over 10–30 % of the flow depth, and a region of little or no shear above. In most

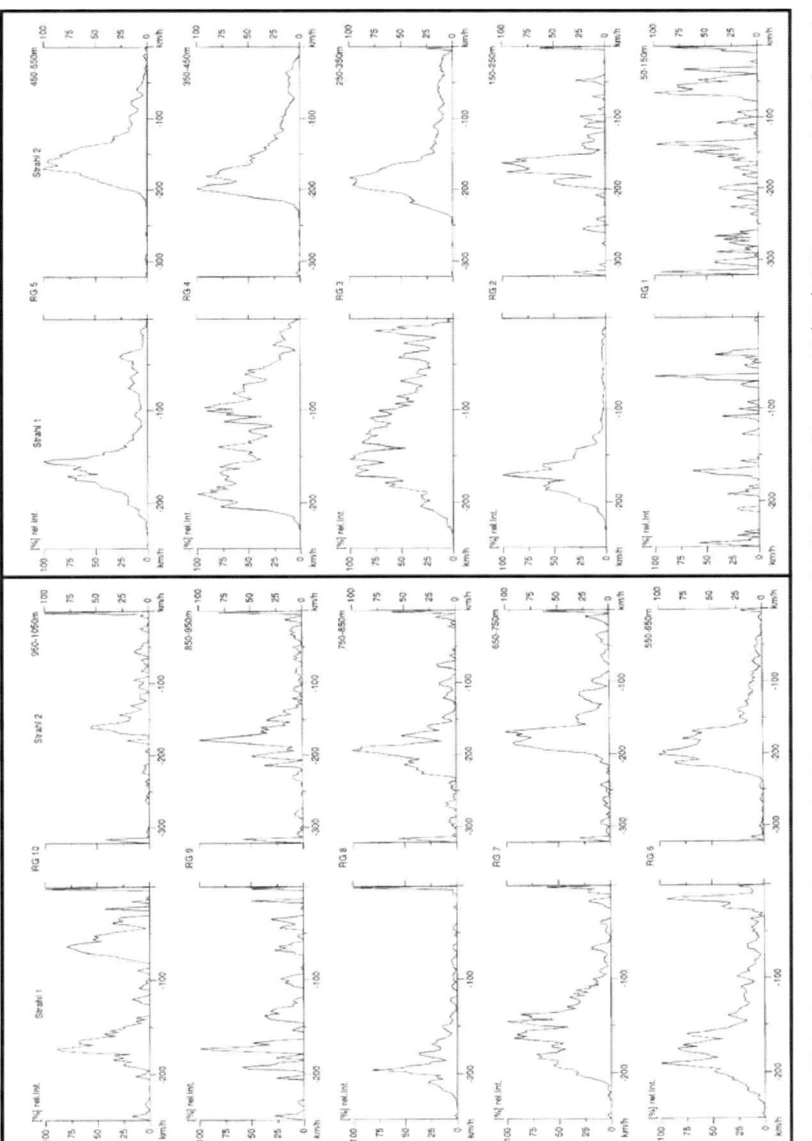

Fig. 7. Velocity spectra from pulsed DOPPLER radar, Vallée de la Sionne, 10/02/1999, approx. 3 s before the avalanche reached the shelter (see Fig. 2). Range gate 1 (RG1) is at 50–150 m from the antenna, RG2 at 150–250 m, etc. "Strahl 1" refers to the C-band radar, "Strahl 2" to the Ka-band radar (most sensitive to the fine suspended particles in the suspension layer). Plots courtesy L. RAMMER

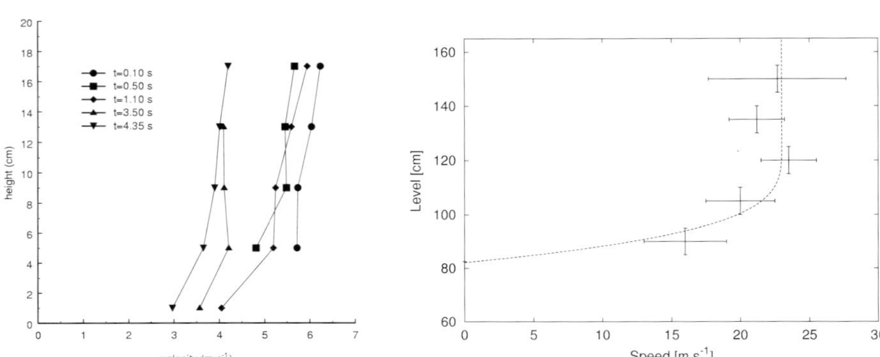

Fig. 8. Vertical profiles of longitudinal velocity from the Revolving Door site (**left**) at different times, and from the Fogas avalanche path (**right**). Obtained through cross-correlations of photoelectric sensors (Revolving Door) and a pair of FMCW radar systems (Fogas). The sliding surface of the Fogas avalanche was about 80 cm above the ground. From [16] (courtesy J. D. DENT) and [36] (redrawn)

cases, the spatial resolution is insufficient to determine whether there is a finite slip velocity[1] or no-slip boundary conditions apply. In the Weissfluhjoch chute measurements, a slip velocity of two thirds of the surface velocity was found shortly after the transition from a rough bottom to a smooth aluminum bottom surface, and the measurements of DENT et al. [16] also appear to favor a significant slip velocity. Clearly, none of the available profiles is compatible with the parabolic profile of a NEWTONian fluid under conditions of stationary open-channel flow. The depth-averaged velocity is 5–15 % smaller than the maximum velocity at the surface. The assumption of a rectangular velocity profile, made explicitly or implicitly in most current dynamical models, thus appears justified as a first approximation, but the shear layer at the bottom leads to a non-negligible transfer of mass from the head to the tail of the avalanche.

No information is presently available on possible differences of the velocity profiles between the head, body and tail of the avalanches, nor are profiles available from a single avalanche at different locations (upper track, lower track, runout zone)[2]. A particularly important open question is whether profiles different from the shear layer–plug flow type occur under normal conditions. Flows on steep chutes (45° or more) with pronounced surface roughness

[1] In wet-snow avalanches, extremely smooth, frozen shear planes parallel to the slope are sometimes found, indicating that the avalanche behaved essentially as a gliding, rigid block in the stopping phase.

[2] The Vallée de la Sionne test site features three pairs of FMCW radar devices just below the starting zone, near the end of a gully and in the upper runout zone. However, none of the events observed so far yielded useful spectra at all six locations, and attempts to obtain velocity profiles have so far been unsuccessful due

or the path segment of a full-size avalanche where the velocity maximum is reached are candidate conditions for finding a more fully fluidized flow regime.

4 Flow Regimes

4.1 Multi-Layer Structure

On the basis of the available experimental evidence, the multi-layered structure of dry-snow avalanches is firmly established. The findings summarized in the previous section strongly suggest the following interpretation:

- Even relatively small dry-snow avalanches may rapidly (i. e., after a few hundred meters) develop a distinct, more mobile and dilute component that flows faster than the dense core. It contains both small snow grains and snowballs of considerable size, the latter usually being constrained to the lower layer of such a powder snow avalanche.
- The distinct flow behavior of this avalanche component favors the view that there is a fairly clear interface between the dense core and the powder snow avalanche rather than a gradual transition.
- As a rule of thumb, the densities in the dense core, the saltation layer and the suspension layer of well-developed, full-size avalanches are about $300, 30$ and $3\,\mathrm{kg\,m^{-3}}$, respectively. Even small avalanches may exhibit this structure [16], but it appears likely that the density of the suspension layer and the size of the snow blocks in it are smaller than in large avalanches.
- Extensive reworking of the overflowed snowcover (snow entrainment at the head, deposition at the tail of the powder snow avalanche) commonly occurs if the snowcover is cold and dry. The entrainment rate may well exceed 50–$100\,\mathrm{kg\,m^{-2}\,s^{-1}}$.

The layering of deposits suggests that the saltation and suspension layers may eventually move ahead of the dense flow. This is confirmed by the pressure measurements at the Shiai-dani [108], Rogers Pass [103] and Vallée de la Sionne sites [102], and to some degree at Ryggfonn [84]. Doppler radar measurements are not able to confirm this because the saltation layer and dense core have the same signature, both containing large particles. However, the more narrow velocity distribution at the avalanche front found in the spectra from Vallée de la Sionne [92] lends indirect support to higher velocities in the saltation layer than in the dense core.

Various observations (e. g. [20,56]) suggest that the dense core of dry-snow avalanches is much more easily deflected sideways at gully bends or upon impact on to a steep opposing slope than the powder-snow part which may climb considerable distances uphill. In this context, it is interesting to note that the (densiometric) Froude number, $\mathrm{Fr}_d = \sqrt{\frac{\rho u^2}{\Delta\rho g h}}$, of the dense

to loss of correlation over the distance of 10 m between the radars (B. Sovilla, pers. comm., September 2002).

avalanche core is typically in the range 5–10, i. e., well in the supercritical regime, while the suspension layer has a Froude number close to 1. This would imply that disturbances do not propagate upstream in the dense core, but two points need to be considered: (i) Obstacles in the avalanche path will not only generate disturbances of the flow height, but also lead to compression of the impacting snow. The compression shock fronts propagate faster than gravity waves and the flow, as evidenced by the pile-up of snow in front of an obstacle. (ii) In the derivation of the Froude number, an earth-pressure coefficient of 1 is tacitly assumed. If the earth-pressure coefficient in the passive state (avalanche contracting) is indeed significantly larger than 1 [100], a modified Froude number based on the normal stresses in the flow direction will be much closer to 1.

4.2 Rheology of the Dense Layer

Qualitative properties. For numerical modeling, the complicated rheological behavior of snow must be simplified to arrive at a tractable model. However, several of these intricacies have a significant effect on the motion of avalanches, especially on the runout distances of avalanches of different masses and on the spatial distribution of pressure in the runout zone. Gubler [32] approached this problem by directly considering the observed physical processes rather than from theoretical preferences for one or the other rheological model, but few of his ideas have since been taken up and put into a more rigorous framework. In the author's view, the following observations need to be reflected in a useful rheological model of avalanching snow:

- Dry-snow avalanches can have quite long runout distances if the flow height is large. The effective friction coefficient can be much smaller (around 0.15 in extreme avalanches) than the measured dry-friction coefficient of snow blocks on snow, which is in the range 0.6–0.8 according to [12].
- In many if not most dry-snow avalanches, a strongly sheared, thin zone at the bottom is overlaid by a deeper layer moving as a plug (or nearly so).
- Avalanche observers often report that the deceleration seems to increase rapidly as the velocity becomes small, even though the slope does not change visibly. At the same time, the snow appears to solidify and to switch from flowing to gliding. Unfortunately, measurements providing hard evidence for this transition are still lacking at present.
- The deposit depth is often substantially higher than the flow depth because snow from the avalanche body piles up on the front that has already been stopped, e. g., at a slope change.
- Very often, a variable amount of snow is deposited along most parts of the avalanche path, presumably from the avalanche tail.

- A full description including the transition to saltation requires density to be variable, both in space and time, but experimental information is largely missing. Densification seems to occur in the dense core and during the stopping phase, while substantial mixing with air accompanies formation of the saltation layer. If that process is disregarded, density may be assumed constant for practical purposes.

Visco-plastic models. DENT and LANG [18] proposed the biviscous BINGHAM model[3], characterized by the stress-strain relationship

$$\dot{\gamma} = \begin{cases} \dfrac{\tau}{\rho\nu'} & \text{if } |\tau| \leq \tau_0 \dfrac{\nu'}{\nu' - \nu}\,, \\ \dfrac{\tau - \operatorname{sgn}(\tau)\,\tau_0}{\rho\nu} & \text{if } |\tau| > \tau_0 \dfrac{\nu'}{\nu' - \nu}\,. \end{cases} \tag{1}$$

τ is the shear stress (in the flow direction, on a plane parallel to the gliding horizon), $\dot{\gamma}$ the corresponding shear rate; τ_0 and ν correspond to the yield strength and the BINGHAM viscosity in the normal BINGHAM model. ν' is the much larger viscosity applicable for stresses below yield. According to [18], the biviscous BINGHAM model with the values $\tau_0/\rho = 2.2\,\mathrm{m}^2\,\mathrm{s}^{-2}$, $\nu = 0.002\,\mathrm{m}^2\,\mathrm{s}^{-1}$ and $\nu' = 0.10\,\mathrm{m}^2\,\mathrm{s}^{-1}$ provides an accurate description of the velocity profile, the runout distance, and the longitudinal deposit profile of their chute experiments with dry snow. Table 1 summarizes (bi)viscous BINGHAM model parameter values estimated from published velocity profiles. For lack of more detailed information, stationary flow conditions are assumed, shear stresses on the upper surface of the dense flow are neglected and snow density and slope angle are estimated. The uncertainties are too large in all measurements for deviations from the BINGHAM rheology to be detectable. Snow conditions have probably varied substantially between experiments, so the differences in yield strength and BINGHAM viscosity between the Revolving Door and Fogas North sites are not surprising, the former having continental climate and the latter experiencing some maritime influence. The very low yield strength and viscosity values inferred from NISHIMURA and MAENO's experiments [82] may be attributed to their technique for preparing the snow sample. Note, however, that the measurements with the higher yield strengths are also the experiments with the larger flow depths—the variation of τ_0 could equally well be related to the MOHR–COULOMB yield criterion.

With regard to the requirements listed on p. 132, the visco-plastic models of the BINGHAM type fulfill the first two points, may or may not satisfy the

[3] In the original BINGHAM rheology, shear flow is possible only if the shear stress exceeds the yield stress, implying that the stress at zero deformation rate is ill-defined. The biviscous model allows to circumvent the numerical problems arising in the conventional Bingham model from this abrupt change from solid to fluid behavior or vice-versa.

Table 1. Analysis of published velocity profiles in terms of the BINGHAM or modified biviscous BINGHAM rheology proposed in [18]. α: slope angle at measurement location; ρ: snow density; $\dot{\gamma}$: mean shear rate over shear layer; τ_0: yield stress; ν: BINGHAM viscosity at high shear rates; ν': viscosity at low shear rates

Location, type of experiment, reference	α	ρ	$\dot{\gamma}$	τ_0/ρ	ν	ν'
		[kg m^{-3}]	[s^{-1}]	[m^2 s^{-2}]	[m^2 s^{-1}]	[m^2 s^{-1}]
Outdoor chute (Montana), film [18]	30°	?	70	2.2	0.002	0.10
Laboratory friction tests [17]	—	?	50–300	1.8	0.004	—
Roughened laboratory chute, fluidized snow; high-speed video [82]	40°	200–300	110–140	0.1–0.3	0.0005 –0.0007	—
Outdoor chute (Hokkaido), video [82]	36°	200–300	160	0.17	0.0006	—
Revolving Door (Montana), optical sensors [16]	35°	?	70–115	0.7	0.002	—
Fogas North (Swiss Alps), FMCW radar [36]	35°	300	56	2.5	0.017	—

third (on which there is little firm information from experiment) and fourth, and do not easily accommodate the fifth point. A pronounced dependence on the depth of the snow slab at relase is the combined result of two effects: For one, the shear stress at the bottom of the flow is bounded from below by τ_0, independent of the slope angle and the flow depth; a thick avalanche can therefore flow on very gentle slopes at low velocities whereas a thin avalanche comes to a stop on quite steep slopes. Assuming a uniform flow depth d and neglecting inertial effects and "pushing" from the tail of the avalanche, the BINGHAM model predicts that the avalanche will stop at a slope angle of

$$\sin \alpha_{\text{stop}} = \frac{\tau_0}{\rho g d} \,. \tag{2}$$

On a qualitative level, this agrees very well with observations: Small avalanches "starve" on steep slopes because a fraction of the thin shear layer is continually left behind the main body and stops rapidly because its depth falls below the threshold defined by (2). Through this process, the flow depth decreases also at the avalanche head; if the mass loss is not offset by snow entrainment, condition (2) will eventually be met even on a steep slope. The flow velocity also depends strongly on the flow depth; assuming that mass loss and entrainment balance, the steady-state surface velocity on a uniform slope is found to be

$$u_{\mathrm{plug}} = \frac{g \sin\alpha}{2\nu} \cdot \left(d - \frac{\tau_0}{\rho g \sin\alpha} \right)^2 . \tag{3}$$

Velocity measurements along the path are now available for several events at the sites of Shiai-dani (Japan), Rogers Pass (Canada), Aulta and Vallée de la Sionne (Switzerland), Ryggfonn (Norway) and the Khibiny mountains (Russia). Comparison of those measurements with the predictions of a BINGHAM model would be very interesting in view of the good agreement found for chute experiments by Dent and Lang [18]. The main impediment in such a comparison lies, however, in the strong effect of snow entrainment, especially in a model of the BINGHAM type: The loss of flow depth due to the shearing of the bottom layer may be more than offset by the fresh snow entering the avalanche, and the model predictions will therefore depend strongly on the entrainment model.

For a complete model, the BINGHAM relationship or its biviscous extension (1) needs to be supplemented with a constitutive relation determining the normal stresses. It appears that the models published so far (depth-integrated hydraulic models) assume isotropy as in a NEWTONian fluid and set the longitudinal stress equal to the hydrostatic pressure. The good agreement of the simulated deposit profile with the experimental one in [18] suggests that this assumption may be reasonable, but further tests would be required to settle this question.

Several extensions of the BINGHAM model are conceivable: A dependence of the yield strength on the normal stress (which increases with depth) would take into account the sintering effects. On the other hand, shearing breaks the bonds between the snow particles and tends to reduce the strength of the material, so the yield strength might be made dependent on the shear rate and/or accumulated shear. Judging from recent modeling of subaqueous debris flows [14,25], such modifications are expected to have a strong influence on the behavior of the flow. However, unless rheological measurements with natural snow, such as those planned in the framework of the EU research project "Avalanche Studies and Model Validation in Europe" (SATSIE), yield clear indications on the shear rate and load dependence of the yield strength and BINGHAM viscosity, this approach would just introduce additional ad hoc parameters.

Granular models. In the quasi-static regime, i. e., at low shear rates and high volumetric particle concentrations, interparticle friction is the dominant mode of momentum transfer in granular materials, and the Coulomb friction law or yield criterion, $\tau = \sigma_n \tan\phi$, is generally accepted to hold to good precision. Here, τ and σ_n are the shear and normal stress at yield, and ϕ is the (static) internal friction angle. If the granular material is more and more rapidly sheared, short-duration collisions between particles account for an increasing fraction of the normal and shear stresses. However, as discussed by SAVAGE and HUTTER [99], experiments show that the stresses grow as the square of the shear rate—the behavior found experimentally by BAGNOLD [2]—only if the particle concentration is held constant. If instead the externally applied normal stress is held constant, the dispersive stresses force the material to dilate, the concentration drops and the stresses remain nearly constant. This is the configuration encountered in dense gravity mass flows. Moreover, the ratio of shear and normal stresses also remains essentially constant and the dynamic internal friction angle is a little smaller but close to the static friction angle. An impressive array of experiments [29,28,49,50,62] have confirmed that the SAVAGE-HUTTER model [99,100] based on this concept predicts the details of the motion of dry granular masses in the laboratory with high precision.

Almost all models used in avalanche hazard mapping today employ a COULOMBian dry-friction term as the dominant dissipation mechanism, even though they were largely inspired by the hydraulics of open-channel flow. Such a friction term is proportional to the effective pressure transmitted by the contacts between grains; as long as dispersive pressures from short-duration grain-grain collisions are neglected, it is therefore proportional to the overburden load and thus to the flow depth. Gravity as the driving force is also proportional to the flow depth. A model with only this dry-friction term would predict runout distances essentially independent[4] of the flow depth (avalanche size), in clear contradiction to observations[5]. By adding a drag term proportional to the square of the velocity and independent of the flow depth or inversely proportional to it, runout distances growing with the flow depth are recovered: In their simplest formulation for a point mass, this class of models obeys the equation

$$\frac{du}{dt} = g(\sin\phi(x) - \mu\cos\phi(x)) - ku^2 \tag{4}$$

with $k = k_0 \cdot (h/h_{\text{ref.}})^{-\alpha}$ and $k_0 = \mathcal{O}(10^{-2})$ for $h_{\text{ref.}} = 1\,\text{m}$. The exponent α is 0 in the PERLA–CHENG–MCCLUNG (PCM) and PERLA–LIED–KRISTENSEN (PLK) models [87,88], 1 in the VOELLMY–SALM (VS) model [96,98,114], and 3 in the NOREM–IRGENS–SCHIELDROP (NIS) model [85,86]. Transforming (4) into a differential equation for the spatial evolution of the specific kinetic energy $u^2/2$, it can readily be solved for the idealized situation of an inclined plane of slope angle ϕ and drop height H, followed by a horizontal plane. The runout distance, R, on the horizontal plane is

$$R = \frac{1}{2k}\ln\left[1 + \left(\frac{\sin\phi}{\mu} - \cos\phi\right)\left(1 - e^{-2kH/\sin\phi}\right)\right]. \tag{5}$$

[4] The flow depth influences the spreading of the mass and thus the location of the distal end of the deposit somewhat.

[5] Of course, the basal friction angle could be prescribed ad hoc to diminish with increasing avalanche size, but this would just highlight the fact that the model does not capture the decisive processes.

In the limit $k \to 0$, the result $R = H \cdot (1/\mu - \cot\phi)$ for a pure COULOMB friction model is recovered. For $k > 0$ and typical avalanche paths, $\exp(-2kH/\sin\phi) \ll 1$, so one obtains

$$R \propto \frac{1}{k_0} \left(\frac{h}{h_{\text{ref.}}} \right)^{\alpha} , \tag{6}$$

thus no flow height dependence in the PCM-type model, linear increase with flow height in the VS-type model, and cubic growth of R with h in the NIS-type model. (In practical applications, the differences in flow-height dependence of the drag term are less important than one might think in the light of the discussion above: The drag coefficient is varied according to the size of the avalanche in the PCM model, thus recovering much of the h-dependence in the other models.)

Statistical analyses of runout distances of a large set of extreme avalanche events in Norway [64,66] show that the runout angle α (inclination of the line from the fracture crown to the toe of the deposit) grows with the steepness of the avalanche track. This observation may be interpreted as indirect evidence of a velocity-dependent drag term: Avalanches in steep terrain reach higher velocities and dissipate more energy per distance traveled than avalanches on gentle slopes. However, as noted in [39, p. 12], avalanches released from steep starting zones tend to have significantly smaller flow heights than avalanches from gentle starting zones. As shown above, the observed size dependence is a strong indication that velocity-dependent drag forces are present in avalanche motion.

In the author's view, the performance of "pure" granular models for snow avalanches can be assessed as follows: (i) At least at the qualitative level, granular flows represent a suitable framework for describing the wide variety of flow regimes that snow avalanches exhibit, from plug flow to partially fluidized flow, saltation and suspension flow. (ii) The observed long runout distances require friction angles well below those measured in dry granular materials. (The same discrepancy is also found in giant rock avalanches and related phenomena.) (iii) The observed strong dependence of the runout distance on the mass and the track slope indicates the presence of an additional, velocity-dependent force that is not proportional to the avalanche mass. It also appears to contradict the shear-rate independence of the ratio of shear and normal stresses found in rapidly sheared dry granular materials under constant normal load.

The difficulties summarized above indicate that snow differs from dry and hard granular materials in certain respects that need to be accounted for in dynamical models of snow avalanches:

1. Snow has *cohesion* that depends strongly on temperature, density and water content. The steep deposit edges and the absence of lateral spreading suggest that cohesion is important in wet-snow avalanches. In dry-snow

avalanches, cohesion is required to explain the appearance of a plug layer within the dense core if the slope angle is larger than the angle of repose.

2. *Comminution and coalescence:* Collisions between snowballs are probably quite inelastic while collisions between snow grains have a much higher restitution coefficient. Large particles will often break up in hard collisions; at low relative velocities, they may stick together instead. This implies that the granulometry of a dry-snow avalanche is constantly changing in the shear layer. Savage and Hutter [99] state that values above approximately 0.1 of the dimensionless number, proportional to the ratio of dispersive pressure and normal load σ_n,

$$R_s = \frac{\rho_p d^2 (U/h_s)^2}{\sigma_n}\,, \tag{7}$$

indicate a granular flow in the inertial regime. ρ_p is the particle density, d the particle diameter, h_s the thickness of the shear layer and U the velocity difference across it. The quadratic dependence of R_s on the particle size implies that comminution may cause an avalanche to revert from the inertial regime to the frictional regime. For the velocity profile from the northern Fogas path [36], one finds that the inertial regime should have prevailed as long as most particles were at least 1 cm in diameter.

3. *Excessive pore pressure:* As will be discussed in more detail in Sect. 5, rapid partial collapse of the snowcover under the load of the avalanche flowing over it may generate a considerable air overpressure in the pores underneath the avalanche and at the interface between snowcover and avalanche. This effect may significantly reduce the effective pressure (the normal stress transmitted by grain contacts) and thus the apparent friction force on the avalanche (see (8c) and (8d) below). It depends sensitively on the snowcover properties, avalanche velocity and density, and on the location within the avalanche. Reducing the dry-friction coefficient in conventional models may capture the effect of fluidization through pore pressure with regard to the runout distance, but velocities in the track and the final deceleration will be underestimated; moreover, a significant fraction of the wide variation of the apparent friction coefficient from one avalanche event to another may be due to pore pressure variations.
Note that it is *not* suggested that excessive pore pressure is the sole or dominant mechanism reducing friction in large avalanches—extremely low effective friction has also been inferred for giant rock avalanches on Mars and the Moon where interstitial-fluid effects are completely negligible. However, it is a mechanism that may be non-negligible in snow avalanches and deserves further study.

4. *Lubrication through melting:* During the descent of a sizeable avalanche, on the order of 5 kJ kg^{-1} of potential energy per unit mass is converted into heat absorbed by the avalanching snow. As most of the dissipation occurs in a third or less of the total mass, this would correspond to a uniform temperature rise of some 7°C in the shear layer. Considering

the low heat conductivity of snow particles and the short duration of the flow, some melting at the surface of the snow particles will likely occur in most cases. This will reduce interparticle friction in collisions and sliding contacts as long as the shear rates are high, but increase the internal friction angle significantly if shearing is sufficiently slow to allow some sintering of particles. This mechanism has been considered for dense-flow avalanches by HUTTER [47], and petrographic evidence for its occurrence in large rockfalls has been cited by ERISMANN [22,23].

Considerable work will be needed to quantify the effects mentioned above for typical conditions occurring in dry-snow avalanches. Tests with a large field rheometer, planned by CEMAGREF Grenoble (M. NAAIM, pers. comm., 2001) as part of the EU project SATSIE, will hopefully soon determine whether the ratio of shear and normal stresses in snow under constant normal stress is more or less independent of the shear rate as it is in dry granular materials. Excessive pore pressures are a potentially very powerful mechanism for dramatically reducing friction in fast dry-snow avalanches.

A variety of rheological models that are capable of describing the intricate behavior of granular materials and avalanching snow are discussed by HUTTER and RAJAGOPAL [51]. For illustrational purposes, the NIS model [85,86] is selected here because it has been routinely and successfully been used in avalanche consulting in Norway and elsewhere. Its rheological basis is a modified and enhanced version of the CRIMINALE–ERICKSEN–FILBEY (CEF) fluid [13][6]. In the constitutive equations (written for the special case of flow in the x-direction and shear only in the x-z-plane for the sake of simplicity)

$$\sigma_x = -p_e - p_u - \rho(\nu_2 - \nu_1)\dot{\gamma}^r \,, \tag{8a}$$

$$\sigma_y = -p_e - p_u \,, \tag{8b}$$

$$\sigma_z = -p_e - p_u - \rho\nu_2\dot{\gamma}^r \,, \tag{8c}$$

$$\tau_{xz} = \tau_c + p_e \tan\phi + \rho m\dot{\gamma}^r \,, \tag{8d}$$

$$\tau_{xy} = \tau_{yz} = 0 \,, \tag{8e}$$

$\dot{\gamma} = \partial_z u_x(x,z,t)$, p_e is the effective pressure (transmitted by the grains), p_u is the pore pressure[7], τ_c is the cohesion, ν_1, ν_2 are normal-stress viscosities, and m is a shear-stress viscosity. Finally, $r = 1$ describes a linear viscous material (presumably more appropriate for wet-snow avalanches and debris flows) while $r = 2$ will be chosen for a granular material.

[6] In the actual implementation of the NIS model, the cohesion and pore pressure are set to zero and the velocity profile $u_x(z)$ is assumed to be that of a steady-state solution. As shown in [65], this makes the model very similar to one-dimensional depth-integrated versions of the VOELLMY-SALM and PCM models; the main difference is that the drag term is proportional to the inverse square of the flow height and centrifugal forces in bends are taken into account.

[7] Tensile stresses are positive in the convention used here.

Equations (8c) and (8d) show that the fraction of the overburden weight supported by a non-vanishing pore pressure does not contribute to the shear stresses (the shear stress in the air is neglected). Moreover, if there is no excessive pore pressure ($p_u = 0$), substitution of (8c) in (8d) yields

$$\begin{aligned} \tau_{xz} &= \max(0, |\sigma_z| - \rho\nu_2\dot{\gamma}^r) + \rho m\dot{\gamma}^r \\ &= \begin{cases} |\sigma_z|\tan\phi + \rho(m - \nu_2\tan\phi)\dot{\gamma}^r & \text{if } \dot{\gamma}^r \leq \frac{|\sigma_z|}{\rho\nu_2}\,, \\ \rho m\dot{\gamma}^r & \text{if } \dot{\gamma}^r > \frac{|\sigma_z|}{\rho\nu_2}\,. \end{cases} \end{aligned} \tag{9}$$

At $\dot{\gamma} \geq \left(\frac{|\sigma_z|}{\rho\nu_2}\right)^{1/r}$, the total weight of the overburden snow is supported by the dispersive pressure, and the growth rate of the total shear stress with the shear strain rate changes abruptly. If $m/\nu_2 < \tan\phi$, the effective friction first diminishes with increasing shear rate from $|\sigma_z|\tan\phi$ to a minimum value of $|\sigma_z| m/\nu_2$ to m/ν_2. This allows a high friction coefficient at low velocities and significant fluidization effects at high velocities. Note, however, that NOREM, IRGENS and SCHIELDROP [85,86] suggest values for which $m/\nu_2 > \tan\phi$, on the basis of experiments on dry granular materials [101]; it is thus of great importance to carry out definitive rheological tests on snow at high shear rates. Finally, with $r = 1$ and $\phi = 0$, one recovers the BINGHAM rheology. HUTTER [46] has investigated ways of constructing an avalanche model based on constitutive equations similar to those of the CEF fluid, but not encumbered by the drastic simplifications made in the present implementation of the NIS model. The resulting equations are too complex for a practically usable model, so suitable approximations retaining the essential features of the system with a variable-depth plug layer will have to be sought.

4.3 The Saltation Layer

Which forces determine the dynamics of the saltation layer? At densities in the range 10–100 kg m^{-3}, inferred from the analysis of impact signals [102] and the deposits of observed powder snow avalanches (see Sect. 5.1) and corresponding to particle volume concentrations[8] c of 1–15 %, neither are particles in continuous contact nor are collisions negligible. Basically, the saltation layer is in the inertial regime as identified by BAGNOLD [2]: The BAGNOLD number, describing the ratio of shear stresses from grain collisions to viscous shear stresses in the fluid phase, can be estimated as

$$\mathrm{Ba} = \frac{c\rho_s d^2\dot{\gamma}^2}{(1-c)\rho_a\nu_a\dot{\gamma}} \approx (3\text{–}50)\cdot 10^3\,, \tag{10}$$

[8] The density of very small snow particles is that of ice (917 kg m^{-3}), but bigger snow blocks have densities of $\rho_s = 400$–600 kg m^{-3}. While collisions of small snow grains are fairly elastic, collisions of large blocks are expected to be very inelastic due to compression, break-up and coalescence.

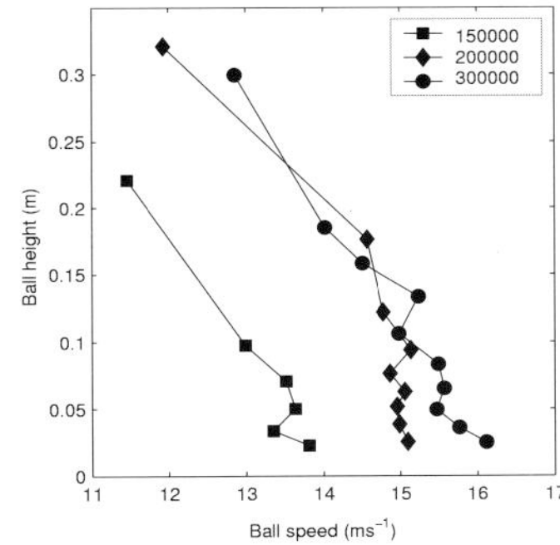

Fig. 9. Ping-pong ball avalanches at the Miyanamori ski jump. (**Left**) Side view of a flow with 550 000 balls. Note the dilute "saltation layer" above and behind the head. (**Right**) Vertical profiles of downslope velocity measured in avalanches of 150 000, 200 000 and 300 000 balls. From [72] with permission, courtesy J. McElwaine.

with ρ_s and ρ_a the snow-particle and air densities respectively, $\nu_a \approx 2 \cdot 10^{-5}\,\mathrm{m^2\,s^{-1}}$ the air viscosity[9], and $\dot{\gamma} = 10$–$40\,\mathrm{s^{-1}}$ the shear rate. Bagnold numbers above 200 indicate strong dominance of the inertial forces. Turbulent shear stresses in the interstitial air are an order of magnitude smaller than the inertial shear stresses from particle collisions. The capacity of the interstitial air to buffer snow-particle interactions is characterized by the Darcy number proposed by Iverson [57],

$$\mathrm{Da} = \frac{\rho_a \nu_a}{c \rho_s \dot{\gamma} k} \approx (2\text{–}20) \cdot 10^{-8}\,\mathrm{m^2} \cdot \frac{1}{k}\,, \tag{11}$$

where k is the permeability of the saltation layer and is probably much larger than $10^{-8}\,\mathrm{m^2}$, a typical value for a snowpack with higher density and much smaller pores. According to Iverson, pore pressure effects were strongly felt at $\mathrm{Da} \geq 1000$, but they can be safely neglected in the saltation layer with $\mathrm{Da} < 1$. (The situation may be significantly different if an approaching avalanche produces large overpressures in the snowpack, see the discussion in Sect. 5.)

The estimates presented above characterize the saltation layer as a granular flow in the inertial regime, assigning a rather limited role to the interstitial air. Yet, the air must play a decisive role in (at least) the formation of the layer if the dense flow usually consists of a thin shear zone at the bottom and a plug flow above it with limited fluctuation velocity (low granular temperature); this question will be taken up in Sect. 5. However, even the well-known ability of avalanches to maintain the saltation layer over a substantial distance

[9] Significantly lower Bagnold numbers are found, however, if the much denser and more viscous mixture of air and fine snow grains is considered as the fluid interacting with the larger snow particles.

beyond the stopping point of the dense flow and on to opposing slopes challenges the conclusions reached from dimensional analysis: Experimentally, it has been found rather difficult to produce fully fluidized flows on slopes below 45° with granular materials comparable to dry snow (see also the theoretical and numerical study by HUTTER et al. [52] which arrived at similar conclusions). Images and measurements from ping-pong ball "avalanches" (Fig. 9) studied by a Japanese group [72,80] show that a layer with similar properties as the saltation layer in snow avalanches is produced in those flows. While the ping-pong balls have a density of only 80–90 kg m^{-3} and a significantly higher restitution coefficient than snow balls, typical velocities in a large dry-snow avalanche are three to four times higher. This implies that the aerodynamic lift forces and the energy per unit mass transmitted in particle collisions are about an order of magnitude larger in the snow avalanches.

The energy dissipation in snow particle collisions suggests a mechanism regulating the density in the saltation layer. Although a consistent mathematical formulation is missing as yet, the main aspects may be described as follows:

- Particle collisions within the saltation layer are partly responsible for preventing collapse of the layer, despite their substantial inelasticity. At low densities approaching those in the saltation layer of snowdrift, rebound from the snowcover surface or ejection of particles upon impact may also contribute to maintain the layer height. (This latter mechanism is invoked in the saltation layer model of powder snow avalanches presented in [53,55].)
- Grazing collisions with the surface of the dense flow or the snowcover are crucial for balancing the particle acceleration induced by the slope-normal component of gravity. Again, the collisions are quite inelastic. The plastic and elastic properties of the snowcover or dense-flow layer lead to steeper ejection than impact angles, as observed in snowdrift [61].
- The energy dissipated in the collisions is supplied by the downslope gravitational force on the jumping particles. If the density is above an equilibrium value, less kinetic energy is gained by the particles between two collisions than they dissipate during the collision, i. e., the granular temperature drops. As a consequence, the layer height diminishes. At the same time, net settling of particles from the saltation layer will reduce the density and increase the mean free path of the snow particles, counteracting the subsiding tendency.
- The maximum sustainable density in the saltation layer is expected to grow with the slope angle and to decrease with increasing particle size, mainly because the restitution coefficient of the snow particles and the bed (dense-flow layer or snowcover) diminishes.

So far, the only information available on the formation of the saltation layer isfrom video tapes, which suggest that ploughing at the front (cf. Sect. 5.2) may be a major contributing factor.

4.4 Suspension Layer

An important amount of experimental information on laboratory turbidity currents is now available (see the classic monograph by Simpson [109], the papers [6,7,41,45,59,104,105] and references therein) and can—with the requisite caution—be transferred to the suspension layer of snow avalanches. The processes inside the suspension layer are in many respects simpler than in the denser saltation and dense-flow layers. Due to the low volumetric concentration (generally less than 1 %, typically around 0.1 % in the runout zone) and the moderate to low Stokes number (the ratio of the particle relaxation time and the characteristic time scale of the flow) of the particles, particle collisions are expected to be negligible. For practical purposes in hazard mapping, the particle-size distribution found in real powder snow avalanches may be replaced by a uniform particle size with the same mean settling velocity [60]. The relative motion of snow particles and air can be modeled as the superposition of this uniform settling velocity and the turbulent diffusion of the particles. As explained in the Introduction, this work will not be reviewed here; only the aspect most directly related to the flow regime of the suspension layer will be discussed, namely turbulence.

Turbulence is of paramount importance in powder snow avalanches as it brings particles from the saltation layer or the snow cover into suspension, maintains the particle suspension and governs air entrainment, the main decelerating process in these flows. The balance of turbulent energy plays a similar role in the suspension layer for closing the field equations as the constitutive relations do in the dense-flow layer. Reasonably accurate turbulence modeling, including the effect of the particles on the turbulence, is therefore important. Direct turbulence measurements at several heights inside full-size powder snow avalanches would be very useful, but require significant effort for sensor development and installation and may not be realized for some time to come. Detailed turbulence measurements in laboratory turbidity currents with a sufficiently high particle load are feasible with modest effort [3,8] and yield valuable information for model developers.

Turbulence production in suspension gravity mass flows occurs at the largest scales, mainly in the zones of intense shear at the bottom and upper flow surfaces. The corresponding structures have sizes of several meters to tens of meters. With regard to these structures, the particles have a very low Stokes number, implying that they follow the interstitial fluid almost perfectly. At the scales of turbulence production, the snow–air mixture thus behaves as a simple fluid with variable density. For example, in the popular k-ϵ turbulence model (see [93] for an overview and references), the shear production term P in the transport equation for the turbulent energy density takes the form

$$P = \tau_{ij}(\partial_i u_j + \partial_j u_i)/2 = \rho\nu_{\text{eff.}}(\partial_i u_j + \partial_j u_i)\partial_i u_j \ , \tag{12}$$

where $\nu_{\text{eff.}}$ is the effective kinematic viscosity combining the (small) molecular viscosity and the turbulent viscosity. (A basically similar, but more elaborate term is obtained in models dealing directly with the transport equations for the components of the stress tensor.) ρ is the *mixture* density, which may be an order of magnitude larger than the density of air in the densest regions of the particle cloud.

As the snow particles fall downward under the action of gravity and a concentration gradient arises, large turbulent eddies transport more particles upward than downward, thereby converting turbulent energy (at large scales) into potential energy. The turbulent energy balance will therefore contain a gravitational term:

$$G = (\rho_p - \rho_f) g_i j_i^{(c)} = (\rho_p - \rho_f) g_i \frac{\nu_{\text{eff.}}}{\sigma_c} \partial_i c \,. \tag{13}$$

ρ_p and ρ_f are the particle and fluid densities, respectively; g_i is the vector of gravitational acceleration, $j_i^{(c)}$ is the turbulent diffusive particle flux in terms of volumetric concentration, c, and $\sigma_c \approx 1.3$ is the turbulent Schmidt number characterizing the efficiency of turbulent particle diffusion relative to momentum diffusion.

At the same time, the falling particles produce turbulence at very small scales by forming wakes, at the rate $(\rho_p - \rho_f) g_i w_i^{(s)}$, where $w_i^{(s)}$ is the mean settling velocity of the particles. That turbulent energy is, however, soon dissipated at the molecular level. In particular, it does not contribute to turbulent particle diffusion and to the effective viscosity. From this consideration, the dominant effects of the particles on the turbulent energy balance are to enhance shear production in proportion to their mass fraction, to extract energy at intermediate and large scales in proportion to the work associated with the turbulent particle diffusion against the gravitational field, and to feed energy into the sub-millimeter scales at a rate proportional to their buoyant mass and settling velocity, effectively enhancing molecular dissipation.

5 Snow Entrainment

For lack of space and scarcity of confirmed results, the discussion of snow entrainment will be kept shorter than this fascinating and important topic deserves in the author's opinion. However, prospects for significant progress are good as several projects on snow entrainment are currently under way.

5.1 Observations and Measurements

Global and local mass balance from surveys. Surveys of spontaneous avalanches have for a long time indicated that the deposit mass may exceed the release mass of an avalanche by a significant factor, but precise numbers were not available. Of the events described in [56], the small Seewis avalanche and the medium-to-large Albristhorn avalanche were estimated to have doubled their mass from release to runout whereas the mass of the extremely large Scex Rouge avalanche probably increased by a little less than a factor of two.

Photogrammetric surveys of the 1999 avalanches at the Vallée de la Sionne test site [113] showed that the first and smallest event may not have had a large net erosion, probably due to a larger spontaneous avalanche that preceded it and eroded and compacted most of the fresh snow. The second event was significantly larger and quadrupled its mass from release to runout. Over large parts of the path, it eroded the entire fresh snow. Even on the opposite slope, 50 m (vertical) above the valley floor, the powder snow avalanche eroded 2 m of snow (and deposited similar masses in irregular patterns). The giant last event again eroded almost all the fresh snow, thereby tripling its initial mass. It even remobilized some of the highly compacted deposits of the preceding avalanches in the gently sloped runout zone.

Sovilla and collaborators made detailed mass balance measurements along the Monte Pizzac avalanche path in the Venetian Alps in the winters 1997/98 to 1998/1999 [110]. They found that all four surveyed avalanches, although differing greatly in initial mass and water content as well as speed and runout distance, eroded[10] 50–200 $\mathrm{kg\,m^{-2}}$ in the steep upper part of the path with slopes generally 35° or more. In all cases, erosion dropped sharply where the slope becomes more gentle, but continued at a sustained rate of 5–75 $\mathrm{kg\,m^{-2}}$, and with higher values in short segments of the gully with slope angles up to 45°. All avalanches attained their maximum moving mass—about twice the initial mass for the smallest and seven times for the largest avalanche—near the middle of the path where the slope angle drops below 35°. Except for the smallest event, which met a rough bed surface in early winter and lost mass from the beginning, appreciable deposition began also near this point. The erosion per unit area correlates with the initial mass (which is also a measure for the amount of fresh, erodible snow in the path). For each avalanche, there is a positive correlation between the speed and the erosion per unit area at different locations. However, the largest two events—one dry, the other wet—with quite similar initial masses also showed similar erosion per unit area in the upper half of the path, even though the wet-snow avalanche was two to three times slower than the dry-snow flow.

Measurements of local erosion rates. The FMCW radar systems initially installed in the Fogas avalanche path and later at three locations (upper and lower track, upper runout zone) at Vallée de la Sionne, provide information on the time evolution of the erosion rate and the flow depth, see Fig. 3. Unfortunately, no comprehensive account has been published to date of these

[10] The values given in this paragraph indicate the *total erosion* (in $\mathrm{kg\,m^{-2}}$) at a fixed location, i. e., the total mass of snow per unit area (parallel to the slope) that was eroded during the entire passage of the avalanche over that location. Below, the variation of the *erosion rate* (in $\mathrm{kg\,m^{-2}\,s^{-1}}$) over time will be discussed; by this term, one understands the mass of snow eroded per unit area and unit time.

measurements, so inferences have to be drawn on the basis of a preliminary report [20]. Also it has not been possible so far to obtain velocity profiles from cross-correlations of the signals from the radar pairs at Vallée de la Sionne, limiting the possibilities for dynamical analysis.

The three artificially released dry-snow avalanches in the winter 1999 produced fairly similar patterns in the profiling radar plots (Fig. 3):

- When the front arrives, the uppermost layer of the snowcover is removed within much less than 1 s. This effect will be called "ploughing". There are uncertainties associated with the detection of the surface of the undisturbed snowcover, but typical ploughing depths appear to be 0.8–1.0 m for all avalanches and all locations with valid measurements. The exception is the first (and smallest) of these avalanches at the middle radar location (at the end of a gully) where probably 2 m of snow were ploughed away. Given the front speeds in the range 40–70 $\mathrm{m\,s^{-1}}$, 8–14 $\mathrm{t\,m^{-1}\,s^{-1}}$ were incorporated into these avalanches per unit width[11].
- A phase lasting 1–10 s follows in which snow is eroded much more gradually. The erosion rate may be roughly constant or diminish with time, as in Fig. 3 from the uppermost radar location (upper track). In this case, the estimated erosion rate drops from 75 $\mathrm{kg\,m^{-2}\,s^{-1}}$ (corresponding to a lowering of the interface between avalanche and snowcover at 0.3 $\mathrm{m\,s^{-1}}$) or to about one third of this rate. Some 600–700 m down the path, at the end of a gully, the same avalanche eroded about 1.3 m in only 1.3 s, or 250 $\mathrm{kg\,m^{-2}\,s^{-1}}$. The second, somewhat smaller avalanche had erosion rates of 50–100 $\mathrm{kg\,m^{-2}\,s^{-1}}$ over 5 s at the middle radar location and 50–70 $\mathrm{kg\,m^{-2}\,s^{-1}}$ over 4 s another 600–700 m downslope in the upper runout zone.
- After this erosive phase, the avalanche body flows on a well-defined glide plane without noticeably eroding or depositing. This phase lasted from 30 s to over one minute in the two largest events.
- Formation of the deposits occurred quite abruptly in all cases, but the deceleration before the final stop is not known without velocity profiles. The deposit heights are small in the single measurement at the uppermost location (Fig. 3) and between 1 and 2 m at the lower two locations.

Interestingly, many of the radar plots suggest that all or most of the entrainment occurred *before* the densest parts of the avalanches arrived at the location. This interpretation is supported by some of the impact-pressure data (cf. Fig. 5), but it is clearly too early to ascribe general validity to such an interpretation, and it certainly does not apply to wet-snow or humid avalanches. Measurements on a wet snow avalanche running out immediately below the lowest radar location showed a pronounced ploughing effect (ploughing depth probably around 2 m) at a front velocity of only 4 $\mathrm{m\,s^{-1}}$, without noticeable gradual erosion afterwards.

[11] This quantity is the instantaneous erosion rate *integrated* over the length of the avalanche.

Mass transfer to the powder snow avalanche. For the 1995 avalanches described in [56], separate estimates of the masses deposited by the dense flow and the powder snow avalanche (saltation and suspension layer combined) were made, albeit with very large uncertainty due to the limited resources for the surveys). Using the mean values so obtained, 25–35 % of the total deposits, corresponding to about 50 % of the initial mass, are from the powder snow part. The corresponding figures for the much larger Albristhorn avalanche are 10–15 % and about 25 %, respectively. The lower mass fraction of the powder snow part in the Albristhorn avalanche is likely a consequence of the higher strength of the snow cover in the track and runout zone at release time. At Scex Rouge, steep and rocky cliffs may have favored snow suspension despite elevated temperatures and moist snow below the release area: The powder snow deposits accounted for 25–35 % of the total deposits and 40–50 % of the initial mass.

One may also obtain a crude estimate of the rate at which snow was transferred to the saltation and suspension layer from the dense core or the snow cover. For the small Seewis avalanche, the mass of the powder snow avalanche deposit was estimated at 1000–2000 t, some of which was entrained and redeposited in the runout zone after separation from the dense core. We will thus assume a mass of 1000 t at the separation point about 800 m from the lower end of the release zone. The area swept over by the avalanche has an average width of 150 m, so the suspended mass per unit width was about $7\,\mathrm{t\,m^{-1}}$. Dividing this by the path segment length of 800 m, one obtains an average net suspension of $8\,\mathrm{kg\,m^{-2}}$. Assuming a mean avalanche length of 250 m and a flow time of 30 s from release to the gully bend, the mean entrainment rate is about $1\,\mathrm{kg\,m^{-2}\,s^{-1}}$. However, as will be discussed later, it is very likely that suspension occurs mainly in the front region of the avalanche, as in subaqueous debris flows in the laboratory [38], and that the suspension rate is an order of magnitude larger there. The corresponding numbers for the much larger Albristhorn avalanche are a net suspension of $40\,\mathrm{kg\,m^{-2}}$ and a mean suspension rate of $2\text{–}5\,\mathrm{kg\,m^{-2}\,s^{-1}}$. Similar numbers are obtained for the giant Scex Rouge avalanche. For saltation layer heights in the range 1–5 m as suggested by various pressure measurements, these estimates point towards the same density range of $10\text{–}100\,\mathrm{kg\,m^{-3}}$ for the saltation layer as was obtained from the impact analysis in [102].

5.2 Conjectured Mechanisms

In [55], four conceivable entrainment mechanisms in dense-flow avalanches were briefly sketched. In the case of "ploughing", the dense core slides on a horizon inside the snowcover (typically the upper surface of the old snow) and shears the fresh snow off at the very front of the avalanche. The entrained snow may be mixed into the avalanche body or pushed in front of the avalanche. An alternative entrainment mechanism at the front was termed "gobbling" and invokes the conveyor-belt effect of the shear layer, by which the plug layer (consisting mainly of large snow blocks) is transported to the avalanche front and dumped onto the snow cover. The latter is thereby broken up and mixed into the avalanche. The "chipping" mechanism designates a more continuous erosion of single snow grains or small snow particles along the bottom of the avalanche, similar to the most common erosion mechanism in a river flowing over a cohesive bed. In rivers or circulating-flume experiments, intermittent detachment of entire chunks of bed material may be observed at high shear stresses and under suitable conditions; the corresponding mechanism in dense-flow avalanches has been termed "ripping" in [55]. The radar data lend support to the conjecture that commonly ploughing at the avalanche front as well as chipping at the bottom of the flow occur. At Vallée de la Sionne, both mechanisms contributed similarly to the total entrainment. It should be noted, however, that the radar data alone cannot discriminate between ploughing and gobbling, or between chipping and ripping.

Considerations concerning the ploughing mechanism. Where is the ploughed snow mixed into the avalanche? Given the general direction of motion of the avalanche, the simplest conjecture would be that it is compressed and piled up in front of the avalanche and so is added directly at the front. This can be treated in analogy to the shock tube problem in aerodynamics. Since the flow depth of the dense core does not grow very much during the event, mass continuity shows at once that the front would have to move at a significantly higher speed than the avalanching snow itself. However, this consequence is clearly contradicted by the measurements of the front and internal velocities. One has to conclude that at least a substantial portion of the ploughed snow is transported towards the body of the dense flow or quickly suspended (the suspension layer grows in density and height during the flow).

Video footage of the early stage of dry-snow avalanches (before everything is shrouded in the bulging powder snow cloud) is indeed strongly indicative of a strong upwelling flow at the front that is quickly deflected towards the body of the avalanche by the air displaced by the front and flows over the avalanche. Small snow grains may become suspended in the turbulent air flow, but larger snow blocks will fall back onto the body of the avalanche. (This mechanism has been observed in subaqueous laboratory debris flows, except that the particles did not come from ploughing of the bed but were ripped off the front by the large hydrodynamic forces.) It is also likely that a fraction of the ploughed snow is overflowed by the head of the avalanche and mixed into the shear layer.

The abrupt transition from rest to motion visible in the FMCW radar data shows that the rupture through the depth of the ploughed layer takes place in 0.1 s or less. Within this fracture time, the front moves at most 5–7 m, perhaps even significantly less. At the measured ploughing depths, the angle ψ between the rupture front and the slope must be at least 8°, but it may well be larger. A naïve static analysis neglecting the acceleration needed to move the snow out of the way of the approaching avalanche would suggest an inclination angle closer to $45° - \phi/2 \approx 30°$. A detailed analysis of the stresses and flow patterns near the avalanche front and/or video observations at chutes appear to be required to clarify this issue.

Does excess pore pressure play an important role? The question whether non-negligible excess pore pressure may develop in dry-snow avalanches naturally ties into the discussion of the ploughing mechanism. In areas with continental and transitional climate and at altitudes where avalanches are frequent, new snow during and shortly after a snowstorm has a volumentric concentration of 10–25 %. A medium to large avalanche exerts a normal stress between 2 and 10 kPa on the snowcover. Under the sudden onset of such a considerable load, accompanied by shear stresses, the texture of the new-snow layer is likely to collapse. Consequently, excess pore pressure p_u somewhat smaller yet similar in magnitude to the load will develop, reducing the effective pressure transmitted between the grains of the snow cover and the avalanche. It is thus conceivable that at least the front of the avalanche might be "aeroplaning" much in the way hydroplaning has been observed in subaqueous debris flows in the laboratory [40,73,74].

One may expect that the permeability is considerably larger in the fresh snow with its low density than in the somewhat compacted avalanching snow. Accordingly, the excess pore pressure initiates an air flow through the snow cover immediately in front of the avalanche. The compressed air in the snow has to be accelerated to a velocity u_{escape} that is somewhat larger than the avalanche front velocity, inducing a pressure drop $\Delta p = (1/2)\rho u_{\text{escape}}^2 \approx 0.5$–1 kPa. If the tensile strength, s_t, of the snow is smaller than the remainder, $p_u - \Delta p$, the snowcover should be disrupted to a depth proportional to $p_u - \Delta p - s_t$.

It appears thus that the air in the snowcover may play a crucial role both for reducing the friction forces in the head of the avalanche and creating the saltation and suspension layers. Again, a much more detailed analysis will be needed to verify that the qualitative considerations given above are correct and to achieve a mathematical description that can be incorporated in future models.

6 Outlook: Consequences for Modeling and Future Experiments

6.1 Answers and Open Questions

What do we know about dry-snow avalanches after six decades of experimental studies? Combining observations made on avalanche deposits, measurements at full scale and laboratory experiments, the following statements can be made with a high degree of certainty:

- In general, three layers can be distinguished: A dense-flow layer at the bottom with density on the order of $300\,\mathrm{kg\,m^{-3}}$, the saltation layer (typical height and density are a few meters and about $30\,\mathrm{kg\,m^{-3}}$, respectively) above or ahead of the dense layer, and the suspension layer (density below $10\,\mathrm{kg\,m^{-3}}$, height up to 200 m or more [56] and growing with distance) above and behind of the saltation layer. The latter two layers are formed during the flow.
- The few available velocity profiles through the depth of the dense layer show a plug-like region riding more or less passively on a thin, highly sheared zone at the bottom. It is not known whether a fully fluidized state can occur under appropriate conditions, nor have velocity profiles been measured in the front and in the saltation layer.
- Runout distances and velocities increase markedly with avalanche size, but there is not enough precise data to check if Froude scaling is respected. The highest measured front velocities approached $80\,\mathrm{m\,s^{-1}}$, internal velocities up to $100\,\mathrm{m\,s^{-1}}$ have been measured.
- The saltation layer contains a wide range of particle sizes from fine-grained snow to snowballs up to approximately 50 cm in diameter. It may move faster than the dense layer and has the tendency to move on a straight line at bends. In the runout zone, the largest particles are deposited first.
- Substantial snow entrainment is the rule rather than the exception at least in medium to large avalanches. The moving mass may thereby increase by almost an order of magnitude. At a given location or instant, erosion at the front may coexist with deposition at the tail. Erosion of part or all of the new snow appears to be common, but erosion to the ground is also observed. The eroded snow is thoroughly mixed into the dry-snow avalanche and transported over long distances.
- The mechanisms of snow entrainment are not fully understood yet, but all large dry-snow avalanches investigated at Vallée de la Sionne showed "ploughing", i. e., an almost instantaneous ($\ll 1\,\mathrm{s}$) erosion of part of the snowcover at the front. In most of those avalanches, more gradual erosion occurred after passage of the front, probably the saltation layer.
- Large powder snow avalanches may erode very strongly even beyond the stopping point of the dense flow and on opposing slopes. No measurements of the erosion rate exist for such flows, however.

Despite the important insights into the dynamic structure gained at several well equipped sites over the past twenty-five years, a number of key questions remain unanswered today:

- Does a fully fluidized flow regime of the dense-flow layer exist, either in very steep segments of the avalanche path or at the front of the avalanche from where the saltation and suspension layers are mainly fed?
- How do the velocity profiles change from one location to another? Do the measured profiles correspond to quasi-steady flow conditions?

- Can an adequate description of the rheology of dry-snow avalanches be achieved with either visco-plastic or granular models?
- If the granular behavior is confirmed, which properties of dry granular materials apply to dry-snow avalanches as well? In particular, is the dynamic friction angle independent of the shear rate?
- How does the granular temperature vary across the bottom shear layer, the plug layer, and the saltation layer?
- Is a significant amount of excess pore pressure generated in the collapse of the snowcover overflowed by the avalanche? If so, can it bring the avalanche front to aeroplaning or disrupt and suspend the snowcover just in front of the approaching avalanche? How quickly does it dissipate and by how much does it reduce the apparent friction?
- Which entrainment mechanisms occur and how do they depend on the dynamical flow variables? Under which conditions does a specific mechanism become dominant?
- Is mass transferred into the saltation and suspension layers from the front only or also from the body surface of the dense-flow layer? Does the underpressure generated by the recirculating air flow play a significant role in the transition to the saltation layer?
- How can the regulation mechanisms for the density of the saltation layer, discussed in Sect. 4.3, be formulated in a consistent, mathematical way that is amenable to numerical modeling?
- How do the velocity and density profiles vary in the suspension layer and how do they evolve along the path? How are the sources and sinks of the turbulent energy distributed in the suspension layer?

From a practical point of view, another source of uncertainty in hazard mapping is of at least equal importance: How do the initial conditions of an avalanche in a given path depend on the recurrence period? Investigation of this topic should also be given high priority in the coming years.

6.2 Suggestions for Future Work

Data analysis. Several experimental techniques available today—notably profiling radar, load cells, and all techniques involving cross-correlations—contain a wealth of information that is not easy to unravel because of the complicated nature of avalanche flow. In the author's opinion, a substantial effort at further developing the respective analysis techniques will bring more insight into some of the central questions listed above.

The present work should have made it clear that (i) in most experiments only a rather limited number of quantities was measured and (ii) a systematic comparison of the obtained results, differentiating with respect to snow properties, initial conditions, location of measurements, etc. is missing as yet. The author is convinced that such a synopsis will provide significant new insight and focus the open questions more than was possible in the present paper.

It might also be beneficial to extend model comparisons from a number of more or less well documented spontaneous avalanches outside test sites [4] to measured avalanches at test sites. The measured velocities and flow depths at different locations, the deposit geometry and, where available, the vertical profile of velocity will allow more detailed probing of the rheological assumptions embodied by these models than has been possible hitherto. It would be important now not only to compare almost identical implementations of the VOELLMY-SALM equations but to include models based on different rheological assumptions, notably a visco-plastic model and granular models like the SAVAGE–HUTTER or NIS models and extensions thereof. However, a meaningful comparison may be possible only if entrainment is measured and taken into account in the simulations.

Desirable sensor development. In order to obtain detailed and reliable data of internal quantities of avalanches, some further development of sensors is desirable or even necessary. On the basis of the author's assessment of present sensor performance, experimental needs, technological possibilities and financial considerations, the following sensor types appear most promising: (i) Low-cost FMCW radar with a small aperture angle and high sweeping and sampling rates; (ii) low-cost capacitance sensors for density measurements in all avalanche layers; (iii) long-range scanning laser distometer for measuring the mass balance along the entire avalanche path; (iv) non-clogging sensors for pore pressure measurements; (v) sturdy, non-clogging and direction-insensitive air-pressure sensors for measuring velocities and possibly densities in the suspension layer. This list is tentative and quite subjective.

Experiments to be performed. In the author's opinion, much can still be learned from systematic analyses of avalanche deposits with different terrain characteristics, sizes, and snow properties. In order to make these observations comparable, standard observation procedures should be elaborated and followed throughout. The number and breadth of such analyses will allow to better appraise the general relevance of findings from the necessarily small number of measured avalanches in test sites, covering only a limited range of possible avalanche behavior.

Over the past few years, interest in snow chute experiments has been revived and the experimental techniques have been considerably refined. This opens the possibility for systematic and detailed studies of fundamental mechanisms in avalanche flow, notably the interaction with obstacles and terrain roughness, snow entrainment, details of the flow regimes, etc. In the envisioned development of models with flow-regime transitions, new viscometric experiments on different varieties of snow with appropriately controlled boundary conditions will play a crucial role.

An effort should be made in full-scale experiments to obtain better and more comprehensive data in locations where flow-regime transitions are likely to occur, notably in the transition from the track to the runout zone. Acceleration, velocity profiles, flow depths, impact pressures and shear and normal stresses at the bed combined would allow to characterize the flow regime and to determine the entrainment or deposition rate, the effective friction and the forces transmitted longitudinally through the avalanche body. This will put several aspects of flow models to a stringent test. Another important point is to characterize the initial conditions and snow properties, and to measure the deposit distribution.

Some additional quantities that have rarely or never been measured in snow avalanches would provide very valuable information. Among them, the pore pressure at several heights in the snow cover is of particular interest because of its potentially enormous importance for determining the bed friction and the entrainment rate. Some preliminary experiments on sudden loading of a natural snowcover might be carried out to establish whether the rough assumptions made in Sect. 5 are realistic. Measurements of granular temperature would provide a very useful check of models that incorporate transitions between the frictional and inertial flow regimes. However, they may be very difficult to carry out in large-scale avalanches for technical reasons. Density is another quantity that has been high on the wish list of avalanche researchers for a long time. Capacitance sensors mounted on a high mast would reveal the density profile from the dense core to the powder cloud.

Needed theoretical work. Much work has been undertaken in the past on quite similar implementations of one- and two-dimensional models of the VOELLMY-SALM or PCM type. It has been suggested here that the more complex and comprehensive rheology at the base of the NIS model may be suitable for capturing the essential phenomena of dry-snow avalanches. However, a more general implementation that determines the velocity profile dynamically (possibly on the basis of an additional transport equation for granular temperature) is needed to unlock the potential of the rheological assumptions. As pointed out by HUTTER [46], this may be a formidable task.

Until this is achieved, a fresh look at visco-plastic models of the BINGHAM or HERSCHEL-BULKLEY type and granular models of the SAVAGE–HUTTER type with the extension introduced by GRAY and TAI [27] or ZWINGER et al. [115,116] might be useful for confronting the conceptually unsatisfactory but well-calibrated traditional models with different approaches that have several important properties built into them from the start and have met with surprising success in studies of muddy subaqueous debris flows and laboratory granular flows, respectively.

Acknowledgements

The author wishes to thank K. HUTTER for many discussions as well as the invitation to participate in workshops of the Sonderforschungsbereich 298 and to contribute to the present volume, and N. KIRCHNER for her editorial patience and encouragement. H. Gubler, M. Kern, K. Kristensen, J. McElwaine, M. Naaim, K. Nishimura, L. Rammer, H. Schaffhauser and B. Sovilla provided valuable additional information on published and unpublished experimental results and/or provided figures. The author is very grateful to A. N. Bozhinskiy, C. Fierz, P. Gauer, R. Greve, K. Hutter, F. Irgens, T. Jóhannesson, C. Keylock, R. I. Perla, M. Schaer, B. Sovilla and T. Zwinger for their critical reading and insightful suggestions that significantly improved the manuscript.

References

1. Adam, V., V. Chritin, M. Rossi, and E. van Lancker (1998) Infrasonic monitoring of avalanche activity: what do we know and where do we go from here? *Annals Glaciol.*, **26**, 324–328
2. Bagnold, R. A. (1954) Experiments on a gravity-free dispersion of large solid spheres in a Newtonian fluid under shear. *Proc. R. Soc. London, Ser. A*, **225**, 49–63
3. Banfi, L. (1994) Turbulenzmessungen an Laborstaublawinen. Diploma thesis, ETH Zürich, CH–8092 Zürich, Switzerland
4. Barbolini, M., U. Gruber, C. J. Keylock, M. Naaim, and F. Savi (2000) Application of statistical and hydraulic-continuum dense-snow avalanche models to five real European sites. *Cold Regions Sci. Technol.*, **31**, 133–149
5. Bartelt, P., B. Salm, and U. Gruber (1999) Calculating dense snow avalanche runout using a Voellmy-fluid model with active/passive longitudinal straining. *J. Glaciol.*, **45** (150), 242–254
6. Beghin, P., E. J. Hopfinger, and R. E. Britter (1981) Gravitational convection from instantaneous sources on inclined boundaries. *J. Fluid Mech.*, **107**, 407–422
7. Beghin, P., and X. Olagne (1991) Experimental and theoretical study of the dynamics of powder snow avalanches. *Cold Regions Sci. Technol.*, **19**), 317–326
8. Best, J. L., A. D. Kirkbridge, and J. Peakall (2001) Mean flow and turbulence structure of sediment-laden gravity currents: new insights using ultrasonic Doppler velocity profiling. *In:* W. D.McCaffrey, B. C. Kneller and J. Peakall (eds.), *Particulate Gravity Currents*, Special Publication no. 31, Intl. Assoc. of Sedimentologists. Oxford, UK, Blackwell Science Ltd., pages 159–172
9. Bozhinskiy, A. N., and K. S. Losev (1998) The Fundamentals of Avalanche Science. Mittlg. no. 55. CH–7260 Davos Dorf, Switzerland, Eidg. Institut für Schnee- und Lawinenforschung
10. Bozhinskiy, A. N., and A. N. Nazarov (1998) Dynamics of two-layer slush flows. *In:* E. Hestnes, ed., *25 Years of Snow Avalanche Research, Voss, 12–16 May 1998.* NGI Publ. no. 203. Norges Geotekniske Institutt, N–0806 Oslo, Norway, pages 74–78
11. Briukhanov, A. V., S. S. Grigorian, S. M. Miagkov, M. Ya. Plam, I. Ya. Shurova, M. E. Eglit, and Yu. L. Yakimov (1967) On some new approaches to the dynamics of snow avalanches. *In* Ôura, H., editor, *Physics of Snow and Ice, Proc. Intl. Conf. Low Temperature Science, Sapporo, Japan, 1966. Vol. I, Part 2.* Sapporo, Hokkaido, Japan, Institute of Low Temperature Science, Hokkaido University, pages 1223–1241

12. Casassa, G., H. Narita, and N. Maeno (1989) Measurements of friction coefficients of snow blocks. *Annals Glaciol.*, **13**, 40–44
13. Criminale, W. O. Jr., J. L. Ericksen, and G. L. Filbey (1958) Steady shear flow of non-Newtonian fluids. *Arch. Rat. Mech. Anal.*, **1**, 4120–417
14. De Blasio, F. V., D. Issler, A. Elverhøi, C. B. Harbitz, P. Bryn, and R. Lien (2003) Dynamics and material properties of the giant Storegga slide as suggested by numerical simulations. In preparation
15. Dent, J. D. (1982) *A biviscous modified Bingham model of snow avalanche motion.* PhD thesis, Dept. of Civil Engineering, Montana State University, Bozeman, Montana, U.S.A.
16. Dent, J. D., K. J. Burrell, D. S. Schmidt, M. Y. Louge, E. E. Adams, and T. G. Jazbutis (1998) Density, velocity and friction measurements in a dry-snow avalanche. *Annals Glaciol.*, **26**, 247–252
17. Dent, J. D. and T. E. Lang (1982) Experiments on the mechanics of flowing snow. *Cold Regions Sci. Technol.*, **1** (2), 109–120
18. Dent, J. D., and T. E. Lang (1983) A biviscous modified Bingham model of snow avalanche motion. *Annals Glaciol.*, **4**, 42–46
19. Dufour, F., U. Gruber, P. Bartelt, and W. J. Ammann (2000) Overview of the 1999 measurements at the SLF test-site Vallée de la Sionne. In: *A Merging of Theory and Practice. Proc. Intl. Snow Science Workshop, Big Sky, Montana, U.S.A., October 1st–6th, 2000.* Bozeman, Montana, U.S.A., Montana State University, pages 327–534
20. Dufour, F., U. Gruber, D. Issler, M. Schaer, N. Dawes, and M. Hiller (1999) Grobauswertung der Lawinenereignisse 1998/1999 im Grosslawinenversuchsgelände Vallée de la Sionne. Interner Bericht 732. CH–7260 Davos Dorf, Switzerland, Eidg. Institut für Schnee- und Lawinenforschung
21. Eckart, W., J. M. N. T. Gray, and K. Hutter (2003) Verification of the Savage–Hutter model with PIV measurements. This volume
22. Erismann, T. H. (1979) Mechanisms of large landslides. *Rock Mech.*, **12**, 15-46
23. Erismann, T. H. (1986) Flowing, rolling, bouncing, sliding: synopsis of basic mechanisms. *Acta Mechanica*, **64**, 101–110
24. Förster, M. (1999) Ausführliche Dokumentation ausgewählter Staublawinenereignisse und Bestimmung ihrer Eingangsparameter für die Verifikation von Staublawinenmodellen. Interner Bericht 730. CH–7260 Davos Dorf, Switzerland, Eidg. Institut für Schnee- und Lawinenforschung
25. Gauer, P. (2002) Ormen Lange slope stability assessment. Run out analyses. Intern Rapport 993016–11. N–0806 Oslo, Norway, Norges Geotekniske Institutt
26. Granada, F., O. Marco, and P. Villemain (1995) Utilisation des techniques d'imagerie pour la cartographie des vitesses à la surface d'une avalanche dense. In: *Comptes Rendus du "Colloque Glaciologie et nivologie — état des recherches à la fin du XX$^{\grave{e}me}$ siècle", Grenoble, France, February 15–16, 1995.* Société Hydrotechnique de France
27. Gray, J. M. N. T., and Y. C. Tai (1998) On the inclusion of a velocity-dependent basal drag in avalanche models. *Annals Glaciol.*, **26**, 277–280
28. Gray, J. M. N. T., W. Wieland, and K. Hutter (1999) Gravity-driven free surface flow of granular avalanches over complex basal topography. *Proc. R. Soc. London Ser. A*, **455**, 1841–1874

29. Greve, R., and K. Hutter (1993) The motion of a granular avalanche in a convex and concave curved chute. *Phil. Trans. R. Soc. London, Ser. A*, **342**, 573–604
30. Grigoryan, S. S., N. A. Urumbayev, and I. V. Nekrasov (1982) Experimental investigation of an avalanche air blast [in Russian]. *Data Glaciol. Studies*, **44**, 87–93
31. Gubler, H. (1987) Measurements and modelling of snow avalanche speeds. *In:* B. Salm and H. Gubler, eds., *Avalanche Formation, Movement and Effects (Proc. of the Davos Symposium, September 1986). IAHS Publ. no. 162.* IAHS Press, Institute of Hydrology, Wallingford, Oxfordshire, UK, pages 405–420
32. Gubler, H. (1989) Comparison of three models of avalanche dynamics. *Annals Glaciol.*, **13**, 82–89
33. Gubler, H. U., ed. (1991) *Proc. Workshop on Avalanche Dynamics, May 14–18, 1990, Weissfluhjoch/Davos (Switzerland).* Mittlg. No. 48. CH–7260 Davos Dorf, Switzerland, Eidg. Institut für Schnee- und Lawinenforschung
34. Gubler, H. (1995) Auswertung Lawinenmessungen Lukmanier vom 7./8. Januar 1994. Technical note. CH–7260 Weissfluhjoch/Davos, Switzerland, ALPUG and Eidg. Institut für Schnee- und Lawinenforschung
35. Gubler, H., and M. Hiller (1984) The use of microwave FMCW radar in snow and avalanche research. *Cold Regions Sci. Technol.*, **9**, 109–119
36. Gubler, H., M. Hiller, G. Klausegger, and U. Suter (1986) Messungen an Fliesslawinen. Zwischenbericht 1986. Mittlg. No. 41. CH–7260 Weissfluhjoch/Davos, Switzerland, Eidg. Institut für Schnee- und Lawinenforschung
37. Gude, M., and D. Scherer (1998) Snowmelt and slushflows: hydrological and hazard implications. *Annals Glaciol.*, **26**, 381–384
38. Hampton, M. (1972) The role of subaqueous debris flow in generating turbidity currents. *Sed. Petrology*, **42** (4), 775–793
39. Harbitz, C. B., ed. (1998) A survey of computational models for snow avalanche motion. NGI Report 581 220–1. Norges Geotekniske Institutt, N–0806 Oslo, Norway
40. Harbitz, C. B., G. Parker, A. Elverhøi, J. G. Marr, D. Mohrig, and P. Harff (in press) Hydroplaning of subaqueous debris flows and glide blocks: Analytical solutions and discussion. Accepted for publication in *J. Geophys. Res.*
41. Hermann, F., and K. Hutter (1991) Laboratory experiments on the dynamics of powder snow avalanches in the runout zone. *J. Glaciol.*, **37** (126), 281–295
42. Hestnes, E. (1998) Slushflow hazard—where, why and when? 25 years of experience with slushflow consulting and research. *Annals Glaciol.*, **26**, 370–376
43. Hestnes, E., and F. Sandersen (1998) Slushflow hazard control. A review of mitigative measures. *In:* E. Hestnes, ed., *25 Years of Snow Avalanche Research, Voss, 12–16 May 1998.* NGI Publ. no. 203. Norges Geotekniske Institutt, N–0806 Oslo, Norway, pages 140–147
44. Hopfinger, E. J. (1983) Snow avalanche phenomena and related phenomena. *Ann. Rev. Fluid Mech.*, **15**, 47–76
45. Hopfinger, E. J., and J.-C. Tochon-Danguy (1977) A model study of powder snow avalanches. *J. Glaciol.*, **19** (81), 343–356
46. Hutter, K. (1988) A continuum model for finite mass avalanches having shear-flow and plug-flow regimes. Interner Bericht 651. CH-7260 Weissfluhjoch/Davos, Switzerland, Eidg. Institut für Schnee- und Lawinenforschung

47. Hutter, K. (1991) Two- and three-dimensional evolution of granular avalanche flow—theory and experiments revisited. *Acta Mechanica (Suppl.)*, **1**, 167–181
48. Hutter, K. (1996) Avalanche Dynamics. *In:* V. P. Singh, ed., *Hydrology of Disasters*. Dordrecht, Kluwer Academic Publishers, pages 317–394
49. Hutter, K., and T. Koch (1991) Motion of a granular avalanche in an exponentially curved chute: experiments and theoretical predictions. *Phil. Trans. R. Soc. London, Ser. A*, **334**, 93–138
50. Hutter, K., T. Koch, C. Plüss, and S. B. Savage (1993) Dynamics of avlanches from initiation to runout. Part II. Laboratory experiments. *Acta Mechanica*, **109**, 127–165
51. Hutter, K., and K. R. Rajagopal (1994) On flows of granular materials. *Cont. Mech. Thermodyn.*, **6**, 81–139
52. Hutter, K., F. Szidarovszky, and S. Yakowitz (1987) Granular shear flows as models for flow avalanches. *In:* B. Salm and H. Gubler, editors, *Avalanche Formation, Movements and Effects. Proc. of the Davos Symposium, September 1986.* IAHS Publication No. 162. IAHS Press, Institute of Hydrology, Wallingford, Oxfordshire, UK, pages 381–394
53. Issler. D. (1998) Modelling of snow entrainment and deposition in powder snow avalanches. *Annals Glaciol.*, **26**, 253–258
54. Issler, D. (1999) European avalanche test sites—Overview and analysis in view of coordinated experiments. Mittlg. no. 59. CH–7260 Davos Dorf, Switzerland, Eidg. Institut für Schnee- und Lawinenforschung
55. Issler, D., P. Gauer, and M. Barbolini (1999) Continuum models of particle entrainment and deposition in snow drift and avalanche dynamics. *In:* H.-D. Alber, R. Balean and R. Farwig, eds., *Models of Continuum Mechanics in Analysis and Engineering*, Proc. of a Workshop organized by the Dept. of Mathematics, Technische Universität Darmstadt, Sept. 30 to Oct. 2, 1998. Aachen, Germany, Shaker Verlag, pages 58–80
56. Issler, D., P. Gauer, M. Schaer, and S. Keller (1996) Staublawinenereignisse im Winter 1995: Seewis (GR), Adelboden (BE) und Col du Pillon (VD). Interner Bericht 694. CH–7260 Weissfluhjoch/Davos, Switzerland, Eidg. Institut für Schnee- und Lawinenforschung
57. Iverson, R. M. (1997) The physics of debris flows. *Rev. Geophys.*, **35** (3), 245–296
58. Kawada, K., K. Nishimura, and N. Maeno (1989) Experimental studies on a powder-snow avalanche. *Annals Glaciol.*, **13**, 129–134
59. Keller, S. (1995) Measurements of powder snow avalanches—Laboratory. *Surveys Geophys.*, **16**, 661–670
60. Keller, S. (1996) Physikalische Simulation von Staublawinen – Experimente zur Dynamik im dreidimensionalen Auslauf. Mittlg. no. 141. CH–8092 Zürich, Switzerland, VAW, ETH Zürich
61. Kobayashi, D. (1972) Studies in snow transport of low-level drifting snow. *Contrib. Inst. Low Temp. Sci. Ser. A*, **24**. Sapporo, Japan, Inst. of Low Temperature Science, Hokkaido University
62. Koch, T., R. Greve, and K. Hutter (1994) Unconfined flow of granular avalanches along a partly curved surface. Part II: Experiments and numerical computations. *Phil. Trans. R. Soc. London, Ser. A*, **445**, 415–435
63. Kozik, S. M. (1962) Raschet dvizheniya snezhnykh lavin [Calculation of movement of snow avalanches]. Moscow, Russia, Gidrometeoizdat

64. Lied, K., and S. Bakkehøi (1980) Empirical calculation of snow-avalanche run-out distance based on topographical parameters. *J. Glaciol.*, **26** (94), 165–177
65. Lied, K., A. Moe, K. Kristensen, and D. Issler (2002) Ryggfonn. Full scale avalanche test site and the effect of the catching dam. NGI Report 581200–35. Postboks 3930 Ullevål Stadion, N–0806 Oslo, Norway, Norges Geotekniske Institutt
66. Lied, K., and R. Toppe (1989) Calculation of maximum snow-avalanche run-out. *Annals Glaciol.*, **13**, 164–169
67. Louge, M. Y., R. Steiner, S. C. Keast, R. Decker, J. D. Dent, and M. Schneebeli (1997) Application of capacitance instrumentation to the measurement of density and velocity of flowing snow. *Cold Regions Sci. Technol.*, **25**, 47–63
68. Marco, O. (1995) *Instrumentation d'un site avalancheux.* Études du CEMAGREF: Équipements pour l'eau et l'environnement, vol. 16. B. P. 22. F–92162 Antony CEDEX, CEMAGREF-Dicova
69. Martinet, G. (1991) St-Venant – modèle pour la simulation d'avalanches de neige humide. *Neige et Avalanches*, **53/54**, 34–40.
70. Martinet, G. (1992) *Contribution à la modèlisation numérique des avalanches de neige dense et des laves torrentielles.* PhD thesis, Université J. Fourier – Grenoble I, France
71. McClung, D. M., and P. A. Schaerer (1985) Characteristics of flowing snow and avalanche impact pressures. *Annals Glaciol.*, **6**, 9–14
72. McElwaine, J., and K. Nishimura (2001) Ping-pong ball avalanche experiments. *Annals Glaciol.*, **32**, 241–250
73. Mohrig, D., A. Elverhøi, and G. Parker (1999) Experiments on the relative mobility of muddy subaqueous and subaerial debris flows, and their capacity to remobilize antecedent deposits. *Marine. Geol.*, **154**, 117–129
74. Mohrig, D., K. X. Whipple, M. Hondzo, C. Ellis, and G. Parker (1998) Hydroplaning of subaqueous debris flows. *GSA Bull.*, **110** (3), 387–394
75. Naaim, M., and C. Ancey (1995) Modelisation of dense avalanches. *In:* Brugnot, G., editor, *Université européenne d'été sur les risques naturels. Session 1992: Neige et avalanches.* BP 22, F–92162 Antony CEDEX, CEMAGREF-Dicova, cemagref éditions, pages 173–181
76. Nakamura, T., H. Nakamura, O. Abe, A. Sato, and N. Numano (1987) A newly designed chute for snow avalanche experiments. *In:* B. Salm and H. Gubler, editors, *Avalanche Formation, Movements and Effects. Proc. of the Davos Symposium, September 1986.* IAHS Publication no. 162. IAHS Press, Institute of Hydrology, Wallingford, Oxfordshire, UK, pages 441–451
77. Nishimura, K. (1990) Studies on the fluidized snow dynamics. Contr. Inst. Low Temp. Sci. A 37. Sapporo, Japan, The Institute of Low Temperature Science, Hokkaido University
78. Nishimura, K., and Y. Ito (1997) Velocity distribution in snow avalanches. *J. Geophys. Res.*, **102** (B12), 27'297–27'303
79. Nishimura, K., and K. Izumi (1997) Seismic signals induced by snow avalanche flow. *Natural Hazards*, **15**, 89–100
80. Nishimura, K., S. Keller, J. McElwaine, and Y. Nohguchi (1998) Ping-pong ball avalanche at a ski jump. *Granular Matter*, **1**(2), 51–56
81. Nishimura, K., N. Maeno, and K. Kawada (1987) Internal structure of large-scale avalanches revealed by a frequency analysis of impact forces. *Low Temp. Sci, Ser. A*, **46**, 91–98

82. Nishimura, K., and N. Maeno (1987) Contribution of viscous forces to avalanche dynamics. *Annals Glaciol.*, **13**, 202–206
83. Nishimura, K., N. Maeno, K. Kawada, and K. Izumi (1993a) Structures of snow cloud in dry-snow avalanches. *Annals Glaciol.*, **18**, 173–178
84. Nishimura, K., N. Maeno, F. Sandersen, K. Kristensen, H. Norem, and K. Lied (1993b) Observations of the dynamic structure of snow avalanches. *Annals Glaciol.*, **18**, 313–316
85. Norem, H., F. Irgens, and B. Schieldrop (1987) A continuum model for calculating snow avalanche velocities. *In:* B. Salm and H. Gubler, eds., *Avalanche Formation, Movements and Effects. Proc. of the Davos Symposium, September 1986.* IAHS Publication no. 162. IAHS Press, Inst. of Hydrology, Wallingford, Oxfordshire, UK, pages 363–380
86. Norem, H., F. Irgens, and B. Schieldrop (1989) Simulation of snow-avalanche flow in run-out zones. *Annals Glaciol.*, **13**, 218–225
87. Perla, R. I., T. T. Cheng, and D. M. McClung (1980) A two-parameter model of snow-avalanche motion. *J. Glaciol.*, **26** (94), 197–207
88. Perla, R., K.Lied, and K. Kristensen (1984) Particle simulation of snow avalanche motion. *Cold Regions Sci. Technol.*, **9**, 191–202
89. Perov, V. F. (1998) Slushflows: Basic properties and spreading. *In:* E. Hestnes, ed., *25 Years of Snow Avalanche Research, Voss, 12–16 May 1998.* NGI Publ. no. 203. Norges Geotekniske Institutt, N–0806 Oslo, Norway, pages 203–209
90. Rammer, L. (1992) Dynamische Messungen an Staublawinen. In: *Internationales Symposion INTERPRAEVENT 1992 – BERN*, pages 421–431
91. Rammer, L. (2000a) Avalanche dynamics measurement facility "Großer Gröben", avalanche on Feb. 21st, 2000. *In:* G. Fiebiger, ed., *Proc. IUFRO Intl. Workshop "Hazard Mapping in Avalanching Areas", St. Christoph/St. Anton, Tyrol, Austria, April 2–7, 2000.* A–5027 Salzburg, Austria, International Union of Forestry Research Organisations (IUFRO), pages 211–223
92. Rammer, L. (2000b) Velocity measurements of avalanches by a pulsed Doppler radar. *In:* G. Fiebiger, ed., *Proc. IUFRO Intl. Workshop "Hazard Mapping in Avalanching Areas", St. Christoph/St. Anton, Tyrol, Austria, April 2–7, 2000.* A–5027 Salzburg, Austria, International Union of Forestry Research Organisations (IUFRO), pages 103–120
93. Rodi, W. (1984) *Turbulence Models and Their Application in Hydraulics — State-of-the-Art Report.* NL–2600 MH Delft, The Netherlands, Intl. Assoc. for Hydraulic Research, 2nd edition
94. Sabot, F., M. Naaim, F. Granada, E. Suriñach, R. Planet, and G. Furdada (1998) Study of the avalanche dynamics by means of seismic methods, image-processing techniques and numerical models. *Annals Glaciol.*, **26**, 319–323
95. Salm, B. (1964) Anlage zur Untersuchung dynamischer Wirkungen von bewegtem Schnee. *ZAMP*, **15**, 357–375
96. Salm, B. (1966) Contribution to avalanche dynamics. In: *Scientific Aspects of Snow and Ice Avalanches, Proc. of the Davos Symposium, 5–10 April 1965.*. IAHS Publication no. 69. IAHS Press, Inst. of Hydrology, Wallingford, Oxfordshire, UK, pages 199–214
97. Salm, B., and H. Gubler (1985) Measurement and analysis of the motion of dense flow avalanches. *Annals Glaciol.*, **6**, 26–34.
98. Salm, B., A. Burkard, and H. U. Gubler (1990) Berechnung von Fliesslawinen. Eine Anleitung für Praktiker mit Beispielen. Mittlg. no. 47. CH–7260 Davos Dorf, Switzerland, Eidg. Institut für Schnee- und Lawinenforschung

99. Savage, S. B., and K. Hutter (1989) The motion of a finite mass of granular material down a rough incline. *J. Fluid Mech.*, **199**, 177–215
100. Savage, S. B., and K. Hutter (1991) The dynamics of avalanches of granular materials from initiation to runout. Part I: Analysis. *Acta Mechanica*, **86**, 201–223
101. Savage, S. B., and M. Sayed (1984) Stresses developed by dry granular materials in an annular shear cell. *J. Fluid Mech.*, **142**, 391–430
102. Schaer, M., and D. Issler (2001) Particle densities, velocities, and size distributions in large avalanches from impact-sensor measurements. *Annals Glaciol.*, **32**, 321–327
103. Schaerer, P. A., and A. A. Salway (1980) Seismic and impact-pressure monitoring of flowing avalanches. *J. Glaciol.*, **26** (94), 179–187
104. Scheiwiller, T. (1986) *Dynamics of powder-snow avalanches.* Mittlg. no. 81. CH–8092 Zürich, Switzerland, VAW, ETH Zürich
105. Scheiwiller, T., K. Hutter, and F. Hermann (1987) Dynamics of powder snow avalanches. *Annales Geophysicae*, **5B** (6), 569–588
106. Scherer, D., M. Gude, M. Gempeler, and E. Parlow (1998) Atmospheric and hydrological boundary conditions for slushflow initiation due to snowmelt. *Annals Glaciol.*, **26**, 377–380
107. Schreiber, H., W. L. Randeu, H. Schaffhauser, and L. Rammer (2001) Avalanche dynamics measurement by pulsed Doppler radar. *Annals Glaciol.*, **32**, 275–280
108. Shimizu, H., E. Huzioka, E. Akitaya, H. Narita, M. Nakagawa, and K. Kawada (1980) A study on high-speed avalanches in the Kurobe canyon, Japan. *J. Glaciol.*, **26**, 141–151
109. Simpson, J. E. (1987) *Gravity Currents: In the Environment and the Laboratory.* Ellis Horwood Series in Environmental Science. Chichester, West Sussex, England, Ellis Horwood Ltd.
110. Sovilla, B., F. Sommavilla, and A. Tomaselli (2001) Measurements of mass balance in dense snow avalanche events. *Annals Glaciol.*, **32**, 230–236
111. Tai, Y. C., J. M. N. T. Gray, K. Hutter, and S. Noelle (2001) Flow of dense avalanches past obstructions. *Annals Glaciol.*, **32**, 281–284
112. Tiefenbacher, F., and M. A. Kern (2003) Experimental devices to determine snow avalanche basal friction and velocity profiles. Submitted to *Cold Regions Sci. Technol.*
113. Vallet, J., U. Gruber, and F. Dufour (2001) Photogrammetric avalanche measurements at Vallée de la Sionne, Switzerland. *Annals Glaciol.*, **32**, 141–146
114. Voellmy, A. (1955). Über die Zerstörungskraft von Lawinen. *Schweiz. Bauzeitung*, **73**, 159–165, 212–217, 246–249, 280–285
115. Zwinger, T. (2000) Dynamik einer Trockenschneelawine auf beliebig geformten Berghängen. PhD thesis, Technische Universität Wien, Wien, Austria
116. Zwinger, T., A. Kluwick, and P. Sampl (2003) Numerical simulation of dry-snow avalanche flow over natural terrain. This volume

Numerical Simulation of Dry-Snow Avalanche Flow over Natural Terrain

Thomas Zwinger[1], Alfred Kluwick[2], and Peter Sampl[3]

[1] Department of Computational Engineering, CSC – Scientific Computing Ltd., P.O. Box 405, 02101 Espoo, Finland, e-mail: thomas.zwinger@csc.fi
[2] Institute of Fluid Mechanics and Heat Transfer, E322, Vienna University of Technology, Resselgasse 3, 1040 Wien, Austria, e-mail: alfred.kluwick@tuwien.ac.at
[3] AVL-List GmbH, Advanced Simulation Technologies, Hans-List-Platz 1, 8020 Graz, Austria, e-mail: peter.sampl@avl.com

received 12 Jul 2002 — accepted 3 Dec 2002

Abstract. A physical model for dry snow avalanche flow is presented. The well established model for dense granular flow proposed by Savage and Hutter [21] is adapted to describe the dense, lower part of a snow avalanche. In order to account for the high velocities of snow avalanches, a velocity dependent bed shear stress in addition to the COULOMB law for dry friction is introduced. Since the model has to be applied to arbitrarily shaped topographies, certain simplifying assumptions of geometrical nature must be introduced. In contrast to former numerical implementations of the SAVAGE–HUTTER model based on Finite Difference schemes, a LAGRANGIAN Finite Volume method is formulated using integral balance laws. For the powder snow avalanche forming on top of the denser part a mixture model with a constant slip velocity between ice particles and air is introduced. The k-ϵ turbulence model modified in order to account for buoyancy effects caused by the suspended particles is implemented. The resulting system of equations is solved by applying a Finite Volume scheme on a fixed grid. The transfer of snow mass from the dense flow into the powder snow avalanche is modelled by an analogy to turbulent momentum transfer.

1 Dry Snow Avalanches

Dry snow avalanches consist of cold and thus cohesionless ice particles with air in between. Due to the low internal friction they are able to overcome longer distances and heights than avalanches consisting of moist snow or even slush. From this, dry snow avalanches impose a great hazard to inhabited Alpine areas.

This hazard has to be accounted for in risk assessment. In Austria as well as in Switzerland, avalanche hazard maps have been introduced in order to provide directives for constructions in Alpine regions. While in earlier times the compilation of such hazard maps was entirely based on the expertise of avalanche experts, numerical simulation has been increasingly introduced as a supporting tool in such processes. This is especially important in cases, where previously uninhabited areas have to be assessed. Under such circumstances it is impossible to consult avalanche-chronicles, a fact which leaves the expert, except for possible marks in the vegetation, without any information of avalanche events that may have previously occurred.

The first steps towards a physical avalanche simulation model were taken by Voellmy [26]. He proposed a one-dimensional center of mass model with a friction law based on hydraulic theory. A physical avalanche model that can be applied in a prognostic manner has to comply with the following criteria:

- Compliance with the basic laws of continuum mechanics, i.e., conservation of mass, momentum and energy.
- Only parameters with constant values for all topographies shall appear in the constitutive relations (e.g., constant friction coefficient).
- Distinction between different flow regimes (e.g., dense granular flow, turbulent powder snow avalanche) [13].

The first item in the list above requires no detailed explanation. Naturally, any violation of the conservation laws will automatically lead to unphysical results. While the global mass balance inherently is fulfilled by a mass point model, the internal frictional behaviour has an essential influence on the shape of the avalanche. From this point of view, continuum models certainly are superior to approaches that divide the avalanche into non-interacting discrete blocks of snow.

The second point concerns the objectivity of the simulation results. If the user has to choose friction parameters himself, the simulation results for the same avalanche track and release-scenario may differ, depending on who ran the simulation. We would like to stress the fact that the physical model presented in this article was validated with different recorded avalanche events with a fixed set of friction parameters.

The last of these three postulations contains the inherent problem of physical avalanche simulation: Avalanche dynamics is highly complex. Consequently, the description of all the processes occurring during a dry snow avalanche event in a global physical model is a tedious task indeed. Different flow layers, as depicted in Fig. 1, have to be treated with different models.

A good indicator for the type of flow regime is the particle volume fraction, Φ. In a two-phase flow of particles and gas it is the ratio between the part of a test volume that contains the solid phase and the test volume itself, i.e., $\Phi = \lim_{V \to 0} V_{\text{solid}}/V$. The mixture density, later defined by equation (45), as well as the ratio between the characteristic particle spacing, l, and the radius of the particles, a, as a function of the particle volume fraction are shown in Fig. 2. This figure also shows that granular flow (i.e., the interstitial fluid can be neglected) can be realized only for ratios of l/a smaller than unity. In other words, only if the mean spacing between the grains is smaller than their radius, particle-particle interactions will be mainly responsible for the momentum transport [9]. Application of granular theories (for which particle-gas interactions do not play a dominant role) is valid in a range of $\Phi \approx 0.15 - 0.55$, where the upper value of the volume fraction stands for the closest random packing.

On the other hand, pure suspension flow (for which particle-particle interactions are completely negligible) is to be found for particle volume fractions of the order $\Phi \leq 10^{-3}$. Between these limiting flow regimes, more complex physics has to be applied to describe the dynamics of the flow.

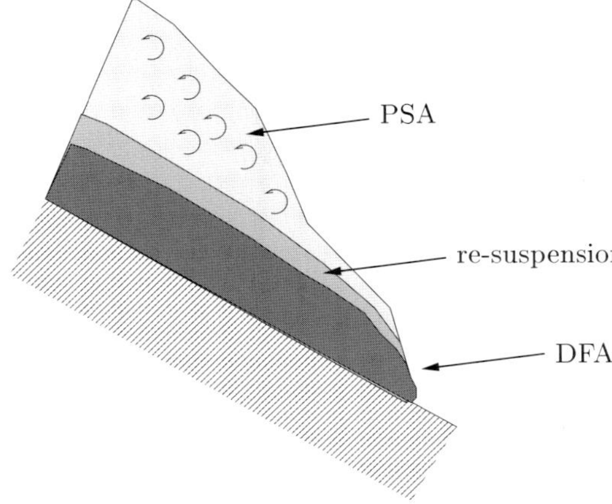

Fig. 1. The layer structure of a dry snow avalanche. The powder snow avalanche (PSA) forms on top of the dense flow avalanche (DFA). Mass transfer from the former into the latter takes place in a small re-suspension layer

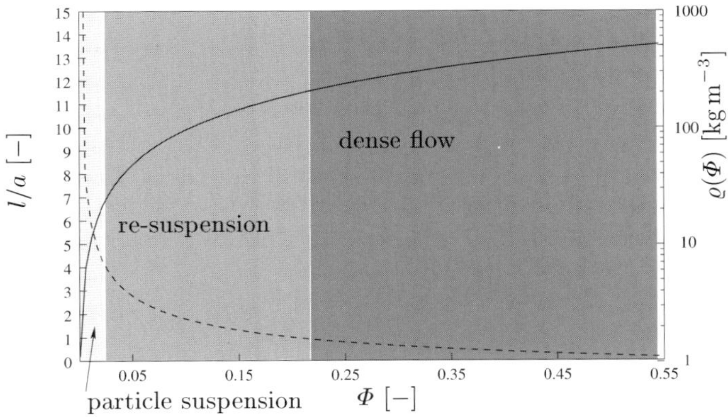

Fig. 2. Mixture density (*solid line, scale on right ordinate*) and ratio between mean distance to particle radius (*dashed line, scale on left ordinate*) of a suspension of spherical mono-disperse ice-particles and air

2 Dense Flow Avalanche Model

Geological materials show a small dependence of the shear stress on the shear rate over a wide range of the shear velocity [21,23]. Hence a relatively simple hydraulic model with a COULOMB-type bed friction law is expected to suffice for the prediction of the main features of dense avalanche flow of such materials over a wide range of velocities. From this point of view we can consider a dense dry snow avalanche to be a deforming granular pile sliding on a small shear layer underneath[1]. In this

[1] As pointed out by Issler [13], presently no experimental evidence on vertical velocity profiles inside dense flow avalanches is available. Nevertheless, the assumption of a shear layer-plug flow type is commonly accepted.

quasi-static deforming pile, a MOHR–COULOMB yield criterion is introduced as a constitutive relation linking normal to tangential stresses. The agreement of the SAVAGE–HUTTER model with laboratory experiments has been demonstrated in channels [8] as well as in unconfined granular flow on inclined surfaces [15]. The obvious advantage of this model over those arising from the kinetic granular theory [14] or even mixture theory is its simple mathematical form; a reduced set of unknown field variables (a two-dimensional velocity profile and the flow depth), given in three equations. The constitutive relations contain two constant and measurable material parameters.

2.1 Governing Equations for Dense Flow Avalanche

The dense flow avalanche is treated as an incompressible single-phase granular continuum. The global conservation of mass for the control volume, $V(t)$, thus reads as

$$\frac{D}{Dt}\int\limits_{V(t)} \varrho_0 dV = \varrho_0 \frac{D}{Dt}\int\limits_{V(t)} dV = 0, \tag{1}$$

with constant flow density, ϱ_0. The substantial derivative is defined by

$$\frac{D}{Dt} = \frac{\partial}{\partial t} + u_i \frac{\partial}{\partial x_i},$$

where u_i stands for the i-th component of the local velocity field. In the momentum balance for the i-th ($i = 1, 2, 3$) direction of the coordinate system,

$$\varrho_0 \frac{D}{Dt}\int\limits_{V(t)} u_i dV = \oint\limits_{\partial V(t)} \sigma_{ij} n_j dA + \varrho_0 \int\limits_{V(t)} g_i dV, \tag{2}$$

the total force acting on the control volume element on the left-hand side consists of the integral of the stress tensor[2], σ_{ij}, over the control volume surface, $\partial V(t)$, and the body force caused by the acceleration due to gravity, g_i. The unit normal, n_j, of the surface $\partial V(t)$ is taken to point outwards of the volume. Restricting ourselves to Lagrangian control volumes, the substantial derivative becomes material $D/Dt \equiv d/dt$. The symbol d/dt stands for the total time derivative. If we further introduce the volume average of a property P, viz.,

$$\overline{P}(\boldsymbol{x}, t) = \frac{1}{V(t)} \int\limits_{V(t)} P dV, \tag{3}$$

the set of governing equations is obtained in their global form

$$\boxed{\frac{dV}{dt} = 0, \qquad \varrho_0 V \frac{d\overline{u}_i}{dt} = \oint\limits_{\partial V(t)} \sigma_{ij} n_j dA + \varrho_0 V \overline{g}_i .} \tag{4}$$

[2] EINSTEIN's convention of summation over same indices is applied throughout the whole text.

Similar to earlier work (e.g. [7]) we introduce the boundary conditions at the free surface and the bottom, which coincides with the slope topography. For a certain position vector, $\boldsymbol{x} = (x_1, x_2, x_3)$, these surfaces are represented by their implicit functions,

$$\begin{aligned} &\text{free surface:} && F_s(\boldsymbol{x};t) = x_3 - s(x_1, x_2; t) = 0, \\ &\text{bottom (i.e, slope):} && F_b(\boldsymbol{x}) = b(x_1, x_2) - x_3 = 0. \end{aligned} \tag{5}$$

For the time being we neglect the lift induced mass flux from the free surface of the dense flow avalanche into the turbulent air above[3]. The free surface is then material, so that

$$\frac{d F_s}{d t} = \frac{\partial F_s}{\partial t} + u_i \frac{\partial F_s}{\partial x_i} = 0. \tag{6}$$

Neglecting the shear stress exerted by the surrounding air, a stress free surface is stipulated,

$$\sigma_{ij} \cdot n_j^{(s)} = 0. \tag{7}$$

At the slope a slip condition without detachment is imposed,

$$u_i \cdot n_i^{(b)} = 0. \tag{8}$$

The normals of the corresponding surfaces are given by the unit vectors in the direction of the gradients of the specific implicit forms (5),

$$n_i^{(s)} = \frac{\partial F_s}{\partial x_j} \cdot \left(\frac{\partial F_s}{\partial x_j} \frac{\partial F_s}{\partial x_j} \right)^{-1/2}, \quad n_i^{(b)} = \frac{\partial F_b}{\partial x_j} \cdot \left(\frac{\partial F_b}{\partial x_j} \frac{\partial F_b}{\partial x_j} \right)^{-1/2}. \tag{9}$$

The dynamic boundary condition is given by the COULOMB law for dry friction with constant bed-friction angle, δ, as proposed by Savage and Hutter [21],

$$\tau^{(b)} = \pm \sigma^{(b)} \tan \delta. \tag{10}$$

Modifications necessary to cover the whole range of shear velocities occurring in snow avalanche flows will be discussed in Sect. 2.4 of this article. Equation (10) represents a relation between the normal stress, $\sigma^{(b)}$, and the shear stress, $\tau^{(b)}$, which later by choice of an appropriate coordinate system will turn out to be the components $\sigma_{33}^{(b)}$ and $\sigma_{13}^{(b)}$, respectively, of the stress tensor at the bottom.

A similar relation between the normal and shear stress is also needed for the bulk of the avalanche. Again, we adopt the proposal of Savage and Hutter [21] and assume simple yielding to occur in the granular pile. This is expressed by a MOHR–COULOMB closure relation,

$$\tau = \pm \sigma \cdot \tan \varphi, \tag{11}$$

where φ is the internal friction angle.

[3] As will be shown later, the averaged volume flux from the DFA into the PSA is sufficiently small that the assumption of a constant control volume for a single time step of the numerical integration is justified.

2.2 Natural Coordinates

The set of equations (4) is exact. The axes of the coordinate system in which this set will further be examined, are assumed to have the directions of the unit vector of the local velocity, the normal of the slope and a third vector resulting from the vector product of the former,

$$\boldsymbol{e}_1 = \frac{\boldsymbol{u}}{||\boldsymbol{u}||}, \quad \boldsymbol{e}_2 = \boldsymbol{e}_3 \times \boldsymbol{e}_1, \quad \boldsymbol{e}_3 = -\boldsymbol{n}^{(b)}. \tag{12}$$

For the sake of simplicity the coordinate system is attached to the slope so that expression (5) for the bottom surface reduces to $x_3 = 0$. We would like to stress the fact that – due to the kinematic boundary condition (8) – equation (12) defines a local orthonormal system, which subsequently will be referred to as the *natural coordinate system* (NCS). Important consequences implied by the introduction of the NCS are:

1. The exterior normal vector at the bottom points exactly in the negative x_3-direction. By means of the KRONECKER symbol this can be expressed as

$$n_i^{(b)} = -\delta_{i3}, \quad (i = 1, 2, 3). \tag{13}$$

2. By applying $(9)_2$, the unit normal vector at the slope is given by

$$\boldsymbol{n}^{(b)} = \frac{1}{\sqrt{(\partial b/\partial x_1)^2 + (\partial b/\partial x_2)^2 + 1}} \cdot \begin{pmatrix} \partial b/\partial x_1 \\ \partial b/\partial x_2 \\ -1 \end{pmatrix}. \tag{14}$$

   Equation (13) then implies that the first partial derivatives of the explicit slope function have to vanish in the NCS, i.e.,

$$\frac{\partial b}{\partial x_i} = 0, \quad (i = 1, 2). \tag{15}$$

3. The velocity components u_2 and u_3 vanish at the slope, $x_3 = b = 0$.

The shape of the control volume will be restricted to prisms as depicted in Fig. 3. The volume content, $V = A_b\,\overline{h}$, is obtained by multiplication of the basal area of the prism, A_b, with the averaged value of the flow-depth,

$$\overline{h} = \frac{1}{A_b} \int\limits_{A_b} [s(x_1, x_2, t) - b(x_1, x_2)]\, dA. \tag{16}$$

The shape of the control volume suggests to split the surface integral in the momentum balance into three distinctive parts,

$$\oint\limits_{\partial V} \sigma_{ij} n_j dA = \oint\limits_{A_s} \sigma_{ij} n_j^{(s)} dA + \oint\limits_{A_b} \sigma_{ij} n_j^{(b)} dA + \oint\limits_{A_h} \sigma_{ij} n_j^{(h)} dA. \tag{17}$$

In view of the dynamic boundary condition (7), the first integral on the right-hand side defined at the free surface vanishes identically. Then, by applying (13), evaluation of the second integral yields

$$\oint\limits_{A_b} \sigma_{ij} n_j^{(b)} dA = -A_b \overline{\sigma}_{i3}, \tag{18}$$

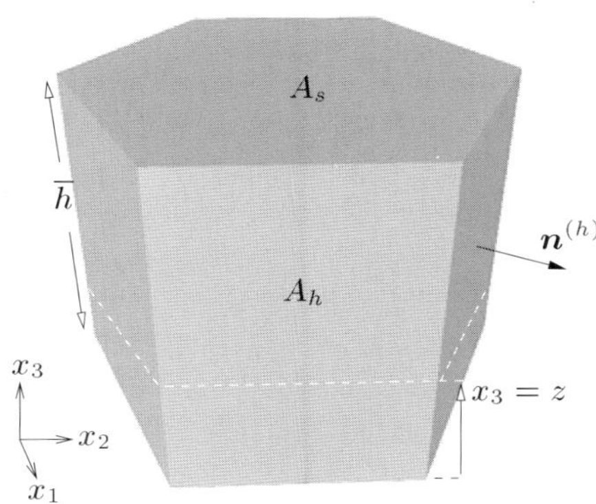

Fig. 3. Prism-like control volume. The surface of the volume can be divided into its basal area A_b, the free surface A_s and the sides A_h

where the bar denotes an average in the sense of (16). Finally, the last right-hand side term of (17) (i.e., the integral over the side faces of the prism) can be rewritten as

$$\oint_{A_h} \sigma_{ij} n_j^{(h)} dA = \oint_{\partial A_b} dl \int_b^s \sigma_{ij} n_j^{(h)}(l) dx_3. \tag{19}$$

Here l stands for the path length of the curve, ∂A_b, bounding the basal area, A_b. Consequently, the following form of the momentum balance is obtained

$$\frac{d\overline{u}_i}{dt} = \frac{1}{\varrho_0 A_b \overline{h}} \Big[\oint_{\partial A_b} dl \int_b^s \sigma_{ij} n_j^{(h)}(l) dx_3 - A_b \overline{\sigma}_{i3} \Big] + \overline{g}_i, \quad (i = 1, 2). \tag{20}$$

Equation (20) is applied to evaluate the acceleration of the control volume in the tangential plane of the slope, i.e., in the directions $i = 1, 2$. The balance in the direction perpendicular to the slope in the direction $i = 3$ will be used to obtain a relation for the vertical distribution of the stress tensor, $\sigma_{33}(x_3)$ for fixed x_1, x_2. To this end, a new average is introduced in which the integral in this direction is not extended from the free surface down to the bottom, but only to a certain level $x_3 = z \geq 0$, as depicted in Fig. 3. We define the average of a property P with respect to the volume given by $V^{(z)} = A_b \cdot (\overline{s} - z)$, similar to (3),

$$\overline{P}^{(z)}(z) = \frac{1}{V^{(z)}} \int_z^{\overline{s}} P \, dx_3. \tag{21}$$

The upper bound of the integral in (21), $\overline{s}$, is the averaged free surface function. As a consequence of the assumptions introduced by the choice of the NCS, i.e., $x_3 = b(x_1, x_2) = 0$, it is equal to the averaged flow-depth, $\overline{h}$, of (16). Taking into account that the stress vector vanishes at the free surface and that the intersecting surface at $x_3 = z$ is parallel to the basal area, A_b, we obtain

$$\begin{aligned} \varrho_0 \cdot (\overline{s} - z) \frac{d\,\overline{u}_3^{(z)}}{dt} =& \varrho_0 \cdot (\overline{s} - z) \overline{g}_3^{(z)} - \overline{\sigma}_{33}(z) \\ &+ \frac{1}{A_b} \oint\limits_{\partial A_b} dl \int\limits_z^s \sigma_{3j} n_j^{(h)}(l) dx_3 . \end{aligned} \tag{22}$$

The kinematic boundary condition at the bottom (8) implies

$$u_3^{(b)} = u_1^{(b)} \frac{\partial b}{\partial x_1} + u_2^{(b)} \frac{\partial b}{\partial x_2}, \tag{23}$$

so that the substantial derivative of $u_3^{(b)}$ is given by

$$\frac{d\,u_3^{(b)}}{dt} = \left(\frac{\partial}{\partial t} + u_1^{(b)} \frac{\partial}{\partial x_1} + u_2^{(b)} \frac{\partial}{\partial x_2} \right) \left(u_1^{(b)} \frac{\partial b}{\partial x_1} + u_2^{(b)} \frac{\partial b}{\partial x_2} \right). \tag{24}$$

As a consequence of the simplifications resulting from the choice of the NCS this relation reduces to

$$\frac{d\,u_3^{(b)}}{dt} = u_1^{(b)\,2} \frac{\partial^2 b}{\partial x_1^2}, \tag{25}$$

which corresponds to the centripetal acceleration caused by the curvature of the slope in the direction of the flow, which is proportional to $\partial^2 b / \partial x_1^2$. The block-like velocity profile in the quasi-static deforming pile allows us to take relation (25) to be valid for all values of the vertical coordinate, i.e.,

$$\frac{d\,\overline{u}_3^{(z)}}{dt} = \overline{u}_1^{(z)\,2} \frac{\partial^2 b}{\partial x_1^2}.$$

Inserting this into (22) yields the normal stress distribution as a function of the vertical coordinate

$$\begin{aligned} \overline{\sigma}_{33}(z) &= \overline{\sigma}_{33}^{(z)} \\ &= \varrho_0 \cdot (\overline{s} - z) \cdot \left[\overline{g}_3^{(z)} - \overline{u}_1^{(z)\,2} \frac{\partial^2 b}{\partial x_1^2} \right] + \frac{1}{A_b} \oint\limits_{\partial A_b} dl \int\limits_z^{\overline{s}} \sigma_{3j} n_j^{(h)}(l)\, dx_3 . \end{aligned} \tag{26}$$

Further simplification is obtained if (26), the lateral momentum balance (20) and the mass balance $(4)_1$ are scaled and assumption of shallowness is imposed.

2.3 Non-dimensional Equations

In complete analogy to the preliminary works using the SAVAGE–HUTTER model (see e.g. Gray et al. [7]) the following set of scalings is introduced:

$$\begin{aligned}
(x_i, dl) &= L \cdot (x_i^*, dl^*), \\
(dx_3, \overline{h}, (\overline{s} - z)) &= H \cdot (dx_3^*, \overline{h}^*, (\overline{s}^* - z^*)), \\
(A_b) &= L^2 \cdot (A_b^*), \\
(t) &= \sqrt{L/g} \cdot (t^*), \\
(\overline{u}_i, \overline{u}_i^{(z)}) &= \sqrt{g\,L} \cdot (\overline{u}_i^*, \overline{u}_i^{*\,(z)}), \\
(\sigma_{ij})_{i \neq j} &= \varrho_0\, g\, H\, \tan\delta \cdot (\tau_{ij}^*), \\
(\sigma_{ii}) &= \varrho_0\, g\, H \cdot (\delta_{ij}\, \sigma_{(i)}^*), \\
(g_i) &= g \cdot (g_i^*) \\
(\partial^2 b / \partial x_i^2)_{i=1,2} &= \frac{1}{R} \cdot (\partial^2 b^* / \partial x_i^{*\,2}).
\end{aligned} \tag{27}$$

Here, dimensionless properties are indicated by a superscripted asterisk. We also split the dimensionless stress tensor into its deviatoric, τ_{ij}^*, and diagonal part, $\sigma_{(i)}^*$.

Since the aspect ratio, $\varepsilon = H/L \ll 1$, representing the ratio between characteristic height, H, and length, L, of dense flow avalanches typically is small, the equations are examined in the limit

$$\varepsilon \to 0, \tag{28}$$

up to an order $\mathcal{O}(\varepsilon^{3/2})$. To this end other dimensionless groups of parameters need to be expressed in terms of the aspect ratio. These are the tangent of the bed-friction angle, $\tan\delta$, and the ratio between the length of the avalanche and the curvature-radius, $\lambda = L/R$. In the following we adopt the scaling proposed by Savage and Hutter [21]

$$\tan\delta = \varepsilon^{1/2}, \qquad \lambda = \varepsilon^{1/2}. \tag{29}$$

Introducing these relations into (26) results in

$$\sigma_{33}^*(z^*) = \sigma_3^*(z^*) = (\overline{s}^* - z^*) \cdot \left[g_3^{*(z)} - \varepsilon^{1/2} \overline{u}_1^{*(z)\,2} \frac{\partial^2 b}{\partial x_1^2} \right] + \mathcal{O}(\varepsilon^{3/2}), \tag{30}$$

in which $\sigma_{33}^*(z^*)$ is expressed as a first order hydrostatic pressure distribution plus a correction of the order $\mathcal{O}(\varepsilon^{1/2})$ caused by centripetal forces due to the slope curvature. Evaluating (30) at the bottom ($z^* = b^* = 0$), we find the normal stress at the slope,

$$\overline{\sigma}_{33}^{*\,(b)} = \sigma_{33}^*(0) = \overline{h}^* \cdot \left[g_3^* - \varepsilon^{1/2} \overline{u}_1^{*\,2} \frac{\partial^2 b^*}{\partial x_1^{*\,2}} \right] + \mathcal{O}(\varepsilon^{3/2}), \tag{31}$$

in first order to be equal to the static overburden of the avalanche. The dimensionless dynamic boundary condition (10) becomes

$$\overline{\sigma}_{13}^{*\,(b)} = -\frac{\overline{u}_i^*}{||\overline{\boldsymbol{u}}^*||} \overline{\sigma}_{33}^* = -\frac{\overline{u}_i^*}{||\overline{\boldsymbol{u}}^*||} \overline{h}^* \cdot \left[g_3^* - \varepsilon^{1/2} \overline{u}_1^{*\,2} \frac{\partial^2 b^*}{\partial x_1^{*\,2}} \right] + \mathcal{O}(\varepsilon^{3/2}). \tag{32}$$

The factor $-\overline{u}_i^* \|\overline{\boldsymbol{u}}^*\|^{-1}$ accounts for the fact that the friction at the bed acts against the current direction of the motion, which by definition of the NCS (12) is $\boldsymbol{e}_1$.

Following the discussions of the vertical normal stress distribution and its value at the slope, we now concentrate on the momentum balance equation (20) in the tangential plane of the slope. In dimensionless form it reads

$$\frac{d\overline{u}_i^*}{dt^*} = \overline{g}_i^* + \varepsilon^{1/2} \frac{\overline{u}_i^*}{\|\overline{\boldsymbol{u}}^*\|} \frac{\overline{\sigma}_{33}^{*\,(b)}}{\overline{h}^*} + \varepsilon \mathcal{B}_i^* + \mathcal{O}(\varepsilon^{3/2}), \quad i = 1, 2. \tag{33}$$

The last term on the right-hand side, in the following referred to as the bisection term, accounts for the hydrostatic forces acting on the side areas of the control volume,

$$\begin{aligned} \mathcal{B}_i^* &= \frac{1}{A_b^* \overline{h}^*} \oint_{\partial A_b} dl^* \int_{b^*(l^*)\equiv 0}^{s^*(l^*)\equiv h^*(l^*)} \sigma_{(i)}^* \delta_{ij} n_j^{(h)} dx_3^* \\ &= \frac{1}{A_b^* \overline{h}^*} \oint_{\partial A_b} dl^* \int_0^{h^*(l^*)} \sigma_{(i)}^* n_i^{(h)} dx_3^*, \quad i = 1, 2. \end{aligned} \tag{34}$$

Identifying the individual terms in (33), we obtain the balance

(net acceleration of control volume)

= (gravitational acceleration) − (bed friction) + (bisection term).

Equation (34) has to be completed by a relation linking the horizontal normal stresses, $\sigma_{(1)}^*$ and $\sigma_{(2)}^*$, to the vertical pressure distribution given by equation (30). In complete analogy to the arguments used by Savage and Hutter [21] we take the diagonal part of the stress tensor as

$$\sigma_{ij}^* = \begin{pmatrix} \sigma_{(1)}^* & 0 & 0 \\ 0 & \sigma_{(2)}^* & 0 \\ 0 & 0 & \sigma_{(3)}^* \end{pmatrix} = (s^* - z^*) \cdot g_3^* \cdot \begin{pmatrix} K_{x_{\text{act/pass}}} & 0 & 0 \\ 0 & K_{y_{\text{act/pass}}}^{(x_{\text{act/pass}})} & 0 \\ 0 & 0 & 1 \end{pmatrix}. \tag{35}$$

The earth pressure coefficients, $K_{x_{\text{act/pass}}}$ and $K_{y_{\text{act/pass}}}^{(x_{\text{act/pass}})}$, are set according to the local state of deformation. The following selection rules[4] (e.g., Gray et al. [7]) apply in the direction of motion,

$$K_{(1)} = \begin{cases} K_{x_{\text{act}}} : & \partial u_1/\partial x_1 > 0, \\ K_{x_{\text{pass}}} : & \partial u_1/\partial x_1 < 0, \end{cases} \tag{36a}$$

and the direction given by the coordinate system axis $\boldsymbol{e}_2$,

$$K_{(2)} = \begin{cases} K_{y_{\text{act}}}^{(x_{\text{act}})} : & \partial u_2/\partial x_2 > 0,\ \partial u_1/\partial x_1 > 0, \\ K_{y_{\text{act}}}^{(x_{\text{pass}})} : & \partial u_2/\partial x_2 > 0,\ \partial u_1/\partial x_1 < 0, \\ K_{y_{\text{pass}}}^{(x_{\text{act}})} : & \partial u_2/\partial x_2 < 0,\ \partial u_1/\partial x_1 > 0, \\ K_{y_{\text{pass}}}^{(x_{\text{pass}})} : & \partial u_2/\partial x_2 < 0,\ \partial u_1/\partial x_1 < 0. \end{cases} \tag{36b}$$

[4] Tai and Gray [24] proposed a modification of equations (36a) and (36b) to renormalise the discontinuities in the coefficients due to changing states of deformation in either of the horizontal directions.

Introducing (35) into (34) yields

$$\mathcal{B}_i^* = \frac{1}{A_b^* \overline{h}^*} \oint_{\partial A_b} g_3^* \, K_{(i)}(l^*) \, n_i(l^*) \frac{h^{*\,2}(l^*)}{2} dl^* . \tag{37}$$

In the Finite Difference formulation, which has been applied in previous implementations of the SAVAGE–HUTTER model (e.g., Gray et al. [7]), (37) appears as the gradient of the hydrostatic pressure, proportional to the gradient of the local flow-depth, $\partial h / \partial x_i$.

Summing up, the governing equations, written in dimensional variables, take the form:

$$\boxed{\begin{aligned} &\frac{d(A_b \cdot \overline{h})}{dt} = 0, \quad \frac{d\,\overline{u}_i}{dt} = \overline{g}_i + \tan\delta \frac{\overline{u}_i}{\sqrt{\overline{u}_j \overline{u}_j}} \frac{\overline{\sigma}_{33}^{(b)}}{\varrho_0 \overline{h}} + \mathcal{B}_i, \ (i = 1, 2), \\ &\overline{\sigma}_{33}^{(b)} = \varrho_0 \overline{h} \cdot \left[g_3 - \overline{u}_1^2 \frac{\partial^2 b}{\partial x_1^2} \right], \quad \mathcal{B}_i = \frac{1}{A_b \overline{h}} \oint_{\partial A_b} g_3 \, K_{(i)}(l) \, n_i^{(h)}(l) \frac{h^2(l)}{2} dl. \end{aligned}} \tag{38}$$

This system represents a set of three equations (conservation of mass and conservation of momentum for the two directions, $\boldsymbol{e}_1$ and $\boldsymbol{e}_2$, of the NCS) for the three unknowns, $\overline{h}$, $\overline{u}_1$ and $\overline{u}_2$. We would like to point out that owing to the choice of the unit normal vector pointing outwards of the control volume, the normal stress component at the bottom, $\overline{\sigma}_{33}^{(b)}$, in general has a negative value. Nevertheless, for the unlikely case of the centripetal acceleration causing a change of the sign, a cut-off for the bed friction term has to be introduced. This accounts for the fact that the SAVAGE–HUTTER model in principle does not include the detachment of dense flow avalanches from the slope terrain.

2.4 Modification of the Bed Friction Law

The dry friction COULOMB law may be applied in granular flows provided the momentum transfer in the shear layer is mainly caused by friction between particles, rather than inter-particle collisions. Thus, it seems to be adequate as a friction law in numerical models of experiments for which the SAVAGE–HUTTER model has been validated (e.g., Koch et al. [15]). Nevertheless, dense flow snow avalanches can reach peak velocities as large as 70 ms^{-1}, far beyond those occurring in laboratory chutes. The friction then will be a function of the shear velocity and thus no longer be independent of the dynamics of the avalanche.

A strong variance of the pauschalgefälle[5] evaluated from data-sets of slope topographies (source: Austrian Federal Ministry for Agriculture and Environment), related to different avalanche events and presented in Table 1 confirms this argument. Applying a simple balance between kinetic and potential energy and the dissipation caused by basal COULOMB friction, the pauschalgefälle is found to be identical to the bed friction angle, δ, and thus a constant. From this point of view the strong difference between several values given in Table 1 clearly disproves the assumption of a constant bed friction angle.

[5] The pauschalgefälle is defined as the angle between the horizontal plane and the straight line connecting the center of mass of the release area with that of the deposition area.

Table 1. Pauschalgefälle of various documented avalanche events in Austria

event	pauschalgefälle	mass [10^6 kg]
Moosbach (Kapl) 1984	21.8°	50
Wolfsgrube (St. Anton) 1988	22.5°	37
Madlein (Ischgl) 1984	24.5°	76
Verwall (Obergurgl) 1986	29.0°	6

Table 2. Model parameters for dry snow avalanche simulations

parameter	symbol	value [unit]
bed friction angle	δ	16.0 [deg]
internal friction angle	φ	35.0 [deg]
dynamic friction coefficient	c_{dyn}	0.022 [–]

Contrary to the approach of combining a Coulomb and a velocity dependent part of the shear stress,

$$\sigma_{13}^{(b)} = \tau_{\text{Coul}} + \tau_{\text{dyn}}(u), \tag{39}$$

as proposed by Gray and Tai [6], we apply a model that applies either the one or the other depending on a criterion for the flow regime (i.e., frictional or collisional momentum transfer) of the shear layer at the bottom of the avalanche. To this end, we still take (11) to be valid, but switch from the quasi-static overburden given by (31), $p_{\text{stat}} = \varrho_0 \overline{h} \cdot \left[g_3 - \overline{u}_1^2 \partial^2 b / \partial x_1^2\right]$, to a velocity dependent normal stress of the form

$$p_{\text{dyn}} = c_{\text{dyn}}\, \varrho_0\, \overline{u}_1^2, \tag{40}$$

if its value exceeds p_{stat}, implying

$$\sigma_{33}^{(b)} = \max(p_{\text{stat}}, p_{\text{dyn}}). \tag{41}$$

The main motivation for this approach was given by the fact that in computational model studies with (41) as dynamic boundary condition the best results could be obtained. With a fixed set of model parameters the deposition area of dense flow avalanches in comparison to field data of several avalanche events listed in Table 1 could be satisfactorily re-calculated. These parameters are listed in Table 2.

From a physical point of view, (41) can be interpreted as an instantaneous transition from a quasi-static friction layer at the bottom of the dense flow avalanche with permanent contact between the particles (Coulomb regime) to a fluidised layer over which the avalanche is sliding[6]. Since in the latter process the momentum transfer is mainly caused by particle-particle collisions, the friction will be

[6] There are reports of field observation indicating a flow regime transition in the deceleration phase of the avalanche in the runout zone [13].

proportional to the square of the rate of deformation. This flow regime has been referred to by Bagnold [2] as the *grain inertia regime*. From this point of view, (40) can be interpreted to arise from a linear vertical velocity distribution inside the friction layer.

3 Powder Snow Avalanche Model

In many avalanche events a powder snow avalanche (PSA) develops on top of the DFA. It consists of ice particles that are lifted from the surface of the DFA or directly from the snow cover by aerodynamic forces. From a physical point of view, the PSA is a suspension consisting of ice particles and air. Furthermore, since characteristic REYNOLDS numbers are typically large, the flow will be turbulent. Because of the particle load, the PSA is heavier[7] than the surrounding, particle free air. Hence, after its formation the PSA is accelerated by its own weight due to gravity and can even decouple from the DFA.

The size of the calculation domain used in the numerical simulations of avalanche events typically lies in the range of a few cubic kilometers. This requires that the governing equations must be treated in their ensemble averaged form since Large Eddy (LES) or direct numerical (DNS) simulation for the time being would by far exceed today's computational capacities.

3.1 Governing Equations for Mean Flow Properties

The conventional mean value of a physical property P is defined by

$$\overline{P} = \lim_{N\to\infty} \frac{1}{N} \sum_{n=1}^{N} P_{(n)}, \tag{42}$$

with $P_{(n)}$ denoting the value of P in the n-th system of an ensemble of N identical systems. The instantaneous decomposition of P into its conventional mean value and fluctuation is given by

$$P = \overline{P} + P'', \quad \overline{P''} = 0.$$

Specifically, we apply the density weighted FAVRE averaging,

$$P = \widetilde{P} + P', \quad \widetilde{P} = \frac{\overline{\varrho P}}{\overline{\varrho}}, \quad \widetilde{P'} = \frac{\overline{\varrho P'}}{\overline{\varrho}} = 0, \tag{43}$$

in order to avoid an artificial source term in the mass balance of the mixture,

$$\boxed{\frac{\partial \overline{\varrho}}{\partial t} + \frac{\partial (\overline{\varrho}\widetilde{u}_i)}{\partial x_i} = 0.} \tag{44}$$

Here the conventionally averaged mixture density, $\overline{\varrho}$, is a function of the ice-particle volume fraction, $\overline{\Phi}$,

$$\overline{\varrho} = \varrho_{\text{ice}}\overline{\Phi}(\boldsymbol{x},t) + \varrho_{\text{air}} \cdot (1 - \overline{\Phi}(\boldsymbol{x},t)). \tag{45}$$

[7] The core of a PSA can show volumetric densities in the range of 10 kg m^{-3}.

The intrinsic densities of the constituents, ϱ_{ice} and ϱ_{air}, are taken to be constant. Density or volume variation of the mixture thus arises from a variable particle volume fraction, $\overline{\Phi}$, only. A separate relation for this unknown is provided in form of the partial mass balance for the ice particles,

$$\boxed{\frac{\partial\overline{\Phi}}{\partial t} + \frac{\partial(\overline{\Phi}\widetilde{u}_i)}{\partial x_i} = -\frac{\partial(\overline{\Phi \cdot u_i'})}{\partial x_i} + u_{\text{sl}}\frac{g_i}{\|\boldsymbol{g}\|}\frac{\partial\overline{\Phi}}{\partial x_i}.} \tag{46}$$

Since the particle volume fraction determines the local density of the mixture, this relation plays a similar role as the energy balance in flows where a coupling between density and temperature field occurs (e.g., natural convection). In the first expression on the right-hand side, in the following referred to as the divergence of the turbulent particle volume flux, $j_i = \overline{\Phi \cdot u_i'}$, a closure relation has to be adapted. The second term is the gradient of a flux caused by a constant slip velocity, u_{sl}, in the direction of the gravitational acceleration, $g_i/\|\boldsymbol{g}\|$. The contribution of the latter to the mixture balances is typically small and will be neglected, i.e., the two constituents are treated to have identical mean values of velocity, $\widetilde{u}_i \equiv \widetilde{u}_{(\text{air})\,i} \approx \widetilde{u}_{(\text{ice})\,i}$, in the momentum and mass balance for the mixture.

The description of the powder snow part of the avalanche is completed with the momentum balance

$$\boxed{\frac{\partial(\overline{\varrho}\widetilde{u}_i)}{\partial t} + \frac{\partial(\overline{\varrho}\widetilde{u}_i\widetilde{u}_j)}{\partial x_j} = -\frac{\partial\overline{p}}{\partial x_i} + \frac{\partial\overline{\tau}_{ij}}{\partial x_j} + \frac{\partial\tau_{ij}^{\text{Rey}}}{\partial x_j} + \overline{\varrho}g_i,} \tag{47}$$

where p denotes the static pressure. The mean values of the viscous stresses, $\overline{\tau}_{ij}$, in general are taken to be negligibly small compared to the turbulent REYNOLDS stresses, $\tau_{ij}^{\text{Rey}} = -\overline{\varrho u_i' u_j'}$. The latter have to be derived again from a turbulence model.

3.2 Turbulence Model

A modified k-ϵ turbulence model [19] is applied to close the system given by (44), (46) and (47). The turbulent viscosity, μ_{tur}, appearing in the closure relation for the turbulent stress components,

$$\tau_{ij}^{\text{Rey}} = \mu_{\text{tur}} \cdot \left(\frac{\partial\widetilde{u}_i}{\partial x_j} + \frac{\partial\widetilde{u}_j}{\partial x_i}\right) - \frac{2}{3}\left(\mu_{\text{tur}}\frac{\partial\widetilde{u}_k}{\partial x_k} + \overline{\varrho}\,k\right)\delta_{ij}, \tag{48}$$

is expressed in terms of the turbulent kinetic energy, $k = \overline{u_i' u_i'}/2$, and its dissipation, ϵ,

$$\mu_{\text{tur}} = C_\mu \overline{\varrho}\frac{k^2}{\epsilon}. \tag{49}$$

This representation is consistent with basic dimensional considerations [19]. In analogy to equation (48), the turbulent particle volume flux is set proportional to the gradient of the mean value of the particle volume fraction,

$$j_i = -(\overline{\Phi \cdot u_i'}) = \frac{\mu_{\text{tur}}}{\varrho(\Phi)\text{Sc}_{\text{tur}}}\frac{\partial\overline{\Phi}}{\partial x_i}. \tag{50}$$

The turbulent SCHMIDT number, $\mathrm{Sc}_{\mathrm{tur}}$, reflects the difference between turbulent mass and momentum diffusivity. In the bulk of the PSA an analogy between turbulent mass and momentum transport is assumed. This is expressed by setting $\mathrm{Sc}_{\mathrm{tur}} = 1$.

For the two unknowns, k and ϵ, in equation (49) further balance laws have to be provided. In the conservation equation for the turbulent kinetic energy,

$$\boxed{\frac{\partial(\overline{\varrho}k)}{\partial t} + \frac{\partial(\overline{\varrho}\widetilde{u}_j k)}{\partial x_j} = -\overline{\varrho}\epsilon + \mathcal{D} + \mathcal{P}_{\mathrm{shear}} + \mathcal{P}_{\mathrm{buo}},} \tag{51a}$$

the net change of k, results from (i) dissipation,

$$-\overline{\varrho}\epsilon = -\mu_{\mathrm{tur}}\overline{\frac{\partial u_i'}{\partial x_j}\frac{\partial u_i'}{\partial x_j}}, \tag{51b}$$

(ii) the diffusive term,

$$\mathcal{D} = \frac{\partial}{\partial x_j}\left(\frac{\mu_{\mathrm{tur}}}{\mathrm{Pr}_k}\frac{\partial k}{\partial x_j}\right), \tag{51c}$$

as well as (iii) production rates due to mean shear,

$$\mathcal{P}_{\mathrm{shear}} = \mu_{\mathrm{tur}}\frac{\partial \widetilde{u}_j}{\partial x_i}\cdot\left(\frac{\partial \widetilde{u}_j}{\partial x_i} + \frac{\partial \widetilde{u}_i}{\partial x_j}\right) - \frac{2}{3}\frac{\partial \widetilde{u}_i}{\partial x_i}\cdot\left(\mu_{\mathrm{tur}}\frac{\partial \widetilde{u}_j}{\partial x_j} + \overline{\varrho}k\right), \tag{51d}$$

and (iv) buoyancy [19],

$$\mathcal{P}_{\mathrm{buo}} = -\frac{\mu_{\mathrm{tur}}}{\mathrm{Sc}_{\mathrm{tur}}\overline{\varrho}}\frac{\partial \overline{\varrho}}{\partial x_i}g_i. \tag{51e}$$

The PRANDTL number for the turbulent kinetic energy, Pr_k, occurring in (51c) represents the ratio between turbulent momentum transfer and the transport of kinetic energy due to turbulent fluctuations. Similarly as for the turbulent SCHMIDT number, we assume an analogy between these two turbulent transport processes to be valid in the bulk of the avalanche by setting $\mathrm{Pr}_k = 1$.

The balance for ϵ reads [19]

$$\boxed{\begin{aligned}\frac{\partial(\overline{\varrho}\epsilon)}{\partial t} + \frac{\partial(\overline{\varrho}\widetilde{u}_j\epsilon)}{\partial x_j} = {} & \frac{\partial}{\partial x_j}\left(\frac{\mu_{\mathrm{tur}}}{\mathrm{Pr}_k}\frac{\partial \epsilon}{\partial x_j}\right) + C_1\frac{\epsilon}{k}\cdot(\mathcal{P}_{\mathrm{shear}} + C_{3\epsilon}\mathcal{P}_{\mathrm{buo}}) \\ & - C_2\frac{\overline{\varrho}\epsilon^2}{k} + C_3\overline{\varrho}\epsilon\frac{\partial \widetilde{u}_i}{\partial x_i}.\end{aligned}} \tag{52}$$

The values of the constants used in the turbulence model (49)–(52) are summarized in Table 3. The multiplication factor of the buoyancy-term in (52),

$$C_{3\epsilon} = 1 - \frac{|\widetilde{u}_i g_i|}{\sqrt{\widetilde{u}_j\widetilde{u}_j}\sqrt{g_j g_j}},$$

Table 3. Constants of the turbulence model

$\mathrm{Sc}_{\mathrm{tur}} = 1$, $\mathrm{Pr}_k = 1$, $C_1 = 1.44$, $C_2 = 1.92$, $C_3 = -0.373$, $C_{3\epsilon} \in [0,1]$, $C_\mu = 0.09$

vanishes if the velocity of the mean flow is aligned with the direction of gravity, i.e., $\widetilde{u}_i g_i = \pm\sqrt{\widetilde{u}_j \widetilde{u}_j}\sqrt{g_j g_j}$, whereas the maximum value 1 is obtained if these two directions are perpendicular with respect to each other, i.e., $\widetilde{u}_i\, g_i = 0$.

4 Re-suspension Layer

As mentioned before, the formation of the powder snow avalanche is caused by the transfer of mass (i.e., ice particles) from the free surface of the DFA into the turbulent air. During this formation process the DFA supplies the PSA with momentum. As the PSA takes on mass, it is accelerated by gravity so that it may even decouple from the DFA. Furthermore ice particles from the resting snow cover can be incorporated into this turbulent mixture flow. The *entrainment* of particles from the loosely packed granular beds or flowing dense granular avalanches into a fluid by aerodynamic forces acting on the particles will subsequently be referred to as *re-suspension*. This process takes place in a thin layer of the powder snow avalanche close to the lower bounding surface. The high REYNOLDS numbers associated with the mean powder avalanche flow imply that this region of re-suspension can be treated as a boundary layer, where high gradients occur.

4.1 Boundary Conditions for the Powder Snow Avalanche

In order to avoid high grid resolutions at solid confinements of the computational domain in high REYNOLDS number turbulent flow simulations, most common CFD (Computational Fluid Dynamics) codes utilise wall functions as boundary conditions for the mean velocity profile [1,25]. The wall distance, y, commonly is made non-dimensional [22] by the dynamic viscosity of air, μ_{air}, the density, $\overline{\varrho}$, and the wall friction velocity, u_{wall},

$$y^+ = y\frac{\overline{\varrho}\, u_{\text{wall}}}{\mu_{\text{air}}}. \tag{53}$$

The wall friction velocity is linked to the local wall shear stress, τ_{wall}, and the density, $\overline{\varrho}$, by the following relation:

$$u_{\text{wall}} = \sqrt{\frac{\tau_{\text{wall}}}{\overline{\varrho}}}. \tag{54}$$

The value of the mean velocity difference, $\Delta\widetilde{u}$, which is parallel to the tangential plane of the slope, is calculated from the law of the wall,

$$\Delta\widetilde{u} = u_{\text{wall}}\frac{1}{\kappa}\ln y^+ + C(k^+), \tag{55}$$

which applies for the non-dimensional wall coordinate in a range of $y^+ \sim \mathcal{O}(10)$ (see e.g. Schlichting and Gersten [22]). The application of the law of the wall in a strict sense is only valid for steady flow conditions. Nevertheless, in the reference frame of the DFA, velocity changes with respect to the surrounding air or PSA are sufficiently slow that the assumption of quasi-stationarity is justified. In (55) κ stands for the VON KÁRMÁN constant, $\kappa = 0.41$. The further constant, $C(k^+)$, depends on the wall roughness, k, whose non-dimensional value is defined as $k^+ = k\overline{\varrho}u_{\text{wall}}/\mu_{\text{air}}$ [22].

The logarithmic law of the wall (55) for single phase shear flows can be derived in many ways; for example by applying asymptotic methods in the high REYNOLDS number limit [16]. In the latter it is a consequence of the matching procedure between the viscous wall layer and the fully turbulent part of the boundary layer. Zwinger [28] extended the classical turbulent boundary layer theory in order to account for particles suspended in the turbulent ambient fluid. He was able to prove that equation (55) also holds for most geophysical gas-solid flows. In order to apply the classical single phase theory, the particle volume fraction in the fully turbulent region of the boundary layer has to be of the same order of magnitude as the primary parameter of expansion. This parameter is given by the inverse of the logarithm of the global REYNOLDS number [16],

$$\mathrm{Re} = \frac{\overline{\varrho}\Delta\widetilde{u}L}{\mu_{\mathrm{air}}} \gg 1.$$

In addition to the mixture density and the viscosity of air, the REYNOLDS number contains the characteristic length of the avalanche, L, and the characteristic velocity difference of the PSA relative to either the DFA or the resting snow cover, $\Delta\widetilde{u}$. The difference between the particle volume fraction at y^+, $\overline{\Phi}(y^+)$, and the flow volume fraction in the DFA, $\Phi_{\mathrm{wall}} = \varrho_0/\varrho_{\mathrm{ice}}$, is determined by the matching procedure and again takes a logarithmic form

$$\Delta\overline{\Phi} = \overline{\Phi}(y^+) - \Phi_{\mathrm{wall}} = -\frac{j_{\mathrm{wall}}}{u_{\mathrm{wall}}}\frac{\mathrm{Sc}^+_{\mathrm{tur}}}{\kappa}\ln y^+ + K(\alpha). \tag{56}$$

The turbulent SCHMIDT number, $\mathrm{Sc}^+_{\mathrm{tur}}$, is the characteristic ratio between the turbulent momentum and mass transport in the re-suspension layer at the surface of the DFA or the snow cover. The constant K depends on the ratio of the material densities, $\alpha = \varrho_{\mathrm{air}}/\varrho_{\mathrm{ice}}$. At sufficiently small wall distances, $y^+ < 1$, i.e., close to the surface of the DFA or the snow cover, the logarithmic terms in (55) and (56) will show large values. The constants therefore may be neglected [18]; this allows computation of the logarithmic terms from both (55) and (56) ,

$$\ln y^+ = -\frac{\kappa}{\mathrm{Sc}^+_{\mathrm{tur}}}\frac{u_{\mathrm{wall}}}{j_{\mathrm{wall}}}\Delta\overline{\Phi} = \frac{\kappa}{u_{\mathrm{wall}}}\Delta\widetilde{u}.$$

Consequently, the turbulent particle volume flux at the wall, j_{wall}, can be expressed in terms of the velocity and volume fraction difference and the wall friction velocity,

$$j_{\mathrm{wall}} = -u^2_{\mathrm{wall}}\frac{1}{\mathrm{Sc}^+_{\mathrm{tur}}}\frac{\Delta\overline{\Phi}}{\Delta\widetilde{u}} = -\frac{\tau_{\mathrm{wall}}}{\overline{\varrho}}\frac{1}{\mathrm{Sc}^+_{\mathrm{tur}}}\frac{\Delta\overline{\Phi}}{\Delta\widetilde{u}}. \tag{57}$$

The turbulent SCHMIDT number has to be determined by experiments. In case of $\mathrm{Sc}^+_{\mathrm{tur}}$ taking on the value 1, complete analogy between the turbulent momentum and mass transport is subsumed.

5 Numerical Implementation of the Coupled Model

Based on the SAVAGE–HUTTER theory, (38), (40) and (41) have been derived to describe the dynamics of the DFA. The ensemble averaged equations for the PSA are given by the mass and momentum balances (44), (46) and (47) as well as the balance equations of the k-ϵ turbulence model (51a) and (52). Due to the different nature of the related flow regimes, the DFA and the PSA are treated in two separate numerical models. The coupling is achieved by applying the expression for the turbulent volume flux (57) from the free surface DFA into the PSA.

These coupled modules are combined in the physical simulation code for dry snow avalanches, SAMOS (**S**now **A**valanche **MO**deling and **S**imulation). SAMOS was developed for the Austrian Federal Ministry for Agriculture, Forestry and Environment in a cooperation between AVL-List, Graz and the Institute for Fluid Mechanics and Heat Transfer of Vienna University of Technology and with consultation from the Austrian Institute for Torrent and Avalanche Control Research, Innsbruck and the Austrian Service for Avalanche and Torrent Control. It is currently used as a tool in avalanche risk assessment in Austria.

The structure of the simulation code is connected to the input of the mountain terrain as well as the initial conditions of the avalanche, i.e., the release area. Hence, we first shall focus on the input data, which the user usually has at her/his disposal.

5.1 Input Data

The quality of the numerical solution is closely affected by the resolution and accuracy of the digital elevation model of the mountain terrain. This data is given by a set $\mathcal{B}$ of $N_b \in \mathbb{Z}^+$ three-dimensional vector components, $\boldsymbol{X}^{(n)} = (X_1^{(n)}, X_2^{(n)}, X_3^{(n)})^T$, describing the surface of the terrain in a General CARTESIAN coordinate System (GCS),

$$\mathcal{B} = \{\boldsymbol{X}^{(1)}, \boldsymbol{X}^{(2)}, \ldots, \boldsymbol{X}^{(n)}, \ldots, \boldsymbol{X}^{(N_b)}\}. \tag{58}$$

In the sense of $s(x_1, x_2)$ in (5), the values $X_3^{(n)}$ can be written as a functionally depending on $X_1^{(n)}$ and $X_2^{(n)}$,

$$X_3^{(n)} = S(X_1^{(n)}, X_2^{(n)}). \tag{59}$$

The avalanche always starts in form of a dense flow avalanche, which originates from an initial slab of snow in the release area that usually disintegrates into smaller sized particles during the first few seconds of the avalanche event [13]. The release area enters the algorithm in form of an initial distribution of a set of $N_{\mathrm{Lag}} \in \mathbb{Z}^+$ LAGRANGE vertex coordinates, $\boldsymbol{X}^{(\alpha)} = (X_1^{(\alpha)}, X_2^{(\alpha)}, X_3^{(\alpha)})^T$,

$$\mathcal{L}(t=0) = \{\boldsymbol{X}^{(1)}, \boldsymbol{X}^{(2)}, \ldots, \boldsymbol{X}^{(\alpha)}, \ldots, \boldsymbol{X}^{(N_{\mathrm{Lag}})}\}_{t=0}. \tag{60}$$

As a consequence of the fact that the avalanche remains attached to the terrain-surface on which it slides down, one of these three coordinates underlies a kinematic constraint. Thus, the component $X_3^{(\alpha)}$ is set to a value, $S^{(\alpha)}$, that is obtained by an interpolation of the function $S(X_1^{(n)}, X_2^{(n)})$ from equation (59) between the known values at the nodes of the slope.

In addition to the LAGRANGE vertex positions, initial volumes, $\mathcal{V}(t=0) = \{V^{(1)}, V^{(2)}, \ldots, V^{(\alpha)}, \ldots, V^{(N_{\mathrm{Lag}})}\}_{t=0}$, are needed as an initial condition for the DFA simulation. In practice this is set by prescribing the snow cover depth, $h_{\mathrm{sc}}^{(\alpha)}$, and density, $\varrho_{\mathrm{sc}}^{(\alpha)}$, for the single parts of the release area. In connection with the shape of the basal area, $A_b^{(\alpha)}$, of the control volume obtained from the initial triangulation of the release area and the assumed flow density, ϱ_0, the initial flow-depths and volumes of the LAGRANGE points

$$\overline{h}^{(\alpha)} = \frac{\varrho_{\mathrm{sc}}^{(\alpha)}}{\varrho_0} h_{\mathrm{sc}}^{(\alpha)}, \quad V^{(\alpha)} = A_b^{(\alpha)} \overline{h}^{(\alpha)}, \tag{61}$$

can be computed.

Triangulations For the simulation the points of the mountain terrain (58) as well as the LAGRANGE vertices (60) need to be combined into a grid. For both sets a triangular grid is used,

$$\begin{aligned} \mathcal{T}_b &= \{T_1, T_2, \dots T_m, \dots T_{M_b}\}, \quad T_m = (m_1, m_2, m_3)^T, \quad M_b \in \mathbb{Z}^+, \\ \mathcal{T}_{\mathrm{Lag}} &= \{T_1, T_2, \dots T_\beta, \dots T_{M_{\mathrm{Lag}}}\}, \quad T_\beta = (\beta_1, \beta_2, \beta_3)^T, \quad M_{\mathrm{Lag}} \in \mathbb{Z}^+. \end{aligned} \tag{62}$$

These triangulations form a set of M_b and M_{Lag} triplets of point indices, T_m and T_β, that describe the triangles formed from the sets $\mathcal{B}$ and $\mathcal{L}$, respectively. Since these two surfaces are geometric entities in $\mathbb{R}^2$, it is sufficient – provided that the mountain terrain can be projected onto the $X_1 - X_2$ plane[8] – to perform these triangulations in the horizontal plane of the GCS only.

In SAMOS, a DELAUNAY triangulation of the corresponding point sets is used (see e.g. Edelsbrunner [4]). It optimises the triangular interconnections of the points such that the global minimal angle is maximised and no vertex lies inside the circumcircle of any triangle. The VORONOI cells are the geometric dual of the DELAUNAY triangulation. A VORONOI cell combines the centers of the circumcircles of the triangles adjacent to a LAGRANGE point. The VORONOI cell in general is a closed, convex polygon. The LAGRANGIAN Finite Volume scheme gains stability if the net flux through the surface of the control volume, i.e., the bisection term $(38)_4$, is evaluated in many directions. Since the resulting prism will have more side faces than one with a triangular basal area, we prefer the VORONOI cell to serve as the basal area of the control volume in our simulation algorithm.

Interpolations of Properties on Triangles Values for properties in points not coinciding with vertices of the triangulations have to be interpolated. For a certain triangle, $T = (m_1, m_2, m_3)^T$, the nodal values of a property P at the boundary points, $\boldsymbol{X}^{(m_1)}$, $\boldsymbol{X}^{(m_2)}$, $\boldsymbol{X}^{(m_3)}$, are given as P_1, P_2 and P_3. It is convenient to define a local coordinate system with its origin $\boldsymbol{X}^{(m_1)}$ and the two directions $\boldsymbol{\Gamma}_1 = \boldsymbol{X}^{(m_2)} - \boldsymbol{X}^{(m_1)}$ and $\boldsymbol{\Gamma}_2 = \boldsymbol{X}^{(m_3)} - \boldsymbol{X}^{(m_1)}$. Any point inside the triangle, $\boldsymbol{X} = \boldsymbol{X}^{(m_1)} + \gamma_1 \cdot \boldsymbol{\Gamma}_2 + \gamma_1 \cdot \boldsymbol{\Gamma}_2$, can be identified by a coordinate pair (γ_1, γ_2), for which the following constraints apply:

$$\gamma_1 \in [0, 1], \quad \gamma_2 \in [0, 1], \quad \gamma_1 + \gamma_2 \leq 1.$$

The interpolated value of the property then takes the form

$$P(\boldsymbol{X}) = (1 - \gamma_1 - \gamma_2) P_1 + \gamma_1 P_2 + \gamma_2 P_3. \tag{63}$$

This interpolation procedure is applied for scalars as well as vector fields on both triangulations, T_b and T_{Lag}.

[8] This is equal to the postulation that the inclination of the slope with respect to the $X_1 - X_2$ plane always lies in the range between 0 and 90 degrees.

5.2 Lagrangian Finite Volume Scheme for DFA

The advantage of the numerical implementation in form of a LAGRANGIAN Finite Volume Scheme is given by the fact that the mesh stretches only over the area of the DFA mass. Compared to the EULERIAN approach, where the entire avalanche track has to be covered by the mesh all the time, this leads to a significant gain of performance of the solution algorithm and, additionally, reduces the needed amount of memory, which meets the requirements of a simulation that can be run on common PC desktop computers.

The clear disadvantage of a LAGRANGIAN scheme is the higher liability to numerical instabilities. These are mainly caused by highly skewed mesh elements that arise from unfavourable constellations of grid points (e.g., a too close proximity of one point to the other) that may occur during the simulation. The closer two LAGRANGE points are, the smaller the basal area of their assigned control volume gets. Because of the incompressibility assumption $(38)_1$ this inevitably leads to a strong local increase of the flow-depth and in turn to unphysically high local gradients of the hydrostatic pressure causing overestimated accelerations. In previous numerical implementations of the SAVAGE–HUTTER model based on the Finite Difference method (e.g., Koch et al. [15]) artificial diffusion terms had been introduced to avoid these proximity instabilities. Due to its more homogeneous spatial distribution of the momentum flux[9] for a single LAGRANGE point the Finite Volume scheme as we apply it, by default, will show a lower liability to this kind of instabilities.

Additionally, the triangulation is updated during the simulation. The VORONOI cell, contrary to a triangular control volume, is assigned to a single LAGRANGE point only. This circumvents re-distribution of the LAGRANGE point volumes, $\mathcal{V}$, which would lead to artificial mass diffusion. The re-triangulation of the LAGRANGE points at each time step takes only a small fraction of the overall computing time and causes no noticeable decrease in the performance of the algorithm.

Natural Coordinates The equations for the DFA (38) have been derived by applying the simplification introduced by the curvilinear coordinates of the NCS (12). As a consequence, a modification of the infinitesimal path element, dl, due to the metric introduced by the curvilinear coordinates should be taken into account. The scaling introduced in $(29)_2$ implies that typical length scales of the avalanche are smaller than the radius of curvature of the slope. Thus, in the direction of the velocity vector the slope can locally be approximated by a circle defined by the local radius of curvature, R. As depicted in Fig. 4, the neglect of the distortion of the path element then is equal to the replacement of the arc, $\hat{l}$, by the secant, l. The length of the arc can be expressed by means of the difference in the inclination angle of the slope , $\Delta\xi$, and the local radius of curvature, r, $\hat{l} = r \cdot \Delta\xi$. From this it follows that $\Delta\xi$ is of the order $\lambda = L/R = \varepsilon^{1/2}$. On the other hand, the length of the secant takes the form $l = 2\,r\tan(\Delta\xi/2)$. Expansion for small angles $\Delta\xi/2 \ll 1$

[9] The bisection term $(38)_4$ is a momentum flux through the side faces of the control volume caused by a gradient of the hydrostatic pressure.

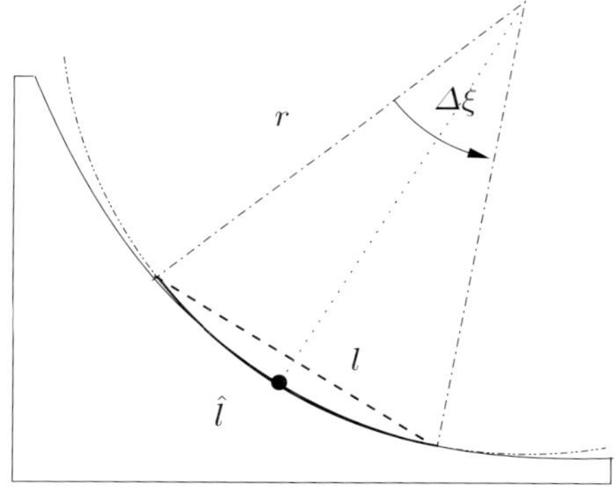

Fig. 4. Approximation of the local arc length by the circle defined by the radius of curvature (*solid thick line*) by the secant (*dashed thick line*)

of the right-hand side leads to

$$\tan(\Delta\xi/2) = \underbrace{\tan(0)}_{=0} + \underbrace{\frac{1}{\cos^2 0}}_{=1}\frac{\Delta\xi}{2} + \underbrace{\frac{\sin 0}{\cos^3 0}}_{=0}\left(\frac{\Delta\xi}{2}\right)^2 + \underbrace{\frac{1+2\sin^2 0}{3\cos^4 0}}_{1/3}\left(\frac{\Delta\xi}{2}\right)^3 + \mathcal{O}\left((\frac{\Delta\xi}{2})^4\right).$$

For the difference between the length of the arc and the secant given in dimensionless coordinates,

$$\hat{l}^* - l^* = \frac{1}{24} r^* \underbrace{\frac{R}{L}}_{=\mathcal{O}\left(\varepsilon^{-1/2}\right)} \underbrace{\Delta\xi^3}_{=\mathcal{O}\left(\varepsilon^{3/2}\right)} = \mathcal{O}(\varepsilon), \tag{64}$$

we obtain an order of magnitude estimation of $\mathcal{O}(\varepsilon)$. Since the path element, dl, appears only in the bisection term, which itself is of the order $\mathcal{O}\left(\varepsilon^{1/2}\right)$, the overall error introduced by replacing the curved slope locally by the tangent is $\mathcal{O}\left(\varepsilon^{3/2}\right)$ and thus compatible with the accuracy of the SAVAGE–HUTTER model.

In terms of the applied coordinates this can be interpreted as a replacement of the curvilinear NCS by local natural CARTESIAN coordinates (NCCS). The transformation of a vector, $\boldsymbol{X}$, from the general CARTESIAN system (GCS) (see Sect. 5.1) into the NCCS, $\boldsymbol{X} \to \boldsymbol{x}$, then is defined by

$$\boldsymbol{x} = \mathbf{A} \cdot \boldsymbol{X}. \tag{65}$$

The rows of the transformation matrix, $\mathbf{A} \in \mathbb{R}^{3\times 3}$ consist of the three vectors (12) defining the axes of the NCS at the position of the LAGRANGE vertex,

$$\mathbf{A} = (\boldsymbol{e}_1, \boldsymbol{e}_2, \boldsymbol{e}_3)^T. \tag{66}$$

As mentioned in Sect. 2.2, these three vectors define an orthonormal system. Consequently, the determinant of the matrix, $\det \mathbf{A} = 1$, has unity value, proving that the transformation (65) is a combination of translations and rotations, that conserves the metric.

Computation of the Bisection Term The bisection term $(38)_4$ for a LAGRANGE vertex, $\boldsymbol{X}^{(\alpha)}$, is given by the sum of the acceleration at all side faces of the assigned VORONOI cell. The basal area of the VORONOI cell is a closed polygon. It is obtained by connecting the circumcircle centers of all triangles, T_β, that contain $\boldsymbol{X}^{(\alpha)}$. Each side of the VORONOI cell bisects one side of these triangles. A bisected triangle side always is the straight connection between $\boldsymbol{X}^{(\alpha)}$ and another LAGRANGE point, $\boldsymbol{X}^{(m_i)}$, $m_i \in T_\beta \wedge \alpha \in T_\beta$, $\alpha \neq m_i$. The net contribution of the neighbour triangles T_β to the acceleration of the LAGRANGE point $\boldsymbol{X}^{(\alpha)}$ is then given by

$$\boldsymbol{\mathcal{B}}(\boldsymbol{X}^{(\alpha)}) = \boldsymbol{\mathcal{B}}^{(\alpha)} = \frac{\varrho_0}{A_b^{(\alpha)} \overline{h}^{(\alpha)}} \sum_{m_i} \boldsymbol{\mathcal{B}}^{(\alpha \to m_i)}. \tag{67}$$

The upper index (α) shall indicate properties evaluated for the VORONOI cell assigned to the LAGRANGE vertex $\boldsymbol{X}^{(\alpha)}$. The index $(\alpha \to m_i)$ indicates a property evaluated at the bisection point $\boldsymbol{X}^{(\alpha \to m_i)} = 1/2 \cdot (\boldsymbol{X}^{(\alpha)} + \boldsymbol{X}^{(m_i)})$.

The separate contributions to the bisection term only have to be computed once for each of the bisection points between two adjacent LAGRANGE vertices that share a triangle of the triangulation, $\boldsymbol{X}^{(\alpha)}$ and $\boldsymbol{X}^{(\gamma)}$. The direction of the normal has to be set according to the orientation of the VORONOI cell

$$\boldsymbol{n}^{(h)}(\boldsymbol{X}^{(\alpha \to \gamma)}) = \begin{cases} (\boldsymbol{X}^{(\gamma)} - \boldsymbol{X}^{(\alpha)}) \cdot \|\boldsymbol{X}^{(\gamma)} - \boldsymbol{X}^{(\alpha)}\|^{-1} & \text{for } \boldsymbol{X}^{(\alpha)}, \\ (\boldsymbol{X}^{(\alpha)} - \boldsymbol{X}^{(\gamma)}) \cdot \|\boldsymbol{X}^{(\alpha)} - \boldsymbol{X}^{(\gamma)}\|^{-1} & \text{for } \boldsymbol{X}^{(\gamma)}. \end{cases} \tag{68}$$

The contribution from a single bisection to either the net balance of the LAGRANGE vertex $\boldsymbol{X}^{(\alpha)}$ or $\boldsymbol{X}^{(\gamma)}$ then is given by

$$\boldsymbol{\mathcal{B}}^{(\alpha \to \gamma)} = \boldsymbol{n}^{(h)}(\boldsymbol{X}^{(\alpha \to \gamma)})\, l_{\mathcal{B}}\, K_{(i)}\, g_3 \frac{h(\boldsymbol{X}^{(\alpha \to \gamma)})}{2}. \tag{69}$$

The local flow-depth is obtained by the arithmetic mean of the values at the end points of the bisected triangle side,

$$h(\boldsymbol{X}^{(\alpha \to \gamma)}) = \tfrac{1}{2} \cdot (h(\boldsymbol{X}^{(\alpha)}) + h(\boldsymbol{X}^{(\gamma)})). \tag{70}$$

If two triangles, T_{β_1} and T_{β_2}, share the bisected side, i.e., the side is not at the boundary of the triangulation, the length of the VORONOI cell side is given by the distance between the center of the circumcircles of the triangles, $l_{\mathcal{B}} = \|\boldsymbol{C}^{(\beta_1)} - \boldsymbol{C}^{(\beta_2)}\|$. If the triangle, T_{β_1}, is at the edge of the domain this length is taken to be the distance between the center of the circumcircle and the bisection point itself, $l_{\mathcal{B}} = \|\boldsymbol{C}^{(\beta_1)} - \boldsymbol{X}^{(\alpha \to \gamma)}\|$.

Discretisation of the Momentum Balance We now turn our attention to the integration of the momentum balance $(38)_2$. In consideration of the non-linearity of the highly coupled equations we chose an explicit scheme for time integration. Taking a time step from $t_{(n)}$ to $t_{(n+1)} = t_{(n)} + \Delta t$, the momentum balance takes the form

$$\frac{\overline{\boldsymbol{u}}_{(n+1)}^{(\alpha)} - \overline{\boldsymbol{u}}_{(n)}^{(\alpha)}}{\Delta t} = \overline{\boldsymbol{g}}_{(n)}^{(\alpha)} + \frac{\overline{\boldsymbol{u}}_{(n+1)}^{(\alpha)}}{\|\overline{\boldsymbol{u}}_{(n)}^{(\alpha)}\|} \frac{\tan\delta\, \overline{\sigma}_{33}^{(b)}}{\varrho_0 \overline{h}_{(n)}^{(\alpha)}} + \boldsymbol{\mathcal{B}}_{(n)}^{(\alpha)}. \tag{71}$$

The normal stress at the bottom,

$$\overline{\sigma}_{33}^{(b)} = \max\left[\varrho_0\,\overline{h}_{(n)}^{(\alpha)} \cdot \left(g_{3\,(n)}^{(\alpha)} - \left(\overline{\boldsymbol{u}}_{(n)}^{(\alpha)}\right)^2 \frac{\partial^2 b_{(n)}^{(\alpha)}}{\partial x_1^2}\right), c_{\text{dyn}}\varrho_0 \cdot \left(\overline{\boldsymbol{u}}_{(n)}^{(\alpha)}\right)^2\right], \tag{72}$$

is inserted according to the friction regime evaluated for the LAGRANGE vertex using (41). The velocity in the numerator of the fraction in the second term of (71) is taken at $t_{(n+1)}$. This provides higher numerical stability, since friction is also introduced in the momentum balance in the direction perpendicular to the velocity, $\overline{\boldsymbol{u}}_{(n)}^{(\alpha)}$. If we solve (71) for the velocity at the new time step, we obtain

$$\overline{\boldsymbol{u}}_{(n+1)}^{(\alpha)} = \left[\overline{\boldsymbol{u}}_{(n)}^{(\alpha)} + \Delta t \cdot \left(\overline{\boldsymbol{g}}_{(n)}^{(\alpha)} + \boldsymbol{\mathcal{B}}_{(n)}^{(\alpha)}\right)\right] \cdot \left[1 - \Delta t \frac{\tan\delta\,\overline{\sigma}_{33}^{(b)}}{\varrho_0 \|\overline{\boldsymbol{u}}_{(n)}^{(\alpha)}\|}\right]^{-1}. \tag{73}$$

Since this scheme is explicit, it is only conditionally stable for all time step sizes.

Computation of New Lagrange Vertex Position and Flow-Depths

New positions for the LAGRANGE vertices, $\boldsymbol{X}_{(n+1)}^{(\alpha)}$, are evaluated in a straightforward manner. The distance obtained from the multiplication of the previously evaluated velocity vector and the time step simply is added to the position of the former time step,

$$\boldsymbol{X}_{(n+1)}^{(\alpha)} = \boldsymbol{X}_{(n)}^{(\alpha)} + \overline{\boldsymbol{u}}_{(n+1)}^{(\alpha)} \cdot \Delta t. \tag{74a}$$

Since this in general will violate the kinematic constraint given by (8), the vertical component of the position vector has to be corrected in order to set the LAGRANGE point back onto the surface topography

$$X_{3\,(n+1)}^{(\alpha)} = S^{(\alpha)} = S(X_{1\,(n+1)}^{(\alpha)}, X_{2\,(n+1)}^{(\alpha)}). \tag{74b}$$

A new triangulation is performed on the updated LAGRANGE vertices. Consequently, new basal areas, $A_{b\,(n+1)}^{(\alpha)}$, of the assigned VORONOI cells have to be evaluated. Additionally, the VORONOI cell volumes, $V_{(n+1)}^{(\alpha)}$, are corrected with respect to the re-suspension flux from the DFA into the PSA given by equation (57). This is the only part of the DFA module that needs to communicate with the PSA code. After performing these updates and corrections the new flow-depths are obtained by applying

$$\overline{h}_{(n)}^{(\alpha)} = V_{(n+1)}^{(\alpha)} / A_{b\,(n+1)}^{(\alpha)}. \tag{75}$$

All steps of the numerical integration of the SAVAGE–HUTTER model are presented in the flowchart given in Table 4.

5.3 Eulerian Finite Volume Scheme for PSA

The shallowness approximation of the DFA (28) does not apply to the PSA. Hence the equations of motion for the PSA must be solved on a three-dimensional domain. To this end, the yetexisting simulation code FIRE™ [1] from AVL-List has been adapted for the PSA part. This code solves the REYNOLDS averaged NAVIER–STOKES equations (RANS) by applying the Finite Volume Method.

Table 4. Flowchart of the time integration loop for the DFA module

Input from main program: all topographic input data of mountain terrain, release area vertices, physical and numerical parameters	
start time integration loop	
	update of Lagrange vertex triangulation, $\mathcal{T}_{\mathrm{Lag}}$
	computation of new Voronoi cell areas, $A_{(b)}$
	correction of x_3 component of Lagrange vertex velocity, u_3; equation (23)
	solution of mass balance; equation (75) corrected by re-suspension flux (57)
	interpolation of topographical data at Lagrange point (slope normals, slope curvature)
	computation of velocity gradients in Lagrange triangles
	computation of bisection terms, $\boldsymbol{\mathcal{B}}^{(\alpha\rightarrow\gamma)}$, for all triangle connections; equation (67) – (70)
	solution of momentum balance; equation (73) with (72)
	relocation of Lagrange vertices; equation (74a)
	correction of Lagrange vertices; equation (74b)
end of time integration loop	
return to main program	

Structured Grid The discretisation of the domain is given by a grid consisting of structured layers of hexahedral volume elements above the terrain surface, as shown in Fig. 5. It covers a region along the sturzbahn of the avalanche[10] and extends a few hundred meters above the mountain terrain, which corresponds to the typical vertical extents of the PSA. The grid is constant with respect to space and time. Typical cell dimensions in the directions parallel to the tangential plane of the mountain terrain are 10 meters. The vertical cell spacing varies from 10 meters at the upper boundary to sub-meter values close to the slope/DFA. For simulations of large scale avalanche events a domain in the range of a few cubic kilometers has to be divided into discrete cells, leading to a total number of grid cells in the range of 10^5 to 10^6. This makes the solution algorithm of the PSA simulation module a task by far more time consuming than the SAVAGE–HUTTER model for the DFA. Since the DFA is assumed to be very shallow, no time dependent modification of the lower PSA grid cells is accounted for, i.e., topographical changes of the mountain terrain caused by dense flowing or deposited snow are neglected.

---

[10] Since the sturzbahn is a result of the simulation it is of course not known beforehand. It is the task of the user to estimate the position and extent of the PSA grid.

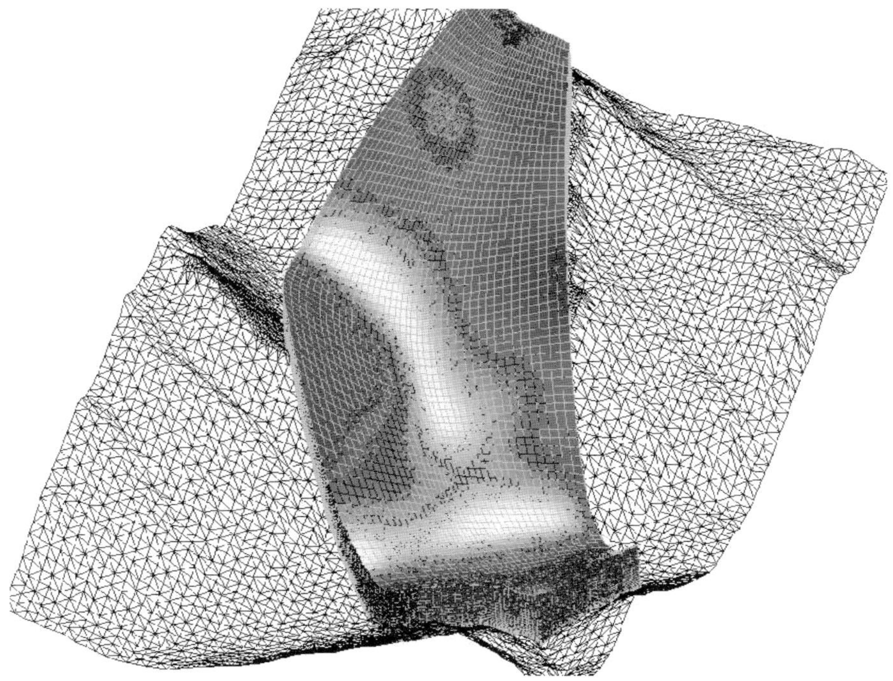

Fig. 5. The triangulated mountain terrain and the superimposed hexahedral structured grid for the PSA simulation

Discretisation of PSA Equations The balance equations of mass (44), particle volume fraction (46), momentum (47), kinetic turbulent energy (51a) as well as dissipation (52) can be written in a general form as

$$\frac{\partial}{\partial t}\left(\overline{\varrho}\boldsymbol{U}\right) + \frac{\partial}{\partial x_j}\left(\overline{\varrho}\widetilde{u}_j\boldsymbol{U} - \boldsymbol{D}\frac{\partial \boldsymbol{U}}{\partial x_j}\right) = \boldsymbol{S} + \boldsymbol{P}. \tag{76}$$

Here, $\boldsymbol{U}$ is the vector of the unknown field variables, $\boldsymbol{D}$ stands for the vector of the turbulent exchange coefficients and $\boldsymbol{S}$ denotes the source vector,

$$\begin{aligned}
\boldsymbol{U} &= (1,\ \widetilde{u}_1, \widetilde{u}_2,\ \widetilde{u}_3,\ \Phi/\overline{\varrho},\ k,\ \varepsilon)^T,\\
\boldsymbol{D} &= \mu_{\mathrm{tur}} \cdot (0,\ 1,\ 1,\ 1,\ 1/(\overline{\varrho}\mathrm{Sc}_{\mathrm{tur}}),\ 1/\mathrm{Pr}_k,\ 1/\mathrm{Pr}_k)^T,\\
\boldsymbol{S} &= (0,\ u_{\mathrm{sl}}\frac{g_i}{\|\boldsymbol{g}\|}\frac{\partial\overline{\Phi}}{\partial x_i},\ \overline{\varrho}g_1,\ \overline{\varrho}g_2,\ \overline{\varrho}g_3,\ \mathcal{P}_{\mathrm{shear}} + \mathcal{P}_{\mathrm{buo}} - \overline{\varrho}\epsilon,\ S_7)^T,\\
S_7 &= C_1\frac{\epsilon}{k}\cdot(\mathcal{P}_{\mathrm{shear}} + C_{3\epsilon}\mathcal{P}_{\mathrm{buo}}) - C_2\frac{\overline{\varrho}\epsilon^2}{k} + C_3\overline{\varrho}\epsilon\frac{\partial\widetilde{u}_i}{\partial x_i}.
\end{aligned}$$

The vector $\boldsymbol{P}$ contains the pressure gradient and shows non-zero elements only in the components corresponding to the momentum equation

$$\boldsymbol{P} = (0,\ 0,\ -\partial\overline{p}/\partial x_1,\ -\partial\overline{p}/\partial x_2,\ -\partial\overline{p}/\partial x_3,\ 0,\ 0)^T.$$

Relation (76) is integrated with respect to time and space, yielding

$$\frac{1}{\delta t}\int_{t_n}^{t_n+\delta t}\left[\int_V \frac{\partial}{\partial t}\left(\overline{\varrho}\boldsymbol{U}\right)dV + \oint_{\partial V}\left(\overline{\varrho}\widetilde{u}_j\boldsymbol{U} - \boldsymbol{D}\frac{\partial \boldsymbol{U}}{\partial x_j}\right)\cdot n_j\,dA\right]dt$$

$$= \frac{1}{\delta t}\int_{t_n}^{t_n+\delta t}\left[\int_V\left(\boldsymbol{S}+\boldsymbol{P}\right)dV\right]dt. \quad (77)$$

Here the control volume, V, with its bounding surface, ∂V, is given by a hexahedral cell of the discretisation shown in Fig. 5. The time step size for the PSA simulation, δt, has to be an integer multiple of the time step size for the DFA simulation, Δt. The integral over the side faces of a hexahedral cell can be replaced by the sum of the six side face areas,

$$\oint_{\partial V} dA \approx \sum_{k=1}^{6} A_k. \quad (78)$$

The spatial integral of the terms on the right-hand side in (77) is approximately written as

$$\int_V \left(\boldsymbol{S}+\boldsymbol{P}\right)dV \approx V\cdot\left(\overline{\boldsymbol{S}}+\overline{\boldsymbol{P}}\right). \quad (79)$$

Spatial derivatives are resolved applying the first order upwind discretisation scheme. In consideration of numerical stability a fully implicit time discretisation scheme is applied. The temporal derivatives of the spatially integrated functions $F(u_i,\boldsymbol{U})$ in the first term on the left-hand side of (77) are evaluated at the new time level $t_{n+1} = t_n + \delta t$,

$$\frac{1}{\delta t}\int_{t}^{t+\delta t} F(u_i,\boldsymbol{U})\,dt = F(u_i,\boldsymbol{U})_{t+\delta t}. \quad (80)$$

Recapitulating, the general transport equation (76) in its linearised form is given by

$$A_p\boldsymbol{U}_c = \sum_n A_n\boldsymbol{U}_n + \overline{\boldsymbol{S}} + \overline{\boldsymbol{P}}, \quad (81)$$

where the subscript c stands for values evaluated at the center of the cell volume and n denotes values at the 18 points at the corners of the hexahedron as well as face centers. The coefficients A_p and A_n result from the spatial and temporal discretisation schemes (78), (79) and (80), [1].

Pressure–Density Coupling In the momentum equations the pressure enters as an additional variable that needs to be treated separately. The mass balance describes the spatial and temporal evolution of the mixture density, $\varrho(\Phi)$. The density

Table 5. Flowchart of SIMPLE iteration loop (PSA module)

provide initial guess for pressure, p^*, and density, ϱ^*	
start iteration loop	
	solve momentum equations (components 3-5 in (81))
	solve transport equation for pressure correction, p'
	correction of the approximated values, $\widetilde{u}_i, \overline{p}$ and $\overline{\varrho}$ ((82a) and (82b))
	solve equation for turbulent kinetic energy and dissipation (components 6 and 7 in (81))
	check for convergence
end of iteration loop	
next time step	

is a dependent variable, since it is determined by the particle volume fraction, Φ, for which a balance equation (46) already exists. By introducing a relation that links the pressure to the density, the mass balance shall act as the corresponding evolution equation for the pressure. Anyhow, if dealing with incompressible constituents of the mixture, no direct relation between the density and the pressure can be drawn. This problem – similar as in incompressible single phase flows – is taken care of by a numerical scheme, commonly termed SIMPLE algorithm [17], that links the momentum to the pressure equation. In this method the variables appearing in the momentum and mass balances, u_i and p, are expressed in terms of an approximation and a correction,

$$\widetilde{u}_i = u_i^* + u_i', \quad \overline{p}_i = p^* + p'. \tag{82a}$$

The relation between density and pressure is obtained by linking the correction values of these two variables,

$$\overline{\varrho} = \varrho^* + \frac{\partial \overline{\varrho}}{\partial \overline{p}} p'. \tag{82b}$$

In our case the density depends only on the particle volume fraction, but not on pressure. Thus the second term on the right-hand side is always zero. If (82a) and (82b) are inserted into (77), a transport equation for the pressure correction, p', can be obtained [1]. Decoupling the pressure gradients from the momentum equation ("checker board" effect [17]) is avoided by evaluating the pressure gradients normal to the cell faces by the difference between the values of the adjacent cells. For one time step the equations have to be iterated until convergence is obtained, i.e., the change in the correction values falls below a given threshold. The numerical scheme for a single time step is depicted in Table 5.

Numerical Treatment of Boundary Conditions The boundary conditions for the velocity as well as the flux of the particle volume fraction at the mountain terrain, and the PSA/DFA interface, respectively, have already been discussed and are given by (55) and (57).

At the free boundaries of the domain, i.e., at those cell faces at the boundary of the PSA grid not adjoining the mountain terrain, the static pressure is set to the value of the atmospheric pressure,

$$p = p_{\mathrm{atm}}. \tag{83}$$

Where flow leaves the domain, the values of all other field variables are obtained from the solution, applying the VON NEUMANN type of condition, $(\partial \boldsymbol{U}/\partial x_i)\cdot n_i = \mathbf{0}$. Here n_i denotes the normal of the side faces at the free boundary of the computational domain. This relation is equal to a free outlet condition. Consequently, if the PSA leaves the computational domain, the flow is not restricted by the boundary. Where flow enters the domain, the velocity is prescribed in the same way. All other values are set to fixed "environment-values", with the particle volume fraction set to zero. Since PSA mass may leave the domain, global mass conservation in general will be violated. From this point of view it is important to dimension the computational domain (i.e., the PSA grid) sufficiently large in order to be able to track the entire PSA mass.

6 Model Validation and Discussion of Applicability

Field data that would provide all necessary information for the complete reconstruction of the dynamics of a dry snow avalanche are presently not available [13]. The records of real avalanche events typically contain outlines of the deposition area and the position of the fraction line of the release area. Sometimes, the expert is also provided with information on the snow cover depth and density of the release area and an estimation on the volume of the snow deposition. The ascertained damage on structural elements, such as bridges or buildings, sometimes allows to make rough estimations of the impact pressures and consequently, assuming a certain flow density, also of the flow velocities of the avalanche. No information of critical dynamical properties – for instance the mass balance during the event – can be reconstructed afterwards. On the other hand, results obtained in laboratories, such as water tank experiments for powder snow avalanches, do not to all extent satisfy the scaling laws [12]. This prohibits the exact reproduction of real world avalanches on a smaller scale.

All these facts make validation of avalanche simulation models a very difficult task. Thus comparison of simulation results with observations of real avalanche events has to be seen under the aspect that the latter contain a high degree of uncertainty.

6.1 Simulation of a Real Avalanche Event

In Austria data of a series of catastrophic avalanches were recorded and analysed [11]. From the modeler's point of view, no matter how limited the accuracy of the data may be, such avalanche records are most valuable material. One of these well documented events is the Madlein avalanche[11] of 1984 in the community Ischgl

[11] Named after the canyon, *Madleinbachgraben*, through which the avalanche approached the village.

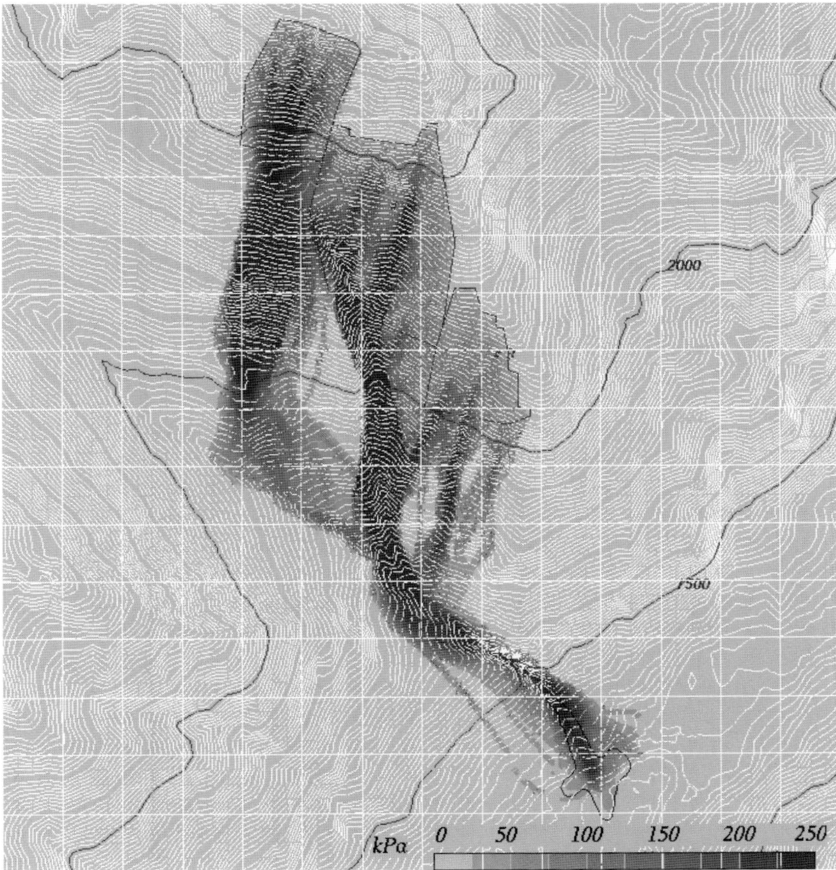

Fig. 6. Peak impact pressure of the DFA along the whole avalanche path. The values are given in units of kPa. Contour lines of the terrain are shown every ten meters. The grid size of the overlayed raster is 200 m

(valley of Paznaun, Tirol). An estimated snow mass of 76 kilotons was released at an altitude between 800 and 1400 meters above the valley. Approximately 100 seconds after triggering, the snow avalanche reached the village where it caused one casualty, personal injury and considerable material damage [10,11].

The numerical grid of the mountain terrain applied in the DFA simulation and the EULERIAN mesh for the PSA are depicted in Fig. 5. The visualisation of the maximum impact pressure in the DFA part along the avalanche path (see Fig. 6) corresponds well with the damage that has been observed on vegetation and constructions.

Since the PSA develops while the avalanche moves it shows maximum impact pressures way down the mountain slope. From the picture displayed in Fig. 7 it follows that impact pressures of the PSA are in the range of 10 kPa and thus an order of magnitude less as compared to the DFA. In this particular event the impact

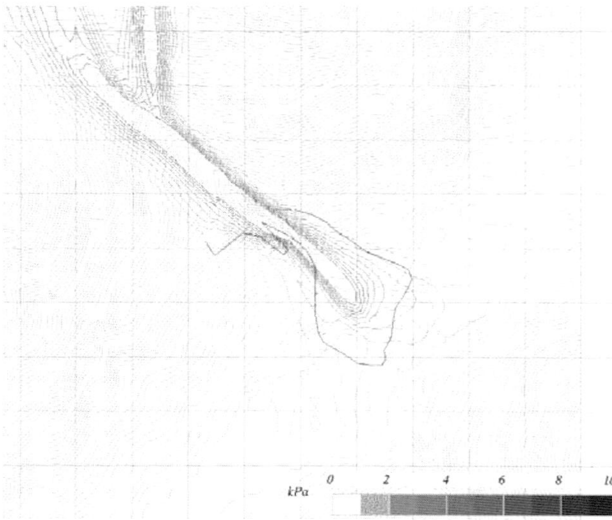

Fig. 7. Contour lines of the computed impact pressure of the PSA in the deposition area. The values are given in units of kPa. The margin of the area where damage caused by the PSA impact has been observed [11] is depicted by the black thick line. Contour lines of the terrain are shown every ten meters. The grid size of the overlayed raster is 200 m

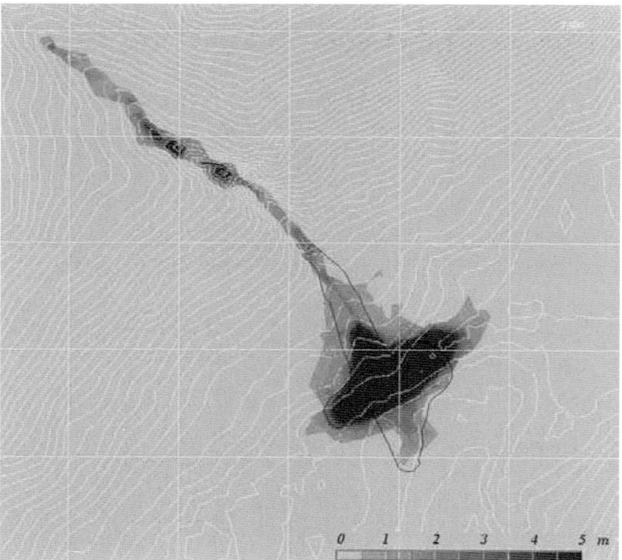

Fig. 8. Contour plot of the computed flow depth of the DFA in the deposition area. The values are given in units of meters. The margin of the observed snow deposition [11] is depicted by the black polygon. Contour lines of the terrain are shown every ten meters. The grid size of the overlayed raster is 200 m

of the DFA mainly caused the damage observed in the village, whereas the PSA destroyed large forest areas along the lower parts of the avalanche track.

The computed results for the resting avalanche of the DFA simulation in Fig. 8 corresponds well with the observed deposition area in the village of Ischgl [11].

6.2 Limitations of the Avalanche Model

Certain preliminary assumptions concerning the physical behaviour of the flow in the different flow regimes have entered the models derived in Sect. 2 and 3. These assumptions involve some limitations when it comes to applications.

The SAVAGE–HUTTER model for the DFA uses the shallowness assumption (28). The error introduced by the expansion of the governing equation is of order $\mathcal{O}(\varepsilon^{3/2})$. Thus, the accuracy will decrease with an increasing ratio of the flow depth to the extension of the avalanche in the directions parallel to the tangential plane of the slope. Anyhow, the DFA typically is shallow enough to impose no major restrictions of applicability.

The scaling introduced by $(29)_2$ implies a typical ratio between the length of the avalanche and the local radius of curvature of the mountain terrain, $\lambda = L/R$, in the order of $\mathcal{O}(\varepsilon^{1/2})$. In general this constraint will be satisfied by most topographies. Nevertheless, we want to mention the fact that the SAVAGE–HUTTER model cannot deal with too sharp edges of the slope along the sturzbahn. Additionally, in such situations the centripetal forces appearing in $(38)_3$ might change the sign of the normal stress causing the DFA to detach from the slope, which would violate a principle assumption of the SAVAGE–HUTTER model. As shown in (64), the scaling introduced by $(29)_2$ is also a necessary condition for the approximation of the curvilinear coordinates of the NCS (12) by local CARTESIAN coordinates (NCCS).

Another shortcoming of the SAVAGE–HUTTER model is the incompressibility assumption applied in the mass balance (1). While it does not introduce major errors for dense flow avalanches in motion, this assumption certainly does not hold in situations where high pressure inside the granular material occurs, e.g., during sudden deceleration due to impact on objects. From this point of view, values of impact pressures on constructions obtained from the DFA simulation should be interpreted as qualitative rather than quantitative results.

With respect to the re-suspension, a net loss of mass of the DFA at its free boundary should be taken into account. This would introduce a surface flux at this boundary in the Finite Volume Scheme of the DFA module. Anyhow, as depicted in Fig. 9, this flux is in the range of 10^{-5} $\mathrm{m\,s^{-1}}$. Thus, for a single time step the assumption of a constant LAGRANGE volume in the SAVAGE–HUTTER model is certainly justified[12].

The boundary condition in form of the turbulent particle volume flux at the wall (57) includes only basic aspects of the very complex physics. Its validity and modifications – for instance with respect to a saltation layer – deserves closer investigation.

In many avalanche events, direct entrainment from the resting snow cover into the avalanche is of great importance. A simple model accounting for the incorporation of snow at the front of the DFA based on the conservation laws for mass and momentum has been added to SAMOS [27]. An application of this additional simulation feature in the re-calculation of the Créta Besse avalanche in Vallée de la Sionne (Switzerland) has been presented by Sailer et. al [20]. A detailed theoretical discussion of this model has to be postponed since it would go beyond the scope of this article.

[12] This assumption certainly has to be reconsidered if additionally entrainment from the resting snow cover into the DFA is taken into account.

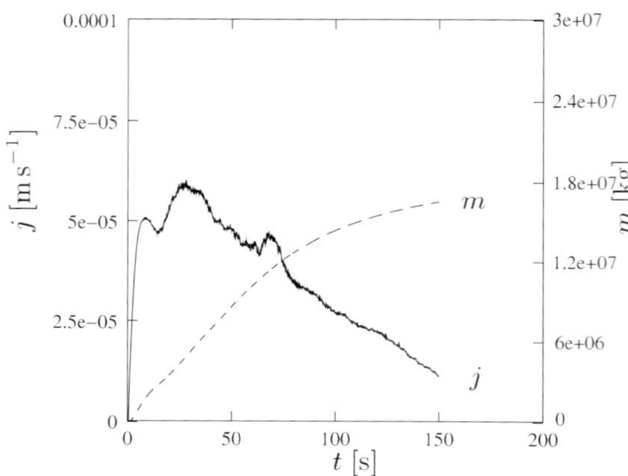

Fig. 9. Averaged particle volume flux, j (*scale on left ordinate, solid line*) and resuspended mass, m (*scale on right ordinate, dashed line*) vs. time of the Madlein avalanche simulation. In total approximately 22% of the DFA mass is transferred into the PSA

The PSA simulation demands a large amount of computational resources, which, with respect to simulation times, imposes limitations on the spatial resolution of the computational domain. The k-ϵ turbulence model is known to have its limitations in unconfined flows and in flows with large extra strains (e.g., strongly curved boundaries, swirling flows) [25]. Nevertheless, the fact that the fully three-dimensional RANS equations are solved for the mixture certainly makes this model superior to depth integrated approaches based on the theory of gravity currents and thermals (e.g., Beghin, Hopfinger and Britter [3]; Fukushima and Parker [5]).

7 Conclusions

A coupled physical model for dry snow avalanches has been presented. As a consequence of the different flow regimes related to the dense flow avalanche (DFA) and the powder snow avalanche (PSA), these two parts of the avalanche have been treated separately. The SAVAGE–HUTTER theory for shallow dense granular flow has been applied to derive the governing equations for the DFA. The model for the PSA is given by the REYNOLDS averaged equations for a particle-gas suspension. A relation for the transfer of snow mass from the former into the latter based on an analogy between turbulent momentum and mass transport has been introduced.

The restriction in applicability of the resulting model imposed by preliminary assumptions and analytical as well as numerical approximations has been discussed in detail. With respect to the inaccuracy of field data, computed results, such as avalanche paths, impact pressures, margins and shapes of avalanche depositions, show excellent agreement with observations of real avalanche events.

Nevertheless, countless open questions concerning the detailed physical processes of avalanche dynamics remain to be answered. The theory of dispersed multiphase flows on the one hand and experimental efforts – in the field as well as in the laboratory – are challenged to provide these answers. This inevitably will lead to enhanced numerical models in the future and the demand for improved numerical techniques that meet the criteria of these new computational implementations.

During recent years, physical avalanche models, like the one presented in this article, have gained importance as useful tools in risk assessment. As construction and planning of industrial flow processes have become unimaginable without the tools provided by CFD (Computational Fluid Dynamics), computational simulation might play a similar role in future risk management. This should be a major motivation for oncoming theoretical, numerical as well as experimental investigations in the related scientific fields.

Acknowledgment

Thomas Zwinger would like to thank AVL-List GmbH for their financial support of his PhD. thesis.

References

1. AVL List (2001) FIRE v7.3 Theory (volume 3). AVL List GmbH Graz
2. Bagnold R.A. (1954) Experiments on gravity-free dispersion of large solid spheres in a Newtonian fluid under shear. Proc. R. Soc. London **225A**, 49–63
3. Beghin P., Hopfinger E.J., Britter R.E (1979) Gravitational convection from instantaneous sources on inclined boundaries. J. Fluid Mech. **107**, 407–422
4. Edelsbrunner H. (1987) Algorithms in combinatorial geometry. EATCS monographs on theoretical computer science **10**, Springer, Heidelberg
5. Fukushima and Parker (1990) Powder Snow Avalanches: Theory and application, J. Glaciology **36** (123), 229–237
6. Gray J.M.N.T., Tai Y.C. (1998) On the inclusion of a velocity-dependent basal drag in avalanche models. Annal. Glac. **26**, 277–280
7. Gray J.M.N.T., Wieland M., Hutter K. (1999) Gravity-driven free surface flow of granular avalanches over complex basal topography. Proc. R. Soc. Lond. A **455**, 1841–1874
8. Greve R., Hutter K. (1993) Motion of a granular avalanche in a convex and concave curved chute: experiments and theoretical predictions. Phil. Trans. R. Soc. London **A 342**, 573–600
9. Haff P.K. (1983) Grain flow as a fluid mechanical phenomenon. J. Fluid Mech. **134**, 401–430
10. Hagen G. (1997) Endbericht über das AVL-Staublawinenmodell. Forsttechnischer Dienst für Wildbach- und Lawinenverbauung (Avalanche and Torrent Control, Forestry Service), Sektion Vorarlberg, Bregenz.
11. Hufnagl H. (1988) Ergebnisse einer rechnerischen Auswertung von fünf Lawinen des Katastrophenwinters 83/84. Proc. Interpraevent 88 **3**, 227–249
12. Hutter K. (1996) Avalanche Dynamics. In V.P. Singh, editor, Hydrology of Disasters, 317–394, Kluwer, Dordrecht
13. Issler D. (2003) Experimental Information on the Dynamics of Dry-Snow Avalanches. This volume
14. Johnson P.C., Nott P., Jackson R. (1990) Frictional-collisional equations of motion for particulate flows and their application to chutes. J. Fluid Mech. **210**, 501–535
15. Koch T., Greve R., Hutter K. (1994) Unconfined flow of granular avalanches along a partly curved surface. II. Experiments and numerical computations. Proc. R. Soc. Lond. A **445**, 415–435

16. Mellor G.L. (1972) The Large Reynold Number Asymptotic Theory of Turbulent Boundary Layers. Int. J. Engng. Sci. **10**, 851–873
17. Patankar S.V. (1980) Numerical heat transfer and fluid flow. McGraw-Hill, New York
18. Prandtl L., Oswatitsch K., Wieghardt K. (1984) Führer durch die Strömungslehre. Eighth edition, Vieweg, Braunschweig/Wiesbaden
19. Rodi W. (1980) Turbulence models and their applications in hydraulics. IAHR, Delft
20. Sailer R., Rammer L., Sampl P. (2002) Recalculation of an Artificially Released Avalanche with SAMOS and Validation with Measurements from a Pulsed Doppler Radar. To appear in Natural Hazards and Earth System Science **2**, 3/4, 211–217
21. Savage S.B., Hutter K. (1989) The motion of a finite mass of granular material down a rough incline. J. Fluid Mech. **199**, 177–215
22. Schlichting H., Gersten K. (2000) Boundary Layer Theory. Springer, Heidelberg
23. Stadler R. (1986) Stationäres, schnelles Fließen von dicht gepackten trockenen und feuchten Schüttgütern. PhD. Thesis, University of Karlsruhe, Karlsruhe
24. Tai Y.C., Gray J.M.N.T. (1998) Limiting stress states in granular avalanches. Annals Glaciology **26**, 272–276
25. Versteeg H.K., Malalasekera W. (1995) An Introduction to Computational Fluid Dynamics. The finite volume method. Addison Wesley, Longman, Essex
26. Voellmy A. (1955) Über die Zerstörungskraft von Lawinen. Schweizerische Bauzeitung **73**, 12. English Translation (1964) *On the destructive force of avalanches*, U.S. Department of Agriculture, Forest Service, Alta Avalanche Study Center Translation **2**
27. Zwinger T., Kluwick A. (2000) Schneeaufnahme am Kopf einer Trockenschneelawine. Project report, Inst. for Fluid Mechanics a. Heat Transfer, Vienna University of Technology
28. Zwinger T. (2000) Dynamik einer Trockenschneelawine auf beliebig geformten Berghängen. PhD. Thesis, Vienna University of Technology, Vienna

Particle Image Velocimetry (PIV) for Granular Avalanches on Inclined Planes

Winfried Eckart[1], John Mark Nicholas Timm Gray[2], and Kolumban Hutter[1]

[1] Institute of Mechanics, Darmstadt University of Technology, Hochschulstr. 1, 64289 Darmstadt, Germany, e-mail: {eckart, hutter}@mechanik.tu-darmstadt.de
[2] Department of Mathematics, University of Manchester, Oxford Road, Manchester M20 2SQ, U.K., e-mail: ngray@ma.man.ac.uk

received 22 May 2002 — accepted 15 Oct 2002

Abstract. This paper is concerned with experimental results of sand avalanches flowing down inclined plexiglass chutes with lateral confinement. We introduce the Particle Image Velocimetry (PIV) technique for granular avalanches and discuss the differences in its implementation compared to standard PIV. Surface and side-wall velocity measurements are described at different downslope chute locations and for four different inclination angles. The PIV-system provides a good measuring technique to determine flow velocities of granular avalanches. The flow behaviour for the smallest inclination angle investigated here deviates significantly from the other results as the avalanche reaches a terminal velocity. This gives valuable indications for the right choice of a bed friction law.

1 Introduction

In this paper we are interested in determining the flow behaviour of *granular avalanches* experimentally. A granular avalanche is a (dense) assembly of particles that flows over an arbitrary bed topography. Granular avalanches may occur both as natural events and in industrial applications and are driven by gravity. The flow of such an avalanche is not only influenced by the type of material that is flowing, but also by the type of bed it is flowing over. In nature snow avalanches are initiated by disturbances to weak layers in the snow, and the path that the avalanche takes depends on various factors such as the slope topography, the frictional resistance and the amount of material that is eroded or deposited during the flow. In industry, avalanches are initiated by raising gates or opening valves and they flow down carefully constructed chutes following a controlled path. Here we will investigate experiments with dry sand avalanches that flow down open plexiglass channels with a prescribed width. This flow is *confined* because the avalanche cannot spread in the cross-slope direction (perpendicular to the steepest slope). This flow configuration resembles the situation encountered in industrial applications rather than in natural events.

We are interested in "measuring" quantities with which the flow behaviour of an avalanche can be described. One of the most important quantities is the *velocity* of the avalanche. In particular, we are seeking information about the velocity of particles at the surface, the base and the confining walls of our channel. As will be explained later, the spatial distribution of the velocity of the avalanche at every instant of time provides important information about its flow behaviour and may help us to understand avalanches better.

We use an optical measurement technique called *Particle Image Velocimetry* (PIV) to measure the velocity of "visible particles" (i.e. particles at the surface/base/boundary) of a granular avalanche. The basic idea is as follows: The region of interest (for example a part of the surface of a flowing granular avalanche) is illuminated twice (say at time t and time $t + \Delta t$), and two pictures (called "frame A" and "frame B") are captured by a camera. Comparing frame A and frame B, "displacements" of identifiable points of the moving avalanche are calculated via pattern recognition. The velocity of such a point is simply its displacement divided by the time difference Δt between both frames. Note that we cannot obtain any information about particle velocities *inside* the granular avalanche, because the PIV-system is an optical system and hence can only "see" what our eyes can see. To our knowledge, PIV-systems have not been applied to granular avalanches before. The only work known to us using a PIV-system for granular materials was recently published by Lueptow et al. [8] who have measured the velocity of particles in a shaker. A brief discussion of the PIV setup and how it is modified for granular flows is therefore given in Sect. 3.

The motion of the avalanche on a confined inclined plane is comparatively simple. The flow is dominated by gravitational acceleration and basal friction, and there are no changes in topography to complicate the motion. Whilst, internal pressure gradients are only of significance close to the outlet. Although there are some three-dimensional effects, the bulk flow is quasi two-dimensional and this is often reduced to one-dimension in theoretical models [18] by applying a depth integration procedure. In the experiments the friction of the plexiglass chute can easily be held constant to obtain reproducible results and has the added advantage that one can see through it to measure the velocity of particles at the base and along the sidewalls.

The experimental results may be divided into three groups: Surface velocity measurements, velocity measurements from the side and from the base. We will present results for different downslope chute locations and inclination angles. We are particularly interested in the velocity profile through the depth of the avalanche, which is an important quantity in depth-integrated theories. This can be measured directly along the side-walls of the chute, but, as we shall show, the presence of the wall has a significant effect on the magnitude of the velocity at high speeds. Some indication of the velocity profile through the depth away from the side-walls can be obtained by making simultaneous measurements of the surface and basal velocity in the bulk flow. A detailed knowledge of the velocity profile provides general information about the physics of granular materials. It may, for example, help us to determine whether the granular material behaves more like a fluid or a solid. In the present paper, we will focus on surface velocity measurements. They are the most easy and reliable to obtain. Specifically, we will present results obtained near the outlet of the silo and three different distances downstream from it (65 cm, 185 cm, 285 cm). The measurements for different inclination angles which were made both from the side and the surface may help us to decide whether the motion of the avalanche depends on the inclination angle in a way not included in existing theories. For example, it is well-known that the flow behaviour for inclination angles slightly larger than the bed friction angle is significantly different from experiments where the inclination angle is much larger than the bed friction angle, compare for example results in [5,12,13,18] and [21]. Note that in [17] the motion of a single ball down an inclined line was investigated and a regime with constant mean velocity was found.

2 Particle Image Velocimetry: Theory

We briefly review some basic theoretical aspects of PIV. Many details are taken from [16], and the reader is referred to this monograph for further details. Two functions are of exceptional importance in PIV, firstly, the *image field intensity* and, secondly, the *cross-correlation function*. The PIV system is an optical system that essentially interprets differences in light intensities as a pattern[1]. The function describing this pattern is the image intensity field. Particular spots of brightness are interpreted as a grey-scale pattern which, in general, move. Such a moving pattern can most easily be detected by comparing two consecutive frames captured by a camera. The second task of the system is to decide which is the correct displacement (hence velocity) among a certain set of possible displacements. This can be accomplished by constructing the cross-correlation function between two consecutive brightness distributions. To this end it is necessary to detect both a pattern on frame A and frame B and a difference between the patterns from both frames.

In Fig. 1 the *geometric imaging* arrangement is sketched. The whole region of interest (here the light sheet) in the 3-dimensional physical space with coordinates $\mathbf{X}$ is subdivided into smaller regions, called "interrogation volumes". With the imaging lens this volume is mapped into the "interrogation area" in the 2-dimensional image space with the coordinates $\mathbf{x}$. In the following we will use the word "interrogation spot" for either interrogation volume or interrogation area.

2.1 Image Intensity Field

For each of the interrogation spots, the image intensity field assigns to each point in the image plane a scalar value which reflects the light intensity of the corresponding point in the physical space. This light intensity depends on many factors, in particular the illumination, but also physical properties of the lens and the camera. An exhaustive discussion is beyond the scope of this paper, see [16] for further details. Here it suffices to mention that for a particular experiment or series of experiments these latter influences are constant.

The image intensity fields $I(\mathbf{x},\mathbf{\Gamma})$ at time t and $I'(\mathbf{x},\mathbf{\Gamma}+\mathbf{D})$ at time $t+\Delta t$ are defined as follows:

$$I(\mathbf{x},\mathbf{\Gamma}) = \sum_{i=1}^{N} V_0(\mathbf{X_i})\,\tau(\mathbf{x}-\mathbf{x_i}), \tag{1}$$

$$I'(\mathbf{x},\mathbf{\Gamma}+\mathbf{D}) = \sum_{j=1}^{N} V_0'(\mathbf{X_j}+\mathbf{D})\,\tau(\mathbf{x}-\mathbf{x_j}-\mathbf{d}), \tag{2}$$

[1] To generate a pattern in a transparent fluid which is the common case in PIV, the fluid is seeded with small, non-transparent *tracer particles*.

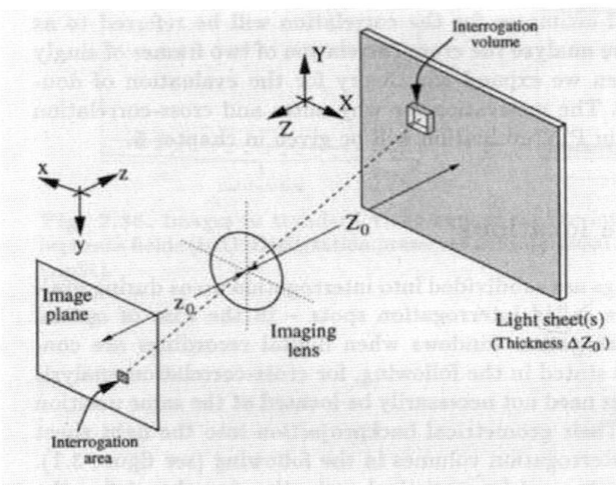

Fig. 1. Sketch of the geometric imaging arrangement (from [16] by permission)

where $\mathbf{\Gamma} = (\mathbf{X_1}, \mathbf{X_2}, ..., \mathbf{X_N})$ denotes the position of the i-th (tracer) particle in the (3-dimensional) physical space, $\mathbf{x}$ the position in the (2-dimensional) image plane, $\mathbf{D}$ the displacement in the physical space ($\mathbf{D} = \mathbf{X_i'} - \mathbf{X_i}$), $\mathbf{d}$ the displacement in the image plane, V_0 the transfer function and τ the point spread function of the lens.

The image intensity field maps the light energy of an individual particle in physical space (i.e. the tracer particles in the observed region of the flow) into an intensity value in the image plane. This function can be viewed as a transformation (or mapping) between two spaces. It is basically a convolution of the transfer function which is responsible for the determination of the "effective" light energy of one particle and the point spread function which describes the optical behaviour of the lens.

What one finally obtains is a field of numbers in the interrogation area for each pixel and each point of time. These numbers may be considered as grey values, and we may interpret the whole set as a *grey scale pattern.* It is in general moving and/or deforming (i.e. a function of time). However, it is common practice to assume that this pattern is moving *rigidly* in the interrogation spot, i.e. all points in this spot are assigned the same velocity.

2.2 Cross-correlation Function

The cross-correlation function $R_{II'}(\mathbf{s}, \mathbf{\Gamma}, \mathbf{D})$ is defined as:

$$R_{II'}(\mathbf{s}, \mathbf{\Gamma}, \mathbf{D}) := \frac{1}{\alpha_I} \int_{\alpha_I} I(\mathbf{x}, \mathbf{\Gamma})\, I'(\mathbf{x} + \mathbf{s}, \mathbf{\Gamma} + \mathbf{D})\, \mathrm{d}\mathbf{x}$$

$$= \frac{1}{\alpha_I} \sum_{i,j} V_0(\mathbf{X_i})\, V_0'(\mathbf{X_j} + \mathbf{D}) \int_{\alpha_I} \tau(\mathbf{x} - \mathbf{x_i})\, \tau(\mathbf{x} - \mathbf{x_j} + \mathbf{s} - \mathbf{d})\, \mathrm{d}\mathbf{x}, \tag{3}$$

where α_I denotes the interrogation area and $\mathbf{s}$ the separation vector in the correlation plane. The function $R_{II'}$ essentially calculates possible displacements by

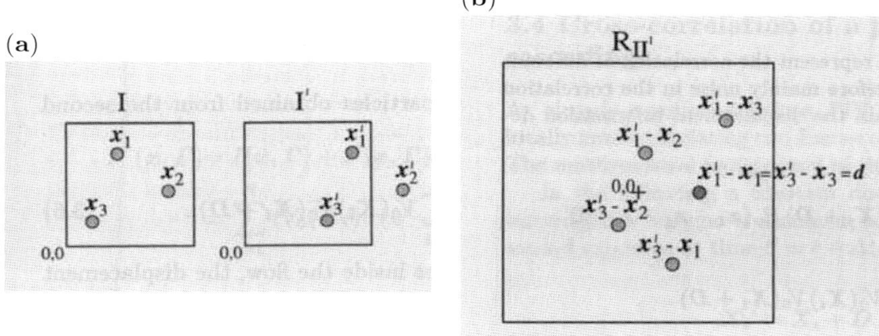

Fig. 2. (**a**) Image intensity fields at time t and $t' = t + \Delta t$ and (**b**) the corresponding cross-correlation (from [16] by permission)

correlating all "grey values" from the first frame (intensity field I) with all "grey values" from the second frame (intensity field I'), where the separation vector is the independent variable that is to be varied. To make this clear, consider the simple example illustrated in Fig. 2. Suppose that three particles are located at $\mathbf{x_1}$, $\mathbf{x_2}$ and $\mathbf{x_3}$ in an interrogation spot at time t, i.e. picture I in Fig. 2 (a). At time $t + \Delta t$, they have moved to the positions $\mathbf{x'_1}$, $\mathbf{x'_2}$ and $\mathbf{x'_3}$ in Fig. 2 (a). In passing we note that $\mathbf{x'_2}$ is not inside the interrogation spot. This is called "loss of pairs". However, the cross-correlation function $R_{II'}$ shown in Fig. 2 (b) is calculated by taking *all* possible displacements *regardless* whether the correct particles are correlated with each other or not, i.e., if the particle at $\mathbf{x_1}$ would have moved to $\mathbf{x'_3}$, then the displacement would be $\mathbf{x'_3} - \mathbf{x_1}$ and so on. Four of the six possible displacements occur only once, but $\mathbf{x'_1} - \mathbf{x_1} = \mathbf{x'_3} - \mathbf{x_3} = \mathbf{d}$ occurs twice. Thus, the displacement $\mathbf{d}$ is the most likely displacement.

$R_{II'}$ may be decomposed into three parts (see KEANE & ADRIAN [7]):

$$R_{II'}(\mathbf{s}, \mathbf{\Gamma}, \mathbf{D}) = R_C(\mathbf{s}, \mathbf{\Gamma}, \mathbf{D}) + R_F(\mathbf{s}, \mathbf{\Gamma}, \mathbf{D}) + R_D(\mathbf{s}, \mathbf{\Gamma}, \mathbf{D}), \tag{4}$$

where R_C is the convolution of the mean intensities (terms $i \neq j$), R_F the fluctuating noise (terms $i \neq j$) and R_D the correlation of identical particle images (terms $i = j$). This means that $R_C + R_F$ produce the errors, whereas R_D is the correct value which is sought. Figure 3 shows a picture of the cross-correlation plane. R_D corresponds to the highest value of $R_{II'}$ in the correlation plane which is called "1st peak". The coordinates s_x and s_y are the separations in the $x-$ and the $y-$direction, respectively. This picture can be read as follows: The most likely displacement is the displacement $d = (s_x(R_D), s_y(R_D))$, where $s_{x,y}(R_D)$ denote the coordinates of R_D in the separation plane.

The quality of this measurement is usually estimated by determining the ratio of the value of the 1st peak divided by the value of the second highest peak (called 1st noise peak, highest peak of the "noise" $R_C + R_F$): A high ratio means high reliability, a low ratio low reliability.

Note that only one displacement vector is calculated within one interrogation spot. The velocity vector is simply the displacement vector divided by the time

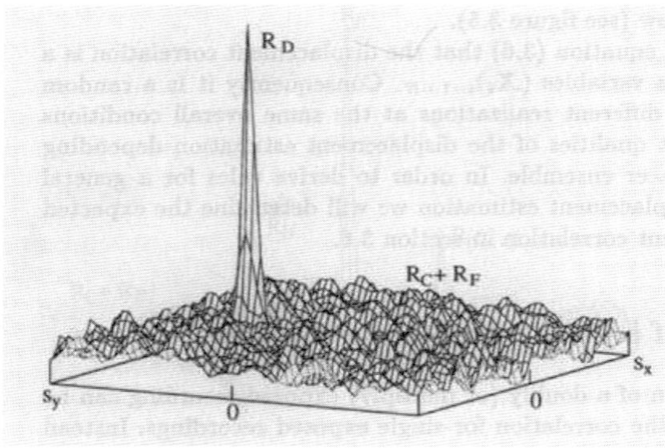

Fig. 3. The cross-correlation plane (from [16] by permission)

delay, Δt, between the two frames. Hence it must be viewed as an average velocity vector of the particles inside the interrogation spot. To increase the spatial accuracy especially for flows with strong displacement gradients, the interrogation spot size must be chosen sufficiently small. Common values are 32 × 32 and 64 × 64. If the camera resolution is 1280 × 1024 and the observed region fits the whole picture, then $1280/32 \times 1024/32 = 1280$ and $1280/64 \times 1024/64 = 320$ velocity vectors are calculated for interrogation spot sizes of 32 × 32 and 64 × 64, respectively.

We may summarise the described procedure as follows:

1. Subdivide the whole image into interrogation spots.
2. Calculate the image intensity fields at time t and $t + \Delta t$ for one interrogation spot.
3. Calculate the correlation between these image intensity fields for one given separation.
4. The cross-correlation plane is built by the correlation values of all possible separations.
5. The most likely displacement is the separation for which the correlation function has the highest value (1st peak).
6. Repeat the procedure for each interrogation spot.

3 Setup and Problems for Granular Materials

In this section, the PIV-setup for granular materials flowing down inclined planes is described. Introducing first the "usual" PIV setup for (transparent) fluids, it will soon become clear that specific problems occur for (generally non-transparent) granular materials. Furthermore, there are some particular problems concerning the use of plexiglass chutes, which will also be discussed in this section.

3.1 Usual PIV Setup for (Transparent) Fluids

Before we describe the PIV setup we used for granular materials, let us briefly review a usual setup in a wind tunnel. A sketch is given in Fig. 4. The heart of each PIV-system is the camera. Together with the lenses this is called the *imaging optics*. For

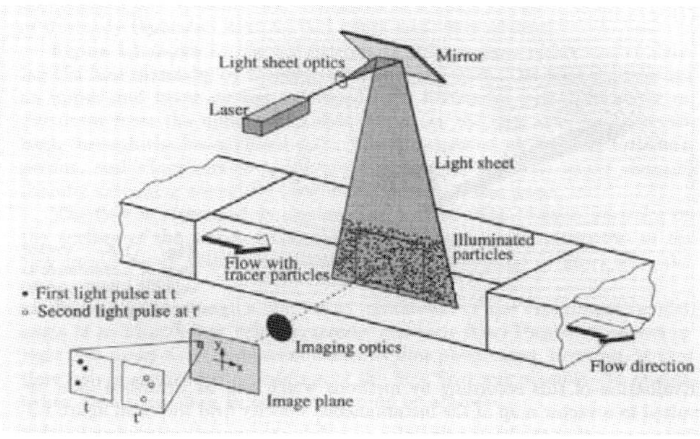

Fig. 4. PIV-setup in a wind tunnel (from [16] by permission)

illumination, a laser including light sheet optics and a mirror are needed. To control and trigger the laser and the camera, a synchroniser and a PC are usually needed. Furthermore, it is convenient to have evaluation and post-processing software. The transparent fluid must be seeded with tracer particles so that the camera and the software can detect some pattern, as described in Sect. 2.

3.2 Setup for Granular Materials

The setup for granular materials differs from that described above as follows. For non-transparent materials, laser light sheets can not be used for illumination. Thus, we employed (computer controllable) flashes which were equipped with a diffusor. In order to guarantee nearly the same brightness on both frames, the intensity and duration of the flashes could be adjusted. From Fig. 4 it is clear that for "usual PIV" the velocity distribution of the tracer particles inside the flow is measured. For "granular PIV", we measure the velocity of particles at the surface or at some boundary. Note that the optical surface structure that is produced by illumination of the surface of the avalanche is already sufficient to detect the motion, i.e. no tracer particles are needed. The flash illumination in connection with the plexiglass chutes that were used may produce different kinds of illumination errors that will be discussed later in greater detail.

We used a system developed by TSI (see www.tsi.com for a detailed description and technical information). It includes two CCD (Charge Coupled Device) cameras of type TSI PIVCAM 13-8, two NIKON super-wide-angle-zoom lenses of type NIKON NIKKOR AF 18-35mm f/3.5D IF, see [15], a synchroniser and a PC including the INSIGHT PIV-software.

For illumination we used either 2 or 4 flashes (with either 1 or 2 cameras) of the type METZ MECABLITZ 60 CT-4, see [14]. Two flashes are needed for each camera. There are two reasons for this. Firstly, it is nearly impossible to trigger and adjust one flash such that the brightness on both pictures would be the same. Secondly, the shutter of the second frame is open for a relatively long time due to technical reasons which would result in distortions if only one flash were used.

(**a**) For surface measurements

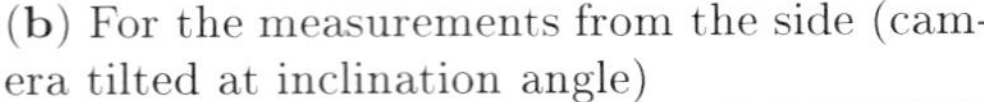

(**b**) For the measurements from the side (camera tilted at inclination angle)

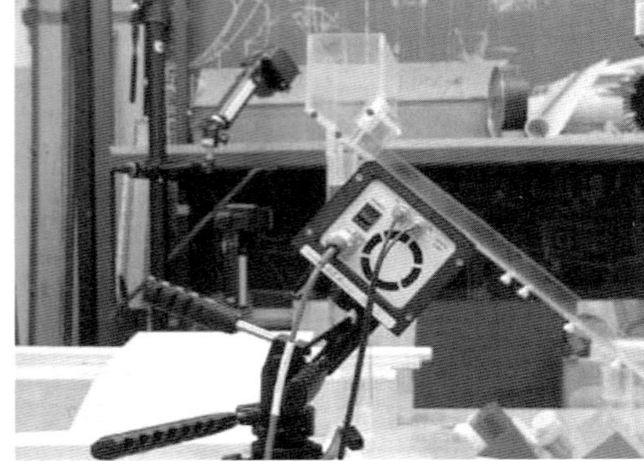

Fig. 5. Setup for granular PIV

The motion can not, in general, be frozen in time and the second picture becomes blurred.

The CCD camera has a resolution of 1,280 × 1,024 pixels and a colour depth of 12 bits (4,096 different grey values). Furthermore, its temporal solution is 4 double frames per second. The time delay between the first and the second frame depends on the range of the velocity values and was chosen to be of the order of 1 msec. in our situation. No polarisation filter was used for the camera. For the measurements made from the side of the chute, the camera was tilted according to the inclination angle used. In this case the down-slope direction is always the positive $x-$direction. Basic features of the setup can be viewed in Fig. 5.

The chutes are made of plexiglass 5 mm thick (Fig. 6). The inclination angles were between 30 and 45 degrees. The length of the short, the long and the extra long chute (measured from the outlet) is 65 cm, 185 cm and 285 cm, respectively. The width is 27.5 cm. The material was manually released from a silo with an outflow height of 25 mm and a volume of approximately 1 litre.

3.3 Experimental Peculiarities Arising for Granular Materials

Errors: As with every measuring system, PIV is subjected to certain measurement uncertainties which cause errors. Some of them are specific for granular avalanches and some are of general relevance. There are two main error sources for granular PIV: Inappropriate surface structure of the avalanche and/or the background and imperfect illumination due to shadows, reflections etc. Both are specific for granular avalanches and do not occur for usual PIV. Because of their exceptional importance, they will be discussed below in greater detail.

Further error sources of general relevance are:

1. Motion perpendicular to the picture plane (2-d measurements),
2. the displacement gradients arising within the interrogation spot: the pattern is not only shifted, but also deforms,

(a) The short chute

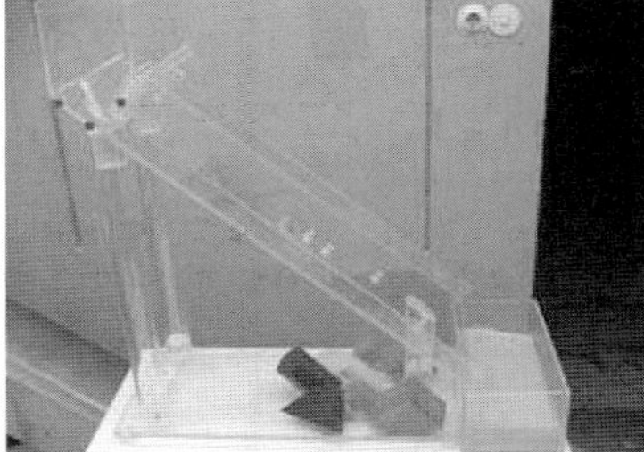

(b) The long chute

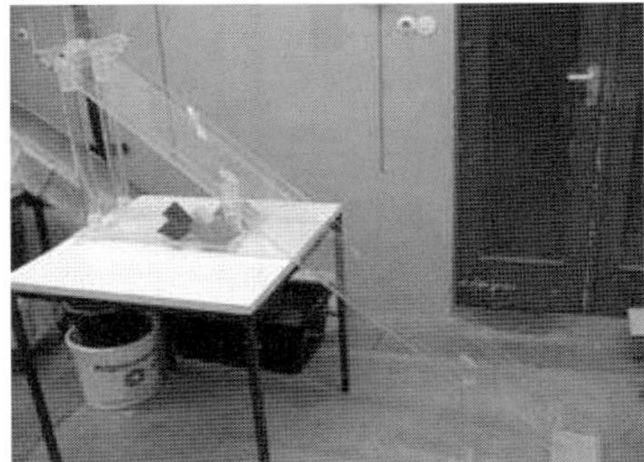

(c) The extra-long chute

Fig. 6. The plexiglass chutes used

3. the angle between the observation plane and the camera is not 90 degrees,
4. the distortion due to the usage of extreme wide angle lenses and short distances between camera and observation plane.

Error no. 1 is only of importance for locations where strong depth gradients arise. However, even for strong depth gradients (which do not occur on our chutes) the influence is rather small in comparison with the other main error sources. Error no. 2 is of some importance at fixed boundaries of the domain and the moving boundaries of the avalanche. Later we will show measurements, where the displacement gradients will cause some error. Further sources of error and errors 3 and 4 (above) are either easy to control or of no importance here.

Beside light reflections, a specific problem for plexiglass chutes is the electrostatic charging. Reflections can in general be avoided by either choosing an appropriate position of the flashes or using indirect flashing. Indirect flashing means, that the flashes are not pointing directly to the chutes, but are directed towards a white paper which reflects the flash light to the chutes. This well-known procedure diffuses the light such that the chutes are in general illuminated with no or very few reflections. Electrostatic charging can be minimised by using an anti-static spray.

Optical Surface Properties The optical surface properties are of extraordinary importance. To illustrate this, we compare velocity measurements of two plates which are at rest. One plate is covered with sand, the other one with a varnish. Whereas the surface of the former is optically quite structured, the surface of the latter is rather homogeneous, see Fig. 7.

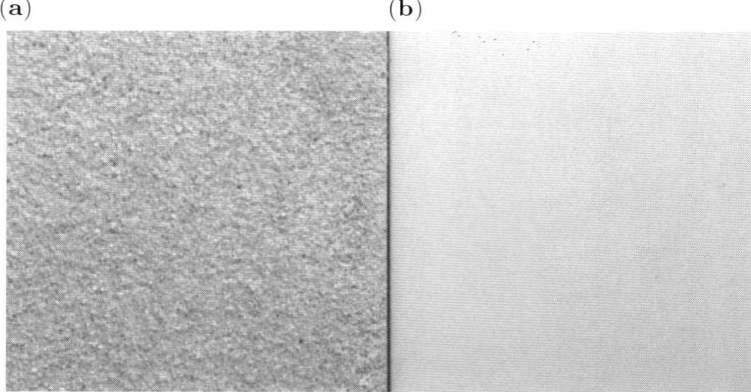

Fig. 7. Two different surfaces, (**a**) sandpaper and (**b**) varnish

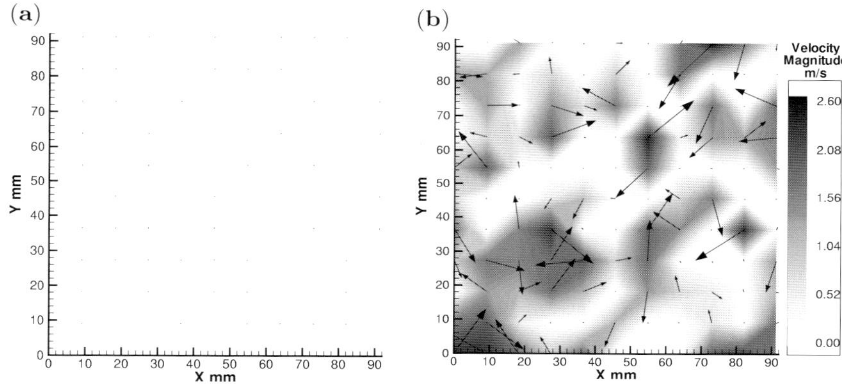

Fig. 8. Measured velocity for, (**a**) the sandpaper and, (**b**) the varnish

The velocity distribution is given in Fig. 8 (recall that the plates are at rest). Whereas the velocity for the sand-plate is measured correctly, the velocity of the varnished plate is not: the in-plane velocities are more or less randomly distributed in magnitude and orientation, which is clearly wrong. With an optically homogeneous (i.e. non-structured) background, erroneous velocity vectors may be produced; which can degrade the overall quality of the data. It is, therefore, recommended to choose a background with optical structure if possible. (This is not possible in the case of plexiglass chutes).

Illumination The second main error source is the illumination. If laser light sheets can be used, the quality of the illumination is generally perfect. We are forced to use other illumination techniques like flashes to illuminate the surface of non-transparent materials. In Figs. 9 and 10, frame A and frame B of one double frame are shown for different positions of flash A and flash B. The object illuminated is sandpaper at rest. To illustrate the effect in a drastic way, the position of the

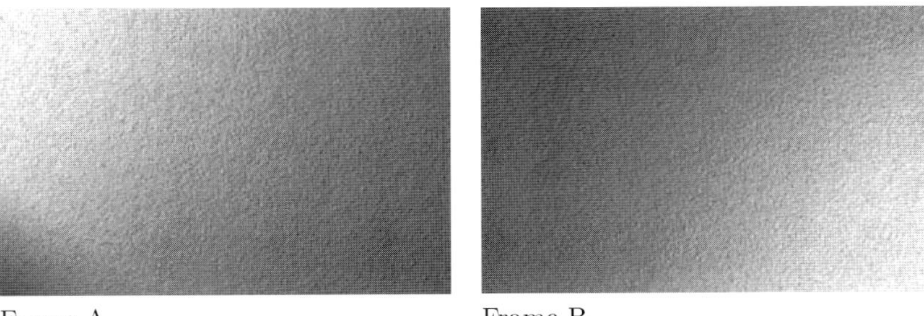

Fig. 9. Flash A upper left corner, flash B lower right corner of the observed region of a sandpaper surface at rest

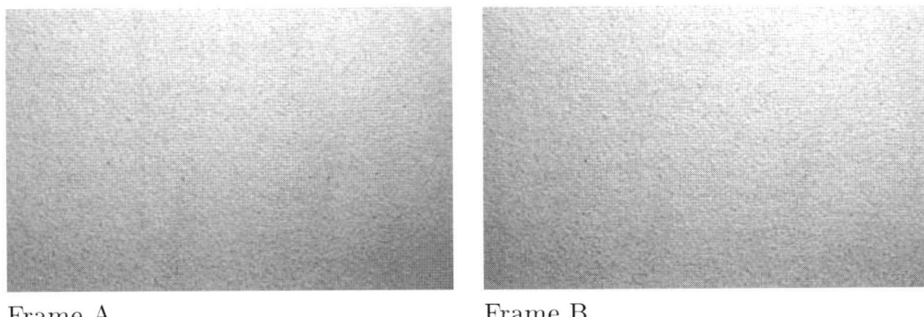

Fig. 10. Flashes as close as possible to each other, illuminating a sandpaper surface at rest

flashes are chosen to exaggerate the errors. It is plain to see that the illumination of subregions on the sandpaper (interrogation spots!) is not the same for frame A and frame B in Fig. 9, whereas it is fairly uniform in Fig. 10. Note that illumination differences between different regions on the sandpaper are of minor importance; crucial is the comparison of the same subregion in frame A with that in frame B.

Figure 11 shows the measured velocity of the sandpaper for the different positions of the flashes. It clearly shows that inappropriate positions of the flashes lead to significant measurement errors. Notice that the error produced here is not of a random nature as was the case in the last paragraph. The direction of the faulty velocity distribution in panel (a) of Fig. 11 reflects the axis of the flash illumination.

Testing the PIV-system In this paragraph a simple method is described to estimate systematic errors of the measured velocity values of the PIV-system. Specifically, the deviation of velocity values of the surface from the bottom at the same spatial locations (influence of the plexiglass) and the deviation of the velocity values from one another in the whole region of either the surface or the bottom is of interest here.

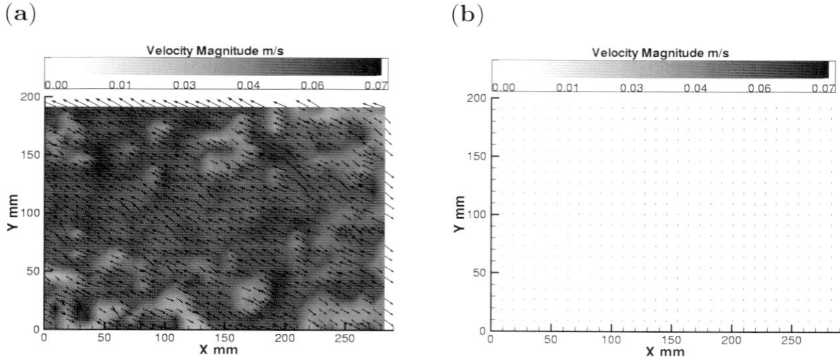

Fig. 11. Measured velocity of sandpaper, (**a**) from Fig. 9 and (**b**) from Fig. 10

Towards this end, a setup with two cameras, a plexiglass chute and a "rigid avalanche" (see Fig. 12) is used. One camera is put below the chute (capturing pictures through the plexiglass) and the other one above it. If the sandpaper is released, double frames of both cameras are taken (simultaneously) from which the frames A are shown in Fig. 13. The measured velocity distribution is given in Fig. 14. Note that the range of the legends is from the lowest (1.833 ms^{-1}) to the highest value (1.898 ms^{-1}) measured and that *raw data* were used, i.e. no filters have been applied to the data shown in Fig. 14. As can be seen, the deviation between the mean surface and the mean bottom velocity is only 0.13%. However, the deviation between the smallest and the highest velocity value in either panel (a) or (b) is approximately 3.5%. This shows once again that surface properties and the illumination which are responsible for the latter deviation are the main error sources, whereas the influence of the plexiglass is negligible. Note, that the results shown in Fig. 14 were obtained under optimal conditions, which means that we cannot expect a local error less than $\pm$ 1.75%, in the experiments considered here.

4 Experimental Results and Discussion

In this section we present experimental data obtained with the PIV-system for a selection of different experiments made from the surface and the side of the channel. In both cases results obtained near the outlet (silo) and in regions downstream sufficiently far away from the outlet have been conducted for four different inclination angles.

One of the tasks in this section is to find indications of a terminal velocity of the avalanche. Another challenging question is the dependence of the flow behaviour on the inclination angle. Different flow behaviour of the avalanche may define a "(phase) transition", where the material behaviour changes significantly, for example from more fluid-like to more solid-like behaviour (compare also [13]). Finally, we are interested in the velocity distribution across the avalanche (or at least the values at the top and the bottom) which may prove or disprove assumptions made for the depth-integration in many theories.

Fig. 12. The "rigid avalanche": A thick piece of (rigid) paper (0.3 mm thickness) with sand (same sand as is used for the avalanches) glued on both sides. The width of it is slightly smaller than the width of the chute

(**a**)

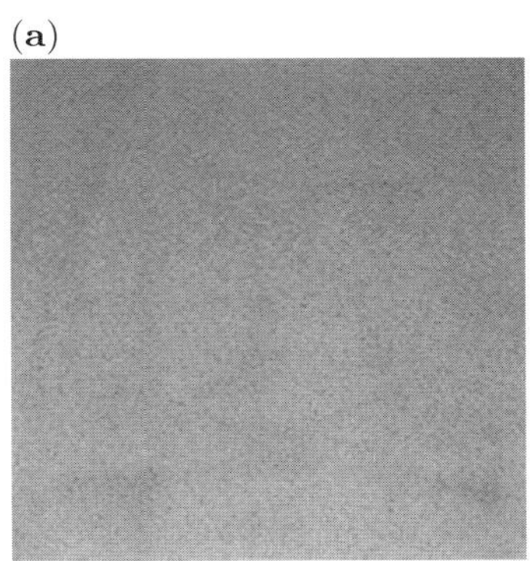

(**b**)

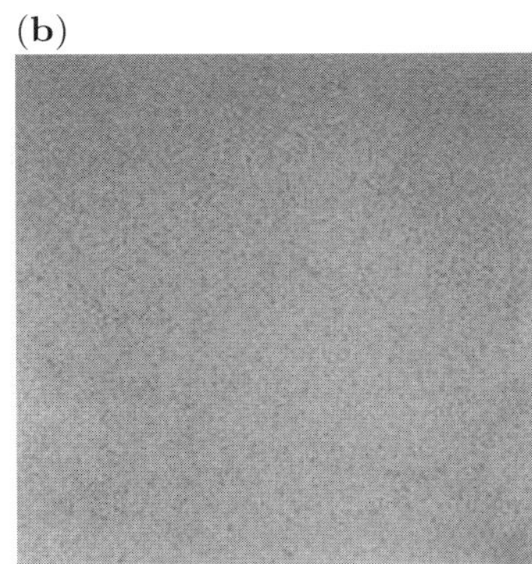

Fig. 13. Frame A of (**a**) the bottom and (**b**) the surface camera (not mirrored). Motion is from top to bottom

(**a**) Mean velocity value: 1.86927 ms^{-1}

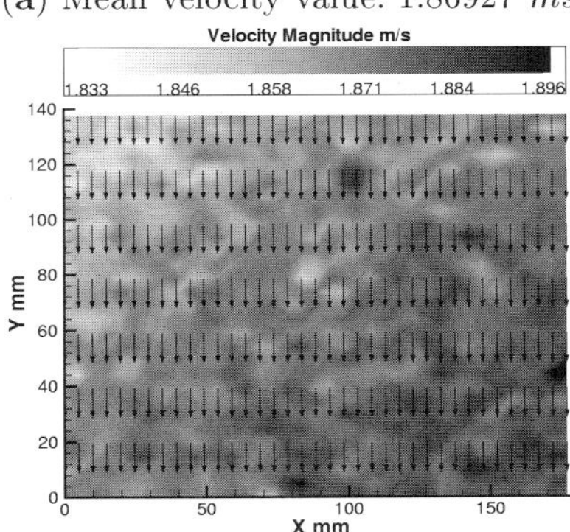

(**b**) Mean velocity value: 1.87178 ms^{-1}

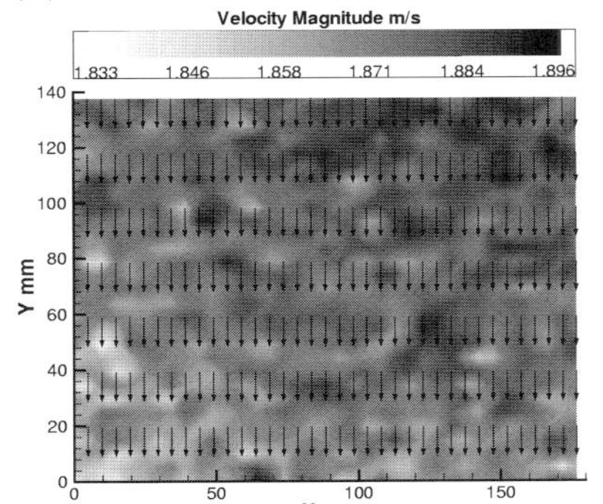

Fig. 14. Measured velocity of (**a**) the bottom and (**b**) the surface of the sandpaper (not mirrored). Motion is from top to bottom

4.1 Remarks on Experimental Procedures

Forces due to electrostatic charging are minimised by applying an anti-static spray before each experiment. Furthermore, each experiment was (immediately) repeated four times to estimate the (statistical) deviation of the velocity field and to ensure nearly the same environmental (humidity) and illumination conditions. It could be established that the absolute values of the velocities in all experiments differ by no more than 5% except for the extra long chute with an inclination angle of 30 degrees (deviation 10%). The results obtained for this particular experiment differ from all the others as will be explained later in greater depth. The variation is likely to originate from the manual release of the avalanche: the initial conditions are not easily kept the same in all four experiments.

Cohesionless sand with a particle diameter of the order of 0.5 mm was used. The particles are of roundish convex shape. The (static) basal friction angle is approximately 20 degrees. No seeding was used: the system was able to detect the velocities of the single-coloured grains without any difficulty, even through the plexiglass. Tests with seeding did not show differences beyond the variations observed in repetitions without seeding.

For the pictures taken from the side, the camera had a distance between 40 and 60 cm to the chute. The visible section corresponds to a rectangular domain of approximately 10 cm × 2 cm. Note, that for all velocity measurements taken from the side through plexiglass, only the velocity *at the boundary* was measured which is in general not equal to the internal velocity away from the boundary. In all these pictures, the avalanche is flowing from left to right. For the pictures taken from the surface, the camera had a distance of 1–2 m to the chute. In this case, the avalanche is always flowing from top to bottom.

At the instant when the material was (manually) released from the silo, a sequence of 10–16 pictures was captured. With the frequency of four double frames per second, a time between 2.5 and 4 seconds could be covered. The interesting sections of the pictures were subjected to the calculation and validation processes to obtain the velocity (done by the software). A cross-correlation algorithm utilising FFT (Fast Fourier Transform) was used[2]. The spot size of the interrogation windows was either 32 × 32 pixels (no oversampling) or 32 × 16 (double oversampling). After filtering possible errors (see later), the processed files were exported to suitable plot programs to post-processing the data. The pictures shown on the following pages are taken from TECPLOT.

As is well-known in PIV, the raw data must be subjected to a *validation procedure* to eliminate the faulty data. This is done by filters. We used two well-known filters, namely the *range filter* and the *standard deviation filter*. With the range filter, the permissible range of velocity values can be adjusted. It was for example mentioned that for the experiments taken from the side of the channel, the avalanche is always flowing from left to right, say in the positive $x-$direction. This clearly suggests that the velocity values in the $x-$direction must remain positive. The standard deviation filter eliminates all velocity values that are not within a certain value from the mean velocity. It can only be applied if no large velocity gradients exist within the viewing range. In our case, the flow field is quite homogeneous, and thus the standard deviation filter is quite effective in eliminating random errors. The blanks were interpolated by means of a neighbourhood filter. However, in any case we restricted the number of interpolated velocity vectors to less than 5% of the total number, although this may produce data which is not very smooth.

---

[2] For further details of the software see: www.tsi.com

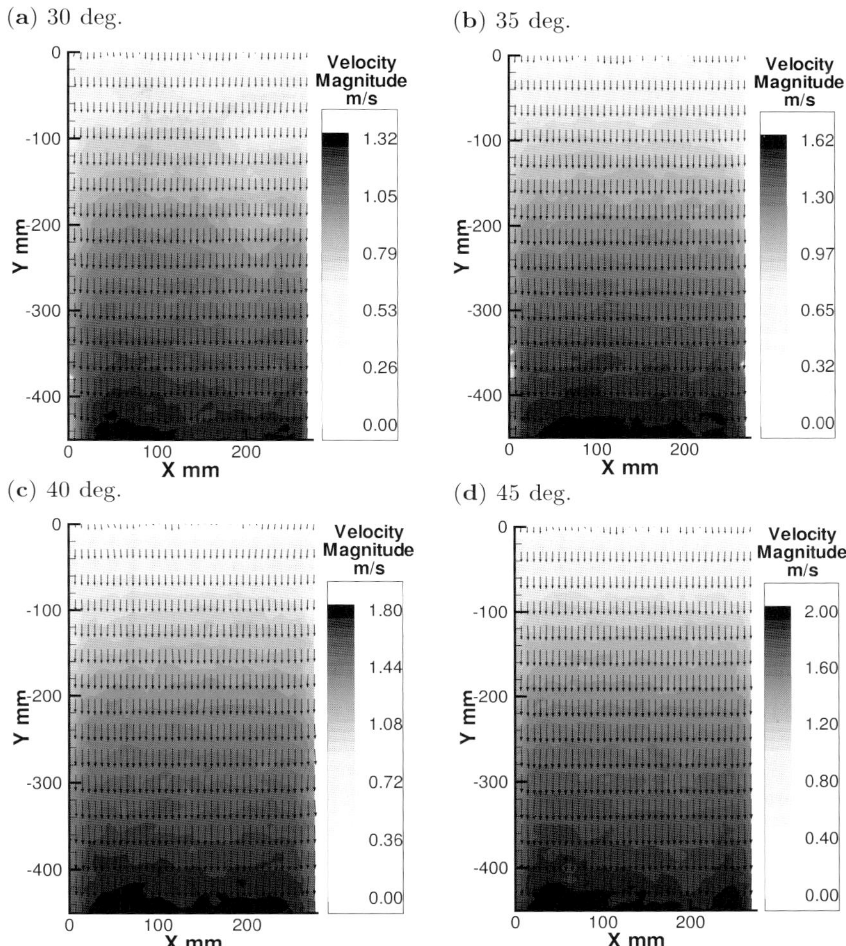

Fig. 15. Measured surface velocity near the outlet ($y = 0$) for four different inclination angles. Motion is from top to bottom

4.2 Results for the Surface

We start our discussion for surface velocity measurementsobtained near the outlet for inclination angles of 30, 35, 40 and 45 degrees, which are presented in Fig. 15. In each panel velocity vectors (in black) and the velocity magnitude is shown, the latter one in shades of grey. The legend is always chosen according to the maximum velocity measured in this particular experiment. The measurements were taken approximately 0.5 s after release, when the whole chute is covered with material. The visible region is 45 cm downstream from the outlet, where the axis $y = 0$ represents the position of the outlet. Here it can clearly be seen that the qualitative behaviour of the flow is the same for all inclination angles investigated. This can be concluded from the relative distribution of the velocity magnitude (i.e. the grey-scale distribution) which is nearly the same for all panels. It is plain to see that the avalanche accelerates downstream, where the maximum velocities of each experiment are reached at the bottom of all pictures. Furthermore, the velocity distribution in the cross-slope direction (i.e. in the $x-$direction) is quite uniform in all panels. From this we conclude that the boundary effects due to the confining side walls are relatively small and restricted to very thin layers in the direct vicinity of these walls.

Next we turn to the end of the long chute, see Fig. 16. Here the velocity was measured at the end of the chute (i.e. $y = 0$ means 185 cm downstream from the outlet), where the visible region corresponds to a domain of 73 cm. The pictures were taken approximately 0.75 s after release. Once again, velocity vectors and the velocity magnitude (in shades of grey) are given in the panels a)–d) for inclination angles of 30, 35, 40 and 45 degrees. First of all it can be seen that the velocity values occurring in Fig. 16 are larger than those in Fig. 15; the avalanche has clearly accelerated further downstream. However, the velocity differences[3] in each panel decrease compared to those in Fig. 15.

The velocity distribution in panel (a) for an inclination angle of 30 degrees differs slightly from that of the other panels. On the one hand, pronounced side boundary layers can be seen where the velocity differs significantly from the velocity in the middle of the chute. The material in these boundary layers has a velocity of approximately 1.35 – 1.5 $m\,s^{-1}$ and seems to accelerate slower or at least with some delay compared to the bulk in the middle of the chute. Recall that the maximum velocity in panel (a) of Fig. 15 was 1.32 $m\,s^{-1}$. The boundary effects for the other inclination angles are less pronounced. On the other hand, there are slightly more "dark" regions in panel (a) than in the other panels which means that the distribution of the velocity magnitude in this panel is slightly different than in the other ones.

Let us now turn to measurements obtained for an extra long chute (3 m), which are presented in Fig. 17. The axis $y = 0$ corresponds to a distance of 285 cm downstream of the outlet, the visible region is 72 cm. Now, a significantly different behaviour can be observed, in particular in panel (a), where the results for an inclination angle of 30 degrees are presented. Firstly, the maximum velocity is not located at the end of the chute as is to be expected, but is more or less randomly distributed in this picture. The velocity magnitude does not change much in this picture except perhaps in the boundary layer. It is worth noting that the maximum velocity in panel (a) is nearly the same as in the panel (a) of Fig. 16 (long chute). From these results we can conclude that a rate-dependent drag law must be included in depth-integrated theories.

The behaviour for panel (a) of Fig. 17 is not seen in the panels (b)–(d). However, the acceleration is significantly decreased in comparison with panels (b)–(d) of Fig. 16 (long chute) which is a hint for a similar behaviour. But to examine this clearly, yet longer chutes are needed.

We may now summarise and clarify the results of Figs. 15, 16 and 17. This is done in Fig. 18, where we present cross-slope-averaged values of the velocity magnitude from the Figs. 15, 16 and 17.

[3] Note that the grey-scale distribution in Fig. 16 is slightly distorted compared to Fig. 15 (i.e. the mean grey colour is shifted to higher velocities) to be able to show flow details.

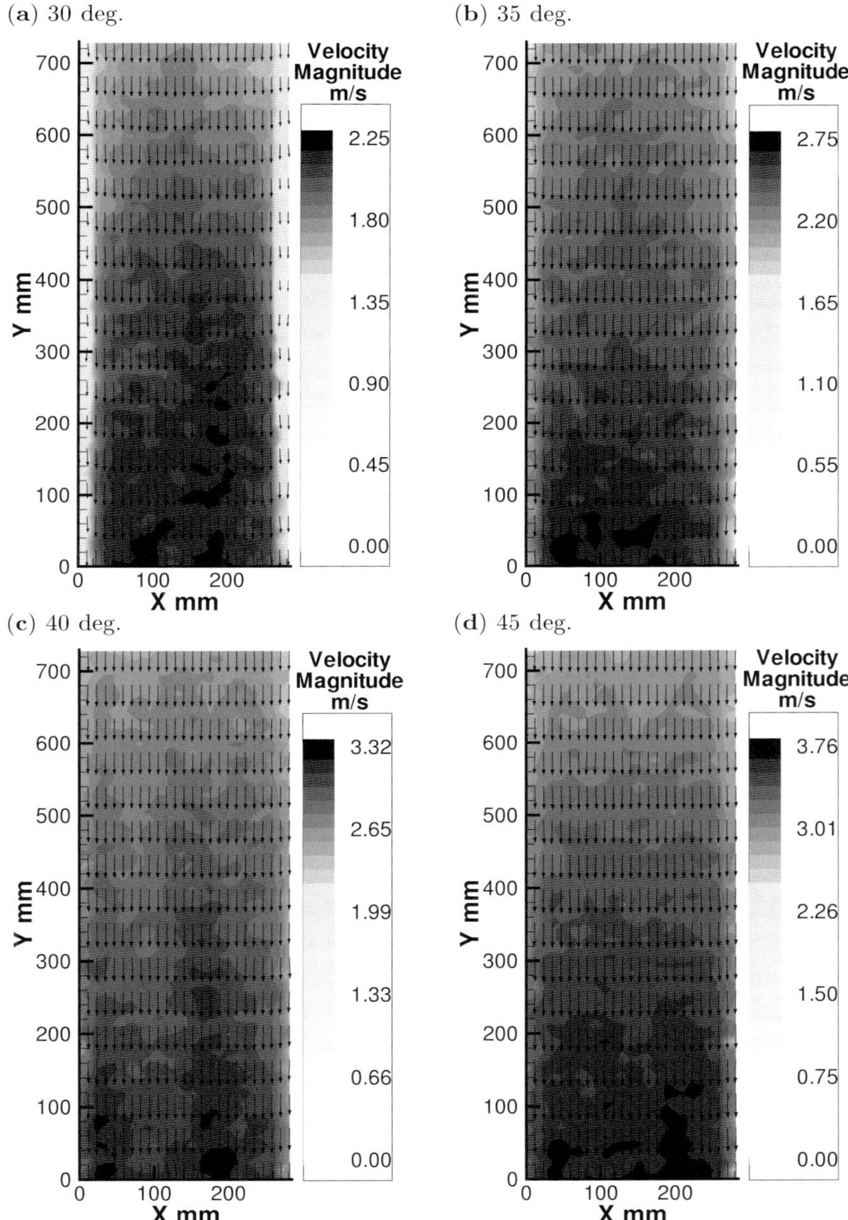

Fig. 16. Measured surface velocity at the end of the long chute ($y = 0$) for four different inclination angles. Motion is from top to bottom

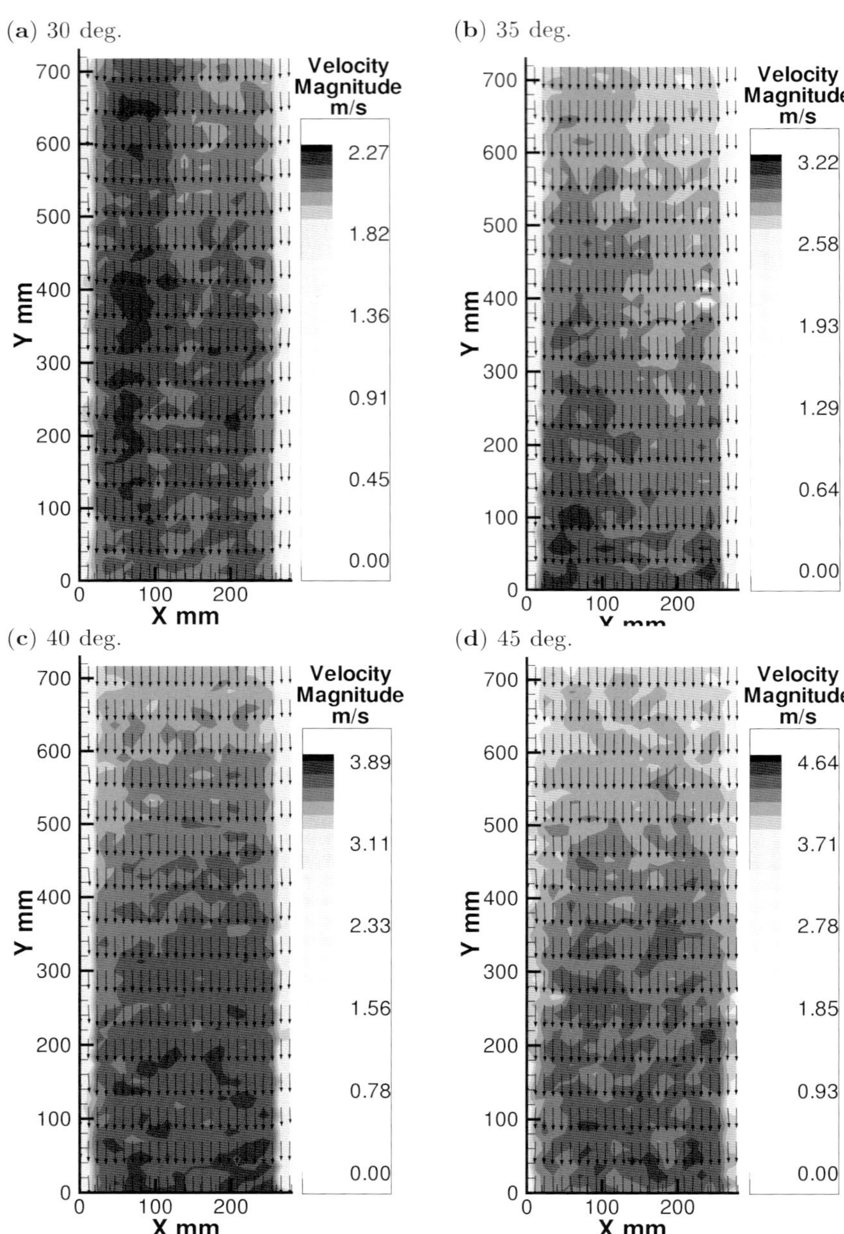

Fig. 17. Measured surface velocity at the end of the extra long chute ($y = 0$) for four different inclination angles. Motion is from top to bottom

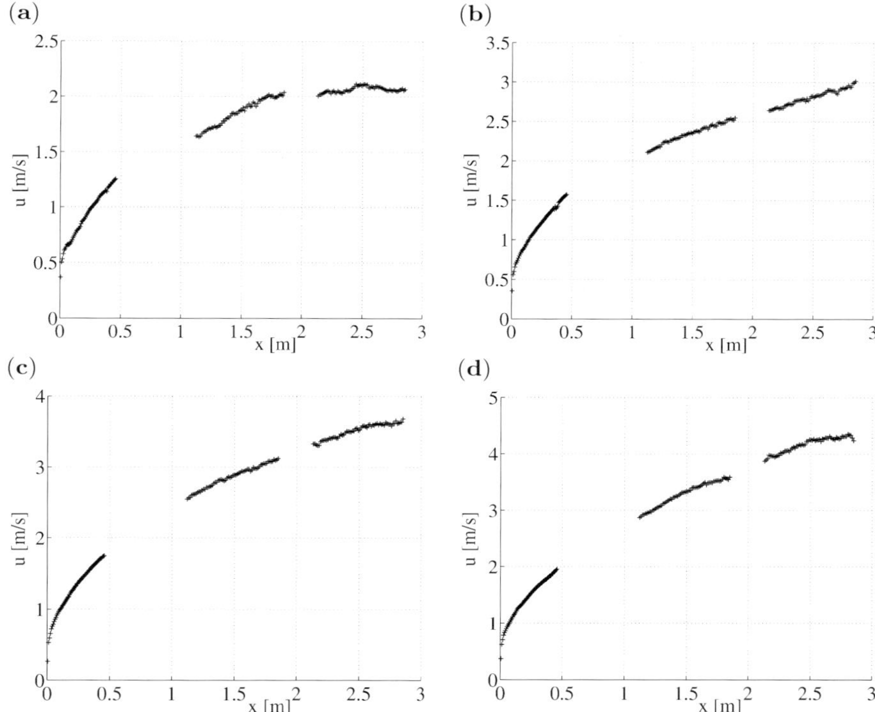

Fig. 18. Cross-slope-averaged velocity values calculated from the experimental results for (**a**) 30, (**b**) 35, (**c**) 40 and (**d**) 45 degrees.

The most remarkable thing that can be seen by comparing panels (a)–(d) of Fig. 18 is that the behaviour for an inclination angle of 30 degrees differs significantly from the other ones. It shows that for travel distances of 2–3 m the avalanche does not accelerate any longer but seems to come to a steady state (terminal velocity). Note that the scale of the y-axis in panels (a)–(d) is chosen according to the maximum velocity for this particular inclination angle. The curves for the inclination angles of 35, 40 and 45 degrees are rather similar in shape.

The boundary layers at the sidewalls have not been explicitly considered so far. One possibility to account for these experimental results would be to assume that the bed friction law depends on the cross-slope coordinate. A more sophisticated approach will be discussed in Sect. 5.

4.3 Results for Observations from the Side

In this subsection we present experimental data resulting from pictures made from the side of the chute. We start by investigating the outflow from the silo. Figure 19 shows pictures made from the side of the chute (through the plexiglass wall) for inclination angles of 30, 35, 40 and 45 degrees. The camera distance was 40 cm. The visible region corresponds to a length of approximately 10 cm, where the outlet is

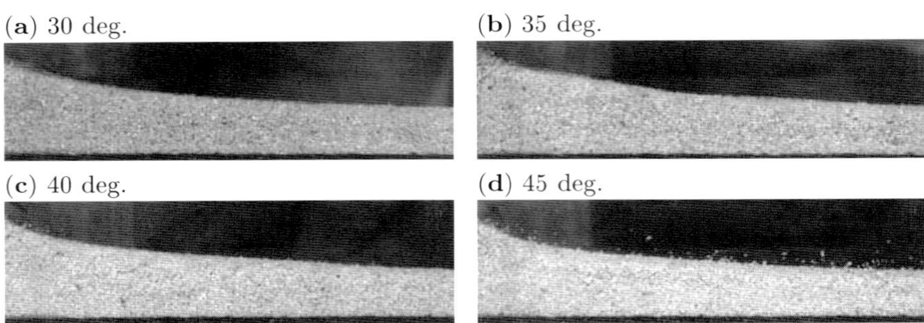

Fig. 19. Pictures from the side of the chute near the outlet for different inclination angles. At the left the material leaves the shutter of the silo; motion from left to right

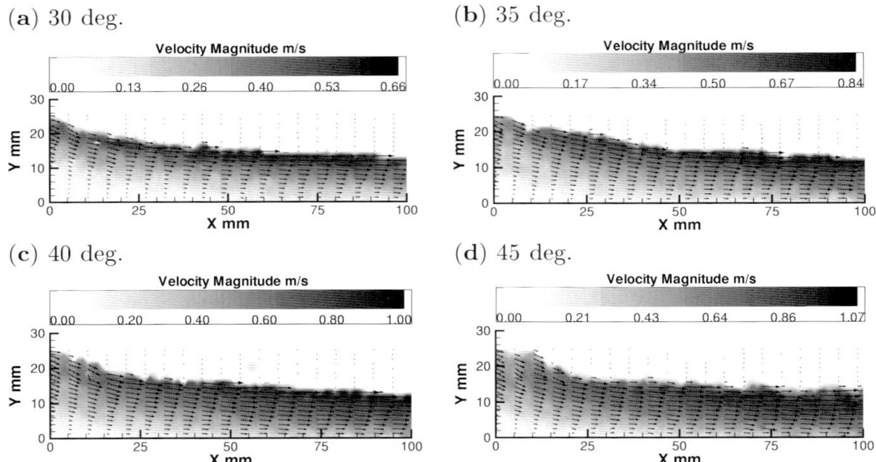

Fig. 20. Velocity measurements from the side at the outlet for different inclination angles

located at $x = 0$. The pictures were captured approximately 0.5 s after release. It can be seen that the qualitative behaviour is the same for all inclination angles. Specifically, the height is decreasing from the left (outlet) to the right margin of panels (a)–(d) in the same manner, i.e. the value of the height 10 cm downstream is nearly the same for all inclination angles.

Figure 20 shows the corresponding velocity measurements. As before, velocity vectors are shown in black, while the velocity magnitude is shown in grey scales. Once again, the legend of each picture is chosen according to the maximum velocity measured for this particular inclination angle.

It can be seen in Fig. 20, that there is a temporal region where the material at the bottom nearly sticks to the bed (henceforth called a "dead zone"), while the material at the top is already flowing: the material is vertically divided into (at least) two regions, where the upper region is slipping over the lower region. This fits

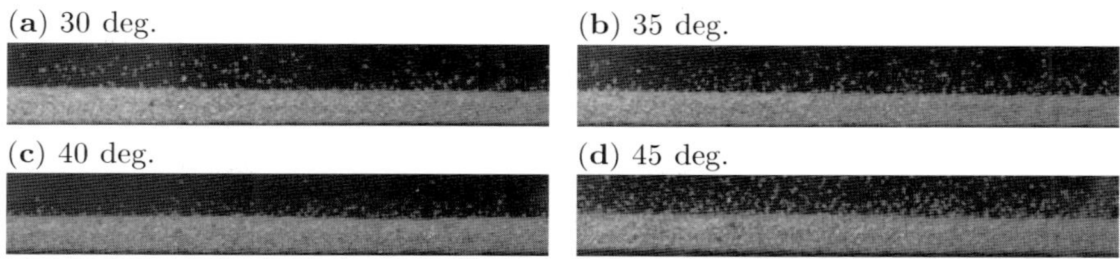

Fig. 21. Pictures from the side near the end of the short chute for four different inclination angles. The left side of the pictures began at approximately 40 cm downstream of the outlet. Motion is from left to right

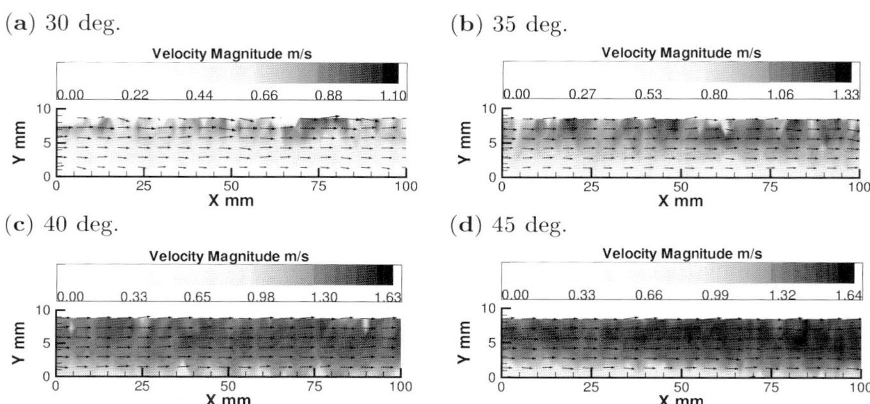

Fig. 22. Velocity measurements from the side near the end of the short chute for four different inclination angles. Motion is from left to right

quite well with the observations and descriptions of creeping flows of soils as shown in [19]. However, the panels of Fig. 20 show only the onset of a flow. The velocity profile is clearly not constant through the depth and shows that additional effects may play a crucial role in this situation. Comparison of the velocity measurements that were taken for different inclination angles show, that the dead zone increases if the inclination angle decreases. Indeed, if we would further decrease the angle, the dead-zone material would remain on the chute after the complete upper layer has flowed down. A similar behaviour is also reported in [12,13].

There is an obvious qualitative interpretation of the fact that the velocity profiles close to the left end show a pronounced dependence across the layer depth with zero velocity at the base and a much more uniform distribution close to the right end. The free surface exerts no shear resistance to the surface particles and so they are accelerated first. It takes some time (i.e. distance from the left to the right) until this information has reached the bottom. This is obviously faster for the steeper chute inclinations than for the shallow ones. The initial shearing is large, but at a distance of 50 to 100 mm from the outlet it is small and the deviation of the velocity profiles from uniformity is not so large.

In Fig. 21, pictures from the side near the end of the short chute are presented. The visible region is once again 10 cm, starting at approximately 50 cm downstream

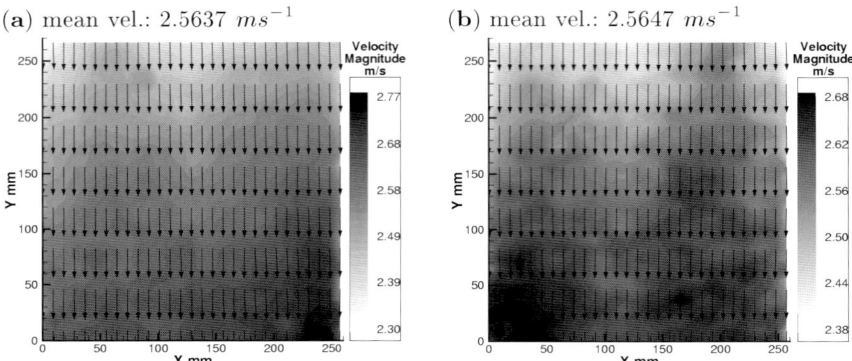

Fig. 23. Simultaneous velocity measurements of (**a**) the base and (**b**) the surface of the avalanche (not mirrored). Motion is from top to bottom

of the outlet. The height gradient is clearly smaller (nearly zero) than near the outlet, as was to be expected.

In the corresponding velocity measurements in Fig. 22 it can be seen that the velocity is, to a good approximation, uniform through the depth. Recall (also from the surface velocity measurements) that the velocity at the confining plexiglass wall is not the same as inside the flow region, hence the velocity distribution through the depth may neither exactly be the same. Because of the small thickness of the granular layer the results, especially at the base and at the free surface, may differ from the real values. As can be estimated from Fig. 21, the depth of the granular layer of 8-9 mm corresponds to at most 20 layers of grains.

4.4 An Experiment with Two Cameras

We now turn to a measurement performed with one camera placed below and the other above the chute, to measure simultaneously at the same location the velocity of the base and the surface, respectively. These measurements serve as additional experimental result to confirm that the velocity profile is approximately uniform throughout the depth.

Figure 23 displays base (panel (a)) and surface (panel (b)) velocity measurements on the long chute for an inclination angle of 40 degrees. The top of the pictures in Fig. 23 corresponds to a location 1.2 m downstream of the outlet. The camera distances were 81.5 cm, resulting in a visible region of approximately 27 cm.

Note that the legends only contain velocities within the range of minimum and maximum velocities in each picture. Furthermore, the mean velocities of the base and the surface are given. They differ from each other up to an amount that is far smaller than the measurement uncertainty. The maximum and minimum velocities in panels (a) and (b) differ only by an amount of 3.5% (which is not far away from the measurement uncertainty). Even in the worst case, the velocity in panel (a) differs only by slightly more than 10% from the velocity in panel (b). From this we conclude that the assumption of a uniform velocity profile through the depth is very reasonable, in particular sufficiently far away from the boundaries.

5 Summary and Conclusions

In this paper we have presented and discussed experimental results of sand avalanches flowing down inclined plexiglass chutes with lateral confinement. Surface, basal and side-wall velocities have been measured with a Particle Image Velocimetry (PIV) system. Although the PIV system was originally not designed for the flow of granular avalanches, it provides a good measuring technique for visible regions of the flow. The major difference to "usual" PIV is the use of flash lights instead of laser light sheets, which cause a more or less inhomogeneous illumination. Careful positioning of the flashes is required to reduce measurement errors. Sand avalanches possess enough surface "structure" for the system to easily detect a moving pattern, i.e. no seeding is needed.

The surface measurements of the avalanche for inclination angles of 35, 40 and 45 degrees showed nearly the same qualitative behaviour. The avalanche accelerates downstream on all chutes and the maximum velocity is higher for larger inclination angles. However, for the smallest inclination angle of 30 degrees a terminal velocity on the longest chute (3 m) could be detected. In this case the avalanche does not accelerate any further. This shows a significantly different behaviour in comparison with the results for the other inclination angles. It strongly suggests the use of a more sophisticated bed friction law than the COULOMB dry friction law which is used in many theories describing avalanche motion [18].

From the measurements made from the side of the channel and the experiment made with two cameras (one below and one above the chute) we concluded that the velocity distribution through the depth is fairly uniform except in a limited region very close to the outlet. Thus, the assumption of a uniform velocity distribution through the depth is reasonable, at least for the flows investigated here.

At high velocities we observed small boundary layers in the vicinity of the confining walls, whilst in the interior the velocity remained uniform across the channel. The boundary layers were more pronounced for smaller inclination angles and for longer distances down the chute. A possibility to take this into account is a *micro-polar constitutive model.* Physically this implies that effects of particle rotations would be taken into account (compare for example [1,9]). Note that the micro-polar theory which was developed by ERINGEN in 1966 introduces internal length scales into the theory. Specifically, this may fit well with the particular flow behaviour of the avalanche observed for the inclination angle of 30 degrees.

Acknowledgements: Thanks to S. P. Pudasaini for many valuable hints. The chutes were built by H. Hoffmann, N. Keller and H. Wiener. Additional electric equipment was provided by K. Polster. Some of the experiments were conducted with the help of S. Ubl and E. Vassilieva.

References

1. Bardet, J.P., Vardoulakis, I. (2001) The asymmetry of stress in granular media. Int. J. Solids Struct. **38**, 353–367
2. Boillot, A., Prasad, A.K. (1996) Optimization procedure for pulse separation in cross-correlation PIV. Experiments in Fluids **21**, 87–93

3. Gray, J.M.N.T., Tai, Y.C. (1998) On the inclusion of a velocity-dependent basal drag in avalanche models. Annals of Glaciology **26**, 277–280
4. Gray, J.M.N.T., Wieland, M., Hutter, K. (1999) Gravity-driven free surface flow of granular avalanches over complex basal topography. Proc. R. Soc. Lond. A **455**, 1841–1874
5. Hutter, K., Greve, R. (1993) Two-dimensional similarity solutions for finite-mass granular avalanches with Coulomb- and viscous-type frictional resistance. J. Glaciol. **39** (132), 357–372
6. Hutter, K., Rajagopal, K.R. (1994) On flows of granular materials. Continuum Mech. Thermodyn. **6**, 81–139
7. Keane, R.D., Adrian, R.J. (1992) Theory of cross-correlation analysis of PIV images. Applied Scientific Research **49**, 191–215
8. Lueptow, R.M., Akonur, A., Shinbrot, T. (2000) PIV for granular flows. Experiments in Fluids **28**, 183–186
9. Mohan, L.S., Nott, P.R., Rao, K.K. (1999) A frictional Cosserat model for the flow of granular materials through a vertical channel. Acta Mech. **138**, 75–96
10. Nohguchi, Y., Hutter, K., Savage, S.B. (1989) Similarity solutions for granular avalanches of finite mass with variable bed friction. Continuum Mech. Thermodyn. **1**, 239–265
11. Pöschel, T., Saluena, C., Schwager, T. (2001) Scaling properties of granular materials. Physical Review E **64**
12. Pouliquen, O. (1999) Scaling laws in granular flows down rough inclined planes. Physics of Fluids **11** (3), 542–548
13. Pouliquen, O., Forterre, Y. (2002) Friction law for dense granular flows: application to the motion of a mass down a rough inclined plane. J. Fluid Mech. **453**, 133-151
14. Product information (2000) "Metz Mecablitz"
15. Product information (2000) "Nikon Nikkor Objektive"
16. Raffel, M., Willert, C., Kompenhans, J. (1998) Particle Image Velocimetry. Springer, Berlin Heidelberg
17. Ristow, G.H., Riguidel, F.-X., Bideau, D. (1994) Different characteristics of the motion of a single particle on a bumpy inclined line. J. Phys. I France **4**, 1161–1172
18. Savage, S.B., Hutter, K. (1989) The motion of a finite mass of granular material down a rough incline. J. Fluid Mech. **199**, 177–215
19. Vulliet, L., Hutter, K. (1988) Viscous-type sliding laws for landslides. Can. Geotech. J. **25**, 467–477
20. Westerweel, J., Dabiri, D., Gharib, M. (1997) The effect of a discrete window offset on the accuracy of cross-correlation analysis of digital PIV recordings. Experiments in Fluids **23**, 20–28
21. Wieland, M., Gray, J.M.N.T., Hutter, K. (1999) Channelized free-surface flow of cohesionless granular avalanches in a chute with shallow lateral curvature. J. Fluid Mech. **392**, 73–100

Existence of Avalanching Flows

Reinhard Farwig

Department of Mathematics, Darmstadt University of Technology,
Schlossgartenstr. 7, 64289 Darmstadt, Germany,
e-mail: farwig@mathematik.tu-darmstadt.de

received 9 Jul 2002 — accepted 10 Sep 2002

Abstract. Avalanches, landslides and debris flows are devastatingly powerful natural phenomena that are far too little understood. These granular matters are mixtures of solid particles and of an interstitial fluid and are easily modelled on the microscopic level by the laws of mechanics. On mesoscopic and macroscopic levels the different scales of the influence of the particles, the fluid and their interaction lead to various models of avalanching flows. In this survey we consider several models of granular materials characterised by height only or by height and momentum, discuss the existence of similarity solutions, existence of arbitrary solutions and particle segregation. The main part concerns the Savage–Hutter equations for dense flow avalanches.

1 Introduction

The number of catastrophes induced by snow avalanches, landslides and debris flows has been increasing during the last decades. The reasons are a possible change of climate with heavy rainfalls, but also the activities of human beings in endangered mountainous regions. Therefore, the determination of runout zones and of endangered regions by analytical and numerical methods for the different types of "avalanches" is of utmost importance. A related physical, but less disastrous behaviour can be observed in the motion of sand dunes, in the pouring of grains leading to free surfaces of stock piles and in hopper flows.

The main feature of the phenomena of *granular materials* is the mixture of solid particles with water or air leading to a behaviour different from that of solids, fluids or gases. On the other hand, the main differences between various kinds of granular flows are due to the small or large fluid-solid interaction, the size and shape of the grains, and due to the predominance of either the solid particles (whilst the influence of the interstitial fluid can be neglected) or of the fluid carrying the small particles.

On the scale of individual grains the behaviour of granular material is described by the laws of classical mechanics. But due to the huge variety of particle sizes, shapes and densities, abrasion of particles, interaction with the fluid, with different layers or with the bed leading to an exchange of particles, it is very difficult to model granular flow on mesoscopic or macroscopic scales. Further typical features are the dilatancy and particle segregation. In everyday life it is observed that after stirring or shaking grains of different size but not necessarily with different specific weight the bigger particles tend to move upwards (*inverse grading*) and to the nose of an avalanche.

As an example consider the different scales of a typical snow avalanche with weight 10^6 kg consisting of snow crystals with radii less than 1 mm, see Table 1. In the lower part of the avalanche sliding on a fluidised layer the solid particles dominate whereas the interstitial air can be neglected. Above this *dense flow avalanche* there may be a *powder snow avalanche* in which turbulent air carries snow crystals. In between there is a thin layer called *resuspension layer* or *saltation layer* feeding the powder snow avalanche, see [28,43].

Table 1. Characteristic parameter ranges in dense flow and powder snow avalanches [28,43]

Physical Quantity		Dense Flow Avalanche	Powder Snow Avalanche
Volume fraction of particles		$0.3 - 0.5$	$1.10^{-3} - 1.10^{-4}$
Density	$[\frac{\mathrm{kg}}{\mathrm{m}^3}]$	$200 - 300$	$0.1 - 10$
Length	[m]	$10 - 1000$	$100 - 2000$
Height	[m]	$0.5 - 20$	$10 - 500$
Speed	$[\frac{\mathrm{m}}{\mathrm{s}}]$	$0 - 70$	$10 - 200$

Although powder snow avalanches have a much lower density than dense flow avalanches, their length, height, velocity and consequently their runout zone is much larger.

This review article is organised as follows. In Sect. 2 we discuss two different models [5,37] of particle segregation. Further we consider stationary solutions of a model of Bouchaud, Cates, Ravi Prakash and Edwards (BCRE model) for (dry) sand piles [3]. Section 3 is devoted to the Savage–Hutter model [40] of dense flow avalanches, its similarity solutions and the mathematical analysis of weak entropy solutions within the theory of systems of conservation laws with source terms. Conclusions and suggestions for further work follow as well as a list of References.

2 On Models of Cohesionless Granular Materials

Section 2 deals with several models of dry, cohesionless granular materials characterised by height only. The existence of stationary solutions and of instationary similarity solutions will be discussed when prescribing different kinds of boundary conditions. We start with two mathematical models for particle segregation.

2.1 Particle Segregation

It is well-known that in granular flows the large particles tend to move upward and to the nose of an avalanche while the small particles concentrate at the bottom and at the rear end of an avalanche. This *inverse grading* can be explained by the percolation effect or the so-called *random fluctuating sieve mechanism* [41]: the probability for a small particle to find a hole in the granular material to fall into is larger than the one for large particles. But since this gravity-induced hole-filling mechanism would lead to a net mass flux downwards, Savage and Lun [41] also propose a *squeezing expulsion mechanism*; by this mechanism the forces exerted by the particles to each other lead to a squeezing of particles in up- or downward direction.

A further discussion of possible reasons for the usual grading and for inverse grading can be found in [22] and in references therein. In addition to percolation effects, to geometrical reorganisation and to segregation driven by convection, inertia or entropy Hong et al. [22] propose a so-called *condensation of hard spheres* as the driving force.

A simple mathematical model for segregation in a mixture of n species has been proposed by Braun [5]. Let $u_i = u_i(x,t)$ denote the concentration of the i–th species, $1 \le i \le n$, in a one-dimensional container $\Omega = (0, L)$ of height $L > 0$. Then the change $\partial u_i/\partial t$ of the concentration u_i is balanced by the negative of the flux $J_i = J_i(u)$ which is the sum of a convectional part $J_i^c = f_i(u)$ and of a diffusional part $J_i^d = -d_i(u)\,\partial u_i/\partial x$ with $d_i(u) > 0$. Defining the diagonal matrix $D(u) = \operatorname{diag}(d_1(u), \ldots, d_n(u))$ we get a system of reaction–diffusion equations

$$u_t - \big(D(u)u_x - f(u)\big)_x = 0\,, \quad u = (u_1, \ldots, u_n)\,. \tag{1}$$

The convective part $f(u)_x$ is related to the random fluctuating sieve mechanism whilst the diffusive term $\big(-D(u)u\big)_x$ accounts for the random effects of collisions and could lead to the squeezing expulsion mechanism. In order to guarantee that particle densities are nonnegative and add up to 1 everywhere in space and time, i.e., that $\sum_{i=1}^n u_i \equiv 1$ and $u_i \ge 0$, we impose the structural conditions

$$d_1(u) = \cdots = d_n(u) =: d(u)$$

as well as

$$\sum_{i=1}^n f_i(u) \equiv 0 \quad \text{and } u_i = 0 \Rightarrow f_i(u) = 0\,. \tag{2}$$

Note that the non-negativity of u_i is a consequence of $(2)_2$ due to well-known maximum principles for parabolic equations, see e.g. [13]. Besides an initial value

$$u(\cdot, 0) = u_0(\cdot) \quad \text{with} \sum_{i=1}^n u_{i0} \equiv 1\,,$$

the flux condition $J\big(u(x,\cdot)\big) = 0$ for $x = 0$ and $x = L$ is used to impose the (non-linear) boundary condition

$$d(u)u_x - f(u) = 0 \quad \text{at } x = 0,\ x = L\,.$$

Then the effect of segregation is reflected by the long-time behaviour of solutions of (1). For $n = 2$ species with concentrations $u := u_1$ and $u_2 = 1 - u_1$ and convectional part $f(u) := f_1(u, 1-u)$ where $f_2 = -f_1$ by (2), the system (1) simplifies to one non-linear parabolic equation

$$\begin{aligned} u_t - \big(d(u)u_x - f(u)\big)_x &= 0 \quad \text{in } (0, L) \\ d(u)u_x - f(u) &= 0 \quad \text{at } x = 0,\ x = L \\ u(\cdot, 0) &= u_0(\cdot)\,. \end{aligned} \tag{3}$$

Theorem 1 [5] *Assume that d and f are twice continuously differentiable.*
(1) *For every prescribed mean concentration*

$$\overline{u} = \frac{1}{L} \int_0^L u(x)\, dx \in [0, 1]$$

the stationary problem

$$d(u)u_x = f(u)\,, \tag{4}$$

cf. (3), *has exactly one solution $u(x)$ with mean value $\overline{u}$.*

(2) *For every initial value $u_0 \in C^0([0, L])$ with mean value $\overline{u} \in [0, 1]$ problem* (3) *has a unique global solution u on $[0, L] \times (0, \infty)$ converging to the stationary solution u of* (4) *with mean value $\overline{u}$ for $t \to \infty$.*

Sketch of Proof (1) A solution u of (4) is defined by the ordinary differential equation

$$\frac{du}{dx} = g(u) := \frac{f(u)}{d(u)} \tag{5}$$

where $f(0) = f(1) = 0$, cf. (2), yields $g(0) = g(1) = 0$. Due to the unique solvability of (5) every solution $u(x)$ of (5) with initial value $u_0 \in [0, 1]$ will exist for all $x \in [0, L]$ and satisfy $u(x) \in [0, 1]$. By the same argument two solutions u_1 and u_2 with $u_1(0) < u_2(0)$ will satisfy $u_1(x) < u_2(x)$ for all $x \in [0, L]$. Since the solution $u = u(\cdot, u_0)$ is a continuous and even a monotonically increasing function of its initial value $u_0 = u(0)$ and since $u_0 = 0$ or $u_0 = 1$ yield $u \equiv 0$ or $u \equiv 1$ respectively, we conclude that the map

$$\overline{u} : [0, 1] \to [0, 1]\,, \quad u_0 \mapsto \overline{u(\cdot, u_0)}\,,$$

is a homeomorphism.

(2) Given an initial value $u_0(x)$ with $u_0(x) \in [0, 1]$ the solution u of (3) exists for all $t > 0$. Then $v = d(u)u_x - f(u)$ satisfies the parabolic equation

$$v_t = a(t, x)v_{xx} + b(t, x)v_x$$

with bounded functions $a = d(u)$, $b = d'(u)u_x - f'(u)$ and vanishing boundary values in $x = 0$, $x = L$. By classical theorems v and $v_x = u_t$ converge to zero for $t \to \infty$. In particular u converges to a stationary solution u_∞ of (3), i.e., u_∞ solves (5). Furthermore (3) easily implies that $\overline{u(\cdot, t)}$ is constant; hence u_∞ is the unique solution of (5) satisfying $\overline{u}_\infty = \overline{u}_0$. □

The proof of Theorem 1(1) is based on topological arguments. Therefore degree theoretical arguments are used in the case of more than two species leading to the existence of at least one stationary solution. Thus uniqueness of a final segregation of particles cannot be guaranteed for more than two species in general. Note that in this model empty space is evenly distributed in the vessel and that compressibility or dilatancy effects are ignored.

A model of particle segregation for polydisperse materials on a free surface has recently been proposed by Pudasaini et al. [37] generalising a model of Prigozhin

[34] for binary mixtures. Consider a granular material consisting of n constituents with different diameters d^i but with same density which are poured with intensity $s^i = s^i(x,t)$, $x \in \Omega \subset \mathbb{R}^2$, $t > 0$, $i = 1, \ldots, n$. The deposited material will form a free surface $y = h(x,t)$ on which the particles flow in the direction of the steepest descent with velocity $-m\nabla h$; here the unknown diffusion coefficient $m \geq 0$ is part of the solution of a related variational inequality, cf. (12), (13) in Sect. 2.2 below. We are looking for the concentration κ_+^i of the ith constituent in the surface mass flux and for its concentration κ_s^i in the heap surface layer. Then the balance equation of mass for the ith constituent assumes the form

$$\kappa_s^i \rho \frac{\partial h}{\partial t} - \operatorname{div}\big(\kappa_+^i m \nabla h\big) = s^i , \tag{6}$$

where ρ denotes the bulk density; cf. (12) below when no particle segregation occurs ($\kappa_s^i = \kappa_+^i$) or for the total balance (sum (6) for $i = 1, \ldots, n$). Particle segregation is modelled by an operator (vector field) $S = (S_1, \ldots, S_n)$ mapping the $(n-1)$–simplex

$$K_+ = \Big\{\kappa_+ = (\kappa_+^1, \ldots, \kappa_+^n) : \ \kappa_+^i \geq 0; \ \sum_{i=1}^n \kappa_+^i = 1\Big\}$$

to the corresponding $(n-1)$–simplex of vectors $\big\{(\kappa_s^1, \ldots, \kappa_s^n)\big\}$ and not depending on the heap geometry or the source distribution but on diameter ratios d^i/d^j. The functions S_i should satisfy certain Hermite interpolation conditions at the vertices of K_+, but more or less are chosen in an *ad hoc* way.

For a one-dimensional flow (flow on an inclined plane with horizontal projection $(0, L)$) with constant point source located at $x = L$ (6) is used to derive that the vector of concentrations κ_+ is the solution of the system

$$\frac{d}{dx}\kappa_+ = \frac{1}{x}\big(S(\kappa_+) - \kappa_+\big), \quad \kappa_+(L) = \kappa_{+0} ,$$

on $(0, L)$. An analogous ordinary differential equation for κ_+ is derived for the flow on a conical heap. For numerical results and the comparison with experimental data we refer to [37].

2.2 Stationary and Similarity Solutions

Consider a granular material such as dry sand poured at a rate $s = s(x,t) \geq 0$ and piling up to form heaps. First, the material builds up without further motion, but eventually the grains start to roll down the faces of the pile when it has reached a critical slope $k = \tan\alpha > 0$. The pile consists of two main parts, the *standing layer* of height $h = h(x,t)$ (and of constant density) and a thin *rolling layer* of relative height $r = r(x,t)$. In the BCRE model, established by Bouchaud et al. [3,4] and modified by de Gennes [8] by omitting diffusion terms, the exchange of grains from the rolling to the standing layer is described by the exchange term

$$\Gamma(h, r) = \gamma r\big(1 - \frac{|\nabla h|^2}{k^2}\big), \quad \gamma > 0 ; \tag{7}$$

thus it is proportional to the thickness $r \geq 0$ of the rolling layer and vanishes iff $r \equiv 0$ or the slope of the bulk equals the critical slope k. Since the grains in the bulk are motionless except for the exchange $-\Gamma$ to the rolling layer, h satisfies the equation

$$h_t = \Gamma(h, r)\,.$$

However, for the rolling layer, there are two source terms s and $-\Gamma$, such that the continuity equation for r reads

$$r_t + \operatorname{div}(vr) = s - \Gamma(h, r)\,,$$

where v is the horizontal projection of the velocity vector of rolling grains. Assuming as in [19] that particles are rolling in the steepest descent direction $-\nabla h$, the term v is modelled by

$$v = -\mu \nabla h$$

with a constant $\mu > 0$. Summarising we arrive at the system of partial differential equations

$$\begin{aligned} h_t &= \gamma r\Big(1 - \frac{|\nabla h|^2}{k^2}\Big), \\ r_t - \operatorname{div}(\mu r \nabla h) &= s - \gamma r\Big(1 - \frac{|\nabla h|^2}{k^2}\Big) \end{aligned} \tag{8}$$

for $(x, t) \in \Omega \times (0, \infty)$ together with the initial conditions $r(x, 0) = 0$ and $h(x, 0) = h_0(x)$, where $h_0(x)$ describes the bottom onto which the granular material is poured. If the domain $\Omega \subset \mathbb{R}^1$ or $\Omega \subset \mathbb{R}^2$ is not the whole space, surplus material drops down at $\partial\Omega$, and if, for simplicity, $h_0 \equiv 0$, we may use the boundary condition ([19–21])

$$h(x, t) = 0 \quad \text{for } x \in \partial\Omega\,. \tag{9}$$

Since $h \geq 0$ close to $\partial\Omega$, the scalar product of ∇h with the exterior normal vector ν on $\partial\Omega$ is nonpositive. For x close to $\partial\Omega$ we see by $(8)_2$ that $r_t = \mu \nabla r \cdot \nabla h + s + O(r)$ indicating that $r(x, t)$ behaves like an outgoing wave near $\partial\Omega$. Thus no boundary value for r may be prescribed.

In the *silo problem* with walls of infinite height at $\partial\Omega$ (implying that no material can leave the silo, see [19–21]), (8) yields the equation

$$\frac{d}{dt} \int_\Omega (h + r)\, dx = \mu \int_{\partial\Omega} r \frac{\partial h}{\partial \nu}\, do + \int_\Omega s\, dx$$

for the balance of the total mass $\int_\Omega (h + r) dx$. Hence $\int_{\partial\Omega} r\, \partial h / \partial \nu\, do = 0$; since r may be arbitrary on $\partial\Omega$, see the discussion above, we get the Neumann boundary condition

$$\frac{\partial h}{\partial \nu}(x, t) = 0 \quad \text{on } \partial\Omega\,. \tag{10}$$

The system (8) is also closely related to an earlier model of L. Prigozhin [33,35] using variational inequalities. In [36] the authors introduce three length scales:

$L_r = \overline{s}/\gamma$ denotes a typical thickness of the rolling layer given a characteristic (mean) source intensity $\overline{s}$,

$L_p = \mu/\gamma$ denotes the mean path of a rolling grain before being trapped in the standing layer,

L denotes the pile size.

Then rescaling variables by

$$x' = \frac{x}{L}\,,\ h' = \frac{h}{L}\,,\ r' = \frac{r}{L_r}\,,\ s' = \frac{s}{\overline{s}}\,,\ t' = \frac{t\overline{s}}{L}\,,$$

and omitting primes ($'$) for the new dimensionless variables and functions, h and $m = (L_p/L)\,r$ solve the system

$$\begin{aligned} h_t &= \frac{1}{(L_p/L)} m\Big(1 - \frac{|\nabla h|^2}{k^2}\Big)\,, \\ \Big(\frac{L_r}{L_p}\Big) m_t - \operatorname{div}(m\nabla h) &= s - \frac{1}{(L_p/L)} m\left(1 - \frac{|\nabla h|^2}{k^2}\right). \end{aligned}$$

Assuming $L_r \ll L_p$ and $L_p/L \to 0$ the second equation implies that

$$\frac{1}{(L_p/L)} m\big(1 - \frac{|\nabla h|^2}{k^2}\big) \approx s + \operatorname{div}(m\nabla h)\,, \tag{11}$$

and consequently that

$$h_t - \operatorname{div}(m\nabla h) \approx s\,.$$

Actually, if $|\nabla h(x,t)| < k$, the term m has to vanish when $L_p/L \to 0$, see (11). Summarising, in the asymptotic limit we obtain the equation

$$h_t - \operatorname{div}(m\nabla h) = s \tag{12}$$

with the restrictions $m \geq 0$, $|\nabla h| \leq k$ and

$$|\nabla h(x,t)| < k \Rightarrow m = 0\,; \tag{13}$$

here $m = m(x,t)$ turns out to be a Lagrange multiplier related to the slope constraint $|\nabla h| \leq k$. By (12), (13) the material may flow uphill only if the slope is critical, for subcritical slopes the flux $-m\nabla h$ even equals 0. Under suitable assumptions and with additional diffusive terms for h and r, this formal analysis is rigorously proved in [36] for a related discretised system with respect to time $t \geq 0$. Furthermore the equation for h and its Lagrange multiplier m is equivalent to a variational problem in the convex set $K = \{\varphi \in H^1(\Omega) : |\nabla\varphi| \leq k \text{ almost everywhere (a.e.)}, \}$:

$$\begin{cases} \text{find } h(x,t) \text{ such that } h(\cdot,t) \in K \text{ for almost all (a.a.) } t > 0, \\ (h_t - s, \varphi - h)_{L^2(\Omega)} \geq 0 \quad \forall \varphi \in K \end{cases}$$

together with an initial condition $h(\cdot,0) = h_0$, see [35].

Note that the original BCRE equations included diffusion terms such as $\varepsilon\Delta r$ in $(8)_2$ leading to a parabolic rather than to a hyperbolic equation for r. However diffusion may lead to grains rolling upwards instead of downwards. The advantage of system (8) is the fact that the exchange between the standing and the rolling layer is easily modelled by the exchange term Γ in (7). The other terms in (8) are just based on the conservation of masses. On the other hand inertia, momenta, longitudinal and lateral pressures as well as density changes are neglected. These effects are incorporated in the Savage–Hutter models for wet snow avalanches, see Sect. 3 below, leading to a highly nonlinear system of conservation laws.

In [19–21], a slightly different constitutive law for the exchange term Γ is used:

$$\Gamma(h,r) = \gamma r\Big(1 - \frac{|\nabla h|}{k}\Big),$$

leading to the avalanche model

$$\begin{aligned} h_t &= \gamma r\Big(1 - \frac{|\nabla h|}{k}\Big), \\ r_t - \operatorname{div}(\mu r \nabla h) &= s - \gamma r\Big(1 - \frac{|\nabla h|}{k}\Big), \end{aligned} \tag{14}$$

together with initial conditions and the boundary conditions (9) or (10) for h. To our knowledge there is no rigorous proof of existence and uniqueness of solutions to (14) up to now. Even the stationary case with $s = 0$ or $s \neq 0$ poses several open problems. One main property and difficulty of the stationary case with $s = 0$, i.e. for the system

$$r\Big(1 - \frac{|\nabla h|}{k}\Big) = 0, \quad \operatorname{div}(r\nabla h) = 0, \tag{15}$$

is the *non-uniqueness* of solutions: Every pair of functions h, r satisfying

$$h \geq 0, \quad |\nabla h| \leq k, \quad r = 0$$

(and even with $|\nabla h| \geq k$ leading to unstable situations) is a solution of (15).

Even in one dimension the boundary value problem

$$|\nabla h| = k \text{ a.e. in } \Omega, \quad h = 0 \text{ on } \partial\Omega,$$

the so-called *eikonal equation* known from geometrical optics, has uncountably many solutions, namely all piecewise linear functions on an interval $\Omega \subset \mathbb{R}^1$ with slope $\pm k$ almost everywhere. However, uniqueness may be obtained in the setting of viscosity solutions of fully nonlinear equations, see [7,12], or when looking for the *maximum volume solution*.

Theorem 2 *Let $\Omega \subset \mathbb{R}^1$ be a bounded open interval or let $\Omega \subset \mathbb{R}^2$ be an open bounded domain. Let the function $\psi \in C^{0,1}(\overline{\Omega})$ describe the bottom topography (bed) and let $\phi : \partial\Omega \to [0,\infty]$ with $\phi \not\equiv \infty$, $\psi \leq \phi$ on $\partial\Omega$, describe the rim (wall) of the container. Then there exists a unique maximum volume solution $h \in C^{0,1}(\overline{\Omega})$ such that*

$$\begin{aligned} &\psi(x) \leq h(x) \text{ in } \overline{\Omega}, \quad h(x) \leq \phi(x) \text{ on } \partial\Omega \\ &\psi(x) < h(x) \text{ for } x \in \Omega \Rightarrow |h|_{C_x^{0,1}} \leq k \\ &\int_\Omega (h - \psi)\, dx = \max . \end{aligned} \tag{16}$$

Here for $x \in \Omega$ the condition $|h|_{C^{0,1}_x} \le k$ means that there exists an open ball B with centre x in Ω such that $|h(y) - h(y')| \le k|y - y'|$ for all $y, y' \in B$. Note that $h(x) = \psi(x)$ iff no granular material lies on the bed at $x \in \overline{\Omega}$. The term $\int_\Omega (h - \psi)\,dx$ measures the total mass (volume) poured onto the bed.

Proof (See [21]) Let

$$\begin{aligned} M = \{h \in C^0(\overline{\Omega}) : h \ge \psi \text{ on } \overline{\Omega}, \quad h \le \phi \text{ on } \partial\Omega\,, \\ h(x) > \psi(x) \text{ for } x \in \Omega \Rightarrow |h|_{C^{0,1}_x} \le k\}\,. \end{aligned}$$

Since Ω is a bounded domain with Lipschitz boundary and since $\psi \in C^{0,1}(\overline{\Omega})$ there exists a constant $K = K(k, \psi)$ such that

$$|h(y) - h(y')| \le K|y - y'| \quad \text{for all } y, y' \in \overline{\Omega} \tag{17}$$

and for all $h \in M$. The assumption $\phi \not\equiv \infty$, i.e., there exists $\xi \in \partial\Omega$ with $\phi(\xi) < \infty$, implies that M is a set of uniformly bounded functions. Furthermore M is closed in $C^0(\overline{\Omega})$. Thus M is bounded in $C^{0,1}(\overline{\Omega})$ and by Arzelà-Ascoli's Theorem [38] even compact in $C^0(\overline{\Omega})$. Since $V(h) = \int_\Omega (h - \psi)dx$ is a continuous functional on $C^0(\overline{\Omega})$ we get the existence of $h \in M$ maximising the volume $V(\cdot)$. Given $h' \in M$ with $V(h') = V(h)$, but different from h, the continuity of h, h' on $\overline{\Omega}$ will lead to the function $\max(h, h') \in M$ with $V(\max(h, h')) > V(h)$ contradicting the maximality of V at h. □

There exists a remarkable analogy [19–21] between (16) and the Dirichlet problem for the Laplacian, i.e.,

$$\Delta u = 0 \text{ on } \Omega, \quad u = g \text{ on } \partial\Omega\,. \tag{18}$$

Under suitable assumptions on $\partial\Omega$ and on g, Perron's method characterises the unique solution u of (18) by sub-harmonic functions:

$$u(x) = \sup\{v \in C^2(\Omega) \cap C^0(\overline{\Omega}) : \Delta v \ge 0 \text{ in } \Omega,\ v \le g \text{ on } \partial\Omega\}\,.$$

Calling a function $h \in C^{0,1}(\overline{\Omega})$ satisfying $(16)_{1,2}$ *subeikonal* we may derive the following result.

Proposition 3 *The solution h of problem* (16) *given by Theorem 2 can be characterised for every $x \in \overline{\Omega}$ by*

$$h(x) = \sup\{g(x) : g \in M\}\,,$$

i.e., $h(x)$ is the supremum and even the maximum of $g(x)$ among all subeikonal functions in (16).

Proof (See [21]) To show that $\tilde{h}(x) := \sup\{g(x) : g \in M\}$ is Lipschitz continuous, fix $y, y' \in \overline{\Omega}$. Then there are sequences $(h_j), (h'_j) \subset M$ such that $h_j(y) \to \tilde{h}(y)$, $h'_j(y') \to \tilde{h}(y')$. Replacing h_j and h'_j by $\max(h_j, h'_j) \in M$ we may assume that $h'_j = h_j$. Then the estimate

$$|h_j(y) - h_j(y')| \le K(k, \psi)|y - y'| \quad \text{for all } j \in \mathbb{N}\,,$$

see (17), yields the desired estimate for $\tilde{h}$ when $j \to \infty$. In particular $\tilde{h}$ is continuous.

To prove $(16)_2$ let $\tilde{h}(x) > \psi(x)$ for some $x \in \Omega$. Having the "maximum" Lipschitz constant $K(k,\psi)$ in mind we find an open ball B with centre x in Ω such that $\tilde{h} > \psi$ on B and that even every $g \in M$ with $g(y) > \big(\tilde{h}(y) + \psi(y)\big)/2$ for some $y \in B$ satisfies $g > \psi$ on B. Given arbitrary $y, y' \in B$ there exists a sequence $(h_j) = (h_j') \subset M$ such that $h_j(y) \to \tilde{h}(y)$, $h_j(y') \to \tilde{h}(y')$. Since the "global" Lipschitz constant of h_j on B is easily seen to be bounded by k for every $j \in \mathbb{N}$, the same holds for $\tilde{h}$ proving that $|\tilde{h}|_{C_x^{0,1}} \leq k$.

Consequently $\tilde{h} \in M$, $\tilde{h}(x) \geq h(x)$ for all $x \in \overline{\Omega}$ and $V(\tilde{h}) \leq V(h)$. If $\tilde{h}(x) > h(x)$, then $\max(\tilde{h}, h) \in M$ would lead to a contradiction to the maximality of $V(h)$. Thus $\tilde{h} \equiv h$. □

The solution h of (16) may also be characterised by transport paths. If for simplicity $\psi \equiv 0$, then for $x \in \overline{\Omega}$

$$h(x) = \inf_{\chi}\{\phi\big(\chi(1)\big) + k\ell(\chi)\}$$

where χ runs through the set of all continuous piecewise linear paths in $\overline{\Omega}$ connecting x with any point $\chi(1) \in \partial\Omega$; here $\ell(\chi)$ denotes the length of χ [21]. In the most elementary case $\psi \equiv 0$ and $\phi \equiv 0$ (no wall), we easily obtain the solution

$$h(x) = k\,\mathrm{dist}(x, \partial\Omega)\,.$$

Note that h will have points or lines in Ω where it is not differentiable; for a discussion of these singular sets for concrete examples and for several general classes of domains, see [19].

Besides the maximum volume solution in Theorem 2 we consider the time-independent standing/rolling layer of thickness h and r, respectively, when granular material is constantly poured onto a flat table $\psi \equiv 0$ with source intensity $s(x)$. For a point source located in y, i.e., formally $s(x) = \delta_y(x)$, the particles pile up to a cone with vertex in y and with slope k, i.e., $h(x)$ equals

$$\Gamma(x,y) = \begin{cases} k\big(\mathrm{dist}(y,\partial\Omega) - |x-y|\big)\,, & |x-y| < \mathrm{dist}\,(y,\partial\Omega)\,, \\ 0, & \text{otherwise}\,. \end{cases}$$

Then, for more general source distributions, we take the maximum (not the sum or integral) of $\Gamma(x,y)$ on $\operatorname{supp} s$, i.e.,

$$h(x) = \max_{y \in \operatorname{supp} s} \Gamma(x,y) = \max_{y \in \Omega} \Gamma(x,y) \cdot \chi_{\operatorname{supp} s}(y)\,. \tag{19}$$

This formula is similar to the solution $u(x) = \int_\Omega G(x,y)f(y)\,dy$ of Poisson's problem $-\Delta u = f$ on Ω, $u = 0$ on $\partial\Omega$ using Green's function $G(x,y)$. In the one-dimensional case $\Omega = (0,\ell)$, $\psi \equiv 0$, $\phi \equiv 0$, problem (14) has a unique stationary solution (h,r). Based on (19) $h(x)$ and also $r(x)$ can be written down explicity; in 2D this problem is not completely solved, see [19].

Finally we consider the silo problem (8) with boundary condition (10), i.e.

$$\begin{aligned} h_t &= \gamma r\Big(1-\frac{|\nabla h|}{k}\Big) && \text{in } \Omega\times(0,\infty)\,, \\ r_t-\operatorname{div}(\mu r\nabla h) &= s-\gamma r\Big(1-\frac{|\nabla h|}{k}\Big) && \text{in } \Omega\times(0,\infty)\,, \\ \frac{\partial h}{\partial \nu} &= 0 && \text{on } \partial\Omega\times(0,\infty)\,. \end{aligned} \tag{20}$$

This instationary hyperbolic system is not yet solved rigorously. In the one-dimensional case exact solutions have been described by Emig et al. ([10]) by parametrising h, r and also x, t in a new coordinate system (μ_1, μ_2). If $s \equiv 0$ and if $\operatorname{div}(\mu r\nabla h)$ is replaced by cr_x, the unknown functions h, r, x, t satisfy a quasi-linear first order 4×4-system of partial differential equations with respect to μ_1, μ_2. This system can be solved 'explicitly' and yields solutions in the parameterised form $h(\mu_1,\mu_2)$, $r(\mu_1,\mu_2)$, $x(\mu_1,\mu_2)$ and $t(\mu_1,\mu_2)$. From these formulae several profiles (h, r) and shock lines can be analysed.

The analysis will become easier in the quasi-stationary case where $s \geq 0$ is independent of t with a mean source intensity

$$\overline{s} = \frac{1}{|\Omega|}\int_\Omega s(x)\,dx > 0\,. \tag{21}$$

In this case, for large t, we expect a *similarity solution*

$$h(x,t) = h_0(x) + \overline{s}t\,, \quad r(x,t) = r(x)\,.$$

Then (20) simplifies to the stationary system

$$\begin{aligned} \overline{s} &= \gamma r\Big(1-\frac{|\nabla h|}{k}\Big) && \text{in } \Omega\,, \\ -\operatorname{div}(\mu r\nabla h) &= s-\gamma r\Big(1-\frac{|\nabla h|}{k}\Big) && \text{in } \Omega\,, \\ \frac{\partial h}{\partial \nu} &= 0 && \text{on } \partial\Omega\,. \end{aligned} \tag{22}$$

Since $\overline{s} > 0$, we conclude from $(22)_1$ that $|\nabla h| < k$ a.e. A simple calculation leads to the highly nonlinear Neumann problem

$$-\operatorname{div}\Big(\frac{\nabla h}{1-|\nabla h|/k}\Big) = \frac{s(x)-\overline{s}}{\mu\overline{s}/\gamma} \text{ in } \Omega, \quad \frac{\partial h}{\partial \nu} = 0 \text{ on } \partial\Omega\,. \tag{23}$$

Proposition 4 *In one space dimension, (22) has a unique similarity solution (up to additive constants in h). For $\Omega = (0,\ell)$ define*

$$U(x) = \overline{s}\,x - \int_0^x s(y)\,dy\,.$$

Then the rolling layer and the slope h_x of the standing layer are given by

$$r(x) = \frac{1}{k\mu}\Big(\frac{\mu k}{\gamma}\,\overline{s} + |U(x)|\Big)\,, \quad h_x(x) = \frac{1}{\mu}\frac{U(x)}{r(x)}\,,$$

respectively.

The proof is given in [19]. Since $\overline{s} > 0$, actually $|h_x| < k$ in $(0, \ell)$. In the two-dimensional case an explicit solution can be found for the disc $\Omega = B_R(0)$ and a point source $s(x) = \delta_0(x)$, see [20]. The general problem in 2D is not yet solved.

3 Existence Results for the Savage–Hutter Avalanche Model

Section 3 deals with the Savage–Hutter model of dense snow avalanches and the construction of its solutions. In Sect. 3.1 we introduce the model together with its basic physical assumptions and emphasise its hyperbolic character. In the one-dimensional plane case we classify the solutions describing avalanches with a similar shape for all times (similarity solutions). The final Sect. 3.3 uses the modern mathematical theory of conservation laws with source terms to construct weak entropy solutions in the one-dimensional case.

3.1 Modelling

The Savage–Hutter equations model the flow of a dense snow avalanche with small aspect ratio on a rough inclined plane or a curved bed by considering the avalanche as a cohesionless granular material in which the interstitial air plays a negligible role. In contrast to the BCRE model of Sect. 2 this model accounts for an exchange of momentum and goes far beyond simple particle models [42,39]; on the other hand it ignores abrasion and exchange of particles between the avalanche and the bed.

In a plane curvilinear coordinate system let x denote the coordinate along the rough incline and let z denote the perpendicular coordinate. Looking for the velocity u of the avalanche and the height h of the free surface the main assumptions of the Savage–Hutter model [40] are as follows:

- The granular material obeys a Mohr–Coulomb-type plastic yield criterion expressed by a constant angle of internal friction ϕ, i.e., given the stress tensor T and the exterior normal vector n on an internal surface the shear traction $S = Tn - n(n \cdot Tn)$ and the normal stress $N = n \cdot Tn$ are related to each other by the formula $|S| = N \tan\phi$. Since shear traction depends on the direction of the velocity vector u,
$$S = -\frac{u}{|u|} N \tan\phi \,,$$
giving rise to a jump discontinuity.
- At the base there exists a very thin fluidised layer (about 10 grain diameters) obeying a Coulomb dry friction law with a bed friction angle $\delta < \phi$, i.e., $S = -(u/|u|)\, N \tan\delta$.
- The longitudinal stress component T_{xx} is related to the perpendicular component T_{zz} by
$$T_{xx} = K_{\mathrm{act/pass}}\, T_{zz} \,,$$
where
$$\left.\begin{array}{l} K_{\mathrm{act}} \\ K_{\mathrm{pass}} \end{array}\right\} = \frac{2(1 \mp \sqrt{(1-\cos^2\phi/\cos^2\delta)})}{\cos^2\phi} - 1 \quad \begin{array}{l} \text{iff } \partial u/\partial x > 0 \,, \\ \text{iff } \partial u/\partial x < 0 \end{array} \tag{24}$$

is the active and passive earth pressure coefficient, respectively. Note that $0 < K_{\text{act}} < K_{\text{pass}}$, where K_{act} applies iff the flow is locally expanding.

- As a major assumption the velocity profile is blunt (except for the fluidised layer): for every $x \in \mathbb{R}$, $t > 0$

$$\int_0^{h(x,t)} u(x,z,t)\,dz = h(x,t)u(x,t), \quad \int_0^h u^2 dz = hu^2 \text{ etc.}$$

  Thus all macroscopic quantities are considered to be z–independent.
- Given a characteristic height H and length L of the avalanche assume that the aspect ratio $\varepsilon = H/L$ is small compared to 1, i.e., $\varepsilon \ll 1$. If the bed is curved with a characteristic radius of curvature R, assume that $L/R = O(\varepsilon^{1/2})$. Finally assume that $\tan\delta = O(\varepsilon^{1/2})$.

Typical values of δ, ϕ and K for glass, quartz, marmor or plastic grains are as follows, see [25]:

$$20^o < \delta < 40^o, \quad 30^o < \phi < 46^o, \quad 5^o < \phi - \delta < 20^o\,,$$

where the bed friction angle also depends on the roughness of the bed. Thus typical earth pressure coefficients are

$$K_{\text{act}} \in (0.7, 0.9), \quad K_{\text{pass}} \in (2.8, 4.6)\,.$$

Ignoring all terms of order higher than ε the Savage–Hutter equations for a thin *two-dimensional* avalanche of height h, velocity $u = (u_1, u_2)$ and momentum hu on a two-dimensional basal profile $z = b(x,y)$ with main downslope direction $(1,0)^T$ take the form [15,27,40]

$$\begin{aligned} \partial_t h + \operatorname{div}(hu) &= 0\,, \\ \partial_t(hu) + \operatorname{div}\big(hu \otimes u + \tfrac{1}{2}\varepsilon h^2 K(\cos\xi)\big) &= hs(u,x)\,, \end{aligned} \tag{25}$$

with the source term

$$s = \sin\xi \begin{pmatrix} 1 \\ 0 \end{pmatrix} - \frac{u}{|u|}\tan\delta\cos\xi - \varepsilon K(\cos\xi)\nabla b\,. \tag{26}$$

Here $\xi = \xi(x)$ is the local inclination angle along the direction $(1,0)^T$ whereas the influence of the curvature has been omitted. Furthermore K denotes the diagonal 2×2-matrix of earth pressure coefficients such that

$$\operatorname{div}\Big(\frac{1}{2}\varepsilon h^2(\cos\xi)K\Big) = \varepsilon h\cos\xi\Big(K_{x,\text{act/pass}}\frac{\partial h}{\partial x}, K_{x,\text{act/pass}}^{y,\text{act/pass}}\frac{\partial h}{\partial y}\Big)^T + \ldots$$

with $K_{x,\text{act/pass}}$ as in (24) and $K_{x,\text{act/pass}}^{y,\text{act/pass}}$ depending on the signs of $\partial u_1/\partial x$ and of $\partial u_2/\partial y$. This term together with the term $\varepsilon hK(\cos\xi)\nabla b$ represents the variation of the normal pressure in x- and y-directions, whereas the first and second term of (26) are due to gravity normalised to 1 and to friction of the avalanche with the bed, respectively. To be more precise in the two-dimensional case, $\varepsilon = H/L$ has to be replaced by a diagonal 2×2 matrix with entries $\varepsilon_x = H/L_x$ and $\varepsilon_y/\varepsilon_{xy} = (H/L_y)/(L_y/L_x)$ for characteristic lengths L_x and L_y.

System (25) is written in the form of a system of conservation laws for (h, hu) with a source on the right-hand side depending on h and u. Looking at the leading terms and ignoring the term containing K, (25) is similar to the shallow water equations and to the Euler equations of gas dynamics. However, besides the fact that there exists no satisfying mathematical theory for systems of conservation laws in more than one space dimension, the jump discontinuity $u/|u|$ and of course the piecewise constant function K depending on signs of ∇u pose new analytical and numerical difficulties. Thus, in every analytical approach – even when looking for similarity solutions, see Sect. 3.2 – K is assumed to be constant.

Of course solutions of (25) may evolve shocks even when the data are smooth. Shocks will mainly occur in the run-out zone when a part of the material has already been deposited. Furthermore shocks can be observed in beautiful experiments on granular matter in rotating drums, see [14].

Proposition 5 *Let $(h, hu) \in \mathbb{R}^3$ be a weak solution of (25) in a domain $\Omega \subset \mathbb{R}^2 \times (0, \infty)$, i.e., for all $\varphi \in C_0^\infty(\Omega; \mathbb{R}^3)$*

$$\iint\limits_{\Omega} \left\{ \begin{pmatrix} h \\ hu \end{pmatrix} \cdot \varphi_t + \begin{pmatrix} hu \\ hu \otimes u + \frac{1}{2}\varepsilon h^2 (\cos\xi) K \end{pmatrix} \cdot \nabla\varphi \right\} dx\, dt$$

$$= -\iint\limits_{\Omega} \begin{pmatrix} 0 \\ hs \end{pmatrix} \cdot \varphi \, dx\, dt\,.$$

Assume that Ω is separated by a smooth, regular surface Γ into two parts Ω_ℓ and Ω_r such that

$$\begin{pmatrix} h \\ hu \end{pmatrix}\Big|_{\Omega_\ell} \in C^1(\overline{\Omega_\ell; \mathbb{R}^3}), \quad \begin{pmatrix} h \\ hu \end{pmatrix}\Big|_{\Omega_r} \in C^1(\overline{\Omega_r}; \mathbb{R}^3)\,.$$

Let $\nu = (\nu_t, \nu_x)$ denote the unit normal vector on Γ directed into Ω_ℓ. Then (h, hu) satisfies the Rankine–Hugoniot jump condition

$$\left[\begin{pmatrix} h \\ hu \end{pmatrix} \right] \nu_t + \nu_x \cdot \left[\begin{pmatrix} hu \\ hu \otimes u + \frac{1}{2}\varepsilon h^2 (\cos\xi) K \end{pmatrix} \right] = 0\,,$$

where as usual $[\cdot]$ denotes the difference of the limits of (h, hu) on Γ taken from Ω_ℓ and from Ω_r, respectively.

Coming back to a one-dimensional avalanche on a basal profile $z = b(x)$, $x \in \mathbb{R}$, let a line of discontinuity Γ be given in parameterised form $\big(\gamma(t), t\big)$. Then $\gamma'(t)$ is the speed of propagation of the discontinuity, and the Rankine–Hugoniot condition takes the simple form

$$\left[\begin{pmatrix} h \\ hu \end{pmatrix} \right] \gamma'(t) = \left[\begin{pmatrix} hu \\ hu^2 + \frac{1}{2}\varepsilon h^2 (\cos\xi) K_{\text{act/pass}} \end{pmatrix} \right]. \tag{27}$$

A more recent generalisation of the Savage–Hutter model considers compressible avalanches of density ρ satisfying a constitutive equation $\rho = \rho(h, u)$, see [16,17].

Since there is no physical evidence for a (monotonically decreasing) dependence on $|u|$, up to now the constitutive equation

$$\rho(h) = h^\alpha, \quad \alpha > 0,$$

has been investigated; see [17] for the mathematically easier case $\alpha = -1/2$. In the one-dimensional case we obtain the system

$$\begin{aligned} \partial_t(\rho h) + \partial_x(\rho h u) &= 0, \\ \partial_t(\rho h u) + \partial_x(\rho h u^2 + \tfrac{1}{2}\beta(x)\rho h^2) &= \rho h s(u,x), \end{aligned} \tag{28}$$

where

$$\beta(x) = \varepsilon K_{\mathrm{act/pass}} \cos\xi(x),$$

$$s = \sin\xi - \varepsilon\cos\xi\, b_x - \frac{u}{|u|}\tan\delta\cos\xi.$$

Assuming an overall constant $K = K_{\mathrm{act/pass}}$ which is physically justified for flows extending globally or contracting globally, it is convenient to introduce new functions to eliminate the x-dependence in the term $\beta(x)\rho h^2$ and to refind the standard form of conservation laws. Let

$$(u_1, u_2) = \left(\left(\frac{\beta}{2\kappa}\right)^{1+\alpha} h^{1+\alpha}, \left(\frac{\beta}{2\kappa}\right)^{1+\alpha} h^{1+\alpha} u \right), \tag{29}$$

where $\kappa = \big(4(1+\alpha)(2+\alpha)\big)^{-1}$, and

$$F(u_1, u_2) = \begin{pmatrix} u_2 \\ u_2^2/u_1 + \kappa u_1^{\frac{2+\alpha}{1+\alpha}} \end{pmatrix}, \quad S_0 = \frac{(1+\alpha)\beta'}{\beta} F + \begin{pmatrix} 0 \\ u_1 s(\frac{u_2}{u_1}, x) \end{pmatrix}. \tag{30}$$

Then (28) takes on the form

$$\partial_t u + \partial_x F(u) = S_0(u,x), \quad u = (u_1, u_2). \tag{31}$$

Note that in (30) also the first component of the source term S_0 is different from zero. However, the Rankine–Hugoniot condition for a shock line Γ with speed of propagation $\gamma'(t)$ has the simple form

$$[u_1]\gamma' = [u_2], \quad [u_2]\gamma' = [F_2(u_1, u_2)], \tag{32}$$

yielding the compatibility condition

$$[u_2]^2 = [F_2(u_1, u_2)] \cdot [u_1] \quad \text{on } \Gamma.$$

By (32) a discontinuity of u_1 or of h with respect to x, say $h(x_\ell, t) > 0 = h(x_r, t)$, is not admissible.

Up to now problems arising from the jump discontinuity in the source terms s and S_0 have been ignored. In Sect. 3.3 we propose to introduce set-valued maps to deal with this discontinuity, see Definition 7 and Remark 8 below.

3.2 Similarity Solutions

Consider the Savage–Hutter model for a one-dimensional incompressible avalanche on a plane moving downwards everywhere, i.e. the system

$$\begin{aligned} \partial_t h + \partial_x(hu) &= 0\,, \\ \partial_t u + u\partial_x u &= \sin\xi - \tan\delta\cos\xi - \beta\, h_x\,, \end{aligned} \tag{33}$$

when $\operatorname{sgn} u = +1$ is constant. Also $K_{\text{act/pass}}$ is assumed to be constant yielding a constant $\beta = \varepsilon K \cos\xi$. In order to discuss the existence of similarity solutions we subtract the motion of the centre of mass. To this end, define

$$u_0(t) = t(\sin\xi - \tan\delta\cos\xi), \quad \tilde{u} = u - u_0(t)$$

and the moving variable

$$\zeta = x - \int_0^t u_0(s)ds\,.$$

Let $g(t)$ denote a typical length of the avalanche at time t, e.g. half the spread of an avalanche with compact support. Now use new coordinates

$$y = \frac{\zeta}{g(t)}, \quad \tau = t$$

in (33) and the notation $(\cdot)'$ and $(\cdot)_y$ for derivatives with respect to τ and y, respectively, to find the system

$$\begin{aligned} h' - y\frac{g'}{g}h_y + \frac{1}{g}(h\tilde{u})_y &= 0\,, \\ \tilde{u}' - y\frac{g'}{g}\tilde{u}_y + \frac{1}{g}(\tilde{u}\tilde{u}_y + \beta h_y) &= 0 \end{aligned} \tag{34}$$

for $(h, \tilde{u})$. Then a solution of the form

$$h(y,\tau) = \ell(\tau)H(y), \quad \tilde{u}(y,\tau) = k(\tau)U(y) \tag{35}$$

is called a *similarity solution* of (34). Owing to the conservation of mass

$$M \equiv \int_{\mathbb{R}} h(\zeta,t)\,d\zeta = \int_{\mathbb{R}} h(y,\tau)g(\tau)\,dy = \ell(\tau)g(\tau)\int_{\mathbb{R}} H(y)\,dy$$

we deduce that

$$\ell = \frac{1}{g},$$

at least when the total mass is finite. It will be seen below that this assumption is not satisfied in general. Inserting $\ell = 1/g$ in (34) yields the system

$$\begin{aligned} H + yH_y - \frac{k}{g'}(HU)_y &= 0\,, \\ U - \frac{g'k}{gk'}yU_y + \frac{k^2}{gk'}UU_y + \frac{\beta}{g^2k'}H_y &= 0\,. \end{aligned} \tag{36}$$

From $(36)_1$ we see that

$$0 = \left(\frac{k}{g'}\right)'(HU)_y\,.$$

Thus either

$$\frac{k}{g'} \equiv \text{const} \quad \text{or} \quad HU \equiv \text{const}\,.$$

Since $k(\tau)$ denotes an overall increase or decrease of the velocity $\tilde{u}$, the change of the characteristic length $g'(\tau)$ has to be proportional to $k(\tau)$. Hence k/g' has to be independent of τ. Actually, if $k/g' \not\equiv$ const, $(36)_1$ would imply that $(HU)_y \equiv 0$ and that $H + yH_y \equiv 0$. These equations yield the general solution $H(y) = c_0/y$, $U = c_1 y$. Then $(36)_2$ can be interpreted as a vanishing linear combination of the functions y and $1/y^2$ with τ-depending coefficients. Now we may conclude that $c_0 = 0$ and consequently $H \equiv 0$ yielding the trivial solution $h \equiv 0$.

In the following assume without loss of generality that

$$k \equiv g'\,,$$

since a constant k/g' different from 1 can be subsumed by the functions H or U, see (35). Then (36) can be written in the simple form

$$\begin{aligned} \big((U-y)H\big)_y &= 0\,, \\ U + \frac{g'^2}{gg''}(U-y)U_y + \frac{\beta}{g^2 g''}\,H_y &= 0\,. \end{aligned} \tag{37}$$

Case 1: $U \equiv y$ In this case the velocity $\tilde{u}(y,\tau)$ is linear in y for every time τ. From $(37)_2$ we derive the equation $g^2 g'' = -\beta H_y/y$. Consequently both sides are constants leading to the identities

$$g^2 g'' = \frac{G_0}{2} \quad \text{and} \quad H(y) = H_0 - \frac{G_0}{4\beta}y^2 \tag{38}$$

with constants H_0, G_0. Let us ignore the elementary case $G_0 = 0$ where $H(y)$ is constant and $g(\tau)$ is linear. If $G_0 \neq 0$, by $(38)_1$ $g'g'' = G_0 g'/(2g^2)$ and consequently

$$g'^2 = -G_0\Big(\frac{1}{g} + \frac{1}{g_0}\Big) \quad \text{with } g_0 \in \overline{\mathbb{R}}\backslash\{0\}\,.$$

If $g_0 = \infty$, then $g \geq 0$ implies that $G_0 < 0$ and $g'g^{1/2} = \pm\sqrt{|G_0|}$. Hence

$$g(\tau) = \big(g(0)^{3/2} + \frac{3}{2}\sqrt{|G_0|}\,\tau\big)^{2/3}, \; H(y) = H_0 + \frac{|G_0|}{4\beta}\,y^2\,; \tag{39}$$

the case $g'g^{1/2} = -\sqrt{|G_0|}$ leads to an unphysical compression of the avalanche or – in other words – to a time reversal in (39). The solution (39), considered only for $|y| < y_0$, defines an avalanche with the shape of an M, called an *M–wave* in [40]. As $\tau \to \infty$,

$$h(y,\tau) \sim c\tau^{-2/3}H(y)\,.$$

If $g_0 \in \mathbb{R}^* = \mathbb{R}\backslash\{0\}$, then $g'(\tau) = \pm\sqrt{|G_0|}\,\sqrt{|g_0+g|}/\sqrt{|g_0 g|}$. First we consider the case when $g_0 > 0$ and $g' > 0$. Then

$$\sqrt{\frac{|G_0|}{g_0}}\tau = \int_{g(0)}^{g(\tau)} \frac{\sqrt{g}\,dg}{\sqrt{g_0+g}} = 2\int_{\sqrt{g(0)}}^{\sqrt{g(\tau)}} \frac{h^2 dh}{\sqrt{g_0+h^2}}$$

$$= \sqrt{g(\tau)}\,\sqrt{g_0+g(\tau)} - g_0 \ln\left(\sqrt{g(\tau)} + \sqrt{g_0+g(\tau)}\right) - C_0 .$$

For $\tau \to \infty$ we deduce the linear behaviour

$$g(\tau) \sim \sqrt{\frac{|G_0|}{g_0}}\,\tau, \qquad \tau \to \infty .$$

Since $g > 0$ and $g_0 > 0$ necessarily imply that $G_0 < 0$, again $H(y) = H_0 + |G_0|\,y^2/(4\beta)$ defines an M–wave on $|y| \le y_0$. But compared to the M-wave above we now get an M-wave with

$$h(y,\tau) \sim \tau^{-1} H(y) \quad \text{as } \tau \to \infty .$$

When $g_0 > 0$ but $g' < 0$, then the differential equation for $g(\tau)$ immediately implies that $g(\tau) \to 0$ and consequently that $\ell(\tau) \to \infty$ in finite time. Thus this case is unphysical.

Next consider the case when $g_0 < 0$, but $g + g_0 \ge 0$ and $g' \ge 0$. Then $G_0 > 0$ and

$$\sqrt{\frac{G_0}{|g_0|}}\tau = 2\int_{\sqrt{g(0)}}^{\sqrt{g(\tau)}} \frac{h^2\,dh}{\sqrt{h^2-|g_0|}}$$

$$= \sqrt{g(\tau)}\,\sqrt{g(\tau)-|g_0|} + |g_0| \ln\left(\sqrt{g(\tau)} + \sqrt{g(\tau)-|g_0|}\right) - C_0 .$$

For $\tau \to \infty$ we deduce the asymptotic behaviour $g(\tau) \sim \tau$. The shape of the avalanche is described by $H(y) = H_0 - G_0 y^2/(4\beta)$ for $|y| \le \sqrt{4H_0\beta/G_0}$ forming a *parabolic cap*, see [40] for the special case $\tau_0 = 0$, $g_0 = -1$, $g(0) = 1$ such that $g'(0) = 0$ (avalanche is starting at rest).

For $g_0 < 0$ and $g+g_0 < 0$, but $g' > 0$, the function $g(\tau)$ is strictly increasing until $g(\tau) \to |g_0|$ where $g'(\tau) \to 0$. For $g(\tau)$ close to $|g_0|$, but less than $|g_0|$, the differential equation for $g(\tau)$ is related to the equation $g'(\tau) = 2\sqrt{|g_0| - g(\tau)}$ showing that $g(\tau)$ actually approaches $|g_0|$ like the parabola $g(\tau) = |g_0| - (\tau_1 - \tau)^2$ as $\tau \to \tau_1-$. Then $k(\tau_1) = g'(\tau_1) = 0$ and $\tilde{u}(\tau_1, y) = 0$ yielding an avalanche at rest! From classical theory of ordinary differential equations it is known that a solution $g(\tau)$ with initial value $g(\tau_1) = |g_0|$ is not uniquely determined for $\tau > \tau_1$. When the avalanche restarts to move and $g(\tau)$ becomes larger than $|g_0|$ for some $\tau > \tau_1$, then we refer to the previous case.

The case $g_0 < 0$, but $g + g_0 > 0$ and $g' < 0$ also leads to an avalanche at rest in finite time. Finally, if $g_0 < 0$, $g + g_0 < 0$ and $g' < 0$, then $g(\tau)$ converges to 0 in finite time. Thus $\ell(\tau) \to \infty$ in finite time leading to an infinite velocity and an unphysical solution.

Case 2: $U \not\equiv y$. In this case $(37)_1$ yields a constant $m \neq 0$, a characteristic momentum, such that

$$H = \frac{m}{U - y}\,; \tag{40}$$

the case $m = 0$ is trivial. Inserting this identity into $(37)_2$ results in

$$U + \frac{g'^2}{gg''}(U-y)U_y - \frac{\beta m}{g^2 g''} \cdot \frac{U_y - 1}{(U-y)^2} = 0\,. \tag{41}$$

Dividing by U, multiplying with $g^2 g''$ and differentiating with respect to y we are led to the equation

$$g'^2 g\Big(\frac{(U-y)U_y}{U}\Big)_y - \beta m\Big(\frac{U_y - 1}{(U-y)^2 U}\Big)_y = 0\,.$$

In order to conclude that $g'^2 g$ is constant we have to exclude the possibility that both terms depending on y vanish. If these terms vanish, we would obtain *two* ordinary differential equations for $U(y)$, leading after elementary calculations, to a contradiction. Thus there exists a constant $c \neq 0$ (since $g > 0$) such that

$$g'\sqrt{g} = c \quad \text{and} \quad g(\tau) \sim \Big(\frac{3}{2}c\tau\Big)^{2/3} \quad \text{as } \tau \to \infty\,,$$

cf. Case 1 with $g_0 = \infty$. Hence $g'' g^2 = -c^2/2$, and (41) yields the differential equation

$$U - 2(U-y)U_y + 2a^3\frac{U_y - 1}{(U-y)^2} = 0\,, \quad a^3 = \frac{\beta m}{c^2} \neq 0\,, \tag{42}$$

which, on defining

$$V(y) = U(y) - y\,, \tag{43}$$

transforms to the equation

$$V_y = \frac{1}{2}\,\frac{V^2(V-y)}{a^3 - V^3}\,. \tag{44}$$

Note that the lines $V \equiv 0$ and $V \equiv a$ as well as points y_0 where $V(y_0) = y_0$ (the 'diagonal') are important in the discussion of local and global properties of solutions of the differential equation (44). Concerning the size of $V_0 = V(0)$ and of a we have to distinguish between several cases.

Since $H(0) = m/U(0) = m/V(0)$ and $\operatorname{sgn} a = \operatorname{sgn} m$, the cases $V_0 < 0 < a$ and $a < 0 < V_0$ lead to unphysical solutions with $H(0) < 0$ and will not be discussed.

Case 2.1: $0 < a < V_0$ *and* $\exists\, y_0 > a : V(y_0) = y_0$ (this case will occur for $V_0 \gg a$).

In this case $V'(0) < 0$ and even $V'(y) < 0$ for all $y \in (-\infty, y_0)$. However, $V'(y_0) = 0$ and $V''(y_0) = y_0^2/[2(y_0^3 - a^3)] > 0$ implying that V has a local minimum at y_0. Since even $V' > 0$ on (y_0, ∞) and $V' < 0$ on $(-\infty, y_0)$, V has a global minimum at y_0. To determine the shape of the corresponding avalanche it is crucial to discuss the asymptotic behaviour of V for $y \to \pm\infty$. By (44) we may exclude that $V(y)$ is bounded for $y \to \infty$. If there exists $y_1 > y_0$ such that $V(y_1) = \frac{1}{2}\,y_1$, the inequality $V'(y_1) = \frac{1}{2}y_1^3/(y_1^3 - 8a_3) > \frac{1}{2}$ leads to a contradiction. Consequently $y/2 < V(y) < y$ for $y > 0$. Defining $w(y) = V(y)/y$ we see that $w(y) \in (1/2, 1)$ and that

$$w' = -\frac{w^2}{2y}\,\frac{(2w-1)(w+1) - 2a^3/(wy^3)}{w^3 - a^3/y^3}\,. \tag{45}$$

If there exists $0 < \delta < 1/2$ such that $w(y) \geq 1/2 + \delta$ for large $y > 0$, then $w'(y) \approx -1/y$ for these y leading to a logarithmic decay $-\log y$. Then finally $w(y)$ will cross the line $w = 1/2 + \delta$ with a negative slope. Thus $w(y) < 1/2 + \delta$ for all large y, even $w(y) \to 1/2$ and $V(y) \sim y/2$ for $y \to \infty$. For $y \to -\infty$ (44) implies that $V(y)$ is unbounded. If there exists $y_2 < -a$ such that $V(y_2) = -y_2$, then $V'(y_2) = -y_2^3/(a^3 + y_2^3) < -1$ leads to a contradiction. Thus $w(y) = V(y)/y < -1$ for all negative y. The possibility that $w(y) \leq w_1 < -1$ for all large $y < 0$ can be excluded since under this assumption $w' \leq \delta/y$ for some $\delta > 0$. Hence $w(y) \to -1$, $V(y) = -y\big(1 + o(1)\big)$ and $U(y) = V(y) + y = o(|y|)$ for $y \to -\infty$. To prove that even $U(y) = O(|y|^{-1/2})$ for $y \to -\infty$ we introduce the auxiliary function $\varphi(y) = |y|^{1/2}U(y)$. By (43), (44)

$$\varphi'(y) = \frac{|y|^{1/2}}{2}\,\frac{w(y)^2(w(y)+1)^2 + O(|y|^{-3})}{w(y)^3(1 + O(|y|^{-3}))}$$

for $y \to -\infty$. Since w is bounded for large negative y, and $1 + w = -U/|y| = -\varphi(y)/|y|^{3/2}$, we obtain that

$$\varphi' = \frac{c(y)}{|y|^{5/2}} - \frac{1}{2|y|^{5/2}|w(y)|}\,\varphi^2\,, \tag{46}$$

where $|c(y)|$ is bounded. Assuming that φ is not bounded for $y \to -\infty$ there exists $y_1 < 0$ such that $\varphi'(y_1)$ is negative. We may even assume that φ is strictly decreasing for $y < y_1$. Thus there are constants $c_1, c_2 > 0$ such that

$$-\frac{c_1}{|y|^{5/2}}\,\varphi^2 \leq \varphi' \leq -\frac{c_2}{|y|^{5/2}}\,\varphi^2\,.$$

However, this differential inequality can be satisfied only for bounded functions. Now the boundedness φ implies that $U(y) = O(|y|^{-1/2})$ for $y \to -\infty$. Summarising the previous results we get for the avalanche characterised by $U = V + y$ and $H = m/(U - y)$ where $m > 0$ that

$$\begin{aligned} 0 < U(y) &\sim \frac{c}{|y|^{1/2}} \quad \text{for } y \to -\infty,\quad U(y) \sim \frac{3}{2}y \quad \text{for } y \to \infty\,,\\ H(y) &\sim \frac{m}{|y|} \qquad\;\; \text{for } y \to -\infty,\quad H(y) \sim \frac{2m}{y} \quad \text{for } y \to \infty\,. \end{aligned}$$

The rate of decay of H for $y \to \pm\infty$ shows that the mass of the avalanche is infinite.

Case 2.2: $0 < a < V_0$ *and* $V(y)$ will not cross or touch the diagonal (this case will occur for $V_0 > a$ close to a).

For $y > 0$ the solution $V(y)$ is strictly decreasing. Since it will not cross the diagonal and cannot cross the line $y \equiv a$, it will approach a in finite "time" $y_1 < a$

with a slope approaching $-\infty$. The behaviour of $V(y)$ for $y \to -\infty$ is the same as in the previous case. Thus the avalanche has the properties

$$\begin{aligned} &0 < U(y) < c/|y|^{1/2} \quad \text{for } y \to -\infty, \quad U(y) \downarrow y_1 + a \quad \text{for } y \to y_1-, \\ &H(y) \sim \frac{m}{|y|} \qquad\qquad \text{for } y \to -\infty, \quad H(y) \uparrow \frac{m}{a} \qquad \text{for } y \to y_1-, \end{aligned}$$

where U and H have infinite slope at $y = y_1$ in which the solution breaks down.

Case 2.3: $0 < V_0 < a$ *and* $\exists\, y_0 > 0 : V(y_0) = y_0$ (this case will occur for small V_0).

Then $V'(0) > 0$, $y_0 < a$, $V'(y_0) = 0$ and $V''(y_0) = -y_0^2/(2(a^3 - y_0^3)) < 0$. Thus V has a local maximum at y_0 and even $V'(y) < 0$ for all $y > y_0$. For $y \to \infty$ the behaviour of $V(y)$ is modelled by the differential equation $V' = -cyV^2$ with a constant $c > 0$ leading to the asymptotic behaviour $V(y) = O(y^{-2})$. Analogously $V(y)$ will tend to 0 as $y \to -\infty$ and $V(y) = O(y^{-2})$. Thus the velocity $U(y)$ has the properties

$$U(y) = y + O(y^{-2}) \quad \text{for } y \to \pm\infty.$$

However the height $H(y) = m/V(y)$ where $m > 0$ diverges as y^2 for $y \to \pm\infty$. Hence the avalanche is unphysical in this case; the "explicit" solutions U, H may be used only locally.

Case 2.4: $0 < V_0 < a$ *and* $V(y)$ will not cross or touch the diagonal (this case will occur for $V_0 < a$ close to a).

Then V is strictly increasing for $y > 0$ until it will reach the level $y = a$ in finite time $y_1 < a$ with slope $+\infty$. For $y \to -\infty$ the behaviour is the same as in Case 2.3. Consequently

$$\begin{aligned} &U(y) = y + O(y^{-2}) \quad \text{for } y \to -\infty, \quad U(y) \uparrow y_1 + a \text{ as } y \to y_1-, \\ &H(y) = O(y^2) \qquad\quad \text{for } y \to -\infty, \quad H(y) \downarrow m/a > 0 \text{ as } y \to y_1-. \end{aligned}$$

This solution is unphysical since it breaks down in finite time and since H is unbounded.

There are four further cases when $a < 0$ and $V_0 < 0$. However, since $W(y) = -V(-y)$ satisfies (44) with a replaced by $-a$, it suffices to refer to Case 2.1–2.4. The corresponding avalanche is described by $U(y) \hat{=} -U(-y)$ and $H(y) \hat{=} H(-y)$.

Case 2.5: $V_0 < a < 0$ *and* $V(y)$ crosses the diagonal. Looking at Case 2.1 we find a solution with the properties ($m < 0$)

$$\begin{aligned} &U(y) \sim \frac{3}{2}y \quad \text{for } y \to -\infty, \quad -\frac{c}{|y|^{1/2}} < U(y) < 0 \quad \text{for } y \to \infty, \\ &H(y) \sim \frac{2m}{y} \quad \text{for } y \to -\infty, \quad H(y) \sim \frac{|m|}{y} \qquad\qquad \text{for } y \to \infty. \end{aligned}$$

Due to the large negative velocities for $y \to -\infty$ the avalanche seems to move upwards everywhere. However, the real similarity avalanche has the height

$$h(x,t) = \frac{1}{g(t)}\, H\Big(\frac{x - \alpha t^2/2}{g(t)}\Big),$$

where the acceleration $\alpha = \sin\xi - \tan\delta\cos\xi$ is assumed to be a positive constant ($\xi > \delta$) and $g(t) \sim \left(\frac{3}{2}ct\right)^{2/3}$ for $t \to \infty$. Moreover, its velocity including the motion of the centre of mass is given by

$$u(x,t) = \alpha t + g'(t)U\Big(\frac{x - \alpha t^2/2}{g(t)}\Big).$$

Since $g'(t) = c/\sqrt{g(t)}$ and $g(t)^{3/2} \sim 3ct/2$ for $t \to \infty$, the asymptotic behaviour of $u(x,t)$ for $t \to +\infty$, i.e. $y \to -\infty$, is given by

$$u(x,t) \sim \alpha t + \frac{3}{2}\,\frac{c}{\sqrt{g(t)}}\,\frac{x - \alpha t^2/2}{g(t)} \sim \frac{\alpha}{2}t + \frac{x}{t}. \tag{47}$$

Hence the physical velocity is positive for large t and approaches half the velocity of the corresponding centre of mass. This behaviour is reflected by the height of the avalanche

$$h(x,t) \sim \frac{m}{|x - \alpha t^2/2|}\begin{cases} 2 & \text{for } x \to -\infty, \\ 1 & \text{for } x \to +\infty. \end{cases}$$

Case 2.6: $a < V_0 < 0$ *and* $V(y)$ crosses the main diagonal. Referring to Case 2.3 we find a solution such that $V(y) = O(y^{-2})$ for $y \to \pm\infty$ and consequently that $U(y) \sim y$, $H(y) = O(y^2)$ for $y \to \pm\infty$.

We drop the two cases analogous to Cases 2.3 and 2.4 when a and V_0 are negative since these solutions fail to exist for large negative y. Finally we mention that even when a is positive there are further solutions existing on a y-semiaxis. E.g., if $a > 0$, consider $y_0 > a$ and $V_0 := V(y_0) > a$ or $< a$, but close to a. Then there exists a solution $V(y)$ for $y > y_0$ evolving an infinite slope for $y < y_0$ where V converges to a; for $y \to +\infty$ $V(y)$ will diverge as $y/2$ or converge to 0 as y^{-2}, see Cases 2.1 or 2.3 for this asymptotic behaviour.

Note that all similarity solutions have been found under the assumption that $\operatorname{sgn} u(x,t) = 1$, i.e., $|\tilde{u}(x,t)| < u_0(t)$. Now the speed of the centre of mass of the similarity avalanche is larger than $\tilde{u}$ iff

$$|g'(t)U(y)| < \alpha t \tag{48}$$

for all admissible y. In several cases, see e.g. the M-wave on a compact interval $|y| \le y_0$, (48) is satisfied for large t since $g'(t)$ is bounded, cf. [40]. For the parabolic cap which has a compact support with respect to y and where $g'(t) \approx t$ a size condition for α, i.e. for the constant slope ξ, has to be assumed [40]. In the important Cases 2.1 and 2.5 condition (48) may be violated locally, but not globally, see (47).

Recall the overall assumption that the earth pressure coefficient $K_{\text{act/pass}}$ was constant, ignoring the fact that the avalanche may be compressed or stretched locally. In [26,32] the authors carefully analyse a parabolic cap solution when even the bed friction angle δ varies either with y or with the centre of mass velocity or with both of them. Although in these cases the height h defines a strict parabolic cap and the velocity u is linear, the equation for the spread g is much more complicated since $g''g^2$ is no longer constant.

Finally we mention that also for a two-dimensional avalanche on an inclined plane there exist parabolic cap solutions starting with a circular support evolving like an ellipse for increasing t, see [24,27]. Let $g_1(t), g_2(t)$ denote the length of the semiaxes of the supporting ellipse in the longitudinal and traverse direction resp., such that the height of the avalanche equals $h(y_1, y_2, t) = (1 - y_1^2 - y_2^2)/(g_1(t)g_2(t)$. Then g_1, g_2 satisfy a second order highly nonlinear system of ordinary differential equations with coefficients depending on the aspect ratios $\varepsilon_1 = [H/L_{x_1}]$ and $\varepsilon_2 = [h/L_{x_2}]$. Numerical results [23,27] show a crucial dependence of the spreads $g_1(t), g_2(t)$ on $\varepsilon_1, \varepsilon_2, \varepsilon_1/\varepsilon_2$ and on the angles ξ and δ.

Note added in proof. When preparing this manuscript we learned that also V. Chugunov, J.M.N.T. Gray and K. Hutter [6] found almost the same set of similarity solutions. However, they use abstract Lie group theory to find invariance properties of (33); then they discuss several cases in more details.

3.3 Existence Results

In this section we present the mathematical analysis of the 2×2-conservation law (25) when the density $\rho \equiv \rho_0 > 0$ is constant and of (28) for a density function $\rho(h) = h^\alpha$, $0 < \alpha < \infty$. For a bed with varying slope $\xi(x)$ we consider the modified height or mass distribution $u_1 = \big(\beta/(2\kappa)\big)^{1+\alpha} \rho h$ and momentum $u_2 = \big(\beta/(2\kappa)\big)^{1+\alpha} \rho h u$, cf. (29) – (31). As indicated in Sect. 3.1 system (25), (28) or (31) will evolve shocks and will allow multiple, even unphysical solutions. Therefore we are looking for suitable (physical) solutions satisfying a sufficiently large set of entropy conditions.

Definition 6 Let $\eta = \eta(u_1, u_2)$, $q = q(u_1, u_2)$ be scalar C^2-functions satisfying

$$\nabla_u \eta(u) \cdot \nabla_u F(u) = \nabla_u q(u) \quad \text{for all } u \in \mathbb{R}_+^* \times \mathbb{R}\,.$$

If η is convex and $\eta(0, \cdot) = 0$, then (η, q) is called a *convex weak entropy-flux pair* (for the flux $F : \mathbb{R}^2 \to \mathbb{R}^2$).

Since the source terms s in (28) and S_0 in (31) have jump discontinuities, it is reasonable – also in view of the striking non-uniqueness of solutions of (15) for sand piles – to use the notion of set-valued maps. Looking at $s = \sin\xi - \varepsilon b_x \cos\xi - (u/|u|)\tan\delta\cos\xi$ we introduce the set-valued sign function

$$\operatorname{sig} u = \begin{cases} [-1, 1] & \text{for } u = 0\,, \\ u/|u| & \text{for } u \neq 0 \end{cases}$$

and

$$\tilde{s}(u, x) = \sin\xi - \varepsilon b_x \cos\xi - \operatorname{sig}(u)\tan\delta\cos\xi\,.$$

Then $\tilde{S}$ is defined by

$$\tilde{S}(u_1, u_2, x) = \frac{(1+\alpha)\beta'}{\beta} F(u_1, u_2) + \begin{pmatrix} 0 \\ u_1 \tilde{s}(u_2/u_1, x) \end{pmatrix}.$$

Finally the system (31) is replaced by the differential inclusion

$$\partial_t u + \partial_x F(u) \in \tilde{S}(u, x) \tag{49}$$

which is made more precise in the following definition.

Definition 7 Given an initial value $u^0 = (u_1^0, u_2^0) \in L^\infty(\mathbb{R})^2$ with $u_1^0 \geq 0$ and $u_2^0/u_1^0 \in L^\infty(\mathbb{R})$ we call a function $u = (u_1, u_2) \in L^\infty\big((0,T) \times \mathbb{R};\ \mathbb{R}_+ \times \mathbb{R}\big)$ a *weak entropy solution* of (49) iff u has the following properties:

(1) there exists $S \in L^\infty_{\mathrm{loc}}\big(\mathbb{R} \times [0,T)\big)^2$ such that

$$S(x,t) \in \tilde{S}\big(u(x,t), x\big) \quad \text{for a.a. } (x,t) \in \mathbb{R} \times [0,T)\,;$$

(2) u is a *weak solution* of (49), i.e., for all $\psi \in C_0^1\big(\mathbb{R} \times [0,T)\big)^2$

$$\int_{\mathbb{R}} \int_0^T (u \cdot \partial_t \psi + F(u) \cdot \partial_x \psi + S \cdot \psi)\, dt\, dx = \int_{\mathbb{R}} u^0 \cdot \psi(x,0)\, dx\,;$$

(3) u satisfies the *entropy inequality*

$$\int_{\mathbb{R}} \int_0^T \big(\eta(u)\partial_t \phi + q(u)\partial_x \phi + \nabla_u \eta(u) \cdot S\phi\big)\, dt\, dx \geq \int_{\mathbb{R}} \eta(u^0)\phi(x,0)\, dx$$

for every test function $0 \leq \phi \in C_0^1\big(\mathbb{R} \times [0,T)\big)$ and every convex-weak entropy flux pair (η, q) for which $\nabla_u \eta(u)$ is locally bounded on $\mathbb{R} \times [0,T)$.

Remark 8 The non-classical part (1) in Definition 7 states the selection of an L^∞_{loc}-function $S(x,t)$ coinciding with $S_0(u_1, u_2, x)$ from (30) when the physical velocity u_2/u_1 does not vanish. In contrast to the usual definition of convex entropy flux pairs the degeneracy of (28) when $u_1 \to 0$ ($h \to 0$) requires to add the condition that $\eta(0, \cdot) = 0$ (weak entropy).

Theorem 9 [16] *Let* $u^0 = (u_1^0, u_2^0) \in L^\infty(\mathbb{R})^2$ *denote an initial value such that* $u_1^0, u_2^0/u_1^0 \in L^\infty(\mathbb{R})$ *and*

$$u_1^0 \geq 0\,,\ u_1^0(x) \to 0\,,\ u_2^0(x) \to 0 \ \text{as } |x| \to \infty$$

and let $\beta \in W^{1,\infty}(\mathbb{R})$ *be given such that* $\beta(x) \geq \beta_0 > 0$. *Then there exists a local weak entropy solution* $u = (u_1, u_2)$ *of (49). If* β *is constant, there exists a global weak entropy solution.*

Sketch of Proof In a first step we consider the viscous approximation

$$u_t + \partial_x F(u) = S_\varepsilon(u,x) + \varepsilon \partial_x^2 u, \quad \varepsilon > 0\,, \tag{50}$$

where S_ε is defined by $\tilde{S}$ via smoothing the jump discontinuity of $\tilde{s}(u,x)$. To prove the existence of classical solutions u^ε and some a priori estimates it is convenient to consider smooth initial values u_ε^0. E.g., we assume that $u_\varepsilon^0 \in C^2(\mathbb{R})^2$, $u_{1\varepsilon}^0 \geq u_1^0 + \varepsilon$. This smoothing and lifting up in addition to the viscous approximation is not contained in the proof in [16], but can easily be included [18]. Standard parabolic theory yields a unique classical solution u^ε in some interval $[0, T_\varepsilon)$. To prove that T_ε can be chosen independently of $\varepsilon > 0$ we apply the theory of invariant regions.

Lemma 10 *Let $u : \mathbb{R} \times [0,T) \to \mathbb{R}_+^* \times \mathbb{R}$ be a classical solution of the system* (50). *Further let $R_i : \mathbb{R}^2 \to \mathbb{R}$ and $M_i : [0,T) \to \mathbb{R}$, $i = 1, 2$, be smooth functions defining the regions*

$$\Sigma(t) = \bigcap_{i=1}^{2} \{u \in \mathbb{R}^2 : R_i(u) \leq M_i(t)\}, \quad t \in [0,T).$$

Assume the following properties:

1. $u(\cdot, 0) \in \Sigma(0)$,
2. $\nabla_u R_i(a)$ *is a left eigenvector of* $\nabla_u F(a)$ *for all* $a \in \partial\Sigma_i(t)$, $t \in [0,T)$,
3. $\Sigma(t)$ *is convex for all* $t \in [0,T)$,
4. $\sup_{x\in\mathbb{R}} \sup_{a\in\partial\Sigma_i(t)} \nabla_u R_i(a) \cdot S_\varepsilon(a,x) \leq M_i'(t)$,

where $\partial\Sigma_i(t) = \partial\Sigma(t) \cap \{a \in \mathbb{R}^2 : R_i(a) = M_i(t)\}$. Then $\Sigma(t)$ is a one-parameter family of invariant regions, i.e., $u(x,t) \in \Sigma(t)$ for all $t \in [0,T)$.

This lemma will be applied using the Riemann invariants $R_\pm(t) = \pm u_2/u_1 + u_1^{1/(2(1+\alpha))}$ of (31) and a function $M(t) = M_\pm(t)$ with initial value $M(0) = \max_{+,-} \|R_\pm(u^0)\|_\infty$ satisfying the nonlinear differential inequality $M'(t) \geq C\big(M(t)^2 + \|\sin\xi\|_\infty\big)$ where $C = C(\|\beta'\|_\infty, \beta_0, \alpha)$. As a conclusion we find a $T_0 > 0$ such that

$$\|u_1^\varepsilon\|_{L^\infty(\mathbb{R}\times(0,T))} + \left\|\frac{u_2^\varepsilon}{u_1^\varepsilon}\right\|_{L^\infty(\mathbb{R}\times(0,T))} \leq C(T)$$

for all $0 < T < T_0$ independent of $\varepsilon > 0$; if the slope ξ is constant, then $T_0 = \infty$ and $C(T)$ is linear in T.

Besides L^2-estimates of u_1^ε and u_2^ε on $\mathbb{R} \times (0, T_0)$ with bounds depending on ε and the crucial non-negativity of u_1^ε, see [16], it is important to have sufficiently many local L^2-estimates of ∇u at hand.

Lemma 11 *Let (η, q) be a convex entropy-flux pair such that $\nabla\eta$ is bounded and $\eta \in C^2$ on $(0,\infty) \times \mathbb{R}$. Further let $u = (u_1^\varepsilon, u_2^\varepsilon)$ be a strong solution of* (50) *such that u^ε, $F(u^\varepsilon)$ and $G_\varepsilon(u^\varepsilon, x)$ are bounded independently of $\varepsilon \in (0,1)$. Then for every bounded set $\Omega \subset \mathbb{R} \times \mathbb{R}_+$ there exists a constant $C(\Omega) > 0$ such that*

$$\varepsilon \iint_\Omega \partial_x u^\varepsilon \cdot \nabla_u^2 \eta(u^\varepsilon) \cdot \partial_x u^\varepsilon \, dx\, dt \leq C(\Omega) \quad \forall \varepsilon \in (0,1).$$

If η is even strongly convex, then a similar estimate holds for $\varepsilon|\partial_x u^\varepsilon|^2$.

Proof Given a solution u^ε of (50) and an entropy-flux pair (η, q) the functions $\eta(u^\varepsilon)$, $q(u^\varepsilon)$ will satisfy the equation

$$\partial_t \eta(u^\varepsilon) + \partial_x q(u^\varepsilon) = \varepsilon\big(\partial_x^2 \eta(u^\varepsilon) - \partial_x u^\varepsilon \cdot \nabla_u^2 \eta(u^\varepsilon) \cdot \partial_x u^\varepsilon\big) + \nabla_u \eta(u^\varepsilon) \cdot G_\varepsilon. \quad (51)$$

Testing with $0 \leq \varphi \in C_0^\infty\big(\mathbb{R} \times [0,\infty)\big)$ such that $\varphi|_\Omega = 1$ will yield the a priori estimate. If η is strictly convex, the estimate $r \cdot \nabla_u^2 \eta(v) \cdot r \geq \delta |r|^2$ with some $\delta > 0$ will prove the second assertion. □

The second main step deals with the limit $\varepsilon \to 0$. Since only very few a priori estimates on u^ε are available and since (49) contains several nonlinear terms, we need to refer to Young measure solutions ([11]) as limits of $\big(u_1^\varepsilon,\ u_2^\varepsilon/u_1^\varepsilon\big)$.

Lemma 12 [11] *Assume that* $(z_k) \subset L^\infty(\Omega)^2$ *is a sequence on* $\Omega \subset [0,T) \times \mathbb{R}$ *with* $z_k(\Omega) \subset K$ *where* $K \subset \mathbb{R}^2$ *is compact and convex. Then there exists a subsequence* $(z_{k_j}) \subset (z_k)$ *and a family of Borel probability measures* $(\mu_{(x,t)})_{(x,t)\in\Omega}$ *on* K *such that for every* $H \in C^0(K)^2$

$$\begin{aligned}&\lim_{j\to\infty} \iint_\Omega H\big(z_{k_j}(x,t)\big)\varphi(x,t)\,dx\,dt \\ &= \iint_\Omega \overline{H}(x,t)\varphi(x,t)\,dx\,dt \quad \forall\, \varphi \in L^1(\Omega)^2 \end{aligned} \tag{52}$$

where

$$\overline{H}(x,t) = \int_K H(\cdot)\,d\mu_{(x,t)} \quad \textit{for a.a. } (x,t) \in \Omega\,.$$

The proof of this famous lemma is based on the Theorem of Banach-Alaoglu [38] applied to the spaces $L^1(\Omega; C^0(K))$ and $L^\infty(\Omega; \mathcal{M}(K))$. When applied to a sequence (z_k) of approximate solutions of a (partial) differential equation this lemma yields a measure-valued solution $\int_K y\,d\mu_{(x,t)}(y)$ in the limit. By these means the notion of strong or weak solutions is generalised to a great extent: there are no longer function values a.e. in Ω, but only *probabilities* of them. E.g., the sequence of Rademacher functions $r_j(x) = \operatorname{sgn}\sin(2^j\pi x)$ on $(0,1)$ attains the values ± 1 with probability $1/2$ for every $j \in \mathbb{N}$. Its limit for $j \to \infty$ in the usual weak sense of $(L^\infty)^*$ is 0, but in the sense of Lemma 12 we get the probability measure $\mu_x \equiv \mu = \frac{1}{2}(\delta_{-1} + \delta_{+1})$ for all $x \in (0,1)$. Then $\overline{H}(x) \equiv \frac{1}{2}\big(H(-1) + H(+1)\big)$ is the limit of $H\big(r_j(x)\big)$ for all $H \in C^0([-1,1])$.

Lemma 12 will be applied to $z^k = \big(u_1^\varepsilon,\, u_2^\varepsilon/u_1^\varepsilon\big)$ where $\varepsilon = 1/k$ and to

$$H_1(z) = \begin{pmatrix} z_1 \\ z_1\,z_2 \end{pmatrix} = \begin{pmatrix} u_1 \\ u_2 \end{pmatrix}, \quad H_2(z) = F(z_1, z_1 z_2)\,.$$

Then for every test function $\varphi \in C_0^1\big(\mathbb{R} \times [0,T)\big)^2$

$$\begin{aligned}&\int_{\mathbb{R}}\int_0^T \big(H_1(z^\varepsilon)\partial_t\varphi + H_2(z^\varepsilon)\partial_x\varphi + H_3^\varepsilon(z^\varepsilon, x)\varphi\big)\,dt\,dx \\ &= \varepsilon \int_{\mathbb{R}}\int_0^T \partial_x u^\varepsilon \cdot \partial_x\varphi\,dt\,dx + \int_{\mathbb{R}} u^0\varphi(x,0)\,dx\end{aligned}$$

where $H_3^\varepsilon(z,x) = S_\varepsilon(z_1, z_1 z_2, x)$ is discontinuous in z. By Lemma 12 there exist Young measures $\mu_{(x,t)}$ for a.a. $(x,t) \in \Omega$ such that

$$\int_{\mathbb{R}}\int_0^T (\overline{H}_1\partial_t\varphi + \overline{H}_2\partial_x\varphi + \overline{H}_3\varphi)\,dt\,dx = \int_{\mathbb{R}} u^0\varphi(x,0)\,dx$$

where $\overline{H}_j(x,t) = \int_K H_j(y)d\mu_{(x,t)}(y)$, $j = 1,2$. However, $\overline{H}_3 \in L^\infty(\mathbb{R}\times\mathbb{R}_+)^2$ cannot be characterised by using the measures $\mu_{(x,t)}$. To deal with functions with jump discontinuities as $\operatorname{sgn} u_1$ or in two dimensions with discontinuities of the type $u/|u|$ a further decomposition of the measures $\mu_{(x,t)}$ is needed, see [29].

To show for the Young measures that $\mu_{(x,t)} = \delta_{z(x,t)}$ for a.a. (x,t) we need the Div-Curl Lemma ([11]) and sufficiently many entropy-flux pairs.

Lemma 13 [11] *Let $\Omega \subset \mathbb{R} \times \mathbb{R}_+$ be bounded, (u^k) a sequence of functions on Ω and (η^1, q^1), (η^2, q^2) weak entropy-flux pairs. Assume that $\eta^i(u^k)$, $q^i(u^k)$ converge weakly in $L^2(\Omega)$ to $\eta^i(u)$, $q^i(u)$ and that $\partial_t \eta^i(u^k) + \partial_x q^i(u^k)$ is relatively compact in $W^{-1,2}(\Omega)$ for $i = 1, 2$. Then, as $k \to \infty$,*

$$\iint\limits_{\Omega} (\eta^1(u^k) q^2(u^k) - \eta^2(u^k) q^1(u^k)) \varphi \, dx \, dt \to \iint\limits_{\Omega} (\eta^1 q^2 - \eta^2 q^1) \varphi \, dx \, dt$$

for every $\varphi \in C_0^\infty(\Omega)$. Furthermore

$$\int\limits_K (\eta^1 q^2 - \eta^2 q_1) d\mu_{(x,t)} = \int\limits_K \eta^1 d\mu_{(x,t)} \cdot \int\limits_K q^2 d\mu_{(x,t)} - \int\limits_K \eta^2 d\mu_{(x,t)} \cdot \int\limits_K q^1 d\mu_{(x,t)} \, .$$

Given an entropy-flux pair (η, q), Definition 6 implies that η satisfies a second order wave equation with respect to u_1, u_2 with (u_1, u_2)-depending coefficients. By [9] every weak entropy η, i.e., $\eta(0, \cdot) = 0$, can be written in the form

$$\begin{aligned} \eta(u_1, u_2) &= \int\limits_{\mathbb{R}} f(\xi) \left(u_1^{\frac{1}{2(1+\alpha)}} - (\xi - \frac{u_2}{u_1})^2 \right)_+^{2+4\alpha} d\xi \\ &= u_1 \int\limits_{-1}^{1} f\left(\frac{u_2}{u_1} + \xi \, u_1^{\frac{1}{2(1+\alpha)}}\right) (1 - \xi^2)^{2+4\alpha} d\xi \, . \end{aligned} \tag{53}$$

In particular, for $f(s) = \frac{1}{2} s^2$, the entropy η coincides up to a multiplicative constant with the mechanical energy $\eta_E = u_2^2/(2u_1) + \kappa' u_1^{\frac{2+\alpha}{1+\alpha}}$. It is easily seen that for $f \in C_0^\infty(\mathbb{R})$ the sequence $\left(\nabla_u \eta(u^k)\right)$ is bounded on Ω and that

$$|r \cdot \nabla_u^2 \eta(u^k) \cdot r| \le C_f r \cdot \nabla_u^2 \eta(u^k) \cdot r \quad \forall r \in \mathbb{R}^2$$

independent of $k \in \mathbb{N}$. A further analysis based on previous a priori estimates shows that the right-hand side of (51) is precompact in $W^{-1,2}(\Omega)^2 + W^{-1,p}(\Omega)^2$ $(p < 2)$ and that the left-hand side of (51) is uniformly bounded in $W^{-1,\infty}(\Omega)^2$. Then by Murat's Lemma $\left(\partial_t \eta(u^k) + \partial_x q(u^k)\right)_{k \in \mathbb{N}}$ is precompact in $W^{-1,2}(\Omega)$ for every weak entropy-flux pair (η, q) generated by $f \in C_0^\infty(\mathbb{R})$, see (53). Hence Lemma 13 may be applied.

This is the starting point to show in a lengthy technical proof [30,31] that the Young measures $\mu_{(x,t)}$ are δ-measures. To be more precise, using the (z_1, z_2)-functions,

$$\mu_{(x,t)} = \begin{cases} \delta_{(z_1, z_2)} & \text{if } z_1 > 0 \\ \delta_{z_1} \times \nu & \text{if } z_1 = 0 \end{cases}$$

where $\operatorname{supp} \nu_{(x,t)} \subset [\liminf_{k\to\infty} z_1^k(x,t),\ \limsup_{k\to\infty} z_1^k(x,t)] \subset\subset \mathbb{R}$. For (u_1, u_2) we conclude that also in the case when $u_1(x,t) = z_1(x,t) = 0$ the measure $\mu_{(x,t)}$ is concentrated in one single point $u(x,t)$. Note that the analysis from [30,31] simplifies to elementary algebraic considerations when $\alpha = -1/2$, see [17]. However, in this case, the constitutive relation $\rho = h^{-1/2}$ seems to be unphysical.

In the final step Lemma 12 implies that $u^k \rightharpoonup u$ weakly in $L^\infty(\Omega)^2$ and using $H(y) = |y|^p$, $1 < p < \infty$, that

$$\iint_\Omega |u^k|^p \, dx\, dt \to \iint_\Omega |u|^p \, dx\, dt$$

for $k \to \infty$. Hence $u^k \to u$ in $L^p(\Omega)^2$ by the Theorem of Radon-Riesz. By similar arguments u can be shown to be an entropy solution of (31). Looking more carefully at $\tilde{S}$ with its jump discontinuity $\operatorname{sig} u$ we may even select a function $S \in L^\infty_{\text{loc}}(\mathbb{R} \times [0,T))^2$ such that u is a solution of (49) in the sense of Definition 7. □

A detailed analysis of the time interval in which a weak entropy solution exists may be performed by the theory of invariant regions, see Lemma 10. In the case when $\rho \equiv \rho_0 > 0$ the following results depending on the behaviour of the slope $\xi(x)$ have been obtained [1,2]:

Theorem 14 *Let a curved base with variable slope angle $\xi(x)$ be given such that for $\beta(x) = \varepsilon K \cos\xi(x)$*

$$\frac{\beta'}{\beta} \in L^\infty(\mathbb{R}).$$

Consider an incompressible avalanche with initial values h_0, u_0 such that

$$\begin{aligned} 0 \le \rho_0 = \beta h_0 \in L^\infty(\mathbb{R}), \quad & m_0 = \beta h_0 u_0 \in L^\infty(\mathbb{R}) \\ \rho_0(x) \to 0, \quad & m_0(x) \to 0 \quad \textit{for } |x| \to \infty. \end{aligned}$$

Finally let $s(x,u) = \sin\xi(x) - \operatorname{sgn}(u)\tan\delta(x)\cos\xi(x)$, where the bed friction angle $\delta(x)$ may depend on $x \in \mathbb{R}$, let $P = (5/8)\,\|\beta'/\beta\|_\infty$, $Q = \|s\|_\infty$ and let $E_0 = \max\left(\|2\rho_0\|_\infty,\ \|u_0^2/2\|_\infty\right)$ measure the initial energy of the avalanche.

1. *If $\beta' = 0$ (constant slope), then the Savage–Hutter equations admit a global weak entropy solution. Furthermore ρ grows at most quadratically as $t \to \infty$ and $u = m/\rho$ grows at most linearly as $t \to \infty$. If in addition $|\xi| \le |\delta|$ for all $x \in \mathbb{R}$, i.e., the slope angle is bounded by the bed friction angle, then ρ and u are uniformly bounded independent of time with a bound depending only on E_0.*
2. *Assume that*

$$\beta' \le 0,\ u_0 > 0,\ \frac{1}{2}u_0^2 \ge 2\rho_0 \ \textit{and}\ \xi \ge \delta\,,$$

i.e., the slope angle $\xi(x)$ is constant or even steepening, the initial velocity is positive and sufficiently large and the slope angle is greater than or equal to the bed friction angle. Then there exists a global weak entropy solution. Furthermore ρ grows at most quadratically and $u = m/\rho$ grows at most linearly as $t \to \infty$. Finally $u \ge 0$ (down slope) and $\frac{1}{2}u^2 \ge 2\rho$.

3. *If the slope is arbitrarily curved* ($\beta' \neq 0$), *then there exists a weak entropy solution on a time interval* $(0, T_{\max})$. *Here it is sufficient to take*

$$T_{\max} = \frac{1}{PQ}\Big(\frac{\pi}{2} - \arctan\sqrt{2E_0P/Q}\Big)\,.$$

In [1,2] the term $\operatorname{sgn}(u)$ in $s(x,u)$ has been smoothed. However, the same results hold in the set-valued formulation, cf. (49), Definition 7, for an incompressible avalanche [18].

4 Conclusions

Several models for avalanching flows were presented. The first part (Sect. 2) dealt with two models of particle segregation and with the so-called BCRE model and some of its variants describing the piling up of granular material by neglecting inertia, pressure and momentum effects. We discussed the existence of stationary solutions when surplus matter drops down at the boundary of the bottom and of similarity solutions of the 1D–silo problem. However, up to now, the instationary problems, first order equations of Hamilton–Jacobi type, are unsolved. It is conjectured that the modern theory of viscosity solutions will help to construct instationary solutions, in particular for arbitrary bottom topographies in 2D, for arbitrary sources, arbitrary boundary conditions and also to prove uniqueness in adequate function spaces.

Section 3 dealt with the Savage–Hutter model for dense snow avalanches. This well-established model accounts for the exchange of momentum and – in the one-dimensional case – leads to a 2×2–system of conservation laws with source term. A further difficulty in the theoretical and numerical analysis is due to jump discontinuities of (1) the earth pressure coefficient depending on the expansive or contractive behaviour of the avalanche and of (2) the shear traction depending on the direction of the flow. In Sect. 3.2 we classified the similarity solutions in the one-dimensional case, i.e. avalanches having a similar shape for all times. In the 2D case only very special similarity solutions are known up to now. The general initial value problem was solved in the final section via the modern theory of weak entropy solutions known from the theory of compressible gas dynamics. We proved the existence of at least one local or global weak solution with height–depending density $\rho = h^\alpha$, $\alpha > 0$. The questions of uniqueness of weak entropy solutions as well as of existence in the two-dimensional case remain open.

References

1. Balean, R. (1997) Existence of Entropy Solution to a One-dimensional Model for Granular Avalanches. Darmstadt University of Technology, Preprint no. 1951
2. Balean, R. (1999) The Flow of a Granular Avalanche Down a Smoothly Varying Slope. In: Models of Continuum Mechanics in Analysis and Engineering. Alber, Balean, Farwig (Eds). Shaker Verlag Aachen, 17–33
3. Bouchaud, J.-O., Cates, M.E., Ravi Prakash, J., Edwards, S.F. (1994) A Model for the Dynamics of Sandpile Surfaces. J. Phys. I France **4**, 1383–1410

4. Bouchaud, J.-O., Cates, M.E., Ravi Prakash, J., Edwards, S.F. (1995) Hysteresis and Metastability in a Continuum Sandpile Model. Pys. Rev. Lett. **74**, 1982–1986
5. Braun, J. (2001) Segregation of Granular Media by Diffusion and Convection. Physical Review E **64**, 011307, 1–7
6. Chugunov, V., Gray, J.M.N.T., Hutter, K. (2003) Group Theoretic Methods and Similarity Solutions of the Savage–Hutter Equations. This volume
7. Crandall, M.G., Lions, P.-L. (1986) On Existence and Uniqueness of Solutions to Hamilton-Jacobi-Equations. Nonlinear Analysis. Theory, Methods and Applications **10**, 353–370
8. de Gennes, P.-G. (1995) Surface Dynamics of a Granular Material. CR Acad. Sci. II-B **321**, 501–506
9. DiPerna, R.J. (1985) Measure-Valued Solutions to Conservation Laws. Arch. Rational Mech. Anal. **88**, 223–270
10. Emig, Th., Claudin, Ph., Bouchaud, J.-P. (2000) Exact Solutions of a Model for Granular Avalanches. Europhys. Lett. **50**, 594–600
11. Evans, L.C. (1990) Weak Convergence Methods for Nonlinear Partial Differential Equations. Amer. Math. Soc.
12. Evans, L.C. (1998) Partial Differential Equations. Amer. Math. Soc.
13. Friedman, A. (1964) Partial Differential Equations of Parabolic Type. Prentice-Hall, Englewood Cliffs, NJ
14. Gray, J.M.N.T. (2001) Granular Flow in Partially Filled Rotating Drums. J. Fluid Mech. **441**, 1–19
15. Greve, R., Koch, T., Hutter, K. (1994) Unconfined Flow of Granular Avalanches Along a Partly Curved Surface. I. Theory. Proc. R. Soc. Lond. A **445**, 399–413
16. Gwiazda, P. (2002) An Existence Result for a Model of Granular Material with Non-Constant Density. Asymptotic Analysis **30**, 43–60
17. Gwiazda, P. (2001) An Existence Result for Balance Laws with Multifunctions. A Model from the Theory of Granular Media. Inst. Appl. Math. Mech., Warsaw University RW01-17
18. Gwiazda, P. (2001) *Private Communication*
19. Hadeler, K.P., Kuttler, C. (1999) Dynamic Models for Granular Matter. Granular Matter **2**, 9–18
20. Hadeler, K.P., Kuttler, C. (2001) Granular Matter in a Silo. Granular Matter **3**, 193–197
21. Hadeler, K.P, Kuttler, C., Gergert, I. (2002) Dirichlet and Obstacle Problems for Granular Matter. SFB 382, Universität Tübingen, Report Nr. 173
22. Hong, D.C., Quinn, P.V., Luding, S. (2001) Reverse Brazil Nut Problem: Competition between Percolation and Condensation. Phys. Rev. Lett. **86**, 3423–3426
23. Hutter, K. (1991) Two- and Three Dimensional Evolution of Granular Avalanche Flow-Theory and Experiments Revisited. Acta Mech. (Suppl.) **1**, 167–181
24. Hutter, K., Greve, R. (1993) Two-Dimensional Similarity Solutions for Finite Mass Granular Avalanches with Coulomb- and Viscous-Type Frictional Resistance. J. Glaciology **39**, 357–372
25. Hutter, K., Koch, T. (1991) Motion of a Granular Avalanche in an Exponentially Curved Chute: Experimental and Theoretical Predictions. Phil. Trans. R. Soc. Lond. A **334**, 93–138
26. Hutter, K., Nohguchi, Y. (1990) Similarity Solutions for a Voellmy Model of Snow Avalanches with Finite Mass. Acta Mech. **82**, 99–127

27. Hutter, K., Siegel, M., Savage, S.B., Nohguchi, Y. (1993) Two-Dimensional Spreading of a Granular Avalanche Down an Inclined Plane. Part I. Theory. Acta Mech. **100**, 37–68
28. Issler, D., Gauer, P., Barbolini, M. (1999) Continuum Models of Particle Entrainment and Deposition in Snow Drift and Avalanche Dynamics. In: Models of Continuum Mechanics in Analysis and Engineering. Alber, Balean, Farwig (Eds). Shaker Verlag Aachen 58–80
29. Kałamajska, A., Gwiazda, P. (2002) On Discontinuity and the Young Measure Characterization. Darmstadt University of Technology, Preprint, to appear
30. Lions, P.L., Perthame, B., Souganidis, P. (1996) Existence and Stability of Entropy Solution for the Hyperbolic Systems of Isentropic Gas Dynamics in Eulerian and Lagrangian Coordinates. Commun. Pure Appl. Math. **49**, 599–638
31. Lions, P.L, Perthame, B., Tadmor, E. (1994) Kinetic Formulation of the Isentropic Gas Dynamics and p-Systems. Commun. Math. Phys. **163**, 415–431
32. Nohguchi, Y., Hutter, K., Savage, S.B. (1989) Similarity Solutions for Granular Avalanches of Finite Mass with Variable Bed Friction. Continuum Mech. Thermodyn. **1**, 239–265
33. Prigozhin, L. (1986) Quasivariational Inequality in a Poured Pile Shape Problem. Zh Vychisl Mat. Mat. Fiz. **26**, 1072–1080
34. Prigozhin, L. (1993) A Variational Problem of Bulk Solids Mechanics and Free–Surface Segregation. Chemical Engineering Sciences **48**, 3647–3656
35. Prigozhin, L. (1996) Variational Model of Sandpile Growth. Euro. J. Appl. Math. **7**, 225–236
36. Prigozhin, L., Zaltzman, B. (2002) On the Approximation of the Dynamics of Sandpile Surfaces. Port. Math. (to appear)
37. Pudasaini, S.P., Mohring, J. (2002) Mathematical Model and Experimental Validation of Free Surface Size Segregation for Polydisperse Granular Materials. Granular Matter **4**, 45–56
38. Rudin, W. (1973) Functional Analysis. McGraw-Hill, New York
39. Salm, B. (1968) On Nonuniform Steady Flow of Avalanching Snow. IAHS Publ. Bern **79**, 19–29
40. Savage, S.B., Hutter, K. (1989) The Motion of a Finite Mass of Granular Material down a Rough Incline. J. Fluid Mech. **199**, 177–215
41. Savage, S.B., Lun, C.K.K. (1988) Particle Size Segregation in Inclined Chute Flow of Dry Cohesionless Granular Solids. J. Fluid Mech. **189**, 311-335
42. Voellmy, A. (1955) Über die Zerstörungskraft von Lawinen. Schweizerische Bauzeitung **73**, 159–162, 212–217, 246–249, 280–285
43. Zwinger, T. (2000) Dynamik einer Trockenschneelawine auf beliebig geformten Berghängen. Dissertation TU Wien, Fakultät für Maschinenbau

Group Theoretic Methods and Similarity Solutions of the Savage–Hutter Equations

Vladimir Chugunov[1], John Mark Nicholas Timm Gray[2], and Kolumban Hutter[3]

[1] Department of Applied Mathematics, Kazan State University, Kazan 420008, Russia, e-mail: chug@ksu.ru
[2] Department of Mathematics, University of Manchester, Oxford Road, Manchester M13 9PL U.K., e-mail: ngray@maths.man.ac.uk
[3] Institute of Mechanics, Darmstadt University of Technology, Hochschulstr. 1, 64289 Darmstadt, Germany, e-mail: hutter@mechanik.tu-darmstadt.de

received 11 Jun 2002 — accepted 12 Sep 2002

Abstract. We consider the spatially one-dimensional time dependent system of equations, obtained by Savage and Hutter, which describes the gravity-driven free surface flow of granular avalanches. All similarity solutions of this system are found by means of group analysis. The family of solutions which are invariant to stretching transformations is investigated in greater detail. Explicit solutions are constructed in three cases and their physical interpretation is given.

1 Introduction

The most general non-dimensional form of the Savage–Hutter theory, which includes both the formulations presented in [6,7], can be obtained by reducing the theory of Gray et al. [2] to one dimension. The time-dependent depth integrated mass and downslope momentum balance equations are

$$\frac{\partial h}{\partial t} + \frac{\partial Q}{\partial x} = 0, \tag{1}$$

$$\frac{\partial Q}{\partial t} + \frac{\partial}{\partial x}(Q^2 h^{-1}) + \frac{\partial}{\partial x}(\beta h^2/2) = hs, \tag{2}$$

where t denotes time, h is the avalanche thickness, Q is the depth averaged downslope volume flux and x is the curvilinear downslope coordinate. The source term s on the right-hand side of (2) is composed of the downslope component of the gravitational acceleration, the Coulomb sliding friction with basal angle of friction δ and gradients of the basal topography height, z^b, above the curvilinear coordinate system. It takes the form

$$s = \sin\zeta - Q|Q|^{-1}\tan\delta\left(\cos\zeta + \lambda\kappa Q^2 h^{-2}\right) - \varepsilon\cos\zeta\frac{\partial z^b}{\partial x}, \tag{3}$$

where ζ is the inclination angle of the curvilinear coordinate to the horizontal and $\kappa = -\partial\zeta/\partial x$ is the curvature. In a typical avalanche the thickness magnitude H^* is much smaller than its length L^*, which is reflected in the small non-dimensional parameter $\varepsilon = H^*/L^*$. The shallowness assumption also requires that typical avalanche lengths are shorter than the radius of curvature R^* of the curvilinear coordinate, which introduces the second non-dimensional parameter $\lambda = L^*/R^*$. The function β is given by

$$\beta = \varepsilon K \cos\zeta, \tag{4}$$

where the earth pressure coefficient K was proposed to take two limiting states K_{act} and K_{pas} associated with extensive ($\partial u/\partial x \geq 0$) and compressive ($\partial u/\partial x < 0$) motions, respectively. Savage & Hutter [6] showed that

$$K_{act/pas} = 2\sec^2\phi\left(1 \mp \{1 - \cos^2\phi\sec^2\delta\}^{1/2}\right) - 1, \tag{5}$$

where ϕ is the internal angle of friction of the granular avalanching material.

Recent experiments suggest that the jump in K at $\partial u/\partial x = 0$ is unrealistic and that a slowly varying function or a constant earth pressure coefficient is more realistic. In this paper it is therefore assumed that K is constant. It shall also be assumed that the bed is flat, $z^b \equiv 0$, and inclined at a constant angle ζ to the horizontal. This implies that the curvature vanishes, $\kappa = 0$, and that β is constant. In addition placing the restriction that $Q > 0$ holds implies that the source term $s = s_0$ is also constant.

Three exact solutions to the system (1), (2) are currently known. These are the parabolic cap similarity solution and the 'M'-wave solutions, derived by Savage & Hutter [6], and the travelling shock wave solution [3] on a non-accelerative slope. In this paper we seek to find further simple solutions of physical interest.

2 Group Analysis and Invariant Solutions

Consider a moving coordinate system

$$\eta = x - s_0 t^2/2. \tag{6}$$

The relative flux $\widehat{Q}$ in the moving coordinate system is then given by

$$\widehat{Q} = Q - h s_0 t. \tag{7}$$

In the new variable η the system (1),(2) may be written in the form

$$\frac{\partial h}{\partial t} + \frac{\partial \widehat{Q}}{\partial \eta} = 0, \tag{8}$$

$$\frac{\partial \widehat{Q}}{\partial t} + \frac{\partial(\widehat{Q}^2 h^{-1})}{\partial \eta} = -\frac{\partial(\beta h^2/2)}{\partial \eta}. \tag{9}$$

If we introduce the relative rate $\widehat{u}$ by the relation $\widehat{u} = \widehat{Q}/h$, then the system (8), (9) takes the form

$$\frac{\partial h}{\partial t} + \frac{\partial}{\partial \eta}(\widehat{u}h) = 0, \tag{10}$$

$$\frac{\partial \widehat{u}}{\partial t} + \widehat{u}\frac{\partial \widehat{u}}{\partial \eta} + \beta\frac{\partial h}{\partial \eta} = 0. \tag{11}$$

These equations coincide with the shallow-water equations considered in [1]. The system (10), (11) admits the symmetry Lie algebra $L_5 \bigoplus L_\infty$ [4]. Using this algebra and the relation $\widehat{u} = \widehat{Q}/h$ we find the following basis of the Lie algebra for (8), (9):

$$X_1 = \frac{\partial}{\partial t}, \quad X_2 = \frac{\partial}{\partial \eta}, \quad X_3 = t\frac{\partial}{\partial t} + \eta\frac{\partial}{\partial \eta},$$

$$X_4 = \eta\frac{\partial}{\partial \eta} + 2h\frac{\partial}{\partial h} + 3\widehat{Q}\frac{\partial}{\partial \widehat{Q}}, \quad X_5 = t\frac{\partial}{\partial \eta} + h\frac{\partial}{\partial \widehat{Q}},$$

$$X_\infty = Z(\widehat{u}, h)\frac{\partial}{\partial \eta} + T(\widehat{u}, h)\frac{\partial}{\partial t},$$

where the functions $Z(\widehat{u}, h), T(\widehat{u}, h)$ are defined by the linear equations

$$\frac{\partial Z}{\partial \widehat{u}} - \widehat{u}\frac{\partial T}{\partial \widehat{u}} + h\frac{\partial T}{\partial h} = 0,$$

$$\frac{\partial Z}{\partial h} - \widehat{u}\frac{\partial T}{\partial h} + \beta\frac{\partial T}{\partial \widehat{u}} = 0.$$

The physical interpretation of the generators is that X_1 and X_2 are time and space translations, X_3 and X_4 are stretching transformations, X_5 is a spatial rotation and X_∞ is the linearising Hodograph transformation.

In this paper we will consider only the invariant solutions with respect to the stretching transformations of η and t (self-similarity solutions). These are generated by the infinitesimal operator (or generator) $X \in L_5$

$$X = (\mu - 1)X_3 + X_4. \tag{12}$$

Any invariant function f solves the partial differential equation $Xf = 0$ [5]. Therefore the basis of invariants is furnished by

$$J_1 = \eta t^{-\mu/(\mu-1)}, \quad J_2 = h t^{-2/(\mu-1)}, \quad J_3 = \widehat{Q} t^{-3/(\mu-1)}, \quad \mu \neq 1,$$

$$J_1 = t, \quad J_2 = h\eta^{-2}, \quad J_3 = \widehat{Q}\eta^{-3}, \quad \mu = 1$$

Consequently, the invariant solution of the system (8), (9) with respect to the group, generated by (12), are defined in the form

$$h = \eta^2 F(t), \quad \widehat{Q} = \eta^3 \Phi(t), \quad \mu = 1, \tag{13}$$

where the functions $F(t)$ and $\Phi(t)$ are found from the equations

$$\dot{\Phi} + 4\Phi^2 F^{-1} + 2\beta F^2 = 0, \quad \dot{F} + 3\Phi = 0, \tag{14}$$

in which the dot denotes differentiation with respect to t, and

$$h = t^{2/(\mu-1)} f(z), \quad \widehat{Q} = t^{3/(\mu-1)} q(z), \quad \mu \neq 1. \tag{15}$$

Moreover, $z = J_1 = \eta t^{-\mu/(\mu-1)}$, and the functions $f(z)$, $q(z)$ satisfy the following system

$$\frac{3}{\mu - 1} q - \frac{\mu}{\mu - 1} q' z + (q^2 f^{-1} + \beta f^2/2)' = 0, \tag{16}$$

$$\frac{2}{\mu - 1} f - \frac{\mu}{\mu - 1} z f' + q' = 0. \tag{17}$$

Here, the prime denotes differentiation with respect to z. Note that (16), (17), and (14) represent ordinary differential equations, and the relative velocity is given by $\widehat{u} = \eta\Phi(t)/F(t)$ for $\mu = 1$ and $\widehat{u} = t^{1/(\mu-1)} q(z)/f(z)$ for $\mu \neq 1$.

3 Qualitative Analysis

An exhaustive analysis of the family of similarity solutions is beyond the scope of this paper. Instead we investigate two cases where the equations may be integrated directly.

3.1 The Case $\mu = 1$

The system (14) may be solved exactly and is easily obtained as

$$\pm t = \begin{cases} \dfrac{3a}{2b^{3/2}} \ln \left[\left(\dfrac{\theta+1}{\theta_0+1} \right) \left(\dfrac{\theta_0 - 1}{\theta - 1} \right) \right] - \dfrac{3}{\sqrt{b}} \left(\dfrac{\theta}{F^{1/3}} - \dfrac{\theta_0}{F_0^{1/3}} \right), & b > 0, \\ \dfrac{2}{\sqrt{a}} \left(F_0^{-1/2} - F^{-1/2} \right), & b = 0, \\ \dfrac{3a}{(-b)^{3/2}} \left(\tan^{-1} \psi - \tan^{-1} \psi_0 \right) + \dfrac{3}{\sqrt{-b}} \left(\dfrac{\psi}{F^{1/3}} - \dfrac{\psi_0}{F_0^{1/3}} \right), & b < 0, \end{cases} \tag{18}$$

where $a = 36\beta > 0$, $\theta = ((aF^{1/3} + b)/b)^{1/2}$, $\psi = (aF^{1/3} + b)/(-b))^{1/2}$, and θ_0 and ψ_0 are the same functions evaluated at $F = F_0$; b, F_0 are constants of integration. For $\mu = 1$ the exact solution of equations (1) and (2) is of the form

$$h = (x - s_0 t^2/2)^2 F(t), \quad Q = h s_0 t + (x - s_0 t^2/2)^3 \Phi(t), \tag{19}$$

where $\Phi(t) = -\dot{F}(t)/3$, and F is implicitly defined by (18). Evidently, the function $F(t)$ has a growing and decaying branch corresponding to the positive and negative roots in (18). The growing branches are particularly interesting as they imply that the avalanche thickness can increase without bound within a finite interval of time for all choices of the parameter b. In case that $b < 0$ holds, the solution degenerates for $F < F^* = (-b/a)^3$.

3.2 The Case $\mu = -2$

Equation (17) can be directly integrated when $\mu = -2$ to give

$$q = 2fz/3 + c_1, \tag{20}$$

where c_1 is an arbitrary constant. Substituting the volume flux from (20), the momentum balance (16) reduces to

$$f' = f^2 (c_1/3 - 2zf/9)(c_1^2 - \beta f^3)^{-1}. \tag{21}$$

For $c_1 = 0$ equation (21) can be integrated to give

$$f(z) = (9\beta)^{-1} z^2 + c_2, \tag{22}$$

where c_2 is an arbitrary constant. The exact solution of (1),(2) in this case is

$$h = t^{-2/3} [(9\beta)^{-1} (x - s_0 t^2/2)^2 t^{-4/3} + c_2], \tag{23}$$

$$Q = h s_0 t + 2t^{-1} [(9\beta)^{-1} (x - s_0 t^2/2)^3 t^{-2} + c_2 (x - s_0 t^2/2) t^{-2/3}]/3. \tag{24}$$

Let us now consider more general solutions for the case $\mu = -2$ when $c_1 \neq 0$ in (21). It is convenient to introduce new variables y, p and ξ for f, q and z by using the scalings

$$f = c_1 y (c_1\beta)^{-1/3}, \quad q = c_1 p, \quad z = 3(c_1\beta)^{1/3}\xi; \tag{25}$$

they transform (21) into a parameter independent form

$$\frac{\partial y}{\partial \xi} = y^2(1 - 2\xi y)(1 - y^3)^{-1}. \tag{26}$$

The avalanche thickness is non-negative; we therefore restrict attention to the domain $y \geq 0$, $-\infty < \xi < \infty$. The solutions of (26) are illustrated in Fig.1. On the line $y = 1$, the gradient $\partial y/\partial \xi$ approaches $\pm\infty$ for all points except one, where $1 - 2\xi y = 0$. Consequently, the point $\xi = 1/2$, $y = 1$ is a singular point. The asymptotic behaviour of the solution in the vicinity of the line $y = 1$ is described by a formula, which can be obtain from (26),

$$y = 1 \pm \sqrt{2[(\xi - \xi_0)(\xi + \xi_0 - 1)]/3}, \quad \xi \to \xi_0, \quad y \to 1. \tag{27}$$

If $\xi_0 = 1/2$, then

$$y = 1 \pm \sqrt{2/3}\,(\xi - 1/2), \quad \xi \to 1/2, \quad y \to 1. \tag{28}$$

The singular point (1/2,1) is a saddle point.

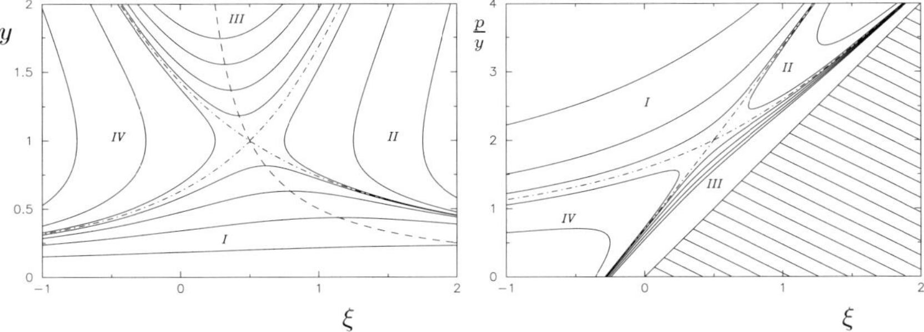

Fig. 1. Left: Four classes of asymmetric thickness, y. Right: Four classes of asymmetric velocity, p/y. Solutions are illustrated in both panels as a function of ξ and labelled *I-IV*. The seperatrix between the solution domains is indicated by the dash-dotted curve, and the dashed line indicates where the first derivative of the thickness vanishes, $\partial y/\partial \xi = 0$. There are no solutions in the hatched region

On the line $y = 1/(2\xi)$ the gradient $\partial y/\partial \xi$ vanishes, as can easily be seen from (26). Therefore, in all points on this line the function $y(\xi)$ assumes an extremum. In the region $y < 1$, $y(\xi)$ has a maximum and when $y > 1$, $y(\xi)$ has a minimum. When $\xi \to \pm\infty$ two asymptotic formulas emerge, first

$$\xi \to \infty, \quad y \to 1/\xi; \quad \xi \to -\infty, \quad y \to -1/(2\xi) \tag{29}$$

for the region $y > 1$, and second

$$\xi \to \pm\infty, \quad y \to \xi^2, \tag{30}$$

for the region $y < 1$.

The curves, which pass through the singular point (1/2, 1), divide the domain into four parts. Each part has its own family of solutions of (26), which are denoted by I, II, III, IV (see Fig.1). The families I-IV together with the relations (6), (7), (15), (20), (25) define the various solutions of the system (1),(2). From a physical viewpoint family I is interesting for the motion of avalanches. Using this family we can construct the solution for various concrete situations.

Let us write the invariant solution for $\mu = -2$ and $c_1 \neq 0$ in the form

$$h(x,t) = (t+t_0)^{-2/3}\frac{c_1}{(c_1\beta)^{1/3}}y\left[\frac{x - s_0(t+t_0)^2/2}{3(c_1\beta)^{1/3}(t+t_0)^{2/3}}\right], \tag{31}$$

$$Q(x,t) = hs_0(t+t_0) + (t+t_0)^{-1}[c_1 + 2(x - s_0(t+t_0)^2/2)h(x,t)/3], \tag{32}$$

where $y(\xi)$ I, II, III, IV and t_0, c_1 are arbitrary constants.

4 Physical Interpretation of the Results

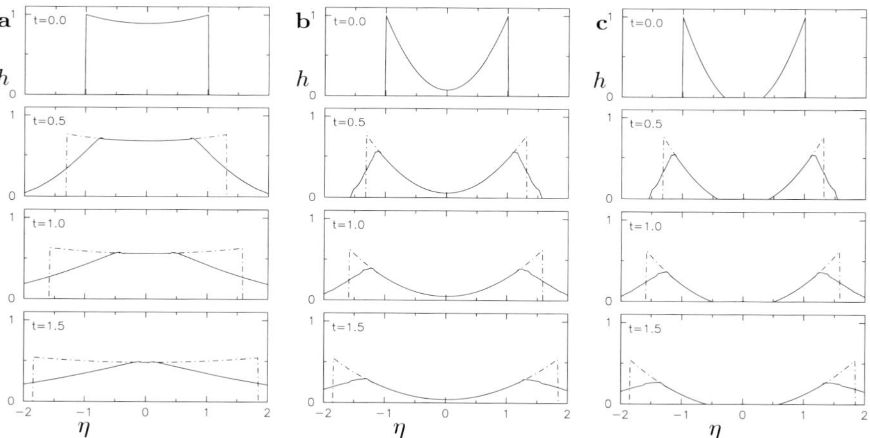

Fig. 2. The temporal and spatial evolution of the avalanche thickness for the cases (**a**) $\beta = 1.0$, (**b**) $\beta = 0.12$ and (**c**) $\beta = 0.1$ are illustrated. The constant $c_2 = 1 - (9\beta)^{-1}$ was used to ensure that $h(\pm 1) = 1$. The solid line shows the computed solution in the accelerating coordinate system η and the dash-dotted line shows the exact solution. Time is measured relative to initial conditions and the M-wave is evaluated at a finite time $t_0 = 1$

4.1 M-waves

The solutions (23), (24) were found by using the separation of variable approach employed by Savage & Hutter [6], who called it an "M"-wave. The name arose from the shape of a truncated solution

$$h(z,t) = \begin{cases} t^{-2/3} f(z), & |z| \leq 1, \\ 0, & |z| > 1, \end{cases} \tag{33}$$

which connected a finite part of the solution (22) with regions of zero thickness and flux on either side. At $z = \pm 1$ there are jump discontinuities in both the thickness and the flux, which should satisfy the mass and momentum jump conditions. They are obtained from (1), (2) as

$$x_j' [\![h]\!] - [\![Q]\!] = 0, \quad x_j' [\![Q]\!] - [\![Q^2 h^{-1} + \beta h^2/2]\!] = 0, \tag{34}$$

where x_j is the position and x_j' is the normal velocity of the discontinuity, the jump bracket $[\![f]\!] = f^+ - f^-$ and $f^\pm = f(x_j \pm 0)$. As we shall show now (33) and (34) cannot hold together. Assuming that on the positive side of the discontinuity $h^+ \neq 0$ and $Q^+ \neq 0$ and that on the negative side $h^- = Q^- = 0$, the jump conditions (34) imply

$$x_j' h^+ - Q^+ = 0, \quad x_j' Q^+ - (Q^+)^2/h^+ - \beta (h^+)^2/2 = 0. \tag{35}$$

Mass balance therefore requires that $x_j' = Q^+/h^+$, and if this is substituted in the momentum jump condition we find that $h^+ = 0$, contradicting our original assumption that $h^+ \neq 0$. The jump conditions are therefore not satisfied by the truncated M-wave solution (33), and expansion fans develop at the jumps.

To understand the collapse of the truncated M-wave in greater detail a series of numerical simulations have been performed using a Total Variational Diminishing Lax Friedrich's scheme [3]. This is a shock capturing method that has been extensively tested against the parabolic cap solution and the travelling shock solution. Figure 2 shows the M-wave (33) and the numerical solutions for the avalanche thickness for various values of β at a sequence of time-steps in the accelerated coordinate system, η. The constant $c_2 = 1 - (9\beta)^{-1}$ was used to ensure that $h(\pm 1) = 1$. In each case the M-wave spreads out laterally, diminishing in height, and close to the discontinuities the shock expands as expected. The overlap domain, where the M-wave (33) and the computed solution are in close agreement, can either expand or contract in the physical domain. For $\beta = 1$ the overlap region decreases with time and the M-wave is destroyed in finite time (Fig.2a). However, for thin avalanches where the aspect ratio $\varepsilon \ll 1$, and hence $\beta \ll 1$, the overlap domain expands in the physical domain despite the collapse close to the discontinuities (Fig.2b). This is because the stretching of the solution is faster than the inward propagation speed of the disturbance from the discontinuities.

For $c_2 < 0$ the M-wave solution (33) contains regions of negative thickness. In this case the M-wave can be linked to the trivial solution $h \equiv 0$ using the jump conditions (34). This time, however, because the thicknesses and fluxes are zero on both sides of the discontinuity, the jump conditions are trivially satisfied. Case (**c**) in Fig.2 shows the evolution of such an M-wave for the case $\beta = 0.1$.

These numerical simulations demonstrate that for shallow avalanches, in which $\beta \ll 1$, the invariant stretching solutions may exist over an expanding region in the physical domain even though they may not satisfy boundary conditions at the ends.

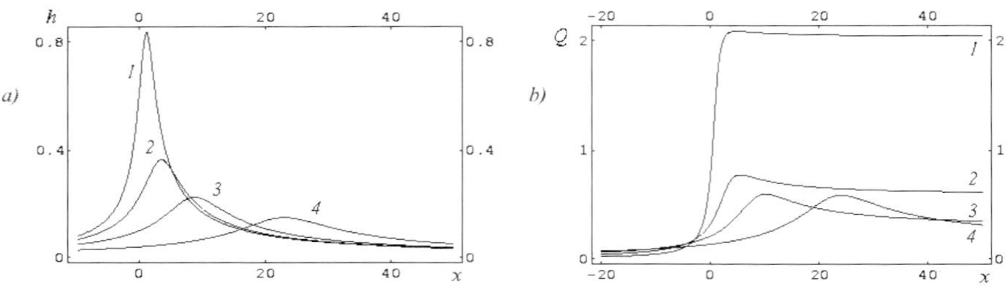

Fig. 3. Evolution of the wave (**a**) and the mass flux (**b**) for (31), (32) (1-$t = 0$, 2-$t = 2$, 3-$t = 5$, 4-$t = 10$)

4.2 Evolution of Case I Waves.

Let us consider a second example: it assumes that at the initial instant the avalanche thickness distribution is described by the formula

$$h_0 = \frac{s_0}{2\beta x_0}\left(\frac{H_0}{3}\right)^2 y\left(\frac{x - x_0}{H_0}\right), \tag{36}$$

where H_0, x_0 are known parameters and the function $y(\xi)$ is one of the family I (Fig.3). This profile has a maximum at $x = x_{max}$. For example, $y(\xi)$ is a function which has the maximum in the point $\xi_{max} = 1$ ($y_{max} = 1/2$). Therefore, $x_{max} = x_0 + H_0$ and $h_{0max} = s_0 H_0^2/(16\beta x_0)$ is the initial amplitude of the wave. The mathematical model for this problem is the Cauchy problem for the system (1),(2) with the initial condition (36).

The solution of this problem is described by the formulas (31), (32) with the constants c_1, t_0 and the function $y(\xi)$ which are defined by comparison of the expression (31) with (36) at $t = 0$. Figure 3a shows the evolution of the wave with time; the calculations were performed with the following values of the parameters: $\zeta = 45^o$, $\delta = 30^o$, $s_0 = 0.2989$, $\beta = 0.1$, $x_0 = 0.1$, $H_0 = 1$. Figure 3b displays the corresponding behaviour of the mass flux.

4.3 Finite-time Singularities

Let us consider the situation when the initial wave begins to grow and moves against the downslope direction (see Fig.4). This is possible when the initial profile of the thickness is described by the expression

$$h_0 = AY\left(\frac{x - x_0}{\lambda_0}\right), \quad Y(\xi) = \frac{1}{y_{max}} y(\xi), \quad y \in I, \tag{37}$$

where A is an amplitude, λ_0 is a typical length of the wave, x_0 is a parameter, and the mass flux is $Q_\infty = -q_\infty(t_0 - t)^{-1} < 0$ on the right side of the wave when $x \to \infty$, and $Q_{-\infty} = 0$ when $x \to -\infty$. In order to construct the solution for this situation we again take the invariant solution (31), (32). It is easy to check that the system (1),(2) admits the transformations

$$Q = -\bar{Q}, \quad h = \bar{h}, \quad t = -\bar{t}, \quad x = \bar{x}, \tag{38}$$

$$Q = \bar{Q}, \quad h = \bar{h}, \quad t = \bar{t} + t_0, \quad x = \bar{x}, \tag{39}$$

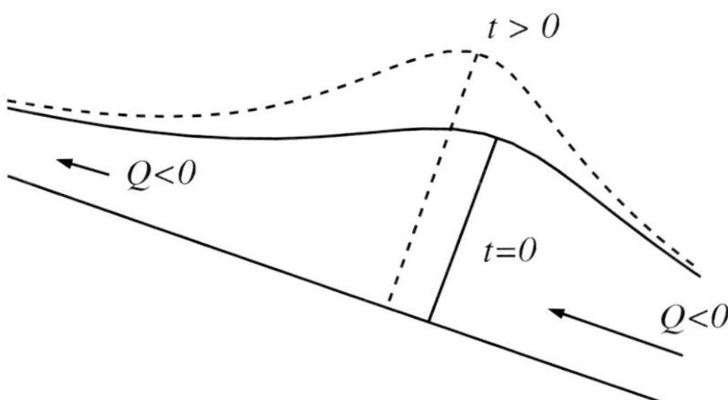

Fig. 4. Schematic diagram of the situation when the initial wave (solid line) begins to grow (dashed line) and moves against the downslope direction

where (38) is a finite group and (39) is the global form of X_1.
Any solution of (1),(2) which is transformed by (38), (39) is again a solution of (1),(2). Therefore, from (31), (32) we have

$$h(x,t) = (t_0 - t)^{-2/3} \frac{c_1}{(c_1\beta)^{1/3}} y \left(\frac{x - \hat{x}_0(t)}{\lambda(t)} \right),$$

$$Q(x,t) = -h(x,t) \left[s_0(t_0 - t) + (t_0 - t)^{-1} \left(\frac{c_1}{h(x,t)} + \frac{2}{3}(x - \hat{x}_0) \right) \right], \tag{40}$$

where $\bar{x}_0(t) = s_0(t_0 - t)^2/2$, $\lambda(t) = 3(c_1\beta)^{1/3}(t_0 - t)^{2/3}$, $y \in I$.
The boundary condition $Q_\infty = -q_\infty(t_0 - t)^{-1}$ defines c_1, namely

$$c_1 = \frac{1}{3} q_\infty .$$

Comparing the first expression of (40) with (37), we find

$$y_{max} = \frac{A\lambda_0}{q_\infty} < 1, \quad t_0 = \frac{1}{3}\sqrt{\frac{\lambda_0^3}{q_\infty \beta}}, \quad \xi_{max} = \frac{1}{2y_{max}}, \quad x_0 = s_0 t_0^2/2, \tag{41}$$

and the finite-time singularity occurs at t_0. For $t > t_0$ the wave no longer exists. The physical meaning is clear. At infinity we have a mass source. The mass of the avalanche grows and moves in the upslope direction. There is a struggle between the supply-rate and the tendency of the avalanche to want to move in the downslope direction, which causes an infinite increase of the amplitude of the wave. Such a situation may develop as an avalanche flows off a steep slope and up an opposing slope, i.e. during run-up. The motion of the wave is obtained from (40),

$$x_{max}(t) = s_0(t_0 - t)^2/2 + 3\xi_{max} \left(\frac{\beta q_\infty}{3} \right)^{1/3} (t_0 - t)^{2/3} .$$

The numerical results are obtained for the following values of the parameters: $s_0 = 0.2989$, $A = 0.083$, $\beta = 0.1$, $\lambda_0 = 1$, $q_\infty = 0.166$. They are illustrated in Fig.5. From (41) we found $y_{max} = 1/2$, $t_0 = 2.587$, $x_0 = 1$.

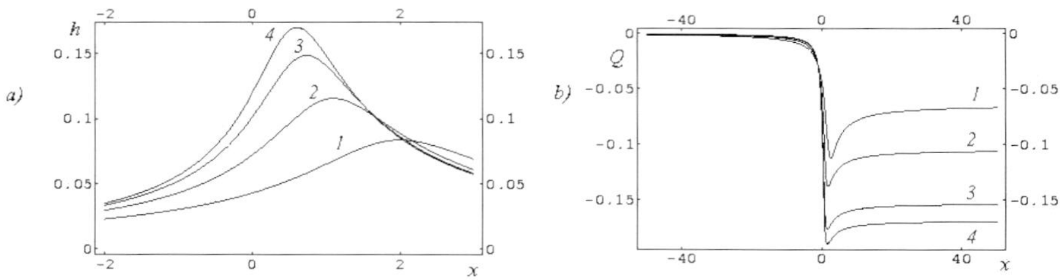

Fig. 5. Behaviour of thickness (**a**) and mass flux (**b**) at four consecutive times prior to the finite-time singularity

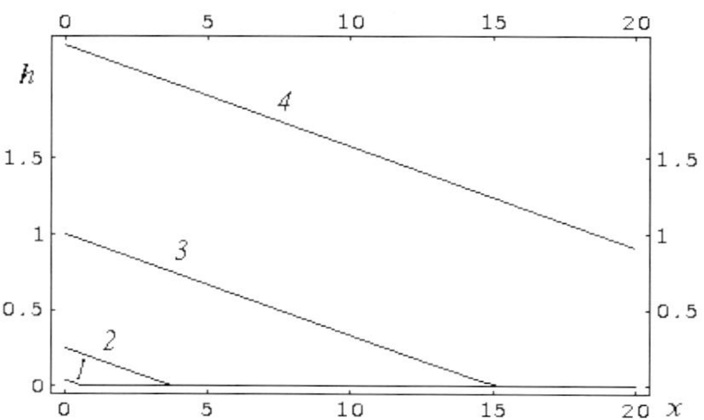

Fig. 6. Dynamics of the thickness of the avalanche for $s_0 = 0.2989,\ h_0 = 0.01$ (1–$t = 0$, 2–$t = 5$, 3–$t = 10$, 4–$t = 15$)

4.4 An Example of the General System when $\mu = 2$

Let us consider the case $\mu = 2$, when the transformations (6) and (15) imply

$$h = t^2 f(z), \quad \hat{Q} = t^3 q(z), \quad z = (x - s_0 t^2/2)t^{-2}. \tag{42}$$

This is a special transformation, as it maps the point $x = 0$ to $z = -s_0/2$ for all time. It is therefore of particular physical interest. We may envisage a problem in which a large source of granular material ($x < 0$) is separated from a grain-free region ($x \geq 0$) by a screen or gate placed at $x = 0$. For $t > 0$, we may construct invariant solutions in which the gate is raised so that the orifice is of size $h_0 t^2$. To solve for the evolution of the free boundary we must integrate the system (16), (17) for $\mu = 2$ subject to the boundary conditions

$$z = -s_0/2, \quad f = h_0, \tag{43}$$

$$f = 0, \quad q = 0. \tag{44}$$

Figure 6 illustrates the dynamics of the thickness of the avalanche for $s_0 = 0.2989$, $h_0 = 0.01$. Curve 1 corresponds to the moment $t = 3$; curve 2 is for $t = 5$; and curves 3, 4 are for $t = 10,\ t = 15$.

Summary

The Savage–Hutter equations for dense granular avalanches have been analysed using group theoretic methods to construct a basis Lie algebra for the system. The generators corresponding to stretching transformations have been used to construct several families of invariant solutions in three special cases, and the solutions have been interpreted physically.

References

1. Aksenov A.V., Baikov V.A., Chugunov V.A., Gazizov R.K., Meshkov A.G. (1995) Lie group analysis of differential equations. Volume 2. Applications in engineering and physical sciences. (Ed.) N.H. Ibragimov. CRC Press
2. Gray J. M. N. T., Wieland M., Hutter K. (1999) Free surface flow of cohesionless granular avalanches over complex basal topography. Proc. Roy. Soc. London A **455**, 1841–1874
3. Gray J.M.N.T., Tai Y.C. (1998) Particle size segregation, granular shocks and stratification patterns. NATO ASI series, H.J. Herrmann et al. (eds.), Physics of dry granular media. Kluwer Academic., 697–702
4. Hydon P.E. (2000) Symmetry Methods for Differential Equations. Cambridge University Press
5. Ibragimov N.H. (2000) Introduction to modern group analysis. Ufa, Russia
6. Savage S. B., Hutter K. (1989) The motion of a finite mass of granular material down a rough incline. J. Fluid Mech. **199**, 177–215
7. Savage S.B., Hutter K. (1991) The dynamics of avalanches of granular materials from initiation to runout. Acta Mech. Part I: Analysis. **86**, 201–223

Part II

Porous and Granular Materials. Fundamentals and Dynamical Processes

Dynamics of Hypoplastic Materials: Theory and Numerical Implementation

Vladimir A. Osinov and Gerd Gudehus

Institute of Soil Mechanics and Rock Mechanics, University of Karlsruhe, Engler-Bunte-Ring 14, 76131 Karlsruhe, Germany
e-mail: {ossinov,gudehus}@ibf-tiger.bau-verm.uni-karlsruhe.de

received 21 Mar 2002 — accepted 22 Apr 2002

Abstract. The paper presents a theoretical study of the dynamic equations and their solutions for a granular material whose mechanical behaviour is described by a constitutive equation of the hypoplasticity theory. Discussed in the paper are well-posedness of the dynamic problem, cyclic shearing of a granular material, the coupling between the transverse and the longitudinal displacement components in the dynamic solutions, the double-frequency effect as a consequence of this coupling, and liquefaction of a saturated granular solid caused by the dynamic cyclic shearing.

1 Dynamic Hypoplasticity. An Overview

Historically, the development of the constitutive modelling of the plastic behaviour of various materials (metals, rocks, soils) began with *elastoplasticity* theories, the main features of which are the decomposition of the deformation into elastic and plastic parts, and the description of the plastic part through the notions of yield surface and flow rule. Subsequently, an alternative way of approaching the problem was found and led to the concept of *hypoplasticity* [39] and to the development of a class of hypoplastic models [40,10,3,38,23] which proved to be able to describe the plastic behaviour of granular materials such as sand, other granular soils, technological powders and others[1]. The advanced versions of hypoplastic equations incorporate the critical-state concept of soil mechanics and the dependence of the stiffness on pressure and density, with the constitutive parameters of a given granular material being valid for a wide range of pressures and densities.

A hypoplastic constitutive equation establishes a relationship between the stress rate and the strain rate in the form

$$\overset{\circ}{\boldsymbol{T}} = \boldsymbol{H}\,(\boldsymbol{D},\, S\,), \tag{1}$$

where $\boldsymbol{T}$ is the Cauchy stress tensor, $\overset{\circ}{\boldsymbol{T}}$ is an objective stress rate (for instance, the Jaumann rate), $\boldsymbol{D}$ is the stretching tensor (the symmetric part of the velocity gradient), and S denotes a set of state variables, that is, physical quantities which are involved in the description of the mechanical behaviour of the material. These

[1] We draw the reader's attention to the fact that the term 'hypoplasticity' is also met with in the literature in another context [6].

state variables (except for the stresses) require additional evolution equations which can be written symbolically in the form

$$\dot{S} = P(\boldsymbol{D}, S), \tag{2}$$

where the dot stands for the rate of change in the state variables. As distinct from elastoplasticity theories, the general form (1), (2) of the description of the plastic deformation does not require (although does not rule out) the introduction of a yield surface and a flow rule, and the decomposition of the deformation into an elastic and a plastic part.

Early versions of hypoplastic equations contained the stress tensor as the only state variable. Then the void ratio was introduced into the equation as an additional state variable. This allowed the critical state concept to be incorporated in the theory in a consistent way, and also made the constitutive parameters independent of the density [10,3,38]. The extended version of hypoplasticity [23] involves the so-called intergranular-strain tensor as a state variable in addition to the stress tensor and the void ratio. This version is able to correctly reproduce the plastic behaviour of a granular material under multi-cycle as well as monotonic loading.

The elaboration of the hypoplasticity theory was accompanied by the use of hypoplastic constitutive equations for the numerical solution of boundary value problems. The present paper is concerned with the dynamic deformation of hypoplastic materials. The study of dynamic problems with hypoplastic constitutive equations was initiated in [24] for the case of small-amplitude transverse waves propagating in a fully saturated material. Solutions to the system of the dynamic equations derived in [24] within the framework of small-amplitude approximation were investigated in more detail in [9,12,13]. Some interesting properties of the solutions were revealed, among which are 'saturation' of the solution away from the boundary for a periodic boundary condition, and a fractal-like dependence of the solution on the boundary data in the class of piecewise-constant functions [12].

Dynamic problems for a dry hypoplastic material were studied in [25,26] with the use of the constitutive equation developed in [10,3]. The analysis of the dynamic equations revealed the possibility of two kinds of ill-posedness of the dynamic problem caused by the loss of hyperbolicity of the dynamic equations; this issue will be discussed below in the present paper. Numerical solutions were obtained and analysed for plane waves propagating in a half-space in the case of simple single-cycle loading at the boundary. The dilatancy effect was shown to result in a coupling between the longitudinal and the transverse components of the displacement in the wave. Two-component waves in a hypoplastic material were also studied in [4] in the small-amplitude approximation.

Constitutive theories in which the state of a granular material is determined only by stress and density are known to fail to correctly describe multi-cycle loading: such equations inevitably produce a too high rate of accumulation of deformation or stress because of the so-called 'ratcheting' effect. The appearance of the extended hypoplastic equation with intergranular strain [23] made it possible to model multi-cycle loading and thus to enhance the capabilities of the theory. Results of the solution of applied geomechanical problems with dynamic hypoplasticity are presented in [27,28], where hypoplasticity was applied to the problem of the earthquake-induced deformation and liquefaction of soil. Some aspects of dynamic hypoplasticity were further discussed in [29]. In particular, it was shown by numerical examples that in the case of cyclic loading the above-mentioned coupling between the displacement components leads to the doubling of the frequency of the induced longitudinal motion as compared to the frequency of the transverse motion.

In the present paper we continue to study dynamic phenomena in hypoplastic granular materials in the light of the results obtained previously. The main issues to be discussed are well-posedness of the dynamic problem, the coupling between the longitudinal and the transverse displacement components, the double-frequency effect as a consequence of this coupling, and wave-induced liquefaction of a saturated granular medium.

2 Hypoplastic Constitutive Equation

For the present analysis we will use the hypoplastic equation developed in [38]. The equation contains the stress tensor $\boldsymbol{T}$ and the void ratio[2] e as state variables, and can be written in the general form as[3]

$$\dot{\boldsymbol{T}} = \mathcal{L}(\boldsymbol{T}, e)\,\boldsymbol{D} + \boldsymbol{N}(\boldsymbol{T}, e)\,\|\boldsymbol{D}\|, \tag{3}$$

where the fourth-order stiffness tensor $\mathcal{L}$ and the second-order tensor $\boldsymbol{N}$ are responsible, respectively, for the linear and the nonlinear parts of the constitutive relation. The term $\|\boldsymbol{D}\| = \sqrt{\mathrm{tr}(\boldsymbol{D}\boldsymbol{D})}$ makes the constitutive equation incrementally nonlinear. Compressive stresses in (3) are taken to be negative. The tensors $\mathcal{L}$ and $\boldsymbol{N}$ are written in rectangular Cartesian coordinates as

$$\mathcal{L}_{ijkl} = \frac{f_b f_e}{\mathrm{tr}(\hat{\boldsymbol{T}}^2)}\Big[F^2 \delta^K_{ik}\delta^K_{jl} + a^2 \hat{T}_{ij}\hat{T}_{kl}\Big], \tag{4}$$

$$N_{ij} = \frac{f_b f_e f_d\, aF}{\mathrm{tr}(\hat{\boldsymbol{T}}^2)}\Big(\hat{T}_{ij} + \hat{T}^*_{ij}\Big), \tag{5}$$

where δ^K_{ij} is the Kronecker delta, and

$$\hat{T}_{ij} = \frac{T_{ij}}{\mathrm{tr}\,\boldsymbol{T}}, \quad \hat{T}^*_{ij} = \hat{T}_{ij} - \frac{1}{3}\delta^K_{ij}. \tag{6}$$

The factor a in (4), (5) is determined by the friction angle φ_c in critical states:

$$a = \sqrt{\frac{3}{8}}\,\frac{(3 - \sin\varphi_c)}{\sin\varphi_c}. \tag{7}$$

The factor F is a function of $\hat{\boldsymbol{T}}^*$:

$$F = \sqrt{\frac{1}{8}\tan^2\xi + \frac{2 - \tan^2\xi}{2 + \sqrt{2}\tan\xi\cos 3\theta}} - \frac{1}{2\sqrt{2}}\tan\xi, \tag{8}$$

[2] Porosity, $e/(1+e)$, could equivalently be used as a measure of density.

[3] For simplicity, in the constitutive equation we write the material time derivative $\dot{\boldsymbol{T}}$ instead of an objective stress rate. Calculations show that for the dynamic problems considered here the difference is unimportant.

where

$$\tan\xi = \sqrt{3}\,\|\hat{\boldsymbol{T}}^*\|, \quad \cos 3\theta = -\sqrt{6}\,\frac{\mathrm{tr}(\hat{\boldsymbol{T}}^{*3})}{[\mathrm{tr}(\hat{\boldsymbol{T}}^{*2})]^{3/2}}. \tag{9}$$

The factors a and F determine the critical-state surface in the stress space.

Three characteristic void ratios are specified as functions of the mean pressure: the minimal possible void ratio, e_d, the critical void ratio, e_c, and the void ratio in the loosest state, e_i. The pressure dependence of these void ratios is postulated in the form

$$\frac{e_i}{e_{i0}} = \frac{e_c}{e_{c0}} = \frac{e_d}{e_{d0}} = \exp\Big[-\Big(\frac{-\mathrm{tr}\,\boldsymbol{T}}{h_s}\Big)^n\Big], \tag{10}$$

with the corresponding reference values e_{i0}, e_{c0}, e_{d0} for zero pressure ($e_{i0} > e_{c0} > e_{d0}$). The constants e_{i0}, e_{c0}, e_{d0} and h_s, n are material parameters.

The factor

$$f_d = \Big(\frac{e - e_d}{e_c - e_d}\Big)^\alpha, \tag{11}$$

where α is a material parameter, tends to unity as the state of the material approaches a critical state. The functions f_e and f_b are defined as

$$f_e = \Big(\frac{e_c}{e}\Big)^\beta, \tag{12}$$

$$f_b = \frac{h_s}{n}\Big(\frac{1+e_i}{e_i}\Big)\Big(\frac{e_{i0}}{e_{c0}}\Big)^\beta\Big(\frac{-\mathrm{tr}\,\boldsymbol{T}}{h_s}\Big)^{1-n}\Big[3 + a^2 - \sqrt{3}\,a\Big(\frac{e_{i0} - e_{d0}}{e_{c0} - e_{d0}}\Big)^\alpha\Big]^{-1},$$

where β is a material parameter. The constitutive parameters of Hochstetten sand [38] are given in Table 1 as an example.

Table 1. Hypoplastic parameters of Hochstetten sand [38]

$\varphi_c[°]$	h_s [MPa]	e_{c0}	e_{d0}	e_{i0}	α	β	n
33	1000	0.95	0.55	1.05	0.25	1.5	0.25

3 Dynamic Equations

Consider a granular body saturated with a fluid. The fluid may be a pure liquid (e.g. water) or a liquid containing a small volume fraction of free gas. The state of the fluid is characterised by the pore pressure $p_l > 0$ and the degree of saturation S_r. According to the effective-stress principle, the total stress is the sum of the effective stress $\boldsymbol{T}$ and the isotropic stress $-p_l\boldsymbol{I}$, where $\boldsymbol{I}$ is the unit tensor, and the effective stress obeys the same constitutive equation as for a dry solid. In the formulation of the dynamic problem we neglect the relative motion of the fluid and solid phases due to seepage, assuming that the permeability of the solid skeleton is sufficiently small. For geomechanical problems relevant to earthquake modelling, this assumption is justified for fine and medium sand.

In the absence of seepage the motion of the medium is described by a single velocity field $\boldsymbol{v}$ for both fluid and solid phases. The equation of motion is written as

$$\operatorname{div}\boldsymbol{T} - \operatorname{grad} p_l + \varrho\,\boldsymbol{g} = \varrho\,\dot{\boldsymbol{v}}, \tag{13}$$

where $\boldsymbol{g}$ is the mass force vector, $\dot{\boldsymbol{v}}$ is the material time derivative of the velocity vector, and ϱ is the bulk density,

$$\varrho = \frac{\varrho_l e S_r + \varrho_s}{1+e}, \tag{14}$$

with ϱ_l and ϱ_s being the true densities of the liquid and solid fractions, respectively (in all numerical calculations below $\varrho_l = 10^3$ kg m^{-3}, $\varrho_s = 2.65 \times 10^3$ kg m^{-3}). The evolution of the effective stress is governed by (3). With incompressible grains, the change in the void ratio is found from the mass balance equation

$$\dot{e} = (1+e)\operatorname{tr}\boldsymbol{D}. \tag{15}$$

The rate of change in the pore pressure is given by

$$\dot{p}_l = -K_f\Big(\frac{1+e}{e}\Big)\operatorname{tr}\boldsymbol{D}, \tag{16}$$

where K_f is the compression modulus of the pore fluid. In the presence of free gas, when $S_r < 1$, the compression modulus K_f depends on the degree of saturation and on the pore pressure, and may therefore vary during the deformation.

We restrict ourselves to plane-wave solutions which depend on only one spatial Cartesian coordinate x_1 and time t, and which have two non-zero components of velocity, v_1 and v_2, with $T_{13} = T_{23} = 0$. For such solutions, equations (13), (3), (15) and (16) produce a system of eight first-order scalar differential equations

$$\frac{\partial v_1}{\partial t} + v_1\frac{\partial v_1}{\partial x_1} = \frac{1}{\varrho}\frac{\partial T_{11}}{\partial x_1} - \frac{1}{\varrho}\frac{\partial p_l}{\partial x_1} + g_1, \tag{17}$$

$$\frac{\partial v_2}{\partial t} + v_1\frac{\partial v_2}{\partial x_1} = \frac{1}{\varrho}\frac{\partial T_{12}}{\partial x_1} + g_2, \tag{18}$$

$$\frac{\partial T_{11}}{\partial t} + v_1\frac{\partial T_{11}}{\partial x_1} = \mathcal{L}_{1111}\frac{\partial v_1}{\partial x_1} + \frac{1}{2}\big(\mathcal{L}_{1112} + \mathcal{L}_{1121}\big)\frac{\partial v_2}{\partial x_1} + N_{11}\,||\boldsymbol{D}||, \tag{19}$$

$$\frac{\partial T_{12}}{\partial t} + v_1\frac{\partial T_{12}}{\partial x_1} = \mathcal{L}_{1211}\frac{\partial v_1}{\partial x_1} + \frac{1}{2}\big(\mathcal{L}_{1212} + \mathcal{L}_{1221}\big)\frac{\partial v_2}{\partial x_1} + N_{12}\,||\boldsymbol{D}||, \tag{20}$$

$$\frac{\partial T_{22}}{\partial t} + v_1\frac{\partial T_{22}}{\partial x_1} = \mathcal{L}_{2211}\frac{\partial v_1}{\partial x_1} + \frac{1}{2}\big(\mathcal{L}_{2212} + \mathcal{L}_{2221}\big)\frac{\partial v_2}{\partial x_1} + N_{22}\,||\boldsymbol{D}||, \tag{21}$$

$$\frac{\partial T_{33}}{\partial t} + v_1\frac{\partial T_{33}}{\partial x_1} = \mathcal{L}_{3311}\frac{\partial v_1}{\partial x_1} + \frac{1}{2}\big(\mathcal{L}_{3312} + \mathcal{L}_{3321}\big)\frac{\partial v_2}{\partial x_1} + N_{33}\,||\boldsymbol{D}||, \tag{22}$$

$$\frac{\partial e}{\partial t} + v_1\frac{\partial e}{\partial x_1} = (1+e)\,\frac{\partial v_1}{\partial x_1}, \tag{23}$$

$$\frac{\partial p_l}{\partial t} + v_1\frac{\partial p_l}{\partial x_1} = -K_f\Big(\frac{1+e}{e}\Big)\frac{\partial v_1}{\partial x_1}, \tag{24}$$

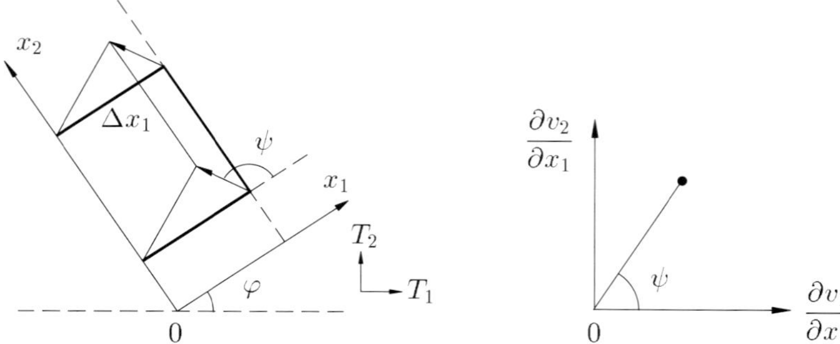

Fig. 1. Definition of the angles φ and ψ

where

$$||\boldsymbol{D}|| = \sqrt{\left(\frac{\partial v_1}{\partial x_1}\right)^2 + \frac{1}{2}\left(\frac{\partial v_2}{\partial x_1}\right)^2} \tag{25}$$

and the coefficients $\mathcal{L}_{ijkl}$, N_{ij} are given by (4), (5). The term $||\boldsymbol{D}||$ and the dependence of the coefficients on the stresses and the void ratio make the system (17)–(24) nonlinear. It is convenient to introduce an angle φ to specify the direction of the wave propagation with respect to the principal axes of the instantaneous stress tensor, Fig. 1. The coefficients $\mathcal{L}_{ijkl}$, N_{ij} may then be viewed as functions of the principal stresses T_1, T_2, T_3 and the angle φ.

4 Wave Speeds and Hyperbolicity

Well-posedness of a dynamic problem requires the system of the governing equations to be hyperbolic. The type and the number of mathematically correct initial and boundary conditions dictated by hyperbolicity are consistent with the conditions which are believed to be correct for a dynamic problem from the physical point of view. If the system loses hyperbolicity, the same set of initial and boundary conditions becomes mathematically incorrect, which means that the response of the body to dynamic loading becomes qualitatively different.

In the general three-dimensional case, when the system of the dynamic equations consists of a large number of scalar equations, the direct evaluation of the characteristic surfaces of the system in rather cumbersome. In such cases the wave speeds can alternatively be found through the eigenvalues of the acoustic tensor $A_{ij} = C_{ikjl} n_k n_l$ calculated for a tangential stiffness matrix C_{ikjl} and a wave propagation direction $\boldsymbol{n}$. Properties of the acoustic tensor and well-posedness of dynamic problems for elastoplastic models are investigated in [14,20,21,33,34,18,19,22,1,2,5]. In relation to a hypoplastic model this question is studied in [24–26].

The system (17)–(24) is nonlinear because of the term $||\boldsymbol{D}||$ and the dependence of the coefficients on the solution. In order to make the system locally quasi-linear and to find its characteristics, we linearise the term $||\boldsymbol{D}||$ about a $\boldsymbol{D}^0 \neq \boldsymbol{0}$ by taking the linear terms of its Taylor series. The velocity gradient has two non-zero components $L_{11} = \partial v_1/\partial x_1$ and $L_{21} = \partial v_2/\partial x_1$. The linearisation of $||\boldsymbol{D}||$ as a function of the two variables L_{11}, L_{21} in the vicinity of L_{11}^0, L_{21}^0 gives

$$||\boldsymbol{D}|| = \sqrt{L_{11}^2 + \frac{1}{2}L_{21}^2} = \frac{L_{11}^0}{||\boldsymbol{D}^0||}L_{11} + \frac{L_{21}^0}{2||\boldsymbol{D}^0||}L_{21} + ... \tag{26}$$

with

$$||\boldsymbol{D}^0|| = \sqrt{\left(L_{11}^0\right)^2 + \frac{1}{2}\left(L_{21}^0\right)^2}. \tag{27}$$

After the linearisation of $||\boldsymbol{D}||$ in (19)–(22), the system can be written in the matrix form

$$\frac{\partial U}{\partial t} + C\frac{\partial U}{\partial x_1} = F, \tag{28}$$

where

$$U = (v_1, v_2, T_{11}, T_{12}, T_{22}, T_{33}, e, p_l) \tag{29}$$

is the column vector of the unknown functions, and

$$C = \begin{pmatrix} v_1 & 0 & -1/\varrho & 0 & 0 & 0 & 0 & 1/\varrho \\ 0 & v_1 & 0 & -1/\varrho & 0 & 0 & 0 & 0 \\ -\kappa_1 & -\kappa_2 & v_1 & 0 & 0 & 0 & 0 & 0 \\ -\kappa_3 & -\kappa_4 & 0 & v_1 & 0 & 0 & 0 & 0 \\ -\kappa_5 & -\kappa_6 & 0 & 0 & v_1 & 0 & 0 & 0 \\ -\kappa_7 & -\kappa_8 & 0 & 0 & 0 & v_1 & 0 & 0 \\ -(1+e) & 0 & 0 & 0 & 0 & 0 & v_1 & 0 \\ K_f(1+e)/e & 0 & 0 & 0 & 0 & 0 & 0 & v_1 \end{pmatrix} \tag{30}$$

is the matrix of the system. The coefficients κ_i in (30) are determined by the constitutive equation (3). They depend on the point of linearisation $\boldsymbol{D}^0$ (through L_{11}^0, L_{21}^0 in (26), (27)) and on the current stresses and void ratio (through $\mathcal{L}_{ijkl}, N_{ij}$ in (19)–(22)). The slopes $\lambda = dx_1/dt$ of the characteristic curves are found from the equation

$$\det\bigl(C - \lambda I\bigr) = 0, \tag{31}$$

where I is the 8×8 unit matrix. The expansion of the determinant (31) shows that one of the roots is of multiplicity four and satisfies the equality

$$(\lambda - v_1)^4 = 0. \tag{32}$$

Four other roots are determined by the expression

$$(\lambda - v_1)^2 = c^2, \tag{33}$$

where

$$c^2 = \frac{1}{2\varrho}\left(\tilde{\kappa}_1 + \kappa_4 \pm \sqrt{(\tilde{\kappa}_1 - \kappa_4)^2 + 4\kappa_2\kappa_3}\right), \tag{34}$$

with $\tilde{\kappa}_1 = \kappa_1 + K_f(1+e)/e$. The difference $\lambda - v_1$ will be referred to as *wave speed.*

As mentioned above, the coefficients κ_i depend on the point of linearisation $\boldsymbol{D}^0$. As seen from (26), these coefficients depend only on the 'direction' of the rate of deformation, $\boldsymbol{D}^0/||\boldsymbol{D}^0||$, which can be specified by an angle ψ, see Fig. 1. This angle determines the instantaneous deformation of an infinitesimally thin layer, i.e the ratio of compression (extension) to shear. With the principal stresses T_1, T_2, T_3, the modulus K_f and the void ratio being fixed, the coefficients κ_i and hence the wave speeds become functions of the angles φ and ψ.

If the quantity (34) is real and positive for both plus and minus signs in front of the square root, there exist four real wave speeds $\pm c_1, \pm c_2$ with

$$c_1 = \frac{1}{\sqrt{2\varrho}} \left(\tilde{\kappa}_1 + \kappa_4 + \sqrt{(\tilde{\kappa}_1 - \kappa_4)^2 + 4\kappa_2\kappa_3} \right)^{1/2}, \tag{35}$$

$$c_2 = \frac{1}{\sqrt{2\varrho}} \left(\tilde{\kappa}_1 + \kappa_4 - \sqrt{(\tilde{\kappa}_1 - \kappa_4)^2 + 4\kappa_2\kappa_3} \right)^{1/2}. \tag{36}$$

Correspondingly, there exist four real characteristics with the slopes $\lambda = \pm c_1 + v_1$ and $\lambda = \pm c_2 + v_1$. In this case the system (17)–(24) is hyperbolic.

If, for the current values of φ and ψ, the term under the square root in (34) becomes negative, formula (34) gives two pairs of complex-conjugate solutions and, accordingly, no real characteristics. In this case, the system (17)–(24) loses hyperbolicity, and the dynamic problem becomes (locally) ill-posed. This type of ill-posedness is called 'flutter instability' [33]. If, for a certain pair φ, ψ, the quantity under the square root in (34) is positive but the term (34) becomes negative when calculated with a minus sign in front of the square root, then two wave speeds become purely imaginary, whereas two other wave speeds remain real. This is another type of possible ill-posedness in addition to the flutter type. Since the absolute value of two wave speeds vanishes before they become imaginary, this type of ill-posedness in called 'stationary discontinuity' [33,14] and is responsible for localisation of deformation in both dynamic and quasi-static cases.

Figures 2a–d show the wave speeds c_1, c_2 as functions of ψ calculated for dry sand ($\tilde{\kappa}_1 = \kappa_1$) with the hypoplastic equation described in Sect. 2. The four figures show the wave speeds for different ratios between the principal stresses and $\varphi = 30°$. If the stress state is hydrostatic or the difference between the principal stresses does not exceed a certain value, there exist two real wave speeds c_1, c_2, and the system is hyperbolic (Fig. 2a,b). The system is hyperbolic as well if the wave propagates along one of the principal stress directions ($\varphi = 0°$ or $90°$) regardless of the ratio between the principal stresses. The higher wave speed, c_1, shown by the upper curves in the figures, is associated with the longitudinal motion: if a purely longitudinal wave propagates in the medium, it propagates with a speed which belongs to this upper spectrum. The lower speed, c_2, is associated with the transverse motion: pure shear waves (e.g. in a fully saturated medium) propagate with speeds which belong to the lower spectrum.

If the ratio between the principal stresses is sufficiently large, there exist pairs φ, ψ for which the speeds c_1, c_2 are complex, and the system loses hyperbolicity, see Fig. 2c. Such pairs form a domain in the (φ, ψ)–plane [25,26]; an example is shown in Fig. 3. If the initial conditions or the current solution fall inside this range of φ and ψ, the problem becomes ill-posed with the flutter type of ill-posedness. A further increase in the ratio between the principal stresses leads to a situation in which the lower wave speed, c_2, vanishes and then becomes purely imaginary, whereas the

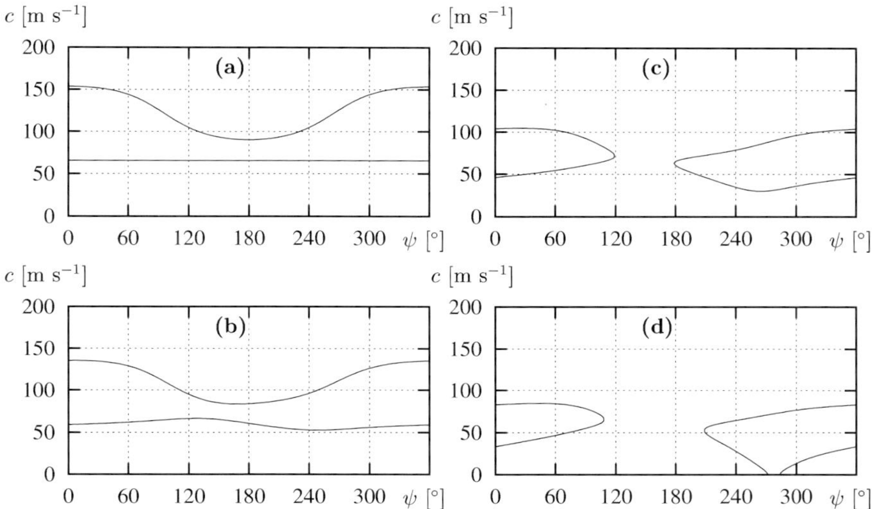

Fig. 2. Wave speeds for dry sand versus angle ψ for $\varphi = 30°$, $e = 0.8$, $T_2 = T_3 = -100$ kPa. **(a)** $T_1 = -100$ kPa, **(b)** $T_1 = -80$ kPa, **(c)** $T_1 = -50$ kPa, **(d)** $T_1 = -30$ kPa

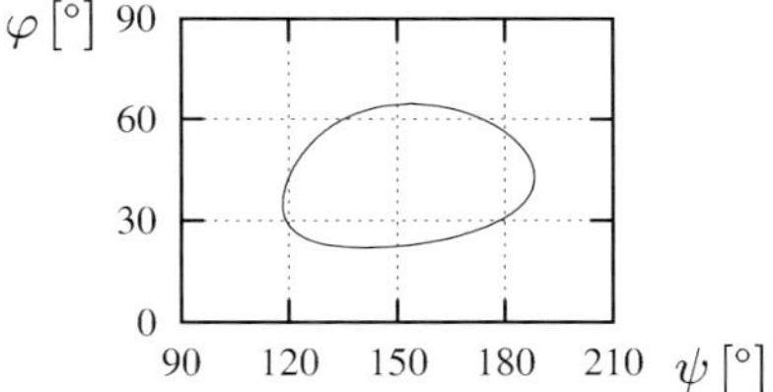

Fig. 3. The domain in which the wave speeds are complex numbers. $e = 0.8$, $T_1 = -50$ kPa, $T_2 = T_3 = -100$ kPa

speed c_1 remains real. In the case shown in Fig. 2d this occurs in the vicinity of $\psi = 280°$ and results in the other type of ill-posedness (stationary discontinuity). As distinguished from the flutter ill-posedness, stationary discontinuity occurs when the difference between the principal stresses is rather high and close to the maximum achievable value for the given granular material with the given density and stress level.

Since stationary discontinuity as a vanishing wave speed is shown to be responsible for the shear band formation, its existence as a type of ill-posedness of a dynamic problem is physically justified. As for the flutter type of ill-posedness, the question still remains open of whether this type reflects actual behaviour of granular materials or should be considered as a shortcoming of the constitutive relation. No clear experimental evidence in support of the flutter ill-posedness has been found so far. At the same time, there seems to be no theoretical reason why

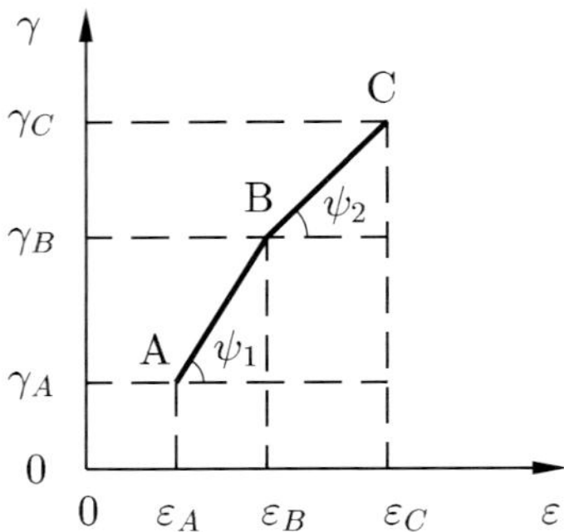

Fig. 4. Experimental evaluation of the wave speeds

flutter has to be excluded for a plastic material. This type of ill-posedness has to be excluded as unreal for an elastic material with a constant stiffness matrix, in which case it would mean the existence of periodic solutions whose amplitudes grow exponentially with time [33] (this explains the use of the term 'flutter' for this type of ill-posedness). This reasoning, however, is not applicable to a plastic material whose stiffness matrix changes during alternating deformation.

Although the flutter ill-posedness is related to a dynamic problem, experimental verification of its existence does not necessarily require dynamic experiments: the problem actually reduces to the verification of the constitutive equation, which can in principle be done in the following way. Consider quasi-static deformation of a layer of granular material as shown in Fig. 1. Let $\varepsilon = \partial u_1/\partial x_1$ and $\gamma = \partial u_2/\partial x_1$ denote the strain components, where u_1, u_2 are the displacements. Subject the body to a deformation path ABC consisting of two successive steps as shown in Fig. 4. The stiffness coefficients κ_1, κ_2 from (30) can then be found as the solution to the system

$$(\varepsilon_B - \varepsilon_A)\kappa_1 + (\gamma_B - \gamma_A)\kappa_2 = \sigma_B - \sigma_A, \tag{37}$$

$$(\varepsilon_C - \varepsilon_B)\kappa_1 + (\gamma_C - \gamma_B)\kappa_2 = \sigma_C - \sigma_B, \tag{38}$$

where $\sigma_A, \sigma_B, \sigma_C$ are the values of the stress T_{11} in the corresponding points.

The coefficients κ_3, κ_4 can be obtained in the same way from the system

$$(\varepsilon_B - \varepsilon_A)\kappa_3 + (\gamma_B - \gamma_A)\kappa_4 = \tau_B - \tau_A, \tag{39}$$

$$(\varepsilon_C - \varepsilon_B)\kappa_3 + (\gamma_C - \gamma_B)\kappa_4 = \tau_C - \tau_B, \tag{40}$$

where τ_A, τ_B, τ_C are the corresponding values of the stress component T_{12}. The wave speeds can then be calculated from (34).

The solvability of the systems (37), (38) and (39), (40) requires the two deformation steps to have different directions: $\psi_1 \neq \psi_2$, Fig. 4. At the same time, however, the difference between ψ_1 and ψ_2 should be small – otherwise the coefficients $\kappa_1, ..., \kappa_4$ obtained from (37)–(40) will characterise 'secant' stiffness which may considerably differ from the required tangential stiffness. In addition, since the wave speeds are to be evaluated for a certain stress state, the deformation steps should also be sufficiently small for the relative stress increments to be small.

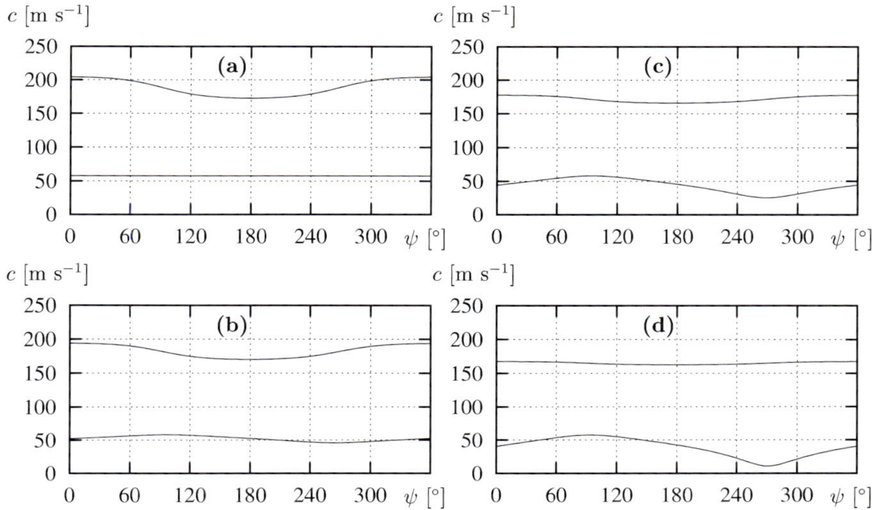

Fig. 5. Wave speeds for a saturated material versus angle ψ for $\varphi = 30°$, $e = 0.8$, $K_f = 20$ MPa, $T_2 = T_3 = -100$ kPa. **(a)** $T_1 = -100$ kPa, **(b)** $T_1 = -80$ kPa, **(c)** $T_1 = -50$ kPa, **(d)** $T_1 = -30$ kPa

The difficulty in performing the described verification is mainly due to the fact that the flutter ill-posedness may occur only in a certain range of the wave propagation directions (angle φ) and the deformation modes (angle ψ), while the capabilities of the conventional experimental techniques are limited to the coaxial or simple-shear types of deformation.

The presence of a fluid in the body changes the wave speeds as compared to a dry body. Figure 5 shows the wave speeds in a hypoplastic material for the same conditions as in Fig. 2 except for the presence of a fluid. The compression modulus of the fluid is taken to be equal to 20 MPa; for a saturated soil it corresponds to a degree of saturation of 0.99 at a depth of 10 m below the water table. Besides the fact that the wave speeds c_2 are higher than in a dry material, neither type of ill-posedness is observed in this case. The presence of a fluid thus has a regularising effect.

5 Hypoplastic Equation with Intergranular Strain

The concept of intergranular strain was introduced in [23] as an extension of the hypoplasticity theory in order to better describe the behaviour of granular materials, in particular, their response to alternating cyclic loading. The concept of intergranular strain as such is not related to a specific hypoplastic equation, and is applicable to any equation of the form (3). The extended constitutive theory contains the so-called intergranular-strain tensor $\boldsymbol{\delta}$ as a new state variable. The equation includes the tensors $\mathcal{L}$ and $\boldsymbol{N}$ from (3), and is written as

$$\dot{T}_{ij} = \mathcal{M}_{ijkl} D_{kl}, \tag{41}$$

where

$$\mathcal{M}_{ijkl} = \left[\rho^{\chi} m_T + (1-\rho^{\chi}) m_R\right] \mathcal{L}_{ijkl} +$$

$$+ \begin{cases} \rho^{\chi}(1-m_T)\mathcal{L}_{ijqs}\hat{\delta}_{qs}\hat{\delta}_{kl} + \rho^{\chi} N_{ij}\hat{\delta}_{kl} & \text{if} \quad \hat{\delta}_{ij} D_{ij} > 0, \\ \rho^{\chi}(m_R - m_T)\mathcal{L}_{ijqs}\hat{\delta}_{qs}\hat{\delta}_{kl} & \text{if} \quad \hat{\delta}_{ij} D_{ij} \leq 0, \end{cases} \tag{42}$$

with $\rho = ||\boldsymbol{\delta}||/R$ and

$$\hat{\delta}_{ij} = \begin{cases} \delta_{ij}/||\boldsymbol{\delta}|| & \text{if} \quad \boldsymbol{\delta} \neq \mathbf{0}, \\ 0 & \text{if} \quad \boldsymbol{\delta} = \mathbf{0}. \end{cases} \tag{43}$$

The intergranular-strain tensor $\boldsymbol{\delta}$ is determined by the evolution equation

$$\dot{\delta}_{ij} = \begin{cases} D_{ij} - \rho^{\beta_r}\hat{\delta}_{ij}\hat{\delta}_{kl} D_{kl} & \text{if} \quad \hat{\delta}_{ij} D_{ij} > 0, \\ D_{ij} & \text{if} \quad \hat{\delta}_{ij} D_{ij} \leq 0. \end{cases} \tag{44}$$

In addition to the constitutive parameters of (3), see Table 1, the extended version involves five parameters R, m_R, m_T, β_r and χ. The values of these parameters used in the present paper for numerical calculations are given in Table 2.

Table 2. Additional constitutive parameters of the extended hypoplastic equation

R	m_R	m_T	β_r	χ
4×10^{-5}	5.0	5.0	0.05	1.5

As distinct from the original hypoplastic equation (3), the extended equation (41) is incrementally linear. More precisely, it is piecewise-linear as it contains two different stiffness matrices for two different sectors in the $\boldsymbol{D}$-space (as for loading and unloading in elastoplasticity). In each sector, there is no dependence of the wave speeds on the angle ψ as in the original version (3). However, since the stiffness matrices depend on the current value of the tensor $\boldsymbol{\delta}$, so also do the wave speeds.

Under monotonic loading with a fixed direction $\boldsymbol{D}/||\boldsymbol{D}||$, the tensor $\boldsymbol{\delta}$ tends asymptotically to $R\boldsymbol{D}/||\boldsymbol{D}||$, and ρ tends to unity [23]. With this asymptotic value of $\boldsymbol{\delta}$ being inserted into (42), equation (41) produces the same constitutive response as the basic equation (3). Thus, during unidirectional deformation the extended equation gradually turns into the original one. This in particular means that the spectrum of the wave speeds of the extended equation includes all wave speeds of the basic equation and, therefore, also exhibits both types of ill-posednesses as revealed above.

The constitutive equation (41) will be used below for the numerical modelling of dynamic cyclic shear under simple-shear conditions. Figure 6 presents an example of quasi-static cyclic shear of dry sand calculated with a constant normal stress $T_{11} = -100$ kPa, initial stresses $T_{22} = T_{33} = -50$ kPa, $T_{12} = 0$ and a void ratio of 0.8. Shown in the figure are the shear stress T_{12}, the principal stress T_{22} and the volumetric deformation $\varepsilon = \partial u_1/\partial x_1$ as functions of the shear strain $\gamma = \partial u_2/\partial x_1$.

Cyclic shear of granular materials, in particular, soils, has been extensively studied experimentally, especially in connection with the modelling of the deformation

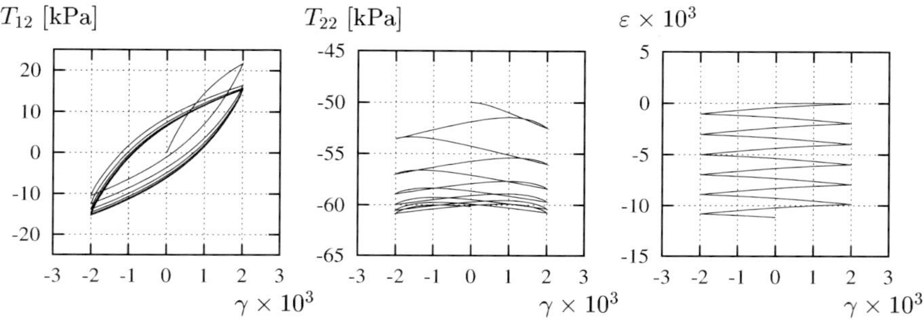

Fig. 6. Cyclic shear of a dry hypoplastic material calculated with equation (41)

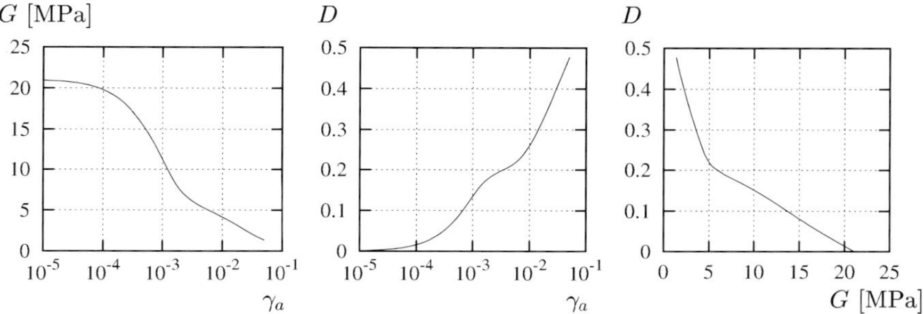

Fig. 7. Secant shear modulus and damping ratio for a dry hypoplastic material

of soil during earthquakes [35,32,8,37,41,11,7,36]. The response of a soil to multi-cycle shear loading is usually quantified through the dependence of the secant shear modulus $G = \tau_a/\gamma_a$ and the damping ratio $D = A/2\pi\tau_a\gamma_a$ on the strain amplitude γ_a [17,16]. Here τ_a is the shear stress amplitude, and A is the area of the hysteresis loop in the (γ, τ)-plane. Figure 7 shows the secant shear modulus and the damping ratio as functions of the strain amplitude calculated with (41) for cyclic simple shear with a constant normal stress $T_{11} = -100$ kPa and the same initial conditions as in Fig. 6. The secant shear modulus and the damping ratio are calculated for the third cycle.

Cyclic shearing of a dry granular material results in its gradual compaction [37,41]. This phenomenon is well described by the present constitutive equation, see the change in ε in Fig. 6. If a granular material is saturated with a fluid, the latter restrains any changes in the volume that would occur in a dry material due to rearrangements of the grains during cyclic shear. In this case, instead of compaction, repeated shearing without drainage is accompanied by the reduction of the effective pressure [35,32,8]. If the shear amplitude is large enough, the shearing may result in the vanishing of the effective pressure and, as a consequence, in the loss of shear stiffness. At the initial stage of the shearing, the effective pressure undergoes small oscillations superimposed on the monotonic decrease in the absolute value. The further evolution depends on the type of loading. In the strain-controlled case, the effective pressure continues to approach zero in a gradual

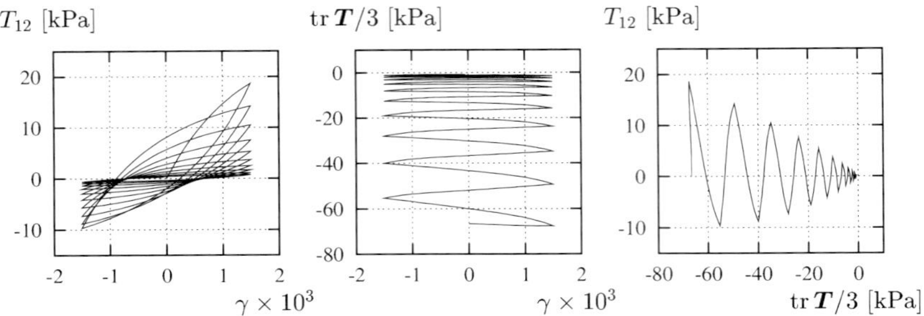

Fig. 8. Strain-controlled cyclic shear of a saturated hypoplastic material

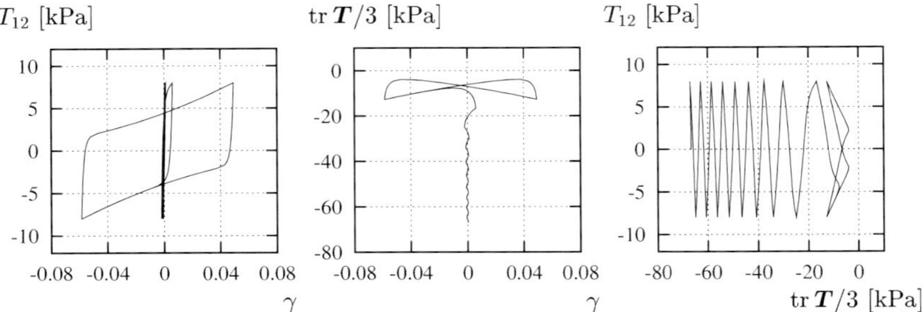

Fig. 9. Stress-controlled cyclic shear of a saturated hypoplastic material

manner. In the stress-controlled case, after a number of cycles, the effective pressure abruptly falls to a low value and then oscillates with a larger amplitude determined by the amplitude of the applied shear stress. The strain amplitude increases as well, reaching several percent after a few cycles ('cyclic mobility' [36]).

Both these cases are correctly modelled by the hypoplastic equation. Figure 8 shows the strain-controlled cyclic shear of a fully saturated material calculated with the same initial state as in Fig. 6. Shown are the shear stress T_{12} and the mean effective stress $\mathrm{tr}\,\boldsymbol{T}/3$ versus the shear deformation γ. After several cycles of loading, the mean effective stress is reduced to zero, thus resulting in the vanishing of the shear stiffness, and the material turns into the state of 'initial liquefaction' [36]. The transition to cyclic mobility during stress-controlled shear is shown in Fig. 9.

6 Numerical Solution of the Dynamic Problem

The system of the governing equations consists of the equation of motion (13), the constitutive equations for stresses (41), intergranular strains (44), pore pressure (16), and the mass balance equation (15). For plane waves propagating along the x_1-axis as shown in Fig. 1, these equations comprise a system of first-order scalar equations which can be written in the form

$$\dot{\boldsymbol{y}} = \boldsymbol{f}\left(\boldsymbol{y}, \frac{\partial \boldsymbol{y}}{\partial x_1}\right), \tag{45}$$

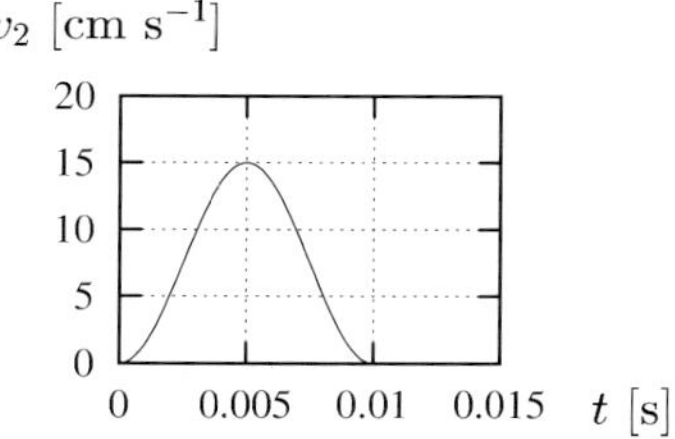

Fig. 10. Boundary condition for the solution shown in Fig. 11

where $\boldsymbol{y}$ is the set of unknown functions $v_1, v_2, T_{11}, T_{12}, T_{22}, T_{33}, \delta_{11}, \delta_{12}, p_l, e$. The system (45) was solved by a finite-difference technique with implicit time integration:

$$\boldsymbol{y}_i^{j+1} = \boldsymbol{y}_i^j + \left[\alpha \boldsymbol{f}(\boldsymbol{y}_i^j, \Delta \boldsymbol{y}_i^j) + (1-\alpha)\boldsymbol{f}(\boldsymbol{y}_i^{j+1}, \Delta \boldsymbol{y}_i^{j+1})\right]\Delta t, \tag{46}$$

where $\boldsymbol{y}_i^j$ stands for the value of $\boldsymbol{y}$ at a spatial point i at time j, $\Delta \boldsymbol{y}_i^j$ is a finite-difference approximation of the derivative $\partial \boldsymbol{y}/\partial x_1$ at the point i calculated at time j, and α is the parameter of the scheme, $0 \leq \alpha \leq 1$. Equations (46) were solved for $\boldsymbol{y}_i^{j+1}$ by successive approximations. The algorithm is described in more detail in [28].

Owing to the nonlinear constitutive behaviour, the propagation of waves in a granular medium is qualitatively different from that in a linearly elastic solid. Two different causes of the nonlinearity can be distinguished. *First*, the stress-strain relation is nonlinear because of the dependence of the stiffness matrix on the current stress state and, in particular, on the stress level. It is known that in certain cases such as the propagation of longitudinal waves this property may lead to the loss of continuity of the solution and to the formation of shock fronts (see e.g. [30]). The *second* source of the nonlinearity is the dependence of the stiffness matrix on the direction of the current deformation, which is responsible for the irreversible (plastic) behaviour of the material. This effect can be isolated and studied separately by considering the simplest case of one-component waves (longitudinal or transverse) with two different stiffness moduli for opposite directions of deformation [24,9,12,13,4].

Granular materials tend to change their volume in shear. In terms of the stiffness matrix, this property (dilatancy) means anisotropy of the stress-strain relation and manifests itself as normal stress effects. In distinction to an anisotropic elastic solid, the volume change in a granular body is irreversible: shear in one direction followed by the shear of the same size in the opposite direction does not restore the volume to its original value. Repeated shearing results in the steady compaction of the material [37,41].

The property of changing the volume in shear influences the dynamics of granular solids. Since shear deformation is accompanied by the displacement of the material in the direction normal to the shear, a transverse wave propagating in a granular medium induces longitudinal displacement and thus gives rise to a longitudinal wave [25,26,29]. Consider a half-space $x_1 \geq 0$ occupied by a dry granular material, and prescribe a boundary condition at $x_1 = 0$ consisting of a purely transverse disturbance $v_2(t)$ for $t \geq 0$ as shown in Fig. 10, with the longitudinal component v_1 being kept equal to zero. The solution to this problem obtained with the constitutive equation (41) is shown in Fig. 11 at two different times. The boundary impulse gives rise to a shear wave propagating into the half-space with a decreasing amplitude. At the same time, the shear wave induces a longitudinal wave which propagates with a higher speed. Since the longitudinal wave is induced by the contraction of the material near the boundary, this wave consists of negative particle velocity, that is, in the direction opposite to the wave propagation. The converse situation where a longitudinal impulse induces a transverse wave is also possible if the wave propagation direction does not coincide with a principal stress direction [26].

A periodic boundary condition for the transverse component results in a periodic motion of the medium in both the longitudinal and the transverse directions. Since one cycle of shearing corresponds to two cycles of densification, the dominant frequency of the induced longitudinal motion turns out to be double the dominant frequency of the transverse motion, the latter being equal to the frequency of the boundary condition [29]. This double-frequency effect is observed in numerical solutions for both a half-space and a finite layer, although in the latter case this effect may be influenced by the reflection from the boundaries and the eigenfrequencies of the layer.

Consider a 30 m thick horizontal layer of dry sand which is set in motion by a sinusoidal boundary condition $v_2(t) = v_0 \sin \omega t$ prescribed for the horizontal component of velocity at the lower boundary of the layer starting from $t = 0$. Such a boundary condition imitates the motion of the rock bed below the sand layer during an earthquake. The upper boundary is assumed to be free of stress. The initial vertical stress T_{11} varies linearly with the depth according to the specific weight of the sand, and $T_{22} = T_{33} = 0.5\,T_{11}$. The problem is solved with $v_0 = 5$ cm s^{-1}, $\omega/2\pi = 4$ Hz and an initial void ratio of 0.8. Figure 12 shows the velocity components at a depth of 18 m for the time interval between 4 and 5 s. Figure 13 shows the spectral amplitudes (the absolute values of the Fourier transform) of the velocity components at the same depth calculated for the time interval from 3 to 6 s. Both figures indicate that the dominant frequency of the induced vertical motion is double the frequency of the boundary condition. A constant negative component of the vertical velocity is caused by the compaction of the sand situated below.

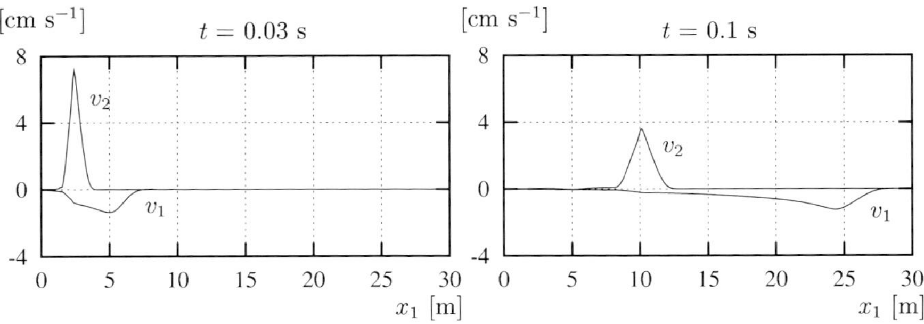

Fig. 11. Transverse and longitudinal waves in a half-space induced by the transverse boundary disturbance shown in Fig. 10. Initial state: $T_{11} = -100$ kPa, $T_{22} = T_{33} = -50$ kPa, $T_{12} = 0$, $e = 0.8$

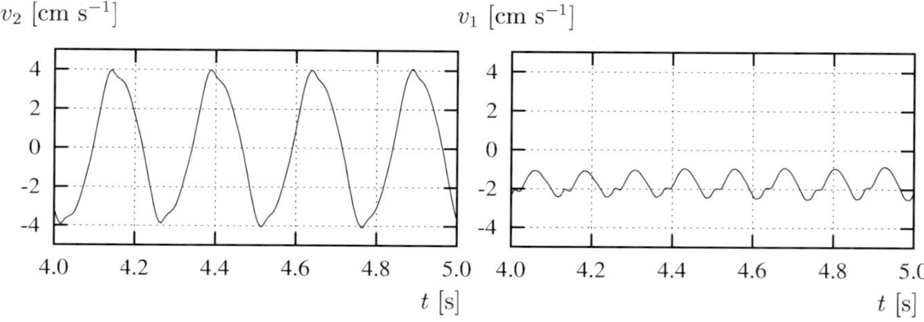

Fig. 12. Horizontal (v_2) and vertical (v_1) velocity components versus time at a depth of 18 m

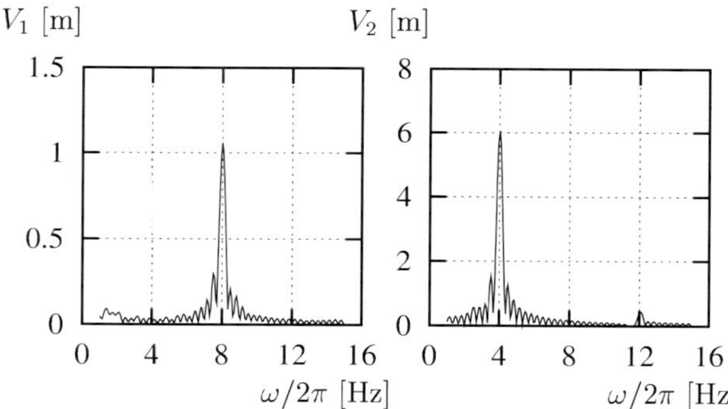

Fig. 13. Spectral amplitudes of the vertical (V_1) and horizontal (V_2) velocity components at a depth of 18 m

7 Earthquake-induced Liquefaction of Saturated Sand

The present hypoplastic model provides an adequate and effective tool for the modelling of the dynamic liquefaction of saturated soil, which is a topical problem in earthquake geotechnics [17,16,15]. In the present study liquefaction is understood as the state of vanishing effective stress ('initial liquefaction' [36]) caused by cyclic shearing. Such states can be reached in dense as well as loose saturated soils. These states do not necessarily lead to an observable failure, but strongly influence the response of the soil mass to the dynamic excitation because of the loss of shear stiffness in the liquefied zones. This plays an important role for the site response analysis.

We will restrict ourselves to a one-dimensional problem for a horizontal layer of saturated sand lying on a hard base (rock). The motion of the sand is assumed to be induced by a plane shear wave coming from below. The influence of the excitation wave is modelled by a boundary condition for the horizontal component of velocity prescribed at the lower boundary of the layer. This one-dimensional problem serves as a good approximation in many practical cases and, being much simpler from the

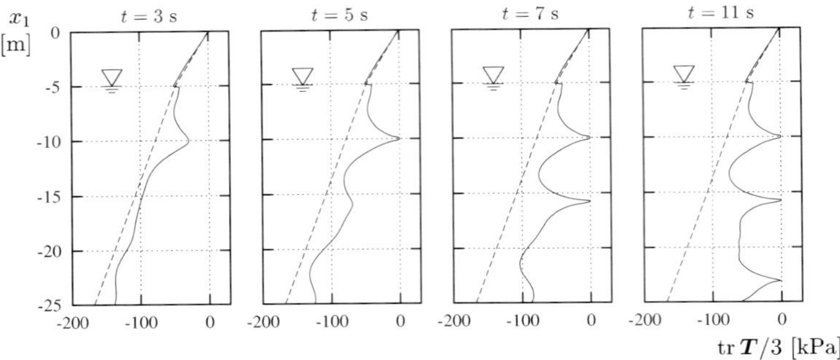

Fig. 14. Evolution of the effective stress in a layer of saturated sand subjected to a sinusoidal shear disturbance at the base. Dashed line: the initial stress

computational point of view than two- and three-dimensional problems, provides wide possibilities for theoretical investigations.

The amplitude of cyclic deformation of the soil during dynamic excitation is different at different depths even if the soil is spatially homogeneous. This is a consequence of both the dynamic nature of the process and the fact that the stiffness of the soil varies with the depth because of the inhomogeneity of the effective stress. A higher shear amplitude produces a higher rate of the reduction of the effective stress. A reduced effective stress means smaller shear stiffness. At the same time, it is reasonable to expect that the reduction of the stiffness at a certain depth will lead to an increase in the shear amplitude at that depth, thus producing a kind of positive feedback. This mechanism is relevant to both dynamic and quasi-static shearing. In the quasi-static case studied in [31], this feedback is shown to manifest itself in instability of slightly inhomogeneous stress states during cyclic shear: a small inhomogeneity increases and eventually evolves into a shear band. A similar situation is observed in dynamic problems: the effective stress develops in such a way that the liquefaction is localised in thin separate zones [27–29].

Figure 14 shows a typical example of the development of the effective stress in a layer of saturated sand calculated with the constitutive equation (41). A 25 m thick layer is subjected to a sinusoidal boundary condition $v_2(t) = v_0 \sin \omega t$ at the lower boundary $x_1 = -25$ m, with $v_0 = 5$ cm s^{-1}, $\omega/2\pi = 4$ Hz, with an initial void ratio of 0.8 and a degree of saturation of 0.99 below the water table. The horizontal effective stresses in the initial state are taken to be equal to half the vertical effective stress. The main feature of the stress pattern is that the liquefaction is localised in narrow zones rather than spreading over a finite thickness. These zones emerge successively from the top to the bottom of the layer. Their number and location depend on the amplitude and the frequency of the boundary condition.

Because of the vanishing shear stiffness and the high hysteretic damping in the liquefied sand, each liquefaction zone makes it impossible for the shear disturbance to propagate through this zone to the above situated part of the layer. After the instant when a liquefaction zone appears, the subsequent motion above this zone proceeds independently of the boundary condition and rapidly decays in spite of the continuing motion of the base. In such a way the emerging liquefaction zones isolate the upper part of the layer from further excitation. The duration of the motion at some depth is shorter than the actual duration of the excitation if a liquefaction zone emerges below that depth.

References

1. An, L., Schaeffer, D.G. (1992) The flutter instability in granular flow. J. Mech. Phys. Solids, **40**(3), 683–698
2. An, L. (1993) Loss of hyperbolicity in elastic-plastic material at finite strains. SIAM J. Appl. Math., **53**(3), 621–654
3. Bauer, E. (1996) Calibration of a comprehensive hypoplastic model for granular materials. Soils and Foundations, **36**(1), 13–26
4. Berezin, Y. A., Osinov, V. A., Hutter, K. (2001) Evolution of plane disturbances in hypoplastic granular materials. Continuum Mech. Thermodyn., **13**, 79–90
5. Bigoni, D., Zaccaria, D. (1994) On the eigenvalues of the acoustic tensor in elastoplasticity. Eur. J. Mech., A/Solids, **13**(5), 621–638
6. Dafalias, Y. F. (1986) Bounding surface plasticity. I: Mathematical foundation and hypoplasticity. J. Eng. Mech., ASCE, **112**(9), 966–987
7. Drnevich, V. P. (1972) Undrained cyclic shear of saturated sand. J. Soil Mech. Found. Div., Proc. ASCE, **98**, SM8, 807–825
8. Finn, W. D. L., Pickering, D. J., Bransby, P. L. (1971) Sand liquefaction in triaxial and simple shear tests. J. Soil Mech. Found. Div., Proc. ASCE, **97**, SM4, 639–659
9. Gordon, M. S., Shearer, M., Schaeffer, D. G. (1997) Plane shear waves in a fully saturated granular medium with velocity and stress controlled boundary conditions. Int. J. Non-Linear Mech., **32**(3), 489–503
10. Gudehus, G. (1996) A comprehensive constitutive equation for granular materials. Soils and Foundations, **36**(1), 1–12
11. Hardin, B. O., Drnevich, V. P. (1972) Shear modulus and damping in soils: design equations and curves. J. Soil Mech. Found. Div., Proc. ASCE, **98**, SM7, 667–692
12. Hayes, B. T., Schaeffer, D. G. (1998) Plane shear waves under a periodic boundary disturbance in a saturated granular medium. Physica D, **121**, 193–212
13. Hayes, B. T., Schaeffer, D. G. (2000) Stress-controlled shear waves in a saturated granular medium. Eur. J. Appl. Math., **11**, 81–94
14. Hill, R. (1962) Acceleration waves in solids. J. Mech. Phys. Solids, **10**, 1–16
15. Ishihara, K. (1993) Liquefaction and flow failure during earthquakes. Géotechnique, **43**(3), 351–415
16. Ishihara, K. (1996) Soil Behaviour in Earthquake Geotechnics. Clarendon Press, Oxford
17. Kramer, S. L. (1996) Geotechnical Earthquake Engineering. Prentice Hall, Upper Saddle River
18. Loret, B., Prévost, J. H., Harireche, O. (1990) Loss of hyperbolicity in elastic-plastic solids with deviatoric associativity. Eur. J. Mech., A/Solids, **9**(3), 225–231
19. Loret, B., Harireche, O. (1991) Acceleration waves, flutter instabilities and stationary discontinuities in inelastic porous media. J. Mech. Phys. Solids, **39**(5), 569–606
20. Mandel, J. (1962) Ondes plastiques dans un milieu indéfini à trois dimensions. Journal de Mécanique, **1**(1), 3–30

21. Mandel, J. (1964) Propagation des surfaces de discontinuité dans un milieu élastoplastique. In: Stress waves in anelastic solids, H. Kolsky and W. Prager (Eds.), Springer, Berlin, 331–340
22. Molenkamp, F. (1991) Material instability for drained and undrained behaviour. Parts 1, 2. Int. J. Num. Anal. Meth. Geomech., **15**, 147–180
23. Niemunis, A., Herle, I. (1997) Hypoplastic model for cohesionless soils with elastic strain range. Mech. Cohesive-frictional Mater., **2**(4), 279–299
24. Osinov, V. A., Gudehus, G. (1996) Plane shear waves and loss of stability in a saturated granular body. Mech. Cohesive-frictional Mater., **1**(1), 25–44
25. Osinov, V. A. (1997) Plane waves and dynamic ill-posedness in granular media. In: Powders and Grains'97, R. P. Behringer and J. T. Jenkins (Eds.), Balkema, Rotterdam, 363–366
26. Osinov, V. A. (1998) Theoretical investigation of large-amplitude waves in granular soils. Soil Dyn. Earthquake Eng., **17**(1), 13–28
27. Osinov, V. A., Loukachev, I. (2000) Settlement of liquefied sand after a strong earthquake. In: Compaction of soils, granulates and powders, D. Kolymbas and W. Fellin (Eds.), Balkema, Rotterdam, 297–306
28. Osinov, V. A. (2000) Wave-induced liquefaction of a saturated sand layer. Continuum Mech. Thermodyn., **12**(5), 325–339
29. Osinov, V. A. (2001) The role of dilatancy in the plastodynamics of granular solids. In: Powders and Grains 2001, Y. Kishino (Ed.), Swets and Zeitlinger, Lisse, 135–138
30. Osinov, V. A. (1998) On the formation of discontinuities of wave fronts in a saturated granular body. Continuum Mech. Thermodyn., **10**(5), 253–268
31. Osinov, V. A. (2002) Cyclic shear of saturated soil: the evolution of stress inhomogeneity. Continuum Mech. Thermodyn., **14**(2), 191–205
32. Peacock, W. H., Seed, H. B. (1968) Sand liquefaction under cyclic loading simple shear conditions. J. Soil Mech. Found. Div., Proc. ASCE, **94**, SM3, 689–708
33. Rice, J. R. (1976) The localization of plastic deformation. In: Theoretical and Applied Mechanics, Proc. 14th IUTAM Congress, W. T. Koiter (Ed.), North-Holland, Amsterdam, 207–220
34. Schaeffer, D. G. (1990) Instability and ill-posedness in the deformation of granular materials. Int. J. Num. Anal. Meth. Geomech., **14**, 253–278
35. Seed, H. B., Lee, K. L. (1966) Liquefaction of saturated sands during cyclic loading. J. Soil Mech. Found. Div., Proc. ASCE, **92**, SM6, 105–134
36. Seed, H. B., Martin, P. P., Lysmer, J. (1976) Pore-water pressure changes during soil liquefaction. J. Geot. Eng. Div., Proc. ASCE, **102**, GT4, 323–346
37. Silver, M. L., Seed, H. B. (1971) Volume changes in sands during cyclic loading. J. Soil Mech. Found. Div., Proc. ASCE, **97**, SM9, 1171–1182
38. von Wolffersdorff, P. A. (1996) A hypoplastic relation for granular materials with a predefined limit state surface. Mech. Cohesive-frictional Mater., **1**(3), 251–271
39. Wu, W., Kolymbas, D. (2000) Hypoplasticity then and now. In: Constitutive modelling of granular materials, D. Kolymbas (Ed.), Springer, Berlin Heidelberg, 57–105
40. Wu, W., Bauer, E., Kolymbas, D. (1996) Hypoplastic constitutive model with critical state for granular materials. Mech. Mater., **23**, 45–69
41. Youd, T. L. (1972) Compaction of sands by repeated shear straining. J. Soil Mech. Found. Div., Proc. ASCE, **98**, SM7, 709–725

Acoustic Waves in Porous Solid–Fluid Mixtures

Krzysztof Wilmański and Bettina Albers

Weierstrass Institute for Applied Analysis and Stochastics, Mohrenstr. 39,
D-10117 Berlin, Germany, e-mail: {wilmansk, albers}@wias-berlin.de

received 25 Jul 2002 — accepted 17 Oct 2002

Abstract. In this article we consider two problems of propagation of weak discontinuity waves in porous materials. In the first part we present basic properties of bulk waves in fully saturated materials. These materials are modelled by a two-component immiscible mixture. We present general propagation conditions for such a model which yield three modes of propagation: P1-, S-, and P2-waves. Then we discuss the dispersion relation and show that results are strongly dependent on the way in which waves are excited. In the second part we present some properties of surface waves. We begin with the classical Rayleigh and Love problems and extend them on heterogeneous materials important in practical applications. Subsequently we proceed to surface waves in two-component porous materials on the contact surface with vacuum (impermeable boundary) and with a liquid (permeable boundary). We show the existence of different modes of surface waves in the high frequency limit as well as the degeneration of the problem in the low frequency limit.

1 Introduction

In this article we present two problems of weak discontinuity waves in porous materials: acoustic waves in saturated media modelled by a two–component continuum, as well as surface waves in such media and their asymptotic properties.

Propagation of acoustic waves in geophysical porous materials plays a particularly important role in testing porous and granular materials because laboratory measurements on such materials usually differ considerably from *in situ* measurements required in practical applications (see e.g. [19]). Most of the theoretical results were obtained within the so–called Biot model (cf. e.g. [3]). They contributed immensely to the understanding of the subject but, simultaneously, there are many very controversial issues related to the application of this model. We mention some of them later in this article.

A particular practical bearing have surface waves. Various theoretical and practical aspects of such waves have been investigated for single-component continua, [4,15,20]. Very little has been done for two-component materials.

During the last decade, acoustics of porous materials was also developed within a different continuous model derived on the basis of modern continuum thermodynamics. This model (in its linear version) is on the one hand simpler than Biot's model; in contrast to the latter it does neither violate the second law of thermodynamics nor the principle of material frame indifference. Moreover it describes changes of porosity by means of an additional microscopic variable. In spite of these differences the number of acoustic modes of propagation and their fundamental properties are the same in both models (see e.g. [21]).

The article is organised as follows: Section 2 contains a review of fundamental properties of P1-, S-, and P2-waves in porous materials. However, we emphasise an aspect of such waves which seems to have been overlooked in the literature. Namely, we demonstrate the dependence of acoustic properties of porous media on the way in which the dynamic disturbance is excited. This way is immaterial for high frequency (short wave) asymptotics determining the speeds of signals in the medium. However, it becomes essential in the limit of low frequencies (long waves) and these are of primary practical importance in soil mechanics and other geophysical applications. As proved by Edelman [7], the monochromatic P2-wave as a solution of an initial value problem does not propagate in the case of low wave-numbers (long waves). It means that such waves do not exist after some impact excitations (chopping, explosions, etc.). Consequently some surface modes of propagation can neither appear in the range of long waves [8]. We return to this problem in the Sect. 3 where we discuss the propagation of surface waves in two limits: high frequency and low frequency. We present results for a poroelastic material with impermeable boundary. Results for both impermeable and permeable boundaries in limits of short waves can be found in [6,9] and for long waves in [8]. Final remarks are given in Sect. 4.

2 Bulk Waves in Two-Component Poroelastic Media

2.1 Field Equations for Two-Component Poroelastic Media

We rely on the model of two-component poroelastic saturated media proposed in a fully nonlinear form in [22,23]. We consider its linear version described by the following fields:

partial mass density of the fluid, $\rho^F(\boldsymbol{x},t)$,
velocity of the fluid, $\boldsymbol{v}^F(\boldsymbol{x},t)$,
velocity of the skeleton, $\boldsymbol{v}^S(\boldsymbol{x},t)$,
symmetric tensor of small deformations of the skeleton, $\boldsymbol{e}^S(\boldsymbol{x},t)$, $\|\boldsymbol{e}^S\| \ll 1$,[1]
porosity, n.

These fields have a purely macroscopic interpretation, and in a theoretical analysis it is not needed to refer to any microscopic quantities related to these macroscopic fields. Certainly, in practical applications such a reference may be necessary. For instance it may be useful to estimate macroscopic elastic parameters in terms of true or drained elastic properties of real materials, partial mass densities in terms of true mass densities or a relative velocity in terms of the filter velocity. In order to keep this article in a reasonable size, we do not discuss this problem here but refer to [13] instead.

[1] The norm of the deformation tensor is usually defined by means of its eigenvalues $\lambda^i, i = 1,2,3: \left(\boldsymbol{e}^S - \lambda^i \mathbf{1}\right)\boldsymbol{k}^i = 0$. Namely $\|\boldsymbol{e}^S\| = \max\left\{|\lambda^1|, |\lambda^2|, |\lambda^3|\right\}$.

For these fields the following field equations hold in the linear model of poroelastic materials[2]:

$$\frac{\partial \rho^F}{\partial t} + \rho_0^F \operatorname{div} \boldsymbol{v}^F = 0, \tag{1}$$

$$\rho_0^F \frac{\partial \boldsymbol{v}^F}{\partial t} + \kappa \operatorname{grad} \rho^F + \beta \operatorname{grad} (n - n_0) + \hat{\boldsymbol{p}} = 0, \quad \hat{\boldsymbol{p}} := \pi \left(\boldsymbol{v}^F - \boldsymbol{v}^S \right), \tag{2}$$

$$\rho_0^S \frac{\partial \boldsymbol{v}^S}{\partial t} - \operatorname{div} \left(\lambda^S \left(\operatorname{tr} \boldsymbol{e}^S \right) \mathbf{1} + 2\mu^S \boldsymbol{e}^S + \beta (n - n_0) \mathbf{1} \right) - \hat{\boldsymbol{p}} = 0, \tag{3}$$

$$\frac{\partial \boldsymbol{e}^S}{\partial t} = \operatorname{sym} \operatorname{grad} \boldsymbol{v}^S, \tag{4}$$

$$\frac{\partial n}{\partial t} + n_0 \operatorname{div} \left(\boldsymbol{v}^F - \boldsymbol{v}^S \right) + \frac{n - n_0}{\tau} = 0, \tag{5}$$

with $\left| \frac{\rho^F - \rho_0^F}{\rho_0^F} \right| \ll 1$. In these equations ρ_0^F, ρ_0^S, n_0 denote constant reference values of partial mass densities, and porosity, respectively. The quantities $\kappa, \lambda^S, \mu^S, \beta, \pi, \tau$ are constant material parameters. The first one describes the macroscopic *compressibility* of the fluid component, the next two are macroscopic *elastic constants* of the skeleton, β is the coupling constant, π is the coefficient of *bulk permeability*, and τ is the *relaxation time*. The construction of (5) is presented, for instance, in Chapt. 10 of the book [23]. For the purpose of this work we assume $\beta = 0$. Then the problem of evolution of porosity described by equation (5) can be solved separately from the rest of the problem, and it does not influence the acoustic waves in the medium. Let us mention that the general case has been considered in earlier papers on the subject (e.g. [24–26]) and it has been shown that coupling effects through β can be neglected in linear models.

2.2 Propagation of Acoustic Fronts in Two-Component Media

We investigate the propagation of an acoustic front carrying weak discontinuities. It is assumed that the front σ_t is given by the relation

$$f(\boldsymbol{x}, t) = 0, \quad \boldsymbol{x} \in \sigma_t \subset \mathcal{B}_t, \quad t \in \mathcal{T}, \tag{6}$$

where the function f is assumed to be at least continuously differentiable with respect to both variables. The symbols $\mathcal{B}_t$, $\mathcal{T}$ denote the current configuration of the medium, and the time interval, respectively. The surface defined by (6) moves with normal speed c and possesses unit normal vector $\boldsymbol{n}$,

$$c := -\frac{\partial f / \partial t}{|\operatorname{grad} f|}, \quad \boldsymbol{n} := \frac{\operatorname{grad} f}{|\operatorname{grad} f|}. \tag{7}$$

Weak discontinuities of fields introduced in Subsect. 2.1 are defined by the following conditions on the surface σ_t (oriented by the field $\boldsymbol{n}(\boldsymbol{x}, t), \boldsymbol{x} \in \sigma_t, t \in \mathcal{T}$):

$$[\![\rho^F]\!] = 0, \quad [\![\boldsymbol{v}^F]\!] = 0, \quad [\![\boldsymbol{v}^S]\!] = 0, \quad [\![\boldsymbol{e}^S]\!] = 0, \tag{8}$$

[2] The so-called principle of separation which yields, for instance, the lack of dependence of the partial stress tensor in the skeleton on the fluid mass density, follows from the second law. This can be changed by adding a dependence on higher gradients (see end of Subsect. 2.2).

where

$$[\![\ldots]\!] := \lim_{\sigma_t^+}(\ldots) - \lim_{\sigma_t^-}(\ldots). \tag{9}$$

Then, according to the Hadamard lemma, the following kinematic compatibility conditions hold

$$\begin{aligned} &\left[\!\!\left[\operatorname{grad}\rho^F\right]\!\!\right] = -\frac{1}{c}R^F\boldsymbol{n}, \quad \left[\!\!\left[\operatorname{grad}\boldsymbol{e}^S\right]\!\!\right] = \frac{1}{2c^2}\left(\boldsymbol{A}^S\otimes\boldsymbol{n}+\boldsymbol{n}\otimes\boldsymbol{A}^S\right)\otimes\boldsymbol{n}, \\ &\left[\!\!\left[\operatorname{grad}\boldsymbol{v}^F\right]\!\!\right] = -\frac{1}{c}\boldsymbol{A}^F\otimes\boldsymbol{n}, \quad \left[\!\!\left[\operatorname{grad}\boldsymbol{v}^S\right]\!\!\right] = -\frac{1}{c}\boldsymbol{A}^S\otimes\boldsymbol{n}, \end{aligned} \tag{10}$$

where

$$R^F := \left[\!\!\left[\frac{\partial\rho^F}{\partial t}\right]\!\!\right], \quad \boldsymbol{A}^F := \left[\!\!\left[\frac{\partial\boldsymbol{v}^F}{\partial t}\right]\!\!\right], \quad \boldsymbol{A}^S := \left[\!\!\left[\frac{\partial\boldsymbol{v}^S}{\partial t}\right]\!\!\right], \tag{11}$$

are the so–called *amplitudes of discontinuity.*

Substitution of the above relations in (1)-(5) evaluated on both sides of the front σ_t yields the conditions

$$R^F = \frac{\rho_0^F}{c}\boldsymbol{A}^F\cdot\boldsymbol{n}, \tag{12}$$

and

$$\begin{aligned} \left(c^2\mathbf{1} - \frac{\lambda^S+\mu^S}{\rho_0^S}\boldsymbol{n}\otimes\boldsymbol{n} - \frac{\mu^S}{\rho_0^S}\mathbf{1}\right)\boldsymbol{A}^S &= 0, \\ \left(c^2\mathbf{1} - \kappa\boldsymbol{n}\otimes\boldsymbol{n}\right)\boldsymbol{A}^F &= 0. \end{aligned} \tag{13}$$

Certainly this is an eigenvalue problem which yields three nontrivial solutions:

$$\begin{aligned} c_{P1} &:= \sqrt{\frac{\lambda^S+2\mu^S}{\rho_0^S}}, \quad \boldsymbol{A}^S\cdot\boldsymbol{n}\neq 0, \quad \boldsymbol{A}_\perp^S := \boldsymbol{A}^S - \left(\boldsymbol{A}^S\cdot\boldsymbol{n}\right)\boldsymbol{n} = 0, \quad \boldsymbol{A}^F = 0, \\ c_{P2} &:= \sqrt{\kappa}, \quad \boldsymbol{A}^F\cdot\boldsymbol{n}\neq 0, \quad \boldsymbol{A}^S = 0, \quad \boldsymbol{A}_\perp^F := \boldsymbol{A}^F - \left(\boldsymbol{A}^F\cdot\boldsymbol{n}\right)\boldsymbol{n} = 0, \\ c_S &:= \sqrt{\frac{\mu^S}{\rho_0^S}}, \quad \boldsymbol{A}_\perp^S\neq 0, \quad \boldsymbol{A}^S\cdot\boldsymbol{n} = 0, \quad \boldsymbol{A}^F = 0. \end{aligned} \tag{14}$$

The first two solutions describe longitudinal P1-, and P2-modes of propagation while the third one is the transversal S–mode in the skeleton. There exists no transversal mode in the fluid: $\boldsymbol{A}_\perp^F \equiv 0$.

The P2-mode is often called Biot's wave. Its theoretical existence is quite natural in the frame of any two–component continuous model even if both components are fluids (a miscible mixture). However, there are problems concerning the practical observation of its propagation if one of the components is solid. The P2-mode has been observed for the first time in an artificial porous material made of sintered glass beads by Plona [14], and in an artificial rock of cemented sand grains by Klimentos & McCann [11], but *in situ* measurements are extremely difficult to perform. The main reason for these difficulties is a very strong attenuation of P2-waves. We discuss this point in some detail further in Subsect. 2.3.

Let us mention that the partial stresses $\boldsymbol{T}^S, \boldsymbol{T}^F$ in the skeleton and in the fluid, respectively, which lead to the above used field equations are not coupled if the constant β is equal to zero. Such a coupling, even though of a different – static – nature, is required in Biot's model commonly used in the wave analysis for porous saturated materials. In the notation of this work such a coupling has the form

$$
\begin{aligned}
\boldsymbol{T}^S &= \lambda^S \left(\operatorname{tr} \boldsymbol{e}^S\right) \mathbf{1} + 2\mu^S \boldsymbol{e}^S - Q \frac{\rho^F - \rho_0^F}{\rho_0^F} \mathbf{1}, \\
\boldsymbol{T}^F &= -\left(\kappa \left(\rho^F - \rho_0^F\right) - Q \operatorname{tr} \boldsymbol{e}^S\right) \mathbf{1},
\end{aligned}
\tag{15}
$$

where Q is Biot's *coupling constant.* Such a model is thermodynamically admissible solely in the case of an additional contribution of the gradient of porosity to the momentum balance equations (2), (3), see [13],

$$
\hat{\boldsymbol{p}} = \pi \left(\boldsymbol{v}^F - \boldsymbol{v}^S\right) - Q \operatorname{grad} n. \tag{16}
$$

In such a case it can be easily shown that the coefficient Q which would give rise to the off–diagonal terms in the eigenvalue problem (13) has an order of magnitude of the pore pressure, i.e. 10^5 Pa in soils and rocks. This must be compared with the elastic constants $\lambda^S, \mu^S, \kappa\rho_0^F$ which are at least of the order 10^8 Pa. Hence, similarly to the assumption that $\beta = 0$, we can leave out this correction in the wave analysis.

The above results do not reveal the attenuation of the waves because the behaviour of amplitudes cannot be determined from the properties of the field equations on the wave front alone. In order to see such effects we have to construct solutions to the field equations. We proceed to do so for monochromatic waves in infinite domains.

2.3 Monochromatic Waves in Two-Component Media

General relations We seek solutions of (1)–(4) in the form of bulk monochromatic waves defined by the following ansatz for harmonic waves:

$$
\begin{aligned}
\rho^F - \rho_0^F &= R^F e^{i(k\boldsymbol{n}\cdot\boldsymbol{x} - \omega t)}, \quad & \boldsymbol{e}^S &= \boldsymbol{E}^S e^{i(k\boldsymbol{n}\cdot\boldsymbol{x} - \omega t)}, \\
\boldsymbol{v}^F &= \boldsymbol{V}^F e^{i(k\boldsymbol{n}\cdot\boldsymbol{x} - \omega t)}, \quad & \boldsymbol{v}^S &= \boldsymbol{V}^S e^{i(k\boldsymbol{n}\cdot\boldsymbol{x} - \omega t)},
\end{aligned}
\tag{17}
$$

where $R^F, \boldsymbol{E}^S, \boldsymbol{V}^F, \boldsymbol{V}^S$ are constant, possibly complex, amplitudes of the disturbance, $\boldsymbol{n}$ denotes the unit vector in the direction of propagation, k is the wavenumber, and ω the frequency of the wave. Both k and ω may be complex.

Straightforward calculations lead to the following compatibility relations with the field equations:

$$
R^F = \frac{k\rho_0^F}{\omega} \boldsymbol{V}^F \cdot \boldsymbol{n}, \quad \boldsymbol{E}^S = -\frac{k}{2\omega} \left(\boldsymbol{V}^S \otimes \boldsymbol{n} + \boldsymbol{n} \otimes \boldsymbol{V}^S\right), \tag{18}
$$

$$
\begin{aligned}
\left(\omega^2 \mathbf{1} - \frac{\lambda^S + \mu^S}{\rho_0^S} k^2 \boldsymbol{n} \otimes n - \frac{\mu^S}{\rho_0^S} k^2 \mathbf{1} + i \frac{\pi\omega}{\rho_0^S} \mathbf{1}\right) \boldsymbol{V}^S - i \frac{\pi\omega}{\rho_0^S} \boldsymbol{V}^F &= 0, \\
-i \frac{\pi\omega}{\rho_0^F} \boldsymbol{V}^S + \left(\omega^2 \mathbf{1} - \kappa k^2 \boldsymbol{n} \otimes n + i \frac{\pi\omega}{\rho_0^F} \mathbf{1}\right) \boldsymbol{V}^F &= 0.
\end{aligned}
\tag{19}
$$

Equations (19) form an eigenvalue problem with a six–dimensional eigenvector $\left(\boldsymbol{V}^S, \boldsymbol{V}^F\right)^T$, and ω^2 are the eigenvalues if k is given. We will consider also a modification of this problem, namely if ω is given.

We can easily separate the components in the direction of the vector $\boldsymbol{n}$, and in the direction perpendicular to this vector. We consider these problems in the subsequent two paragraphs.

Longitudinal Modes of Propagation Scalar multiplication of (19) by the vector $\boldsymbol{n}$ yields

$$\begin{pmatrix} \omega^2 - \dfrac{\lambda^S + 2\mu^S}{\rho_0^S} k^2 + i\dfrac{\pi\omega}{\rho_0^S} & -i\dfrac{\pi\omega}{\rho_0^S} \\ -i\dfrac{\pi\omega}{\rho_0^F} & \omega^2 - \kappa k^2 + i\dfrac{\pi\omega}{\rho_0^F} \end{pmatrix} \begin{pmatrix} \boldsymbol{V}^S \cdot \boldsymbol{n} \\ \boldsymbol{V}^F \cdot \boldsymbol{n} \end{pmatrix} = 0. \tag{20}$$

This two–dimensional eigenvalue problem yields immediately the *dispersion relation*

$$\left(\omega^2 - c_{P1} k^2 + i\frac{\pi\omega}{\rho_0^S}\right)\left(\omega^2 - c_{P2}^2 k^2 + i\frac{\pi\omega}{\rho_0^F}\right) + \frac{\pi^2\omega^2}{\rho_0^S \rho_0^F} = 0. \tag{21}$$

We consider two cases.

1. The frequency ω is real and given. This corresponds to the problem of a harmonic excitation with a given frequency ("boundary value problem").
2. The wavenumber k is real and given. This corresponds to an external impact ("initial value problem", chopping, explosion).

In the first case, (21) can be easily solved for k and we obtain

$$\begin{aligned} k^2 &= \frac{1}{2}\left[\frac{1}{c_{P1}^2}\left(\omega^2 + i\frac{\pi\omega}{\rho_0^S}\right) + \frac{1}{c_{P2}^2}\left(\omega^2 + i\frac{\pi\omega}{\rho_0^F}\right) \pm \sqrt{D}\right], \\ D &:= \left[\frac{1}{c_{P1}^2}\left(\omega^2 + i\frac{\pi\omega}{\rho_0^S}\right) - \frac{1}{c_{P2}^2}\left(\omega^2 + i\frac{\pi\omega}{\rho_0^F}\right)\right]^2 - \frac{4}{c_{P1}^2 c_{P2}^2}\frac{\pi^2\omega^2}{\rho_0^S\rho_0^F}. \end{aligned} \tag{22}$$

The solution with the plus sign corresponds to the longitudinal P1-wave. Correspondingly, the minus sign yields the relation for the P2-wave. In the limit of low and high frequencies one obtains easily

$$\begin{aligned} \text{P1}: \quad & \lim_{\omega\to 0}\frac{\omega}{\operatorname{Re} k} = \sqrt{\frac{\lambda^S + 2\mu^S + \kappa\rho_0^F}{\rho_0^S + \rho_0^F}} =: c_{oP1}, \\ & \lim_{\omega\to\infty}\frac{\omega}{\operatorname{Re} k} = \sqrt{\frac{\lambda^S + 2\mu^S}{\rho_0^S}} \equiv c_{P1}, \\ \text{P2}: \quad & \lim_{\omega\to 0}\frac{\omega}{\operatorname{Re} k} = 0, \quad \lim_{\omega\to\infty}\frac{\omega}{\operatorname{Re} k} = \sqrt{\kappa} \equiv c_{P2}. \end{aligned} \tag{23}$$

These limits were obtained also within Biot's model. Obviously the additional coupling appearing in this model does not influence the result.

In the next two figures we illustrate these results for the following numerical data

$$c_{P1} = 2500\frac{\text{m}}{\text{s}}, \quad c_{P2} = 1000\frac{\text{m}}{\text{s}}, \quad \rho_0^S = 2500\frac{\text{kg}}{\text{m}^3}, \quad \rho_0^F = 250\frac{\text{kg}}{\text{m}^3}. \tag{24}$$

In Fig. 1 we plot the phase velocity $c_{ph} = \dfrac{\omega}{\mathrm{Re}\, k}$ of both longitudinal modes, and in Fig. 2 the attenuation $\gamma = \mathrm{Im}\, k$ is displayed.

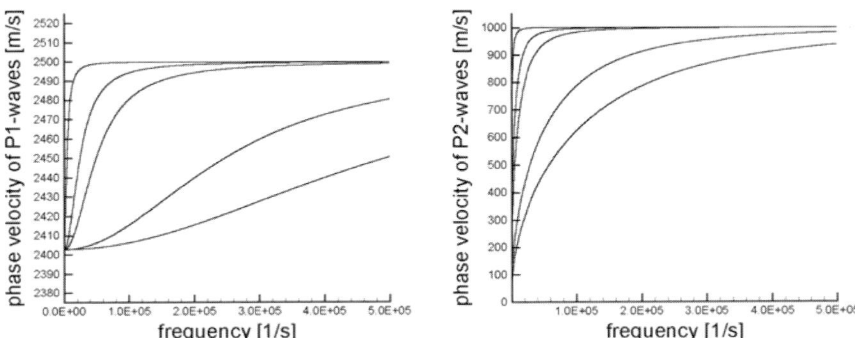

Fig. 1. Phase speed of P1- (left), and P2-waves (right) as functions of frequency ω. The curves correspond to the permeabilities π (from top to bottom): $10^6, 5 * 10^6, 10^7, 5 * 10^7, 10^8$ [kg/(m^{3}s)]

Inspection of Fig. 1 shows that both modes of propagation exist for any frequency of the excitation. The phase speed of P1-waves grows a little from its initial value to the asymptotic speed c_{P1} for $\omega \to \infty$. On the other hand the phase speed of P2-waves is equal to zero for $\omega = 0$ and grows asymptotically to the limit c_{P2} for $\omega \to \infty$. For both modes the growth becomes slower for larger permeability coefficients π which we demonstrate for practically reasonable values of this coefficient.

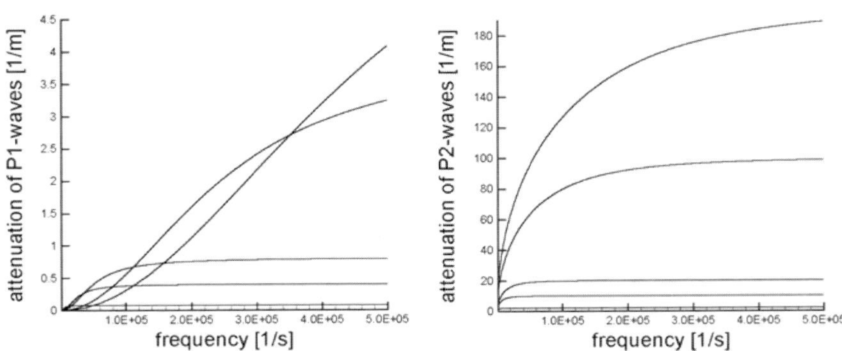

Fig. 2. Attenuation of P1- (left), and P2-waves (right) as functions of frequency ω. The same values of permeability π as in Fig. 1 are used, however, here they are growing from the bottom to the top

It is clear from Fig. 2 that the attenuation of P2-waves is much stronger than the one of P1-waves. This observation justifies the remark made in the Introduction that the strong attenuation of P2-waves causes difficulties in their *in situ* measurements.

The above described properties of monochromatic waves have been discussed in detail in earlier works on this model of poroelastic materials (e.g. [22–26]).

We proceed to present properties of the second case – external impact (initial value problem, chopping). In this case, the wavenumber k is given and real, and the frequency ω is complex. It follows as the solution of the dispersion relation (21). This solution cannot be obtained analytically. Asymptotic analysis for large and small wave-numbers has been performed by Edelman [7], where the existence of the critical wavenumber k_{cr} for the P2-wave has been proven analytically. We present here a few typical numerical examples and use the data given in (24).

In contrast to the above discussed boundary value problem, P2-waves may not exist in the case of the initial value problem. For any chosen real wave number k solutions of the dispersion relation (21) consist of four complex values of ω, which are symmetric with respect to zero. Consequently there are two essential real parts of ω which determine the P1-, and P2-mode. In Fig. 3 we show the real part of ω corresponding to the P2-mode for different values of the permeability coefficient π.

It is seen that for sufficiently small wavenumbers k (i.e. long waves) the real part of ω is equal to zero. Consequently in these ranges the P2-modes contain

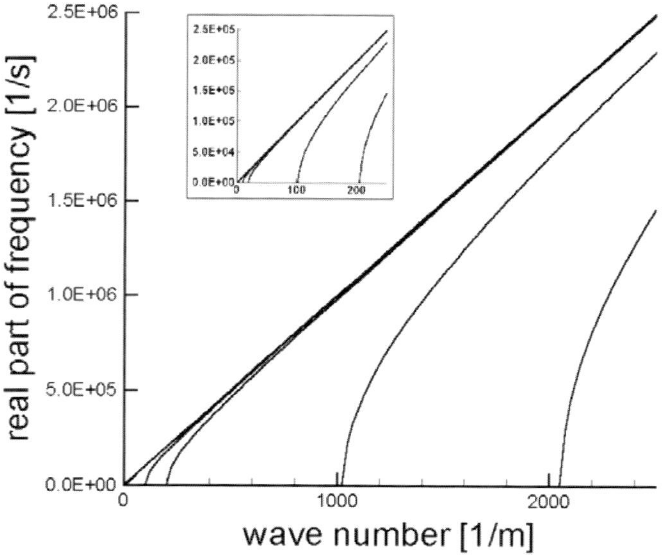

Fig. 3. Real part of the frequency as a function of the wave number for P2-waves. In the figure the curves for $\pi = 10^6$ to 10^9 $[\mathrm{kg}/(\mathrm{m}^3\mathrm{s})]$ are shown. The inlet is the magnification of the large figure for the following values of permeability π : $10^6, 5 * 10^6, 10^7, 5 * 10^7, 10^8$ $[\mathrm{kg}/(\mathrm{m}^3\mathrm{s})]$ growing from the left to the right

only damping and they cannot propagate as waves. The extent of the plateau of the constant real part of frequency changes approximately linearly with π and, for instance, for $\pi = 10^9$ [kg/(m^{3}s)] (see Fig. 3) it reaches the value $k \approx 2050$ [1/m], which corresponds to the wave length ~ 0.05 cm. Obviously from the physical point of view, the P2-wave does no longer exist because the wave length would have to be smaller than the characteristic dimension of the microstructure. However, the minimum length of the wave for smaller permeabilities lies in the physically reasonable range. For instance, for $\pi = 10^7$ [kg/(m^{3}s)] it is approximately 5 cm (see Fig. 3, inlet).

The problem of existence of propagation does not concern the P1-mode. These waves behave similarly to those of the boundary value problem. In Fig. 4 we show their phase speeds for the data (24). The speed grows with k a little and reaches the limit value c_{P1} for $k \to \infty$. As indicated above, the P2-waves do not propagate below a critical value of k which changes with π. We show this behaviour in Fig. 5. In the range of large values of k, the P2-modes propagate and reach the limit value c_{P2} for $k \to \infty$.

Imaginary parts of the frequency ω determine the damping of waves. This attenuation in time behaves differently from the attenuation in space discussed in the first case. For P1-waves (Fig. 6) it grows with increasing wavenumber k (i.e. with the decay of the wave length). However in the range of long waves the damping in media with larger permeability π is smaller than it is for media with smaller permeability. Most likely this is related to the fact that the energy of the wave created by the impact remains longer in the vicinity of the impact if the value of π is larger which, as seen in Fig. 4, yields a lower speed of propagation. The behaviour of the P2-modes is entirely different due to the existence of plateaus. The ranges of these plateaus are visible also in Fig. 7 which illustrates the attenuation of P2-modes. For any value of permeability π the range of small values of k contains solely damping – the frequency ω is imaginary. For larger values of k we see the attenuation of P2-waves. As for the boundary value problem it is much stronger than in the case of P1-waves.

The above described properties of initial value problems have an important influence on the construction of asymptotic solutions in the range of low frequencies. For instance, they lead to an entirely different structure of surface waves than for high frequencies [6,9]. Asymptotic analysis of this problem has been performed by Edelman [8]. We shall discuss some aspects of this problem in Sect. 3 of this work.

Transversal Modes of Propagation Let us introduce the following quantities:

$$\mathbf{V}_{\perp}^{F} := \mathbf{V}^{F} - \left(\mathbf{V}^{F} \cdot \mathbf{n}\right) \mathbf{V}^{F}, \quad \mathbf{V}_{\perp}^{S} := \mathbf{V}^{S} - \left(\mathbf{V}^{S} \cdot \mathbf{n}\right) \mathbf{V}^{S}. \tag{25}$$

Then, from (19), for arbitrary components of the above vectors $V_{\perp}^{F} := \mathbf{V}_{\perp}^{F} \cdot \mathbf{t}$, $V_{\perp}^{S} := \mathbf{V}_{\perp}^{S} \cdot \mathbf{t}$, with $\mathbf{t}$ being any unit vector perpendicular to $\mathbf{n}$,we obtain

$$\begin{pmatrix} \omega^2 - \dfrac{\mu^S}{\rho_0^S} k^2 + i\dfrac{\pi\omega}{\rho_0^S} & -i\dfrac{\pi\omega}{\rho_0^S} \\ -i\dfrac{\pi\omega}{\rho_0^F} & \omega^2 + i\dfrac{\pi\omega}{\rho_0^F} \end{pmatrix} \begin{pmatrix} V_{\perp}^{S} \\ V_{\perp}^{F} \end{pmatrix} = 0. \tag{26}$$

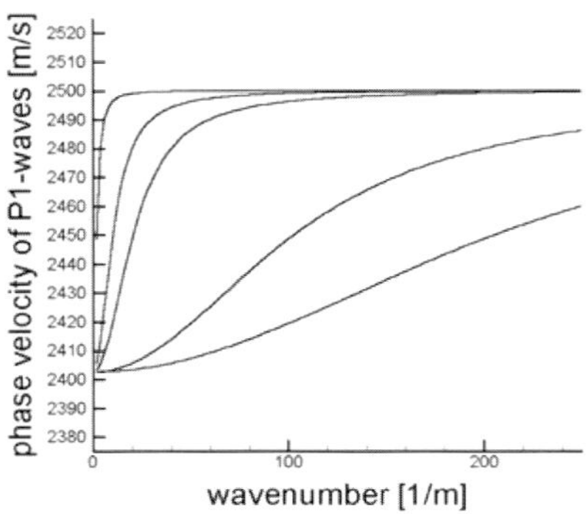

Fig. 4. Phase velocity of P1-waves for permeabilities π (from the left to the right): $10^6, 5*10^6, 10^7, 5*10^7, 10^8$ $[\mathrm{kg}/(\mathrm{m}^3\mathrm{s})]$

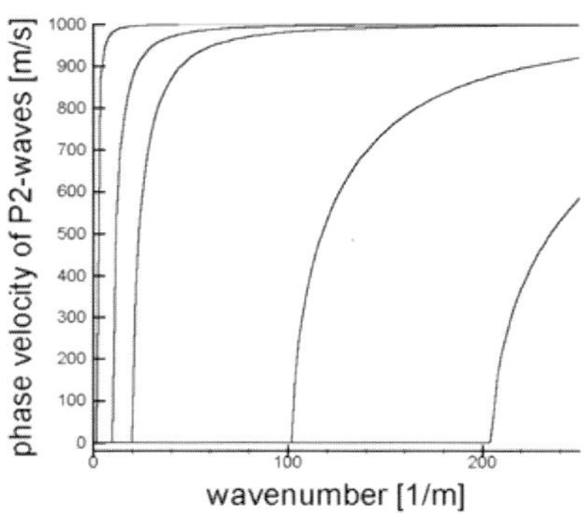

Fig. 5. Phase velocity of P2-waves for permeabilities π (from the left to the right): $10^6, 5*10^6, 10^7, 5*10^7, 10^8$ $[\mathrm{kg}/(\mathrm{m}^3\mathrm{s})]$

This is again an eigenvalue problem which yields the dispersion relation

$$\omega^3 + i\pi\left(\frac{1}{\rho_0^S} + \frac{1}{\rho_0^F}\right)\omega^2 - c_S^2 k^2 \omega - i c_S^2 k^2 \frac{\pi}{\rho_0^F} = 0. \tag{27}$$

As before we calculate limit speeds for high and low frequencies and obtain

$$\lim_{\omega\to 0}\frac{\omega}{\operatorname{Re} k} = \sqrt{\frac{\mu^S}{\rho_0^S + \rho_0^F}} =: c_{oS}, \quad \lim_{\omega\to\infty}\frac{\omega}{\operatorname{Re} k} = \sqrt{\frac{\mu^S}{\rho_0^S}} \equiv c_S. \tag{28}$$

These are relations commonly used in geophysical applications.

We illustrate the solutions of (26) in Figs. 8 and 9 for the data

$$c_S = 1500\,\frac{\mathrm{m}}{\mathrm{s}}, \quad \rho_0^S = 2500\,\frac{\mathrm{kg}}{\mathrm{m}^3}, \quad \rho_0^F = 250\,\frac{\mathrm{kg}}{\mathrm{m}^3}. \tag{29}$$

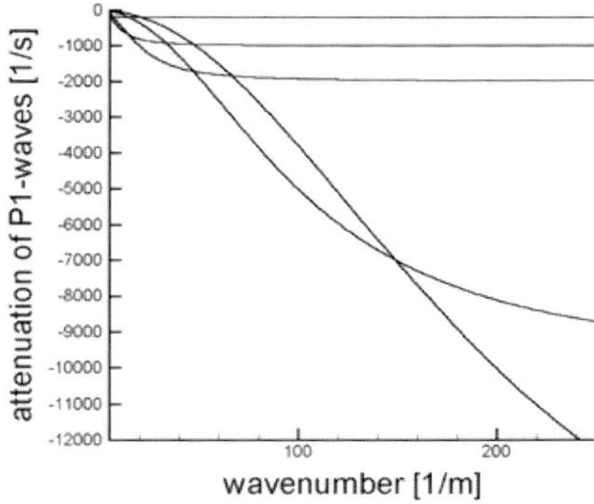

Fig. 6. Attenuation of P1-waves for permeabilities π: 10^6 (the smallest attenuation), $5 * 10^6, 10^7, 5 * 10^7, 10^8$ (the largest attenuation) $[\mathrm{kg}/(\mathrm{m}^3\mathrm{s})]$

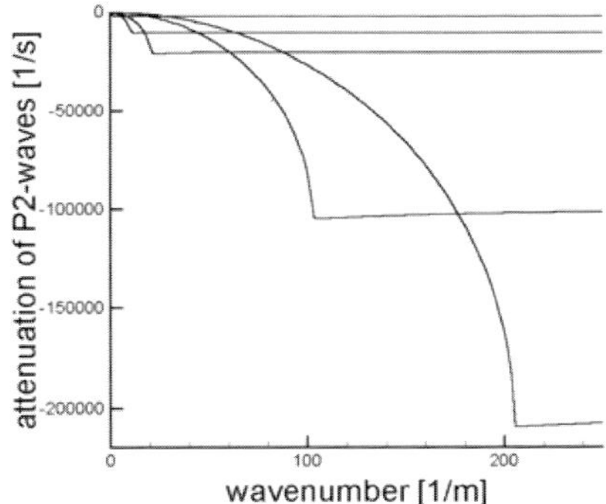

Fig. 7. Attenuation of P2-waves for permeabilities π: 10^6 (the smallest attenuation), $5 * 10^6, 10^7, 5 * 10^7, 10^8$ (the largest attenuation) $[\mathrm{kg}/(\mathrm{m}^3\mathrm{s})]$

For the phase speed we obtain a behaviour quite similar to the one of P1-waves. After the initial growth, the phase speed tends to the limit value c_S for $k \to \infty$. The behaviour of the attenuation is also similar to that of P1-waves. This is shown in Fig. 9.

Group Velocities Propagation of waves with dispersion leads to the effect of the propagation of packages of waves of frequencies from a certain interval in the form of an envelope of harmonic waves whose speed is different from a phase speed of any of the waves belonging to such a package. The speed of a package is called the *group velocity*, and it is locally defined as the derivative of the frequency with respect to the wavenumber. It is well known (see e.g. [18]) that measurements of

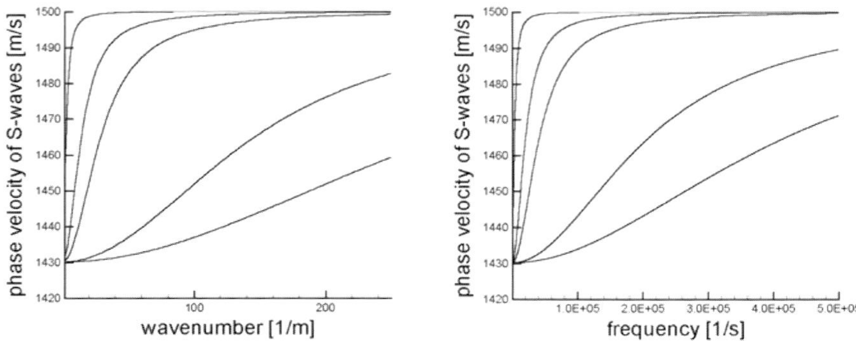

Fig. 8. Phase speed of S–waves for the permeabilities π : $10^6, 5*10^6, 10^7, 5*10^7, 10^8$ [kg/(m^{3}s)]. The left diagram corresponds to the initial value problem while the right diagram to boundary vibrations. The upper curve corresponds to the lowest permeability

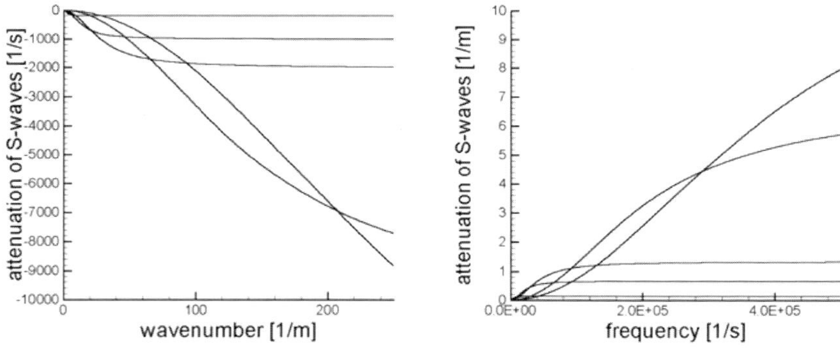

Fig. 9. Attenuation of S–waves for the permeabilities π: $10^6, 5*10^6, 10^7, 5*10^7, 10^8$ [kg/(m^{3}s)]. The left diagram corresponds to the initial value problem while the right diagram to boundary vibrations. The upper curve corresponds to the lowest permeability

speeds may give either a phase speed or a group speed depending on the way of excitation. Therefore it is essential to know the difference between both speeds for an arbitrary frequency.

In the case considered in this article, the group velocity for the boundary value problem (a source vibrating with a given frequency ω) follows by differentiation of (22) with respect to ω as

$$c_{gr} = \operatorname{Re}\left(\frac{dk}{d\omega}\right)^{-1}. \tag{30}$$

The group velocity for the initial value problem (chopping) follows from the dispersion relation by differentiation with respect to k. One obtains easily

$$c_{gr} := \operatorname{Re}\left(\frac{d\omega}{dk}\right) = \operatorname{Re}\left\{2\frac{\omega}{k}\frac{c_{P1}^2 G_2 + c_{P2}^2 G_1}{\left(\frac{\omega^2}{k^2} + c_{P1}^2\right) G_2 + \left(\frac{\omega^2}{k^2} + c_{P2}^2\right) G_1}\right\}, \tag{31}$$

where

$$G_1 := \frac{\omega^2}{k^2} - c_{P1}^2 + i\frac{\pi}{\rho_0^S}\frac{\omega}{k}, \qquad G_2 := \frac{\omega^2}{k^2} - c_{P2}^2 + i\frac{\pi}{\rho_0^F}\frac{\omega}{k}, \tag{32}$$

and the frequency ω is a function of the wavenumber k determined by the dispersion relation (21). In the present notation this relation has the form

$$G_1 G_2 + \frac{\pi^2}{\rho_0^S \rho_0^F k^2}\frac{\omega^2}{k^2} = 0. \tag{33}$$

In Fig. 10 we show in juxtaposition the behaviour of the group velocities for the boundary value problem (left), and for the initial value problem (right). We used the data (24) and made the calculation for a single but representative value of the permeability coefficient $\pi = 10^7 \mathrm{kg}/(\mathrm{m}^3\mathrm{s})$.

The behaviour of the group velocity for the boundary value problem is, as expected, smooth and it approaches the limit of c_{P1}, and c_{P2}, respectively, as $\omega \to \infty$. It exceeds the limit values a little for medium frequencies because the dispersion curves are in this range convex functions of the frequency.

The situation changes dramatically for the group velocity of P2-waves initiated by the initial impact in the range of wave-numbers higher than a critical value (app. 20.556 1/m in the numerical example). After the plateau of the zero velocity (right panel of Fig. 10) the group velocity decays from an *infinite value* to the c_{P2} limit. This infinite critical value is related to the fact that the corresponding dispersion relation (Fig. 3) possesses a vertical tangent in this point. This is a behaviour which is characteristic for *parabolic problems*. Our model is fully hyperbolic but its limit behaviour for P2-waves reminds this of the Darcy's parabolic model of diffusion in which inertial effects in the fluid component are neglected. In this sense the Darcy's model may be considered to be a low frequency approximation of the hyperbolic two-component model.

The above described properties of P2-waves for initial impact have important consequences for the existence of surface waves. We discuss this problem in the next section.

Conclusions for Bulk Waves in Two-Component Media Results presented in the above section show that the simplest possible model of saturated poroelastic materials yields qualitatively the same properties of wave motion as the more sophisticated Biot model. However, in contrast to the latter the model used in this work does not contradict any principal rules of modern continuum thermodynamics. In addition notions such as tortuosity, anisotropic permeability, etc. which may be essential in some practical applications, are not needed in the construction

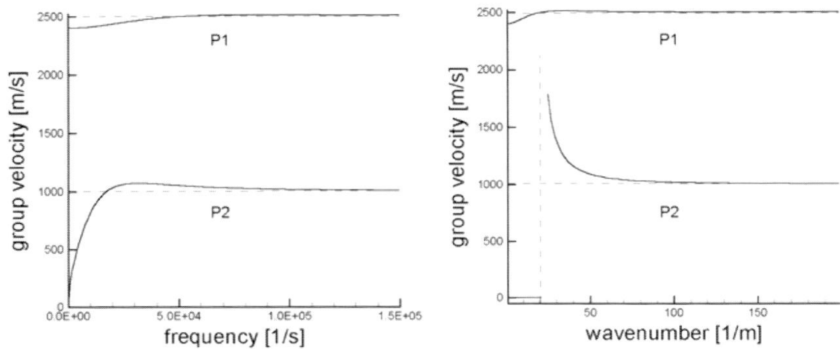

Fig. 10. Group velocities of P1-, and P2-waves for $\pi = 10^7$ [kg/(m^{3}s)]. The left figure corresponds to the boundary value problem while the right one to the initial value problem. The critical value of the wave number equals $k = 20.556$ [1/m]

of all important bulk modes of propagation in spite of claims in the literature on Biot's model.

As the analysis of monochromatic waves shows, the asymptotic behaviour for high frequencies checks with the expectations following from the analysis of singularities of fields. This is independent of the fact whether one controls the propagation by harmonic excitations on the boundary (by a given real frequency ω) or whether one controls an initial condition in which a wave of a particular length (by prescribing a real wavenumber k) is excited.

However, the situation changes if the low frequency limit is considered. This limit is smooth independently of the external control for the classical two modes of propagation – P1- and S-waves. Both wave types have finite phase speeds for $\omega \to 0$ and these are a bit smaller than the speeds of propagation of the corresponding fronts. This is, however, not the case for the P2-mode. This mode behaves like a wave for harmonic excitations on the boundary. The phase speed of this wave goes to zero as $\omega \to 0$. In the vicinity of the zero frequency, it has approximately parabolic character. The behaviour changes entirely in the case of an initial value problem. In the vicinity of the zero wavenumber k (infinitely long waves), the P2-mode has zero phase velocity and it is solely damped. After a plateau of zero velocity whose length depends on the value of the permeability coefficient π, this mode behaves again as a wave. In the limit of high frequencies (short waves) this behaviour is the same as the one of P2-waves excited by harmonic vibrations.

Such behaviour has very important practical bearings. The lack of observation of P2-waves in *in situ* measurements may be related not only to the high attenuation of P2-waves, but also to the nonexistence of these waves for low frequency initial excitations. It is also very important in the analysis of surface waves in the range of low frequencies commonly used in geophysical applications. We will return to this question in Sect. 3.

Let us finally mention that the attenuation properties of all modes are caused by the relative motion of components reflected by the permeability coefficient π.

As the examples presented above clearly show, these properties agree well with the expectations.

3 Surface Waves

3.1 Surface Waves in Single Component Media

Introduction In contrast to bulk waves, surface waves propagate along surfaces and their penetration in the direction perpendicular to the surface decays so fast that amplitudes of disturbances can be assumed to be zero at a depth of a few wave lengths. Consequently, their geometric dispersion is determined in two-dimensional rather than in three-dimensional space, as is the case for bulk waves. Hence, for a point source, the amplitude of surface waves decays as r^2 (and not as r^3 as for bulk waves), where r is the distance from the source. This property is the main reason for destructive actions of surface waves in earthquakes and, simultaneously, it is the reason for their importance in nondestructive testing of soils. The latter means that one can investigate properties of soils to a depth of approximately twice the wave length without the necessity to drill boreholes and to investigate laboratory samples distorted by boring and transport.

In the following subsections, we present some properties of surface waves for single-component heterogeneous elastic media as well as surface waves in two-component porous materials.

Single-component models apply to processes in porous materials in which a relative motion of components is not essential. Examples are quasi-static geophysical processes as well as far-field properties of waves in which an influence of P2-waves no longer appears. An extensive presentation of wave properties under such conditions can be found in books on seismology of the earth (e.g. [1,2,17]).

In the case of surface waves in two-component materials, we consider separately two types of boundary conditions for a far-field approximation of a harmonic boundary source. Finally, we review some properties of surface waves in two-component porous materials initiated by an impact.

Rayleigh Waves in Homogeneous Elastic Materials We begin with a brief reminder of the classical Rayleigh problem. It is a two-dimensional dynamical solution of the boundary value problem for a semi-infinite elastic body described by the equations

$$\rho \frac{\partial^2 \boldsymbol{u}}{\partial t^2} = \operatorname{div} \boldsymbol{T}, \quad \boldsymbol{T} = \lambda \left(\operatorname{tr} \boldsymbol{e} \right) \mathbf{1} + 2\mu \boldsymbol{e}, \quad \boldsymbol{e} := \operatorname{sym} \operatorname{grad} \boldsymbol{u}, \tag{34}$$

where ρ is a constant mass density, and λ, μ denote the Lamé constants.

It is known that the decomposition of the displacement vector $\boldsymbol{u}$ into potential and solenoidal parts, $\boldsymbol{u} = \boldsymbol{u}_L + \boldsymbol{u}_T$, yields the equivalent set of equations

$$\begin{aligned} \frac{\partial^2 \boldsymbol{u}_L}{\partial t^2} &= c_L^2 \Delta \boldsymbol{u}_L, \quad \operatorname{curl} \boldsymbol{u}_L = 0, \quad c_L := \sqrt{\frac{\lambda + 2\mu}{\rho}}, \\ \frac{\partial^2 \boldsymbol{u}_T}{\partial t^2} &= c_T^2 \Delta \boldsymbol{u}_T, \quad \operatorname{div} \boldsymbol{u}_T = 0, \quad c_T := \sqrt{\frac{\mu}{\rho}}, \end{aligned} \tag{35}$$

where $\triangle$ is the Laplace operator. Each of these equations describes a bulk wave in the infinite medium – the first one is a longitudinal wave (called a P-wave in geophysics), and the second one a transversal wave (called S-wave in geophysics).

We seek the solution of the following boundary value problem:

$$\boldsymbol{T}\boldsymbol{n}\big|_{z=0} = 0, \quad \boldsymbol{n} \equiv -\boldsymbol{e}_3, \quad \boldsymbol{u}\big|_{z\to\infty} = 0. \tag{36}$$

The direction $\boldsymbol{n}$ coincides with the negative direction of the z-axis.

We make the following ansatz

$$\begin{aligned} \boldsymbol{u}_L &= A_L e^{-\gamma z} e^{i(kx-\omega t)} \boldsymbol{e}_1 + B_L e^{-\gamma z} e^{i(kx-\omega t)} \boldsymbol{e}_3, \\ \boldsymbol{u}_T &= A_T e^{-\beta z} e^{i(kx-\omega t)} \boldsymbol{e}_1 + B_T e^{-\beta z} e^{i(kx-\omega t)} \boldsymbol{e}_3, \end{aligned} \tag{37}$$

where $A_L, B_L, A_T, B_T, \gamma, \beta, k, \omega$ are constants. Substitution in $(35)_{2,4}$ yields the compatibility conditions

$$\frac{\gamma^2}{k^2} = 1 - \frac{c_R^2}{c_L^2}, \quad \frac{\beta^2}{k^2} = 1 - \frac{c_R^2}{c_T^2}, \quad c_R := \frac{\omega}{k}. \tag{38}$$

Simultaneously, substitution of (37) in $(35)_{1,3}$ leads to the following form of the solution

$$\begin{aligned} B_L &= i\frac{\gamma}{k} A_L \quad \Rightarrow \quad \boldsymbol{u}_L = \left(\boldsymbol{e}_1 + i\frac{\gamma}{k}\boldsymbol{e}_3\right) A_L e^{-\gamma z} e^{i(kx-\omega t)}, \\ B_T &= i\frac{k}{\beta} A_T \quad \Rightarrow \quad \boldsymbol{u}_T = \left(\boldsymbol{e}_1 + i\frac{k}{\beta}\boldsymbol{e}_3\right) A_T e^{-\beta z} e^{i(kx-\omega t)}. \end{aligned} \tag{39}$$

There remains to exploit the boundary conditions $(36)_1$. The second condition is identically satisfied provided the constants γ, β are chosen to be positive. According to $(34)_2$, the boundary conditions for stresses can be written in the form

$$\begin{aligned} \left(c_L^2 - 2c_T^2\right) \frac{\partial \boldsymbol{u}\cdot\boldsymbol{e}_1}{\partial x} + c_L^2 \frac{\partial \boldsymbol{u}\cdot\boldsymbol{e}_3}{\partial z}\bigg|_{z=0} &= 0, \\ \frac{\partial \boldsymbol{u}\cdot\boldsymbol{e}_1}{\partial z} + \frac{\partial \boldsymbol{u}\cdot\boldsymbol{e}_3}{\partial x}\bigg|_{z=0} &= 0. \end{aligned} \tag{40}$$

Finally, substitution of (39) in the above conditions leads to a homogeneous set of equations for the constants A_L, A_T. Consequently, the determinant of this set should be zero, and this gives rise to

$$\mathcal{P}_R := \left(2 - \frac{c_R^2}{c_T^2}\right)^2 - 4\sqrt{1 - \frac{c_R^2}{c_T^2}}\sqrt{1 - \frac{c_R^2}{c_L^2}} = 0. \tag{41}$$

This equation determines the phase speed $c_R = \omega/k$ of the wave described by (39). It is clear that this solution is independent of the choice of the frequency ω. Hence these waves are nondispersive. Their amplitudes decay with depth z in an exponential way. For this reason they are called *surface waves.* They have been discovered by Rayleigh. It can also be shown that (41) possesses a single real positive solution $c_R < c_T$.

The above solution is not the only surface wave solution of classical elasticity. In the next paragraph we show another one discovered by Love.

Waves in a Layer of an Ideal Fluid and Love Waves In order to appreciate the influence of heterogeneities on the propagation of surface waves, we investigate first a simple example of a layer of an *ideal fluid* $-\infty < x < \infty$, $0 \leq z \leq H$. The upper surface $z = H$ is assumed free of loading and the lower surface $z = 0$ is in contact with a *rigid body*. The problem is described by the equations of mass and momentum conservation

$$\frac{\partial \rho}{\partial t} + \rho_0 \operatorname{div} \boldsymbol{v} = 0, \quad \rho_0 \frac{\partial \boldsymbol{v}}{\partial t} = -\operatorname{grad} p, \quad p = p_0 + \kappa (\rho - \rho_0), \tag{42}$$

where ρ_0, p_0 are reference constant values of the mass density and pressure, respectively, and κ denotes a constant compressibility coefficient of the fluid. Simple manipulations lead to the following wave equation for the pressure p,

$$\frac{\partial^2 p}{\partial t^2} = \kappa \Delta p, \quad (x, z) \in (-\infty, \infty) \times (0, H), \tag{43}$$

The solution of (43) must satisfy the boundary conditions:

$$p(x, z = H, t) = 0, \quad v_z (x, z = 0, t) = 0. \tag{44}$$

We seek a monochromatic wave solution of the frequency ω,

$$p = \left(A e^{irkz} + B e^{-irkz} \right) e^{i(kx - \omega t)}. \tag{45}$$

Then the second boundary condition can be replaced by

$$\frac{\partial p}{\partial z} (x, z = 0, t) = 0. \tag{46}$$

Substitution of (45) in (43) yields the compatibility relation

$$r^2 = \frac{c_{ph}^2}{c^2} - 1, \quad c_{ph} := \frac{\omega}{k}, \quad c := \sqrt{\kappa}. \tag{47}$$

Simultaneously, evaluation of the boundary conditions with the ansatz (45) yields the set of homogeneous algebraic relations for the constants A and B, namely

$$A e^{irkH} + B e^{-irkH} = 0, \qquad A - B = 0. \tag{48}$$

The determinant of this set must be equal to zero, so

$$\cos(rkH) = 0. \tag{49}$$

For nontrivial solutions r must be real. This means, however, that the phase velocities c_{ph} are bigger than the speed of propagation c appearing in the wave equation for the pressure (43)($\kappa \equiv c^2$). If we require that waves of the form (45) do exist, then this seems to violate the basic property of the hyperbolic problem. This result is a consequence of the assumption that the foundation of the fluid is a *rigid body* in which all disturbances propagate with an *infinite speed.* As we see further a modification of the boundary condition $(44)_2$ for the case of contact with an elastic body (which we make for the so-called Love waves) eliminates this paradox.

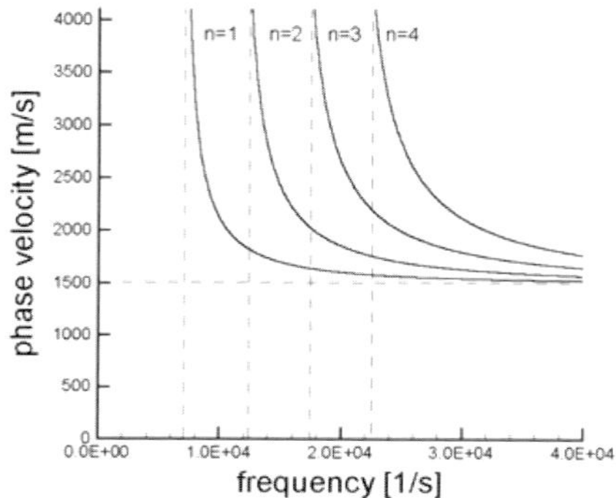

Fig. 11. Phase velocity for a layer of ideal fluid. Numerical data: $c = 1500$ m/s, $H = 1$ m. Modes $n = 1, 2, 3, 4$ are shown

Solution of (49) yields immediately the following relation between the phase speed and the frequency:

$$c_{ph} = \frac{c}{\sqrt{1 - \omega_{cr}^2/\omega^2}}, \quad \omega_{cr} := \left(n + \frac{1}{2}\right)\pi\frac{c}{H}, \quad n = 1, 2, \ldots \tag{50}$$

This relation is illustrated in Fig. 11.

The paradox of infinite phase speeds does no longer appear for surface waves which propagate in an elastic layer over an elastic half-space. Transversal waves in such a system have been described in 1911 by Love. We proceed to briefly present these results. They form the simplest illustration of the problem of surface waves in heterogeneous materials.

Consider the propagation of a wave whose amplitude has only an $\boldsymbol{e}_2$-component $u_2 \equiv \boldsymbol{u} \cdot \boldsymbol{e}_2$ (perpendicular to the (x, z)-plane). The body consists of a layer of thickness H in the z-direction in which the mass density is ρ' and the shear-wave speed is c'_T. This layer is connected to the elastic half-space $z \leq 0$ whose mass density and shear-wave speed are ρ and c_T, respectively.

We seek the solution of the wave equations

$$\begin{aligned} \frac{\partial^2 u'_2}{\partial t^2} &= c_T'^2 \Delta u'_2, \quad 0 < z < H, \\ \frac{\partial^2 u_2}{\partial t^2} &= c_T^2 \Delta u_2, \quad z < 0, \end{aligned} \tag{51}$$

in the form

$$u'_2 = \left(A' e^{iks'z} + B' e^{-iks'z}\right) e^{i(kx - \omega t)}, \quad u_2 = B e^{ksz} e^{i(kx - \omega t)}, \tag{52}$$

i.e. in a form of a monochromatic wave which propagates in the direction of the x-axis with frequency ω, wavenumber k in this direction, and with phase speed $c := \omega/k$. The wave should decay in the z-direction, i.e. s must be positive. We check now whether the ansatz (52) can fulfil (51) and the following boundary conditions:

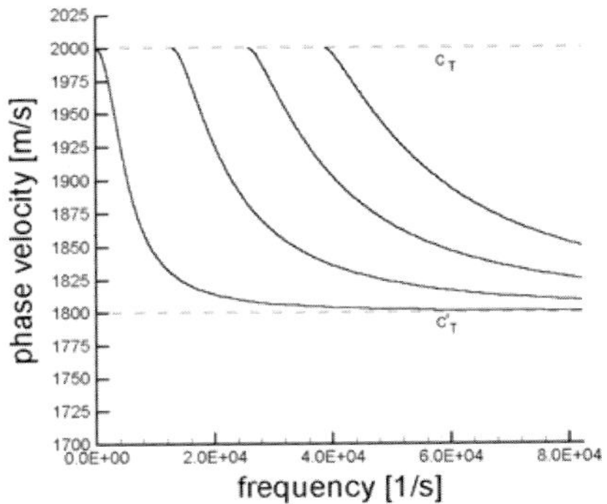

Fig. 12. Fundamental and three higher modes ($n = 1, 2, 3$) of the Love wave

1) shear stress on the plane $z = H$ is equal to zero, i.e.

$$\frac{\partial u_2'}{\partial z}(x, z = H, t) = 0, \tag{53}$$

2) shear stress and displacement must be continuous on the interface $z = 0$, i.e.

$$\begin{aligned} \rho' c_T' \frac{\partial u_2'}{\partial z}(x, z = 0, t) &= \rho c_T \frac{\partial u_2}{\partial z}(x, z = 0, t), \\ u_2'(x, z = 0, t) &= u_2(x, z = 0, t). \end{aligned} \tag{54}$$

Substitution of (52) in (51) yields

$$s'^2 = \frac{c^2}{c_T'^2} - 1, \quad s^2 = 1 - \frac{c^2}{c_T^2}, \quad c \equiv \frac{\omega}{k}. \tag{55}$$

The boundary conditions (54) lead to a homogeneous set of three algebraic relations for the constants A', B', B. Consequently, its determinant must be zero, and this condition yields

$$\omega = \frac{c}{Hs'}\left[\arctan\left(\frac{\rho c_T^2 s}{\rho' c_T'^2 s'}\right) + n\pi\right], \quad n = 1, 2, 3, \ldots, \tag{56}$$

and both s, and s' must be real, i.e.

$$c_T' \le c \le c_T. \tag{57}$$

This is the condition for the *existence* of Love waves. Hence, the Love waves can propagate solely in layers which are *softer than the foundation.* In addition there exist infinitely many modes of propagation whose existence is limited from below by a corresponding critical frequency. All these modes are dispersive because the phase speeds depend on the frequency given by the inverse relation to (56).

In Fig. 12 we show an example of the solution of relation (56) for the following data:

$$c_T = 2000\frac{\mathrm{m}}{\mathrm{s}}, \quad c_T' = 1800\frac{\mathrm{m}}{\mathrm{s}}, \quad \frac{\rho'}{\rho} = 0.8, \quad H = 1\mathrm{m}. \tag{58}$$

Surface Waves in Elastic Heterogeneous Materials Surface waves observed in geotechnics propagate always over a heterogeneous soil. In such a case not only Love waves but also Rayleigh waves possess multiple modes of propagation. They are described by equations following from the momentum conservation law in which one has to substitute Hooke's law with coefficients that are dependent on the depth z. Instead of (34), we then have

$$\rho \frac{\partial^2 \boldsymbol{u}}{\partial t^2} = \mu \triangle \boldsymbol{u} + (\lambda + \mu) \operatorname{grad} \operatorname{div} \boldsymbol{u} + \boldsymbol{e}_3 \frac{d\lambda}{dz} \operatorname{div} \boldsymbol{u} + \frac{d\mu}{dz} \left(\boldsymbol{e}_3 \times \operatorname{curl} \boldsymbol{u} + 2 \frac{\partial \boldsymbol{u}}{\partial z} \right), \quad (59)$$

where ρ, λ, μ are function of z and $\boldsymbol{e}_3$ is the unit vector perpendicular to the boundary (in the direction of the z-axis). A solution for harmonic waves is sought in the form

$$\boldsymbol{u} = \left(u_1 (z, k, \omega) \boldsymbol{e}_1 + u_3 (z, k, \omega) \boldsymbol{e}_3 \right) e^{i(kx - \omega t)}. \quad (60)$$

Substitution of this ansatz in (59) yields the following set of ordinary differential equations

$$\frac{d\boldsymbol{f}}{dz} = \boldsymbol{A}(z) \boldsymbol{f}, \quad \boldsymbol{f} \in \Re^4, \quad \mathbf{A} \in \Re^4 \times \Re^4, \quad (61)$$

where the vector $\boldsymbol{f}$ and the matrix $\boldsymbol{A}$ are defined as follows:

$$\begin{aligned} &\boldsymbol{f} := (u_1, u_3, f_3, f_4)^T, \\ &f_3 := \mu \left(\frac{du_1}{dz} + k u_3 \right), \quad f_4 := (\lambda + 2\mu) \frac{du_3}{dz} - k \lambda u_1, \end{aligned} \quad (62)$$

$$\boldsymbol{A} := \begin{pmatrix} 0 & -k & \mu^{-1} & 0 \\ k\lambda (\lambda + 2\mu)^{-1} & 0 & 0 & (\lambda + 2\mu)^{-1} \\ k^2 \zeta - \omega^2 \rho & 0 & 0 & -k\lambda (\lambda + 2\mu)^{-1} \\ 0 & -\omega^2 \rho & k & 0 \end{pmatrix},$$

$$\zeta := 4\mu \frac{\lambda + \mu}{\lambda + 2\mu}. \quad (63)$$

With this notation the shear stress, τ_{xz}, and the stress component normal to the boundary, σ_z, can be written in the form

$$\tau_{xz} = f_3 e^{i(kx - \omega t)}, \quad \sigma_z = f_4 e^{i(kx - \omega t)}. \quad (64)$$

The set of equations (61) defines a linear differential eigenvalue problem with eigenfunctions $\boldsymbol{f}$. Boundary conditions associated with this problem follow from the requirement that the stress components (64) vanish for $z = 0$, and the eigenvector $\boldsymbol{f}$ vanishes as $z \to \infty$.

Nontrivial solutions of this eigenvalue problem for a given frequency ω exist only for some values of the wavenumber, k, say k_j, $j = 1, \ldots, M$ which are called eigenvalues of the problem. The relation between the frequency and eigenvalues is only known in form of an implicit function

$$\mathcal{D}_R \left(\lambda (z), \mu (z), \rho (z), k_j, \omega \right) = 0; \quad (65)$$

it is called the Rayleigh dispersion relation. Solutions of this relation are real, which means that Rayleigh waves in heterogeneous materials are attenuated. In addition they depend on the frequency which means that Rayleigh waves in heterogeneous materials are dispersive.

Several numerical techniques have been developed to solve the above eigenvalue problem. The most popular and successful is most likely the so-called propagator-matrix method (Thomson-Haskell algorithm). It has been introduced by Kausel as a thin-layer method of discretising the problem in the z-direction (multilayer system). We shall not present here any details of these techniques, and refer to Lai [12] for their presentation and corresponding references.

3.2 Surface Waves in Two-Component Poroelastic Materials

The theory of surface waves in two-component systems differs qualitatively from that for one-component continua. Such waves are produced in linear models by a combination of bulk waves. For a one-component continuum there are two bulk modes of propagation which yield a single Rayleigh wave. For two-component systems we have three bulk modes: P1-waves, P2-waves and S-waves which produce two surface modes in case of an impermeable boundary. For a permeable boundary, i.e. an additional system such as the presence of a fluid in the exterior, there may exist three surface modes, etc.

In this Subsection we consider surface waves in two-component homogeneous poroelastic materials with impermeable boundary. We indicate as well some properties related to the permeable boundary condition. This condition has been proposed by Deresiewicz & Skalak in the 60ies [5]. However, the recent work [13] on problems in which this condition is incorporated indicates some flaws which, so far, have not been successfully removed.

Compatibility Conditions and Dispersion Relation We follow here the presentation in [27]. The solution of the differential eigenvalue problem (74) was presented in a different form in [10]. We seek a solution of the set (1)–(5) in which the displacement vector $\boldsymbol{u}^S$ for the skeleton, and formally the displacement vector $\boldsymbol{u}^F$ for the fluid are introduced. The latter is incorporated solely for technical symmetry of considerations, and it does not have any physical bearing. Then,

$$\begin{aligned} \boldsymbol{u}^S &= \operatorname{grad}\varphi^S + \operatorname{curl}\boldsymbol{\psi}^S, \quad \boldsymbol{v}^S = \frac{\partial \boldsymbol{u}^S}{\partial t}, \quad \boldsymbol{e}^S = \operatorname{sym}\operatorname{grad}\boldsymbol{u}^S, \\ \boldsymbol{u}^F &= \operatorname{grad}\varphi^F + \operatorname{curl}\boldsymbol{\psi}^F, \quad \boldsymbol{v}^F = \frac{\partial \boldsymbol{u}^F}{\partial t}. \end{aligned} \tag{66}$$

where $\varphi^S, \varphi^F, \boldsymbol{\psi}^S, \boldsymbol{\psi}^F$ are the so-called displacement potentials. As the problem is assumed to be two-dimensional we make the following ansatz for solutions harmonic in the x-direction:

$$\begin{aligned} \varphi^S &= A^S(z)\exp\left[i(kx-\omega t)\right], \quad \varphi^F = A^F(z)\exp\left[i(kx-\omega t)\right], \\ \psi_z^S &= B^S(z)\exp\left[i(kx-\omega t)\right], \quad \psi_z^F = B^F(z)\exp\left[i(kx-\omega t)\right], \\ \psi_x^S &= \psi_y^S = \psi_x^F = \psi_y^F = 0, \end{aligned} \tag{67}$$

and

$$\begin{aligned}\rho^S-\rho_0^S&=A_\rho^S(z)\exp\left[i(kx-\omega t)\right],\\ \rho^F-\rho_0^F&=A_\rho^F(z)\exp\left[i(kx-\omega t)\right],\\ n-n_0&=A^\Delta(z)\exp\left[i(kx-\omega t)\right].\end{aligned}\tag{68}$$

Substitution to (1)-(5) leads after straightforward calculations to the following compatibility conditions:

$$\begin{aligned}&B^F=\frac{i\pi}{\rho_0^F\omega+i\pi}B^S,\quad A^\Delta=-\frac{n_0\omega\tau}{i+\omega\tau}\left(\frac{d^2}{dz^2}-k^2\right)\left(A^F-A^S\right),\\ &A_\rho^S=-\rho_0^S\left(\frac{d^2}{dz^2}-k^2\right)A^S,\quad A_\rho^F=-\rho_0^F\left(\frac{d^2}{dz^2}-k^2\right)A^F,\end{aligned}\tag{69}$$

as well as

$$\begin{aligned}&\left[\kappa\left(\frac{d^2}{dz^2}-k^2\right)+\omega^2\right]A^F+\\ &+\left[\frac{n_0\beta\omega\tau}{\rho_0^F(i+\omega\tau)}\left(\frac{d^2}{dz^2}-k^2\right)+\frac{i\pi}{\rho_0^F}\omega\right]\left(A^F-A^S\right)=0,\end{aligned}\tag{70}$$

$$\begin{aligned}&\left[\frac{\lambda^S+2\mu^S}{\rho_0^S}\left(\frac{d^2}{dz^2}-k^2\right)\omega^2\right]A^S-\\ &-\left[\frac{n_0\beta\omega\tau}{\rho_0^S(i+\omega\tau)}\left(\frac{d^2}{dz^2}-k^2\right)+\frac{i\pi}{\rho_0^S}\omega\right]\left(A^F-A^S\right)=0,\end{aligned}\tag{71}$$

$$\left[\frac{\mu^S}{\rho_0^S}\left(\frac{d^2}{dz^2}-k^2\right)\omega^2\right]B^S+\frac{i\pi\rho_0^F}{\rho_0^S(\rho_0^F\omega+i\pi)}\omega^2B^S=0.\tag{72}$$

It is convenient to introduce a dimensionless notation. In order to do so we define the auxiliary quantities

$$\begin{aligned}&c_s:=\frac{c_S}{c_{P1}}<1,\quad c_f:=\frac{c_{P2}}{c_{P1}},\quad \pi':=\frac{\pi\tau}{\rho_0^S}>0,\quad \beta':=\frac{n_0\beta}{\rho_0^Sc_{P1}^2}>0,\\ &r:=\frac{\rho_0^F}{\rho_0^S}<1,\qquad z':=\frac{z}{c_{P1}\tau},\quad k':=kc_{P1}\tau,\quad \omega':=\omega\tau,\end{aligned}\tag{73}$$

where the speeds c_{P1}, c_S, c_{P2} are defined by (14). Further, we omit the prime for typographical reasons. Substitution of (73) in (70)-(72) yields

$$\begin{aligned}&\left[c_f^2\left(\frac{d^2}{dz^2}-k^2\right)+\omega^2\right]A^F+\\ &+\left[\frac{\beta\omega}{r(i+\omega)}\left(\frac{d^2}{dz^2}-k^2\right)+i\frac{\pi}{r}\omega\right]\left(A^F-A^S\right)=0,\\ &\left[\left(\frac{d^2}{dz^2}-k^2\right)+\omega^2\right]A^S-\left[\frac{\beta\omega}{i+\omega}\left(\frac{d^2}{dz^2}-k^2\right)+i\pi\omega\right]\left(A^F-A^S\right)=0,\\ &\left[c_s^2\left(\frac{d^2}{dz^2}-k^2\right)+\omega^2+\frac{i\pi\omega}{\omega+i\frac{\pi}{r}}\right]B^S=0.\end{aligned}\tag{74}$$

This is a differential eigenvalue problem that can be easily solved because for homogeneous materials the matrix of the coefficients is independent of z. Consequently, we seek solutions in the form

$$A^F = A_f^1 e^{\gamma_1 z} + A_f^2 e^{\gamma_2 z}, \quad A^S = A_s^1 e^{\gamma_1 z} + A_s^2 e^{\gamma_2 z}, \quad B^S = B_s e^{\zeta z}, \tag{75}$$

where the exponents $\gamma_1, \gamma_2, \zeta$ must possess negative real parts. Substitution in (74) yields them in the form

$$\left(\frac{\zeta}{k}\right)^2 = 1 - \frac{1}{c_s^2}\left(1 + \frac{i\pi}{\omega + i\frac{\pi}{r}}\right)\left(\frac{\omega}{k}\right)^2, \tag{76}$$

and

$$\begin{aligned}&\left[c_f^2 + \left(c_f^2 + \frac{1}{r}\right)\frac{\beta\omega}{i+\omega}\right]\left[\left(\frac{\gamma}{k}\right)^2 - 1\right]^2 + \left[1 + \left(1 + \frac{1}{r}\right)\frac{i\pi}{\omega}\right]\left(\frac{\omega}{k}\right)^4 \\ &+ \left[1 + c_f^2 + \left(1 + \frac{1}{r}\right)\frac{\beta\omega}{i+\omega} + \left(c_f^2 + \frac{1}{r}\right)\frac{i\pi}{\omega}\right]\left[\left(\frac{\gamma}{k}\right)^2 - 1\right]\left(\frac{\omega}{k}\right)^2 = 0.\end{aligned} \tag{77}$$

Simultaneously the following relations for the eigenvectors are obtained:

$$\boldsymbol{R}^1 = \left(B_s, A_s^1, A_f^1\right)^T, \quad \boldsymbol{R}^2 = \left(B_s, A_s^2, A_f^2\right)^T, \tag{78}$$

where

$$A_f^1 = \delta_f A_s^1, \quad A_s^2 = \delta_s A_f^2, \tag{79}$$

$$\delta_f := \frac{1}{r}\frac{\dfrac{\beta\omega}{i+\omega}\left[\left(\dfrac{\gamma_1}{k}\right)^2 - 1\right] + \dfrac{i\pi}{\omega}\dfrac{\omega^2}{k^2}}{\left(c_f^2 + \dfrac{1}{r}\dfrac{\beta\omega}{i+\omega}\right)\left[\left(\dfrac{\gamma_1}{k}\right)^2 - 1\right] + \left(\dfrac{\omega}{k}\right)^2 + \dfrac{i\pi}{\omega r}\dfrac{\omega^2}{k^2}}, \tag{80}$$

$$\delta_s := \frac{\dfrac{\beta\omega}{i+\omega}\left[\left(\dfrac{\gamma_2}{k}\right)^2 - 1\right] + \dfrac{i\pi}{\omega}\dfrac{\omega^2}{k^2}}{\left(1 + \dfrac{\beta\omega}{i+\omega}\right)\left[\left(\dfrac{\gamma_2}{k}\right)^2 - 1\right] + \left(\dfrac{\omega}{k}\right)^2 + \dfrac{i\pi}{\omega}\dfrac{\omega^2}{k^2}}. \tag{81}$$

The above solution for the exponents still leaves three unknown constants B_s, A_f^2, A_s^1 which must be specified from boundary conditions. This is the subject of the next subsection. However, for technical reasons we solve the problem under the simplifying assumption $\beta = 0$. We have already mentioned in Sect. 2 that this simplification does not change qualitative properties of acoustic waves, and that the quantitative influence is small for practically relevant values of β. In addition we only solve the limit problems in the range of high and low frequencies.

For the *high frequency approximation* , $1/\omega \ll 1$, we immediately deduce from (76) and (77)

$$\begin{aligned}&\left(\frac{\zeta}{k}\right)^2 = 1 - \frac{1}{c_s^2}\left(\frac{\omega}{k}\right)^2, \\ &\left(\frac{\gamma_1}{k}\right)^2 = 1 - \left(\frac{\omega}{k}\right)^2, \quad \left(\frac{\gamma_2}{k}\right)^2 = 1 - \frac{1}{c_f^2}\left(\frac{\omega}{k}\right)^2,\end{aligned} \tag{82}$$

and

$$\delta_f = \delta_s = 0 \quad \Rightarrow \quad \boldsymbol{R}^1 = \left(B_s, A_s^1, 0\right)^T, \quad \boldsymbol{R}^2 = \left(B_s, 0, A_f^2\right)^T. \tag{83}$$

For the *low frequency approximation*, $\omega \ll 1$, (77) becomes singular. It can be written in the form

$$\begin{aligned} & c_f^2\omega^2\left[\left(\tfrac{\gamma}{k}\right)^2 - 1\right]^2 + \omega\left[\omega + i\pi\left(1+\tfrac{1}{r}\right)\right]\left(\tfrac{\omega}{k}\right)^4 + \\ & + \left[\omega\left(1+c_f^2\right) + i\pi\left(c_f^2+\tfrac{1}{r}\right)\right]\omega\left[\left(\tfrac{\gamma}{k}\right)^2 - 1\right]\left(\tfrac{\omega}{k}\right)^2 = 0. \end{aligned} \tag{84}$$

With the substitution

$$W := \omega\left[\left(\frac{\gamma}{k}\right)^2 - 1\right] + \omega\left(\frac{\omega}{k}\right)^2\frac{1+c_f^2}{2c_f^2}, \tag{85}$$

we obtain the following quadratic equation for W,

$$\begin{aligned} & c_f^2 W^2 + i\pi\left(c_f^2 + \frac{1}{r}\right)\left(\frac{\omega}{k}\right)^2 W - \\ & - \left\{i\pi\left[\frac{\left(c_f^2+\frac{1}{r}\right)\left(1+c_f^2\right)}{2c_f^2} - \left(1+\frac{1}{r}\right)\right]\left(\frac{\omega}{k}\right)^4\right\}\omega + O\left(\omega^2\right) = 0, \end{aligned} \tag{86}$$

which, for small ω, can be solved by the regular perturbation method, viz.,

$$W = W_0 + \omega W_1 + O\left(\omega^2\right). \tag{87}$$

After easy calculations we obtain

$$W = \begin{cases} \left[\dfrac{1+c_f^2}{2c_f^2} - \dfrac{r+1}{rc_f^2+1}\right]\left(\dfrac{\omega}{k}\right)^2\omega, \\ -\left[\dfrac{1+c_f^2}{2c_f^2} - \dfrac{r+1}{rc_f^2+1}\right]\left(\dfrac{\omega}{k}\right)^2\omega - i\pi\dfrac{rc_f^2+1}{rc_f^2}\left(\dfrac{\omega}{k}\right)^2 \end{cases}. \tag{88}$$

Bearing (85) in mind, we arrive at the following results for the exponents

$$\begin{aligned} \omega \ll 1: \qquad & \left(\frac{\zeta}{k}\right)^2 = 1 - \frac{r+1}{c_s^2}\left(\frac{\omega}{k}\right)^2, \\ & \left(\frac{\gamma_1}{k}\right)^2 = 1 - \frac{r+1}{rc_f^2+1}\left(\frac{\omega}{k}\right)^2, \\ & \left(\frac{\gamma_2}{k}\right)^2 = 1 - \frac{rc_f^4+1}{c_f^2\left(rc_f^2+1\right)}\left(\frac{\omega}{k}\right)^2 - \frac{i\pi}{\omega}\frac{rc_f^2+1}{rc_f^2}\left(\frac{\omega}{k}\right)^2, \end{aligned} \tag{89}$$

and for the coefficients of amplitudes

$$\delta_f = 1 - \frac{\omega r}{i\pi}\frac{1-c_f^2}{1+rc_f^2}, \quad \delta_s = -rc_f^2\left(1 - \frac{\omega r}{i\pi}\frac{1-c_f^2}{1+rc_f^2}\right). \tag{90}$$

Obviously, due to the singular character of (84), the last contribution to γ_2/k becomes singular as $\omega \to 0$.

Boundary Value Problems for Surface Waves In order to determine surface waves in a saturated poroelastic medium we need conditions for $z = 0$. We discuss in detail the problem in which this boundary is impermeable, i.e. a poroelastic medium is in contact with a vacuum. Boundary conditions then have the form

$$T_{13}\big|_{z=0} \equiv T_{13}^S\Big|_{z=0} = \mu^S \left(\frac{\partial u_1^S}{\partial z} + \frac{\partial u_3^S}{\partial x} \right)\Big|_{z=0} = 0, \tag{91}$$

$$T_{33}\big|_{z=0} \equiv \left(T_{33}^S - p^F\right)\Big|_{z=0}$$
$$= c_{P1}^2 \rho_0^S \left(\frac{\partial u_1^S}{\partial x} + \frac{\partial u_3^S}{\partial z} \right) - 2c_S^2 \rho_0^S \frac{\partial u_1^S}{\partial x} - c_{P2}^2 \left(\rho^F - \rho_0^F\right)\Big|_{z=0} = 0, \tag{92}$$

$$\frac{\partial}{\partial t}\left(u_3^F - u_3^S\right)\Big|_{z=0} = 0, \tag{93}$$

where the first two conditions mean that the surface $z = 0$ is stress-free (far-field approximation), and the last condition means that there is no transport of fluid mass through this surface (impermeable boundary). By u_1^S, u_3^S, we denote the components of the displacement $\boldsymbol{u}^S$ in the direction of the x-axis and z-axis, respectively, while u_3^F is the z-component of the displacement $\boldsymbol{u}^F$.

For a permeable boundary the last condition would not hold. Instead the mass transport through the surface must be specified by a relation to a driving force. According to the proposition of Deresiewicz & Skalak [5] such a driving force is proportional to the difference of the pore pressures on both sides of the boundary. In an earlier paper on surface waves on such boundaries, [9], we have used the condition

$$\rho_0^F \frac{\partial}{\partial t}\left(u_3^F - u_3^S\right) - \alpha\left(p^F - n_0 p_{ext}\right)\Big|_{z=0} = 0, \tag{94}$$

where α denotes a surface permeability coefficient and p_{ext} is an external pressure. This condition relies on the assumption that the pore pressure p and the partial pressure p^F satisfy the relation $p^F \approx n_0 p$ at least in a small vicinity of the surface. In some cases it may be a good approximation and the results presented in [9] agree qualitatively very well with observations. However, the condition seems to be violated on boundaries of granular materials with a relatively small porosity. We leave this issue un-clarified in this work and refer to future research.

Substitution of results of Subsection 3.2.1. in the boundary conditions (91)-(93) yields the following equations for the three unknown constants B_s, A_f^2, A_s^1:

$$\boldsymbol{AX} = \boldsymbol{0}, \tag{95}$$

where

$$\boldsymbol{A} := \begin{pmatrix} \left(\frac{\zeta}{k}\right)^2 + 1 & 2i\frac{\gamma_2}{k}\delta_s & 2i\frac{\gamma_1}{k} \\ -2ic_s^2\frac{\zeta}{k} & \begin{matrix}\left[\left(\frac{\gamma_2}{k}\right)^2 - 1 + 2c_s^2\right]\delta_s + \\ +rc_f^2\left[\left(\frac{\gamma_2}{k}\right)^2 - 1\right]\end{matrix} & \begin{matrix}\left(\frac{\gamma_1}{k}\right)^2 - 1 + 2c_s^2 + \\ +rc_f^2\left[\left(\frac{\gamma_1}{k}\right)^2 - 1\right]\delta_f\end{matrix} \\ i\frac{r\omega}{r\omega + i\pi} & -\left(\delta_s - 1\right)\frac{\gamma_2}{k} & \left(\delta_f - 1\right)\frac{\gamma_1}{k} \end{pmatrix}, \tag{96}$$

$$\boldsymbol{X} := \left(B_s,\, A_f^2,\, A_s^1 \right)^T .$$

This homogeneous set yields the *dispersion relation* $\det \boldsymbol{A} = 0$, determining the ω-k relation. We investigate separately solutions of this equations for high and low frequencies.

High Frequency Approximation In the case of high frequencies, $1/\omega \ll 1$, we have $\delta_s = \delta_f = 0$ and the dispersion relation takes on the form

$$\mathcal{P}_R \sqrt{1 - c_f^2 \left(\frac{\omega}{k}\right)^2} + \frac{r}{c_s^4} \left(\frac{\omega}{k}\right)^4 \sqrt{1 - \left(\frac{\omega}{k}\right)^2} = 0, \tag{97}$$

where

$$\mathcal{P}_R := \left(2 - \frac{1}{c_s^2}\left(\frac{\omega}{k}\right)^2\right)^2 - 4\sqrt{1 - \left(\frac{\omega}{k}\right)^2}\sqrt{1 - \frac{1}{c_s^2}\left(\frac{\omega}{k}\right)^2}. \tag{98}$$

Hence for $r = 0$, (97) reduces to $\mathcal{P}_R = 0$ which is the Rayleigh dispersion relation for single-component continua. Otherwise we obtain the relation identical with the one analysed by Edelman [6] and Edelman & Wilmański [9] in the limit of short waves (i.e. $1/k \ll 1$). Consequently, the conclusions for this case are the same as well. As shown in [5,8], (97) possesses two roots defining two surface waves: a true Stoneley wave which propagates almost without attenuation at a speed slightly smaller than c_f, as well as a generalised Rayleigh wave which is leaky (i.e. it radiates energy to the P2-wave) and propagates with the speed c_R, $c_f < c_R < c_s$.

Low Frequency Approximation A limit of long waves has been recently analysed by Edelman [8], where it has been shown that existence of a critical value k_{cr} of the wavenumber k for initial value problems as described in Sect. 2 for bulk waves yields nonexistence of Stoneley surface waves in the range $0 \le k \le k_{cr}$. This is not the case for boundary value problems in which we consider the limit of small frequencies ω rather than the limit of long waves. In this limit surface waves do exist and we proceed to investigate this problem in some detail.

If we account for (89) and (90) in the condition $\det \boldsymbol{A} = 0$, then we obtain the dispersion relation reflecting a dependence of ω/k on ω. The expansion with respect to $\sqrt{\omega}$ yields an identity in the zeroth order and the following relation for the higher order

$$\left(\frac{\omega}{k}\right)\left\{\left(2 - \frac{r+1}{c_s^2}\left(\frac{\omega}{k}\right)^2\right)^2 - 4\sqrt{1 - \frac{r+1}{c_s^2}\left(\frac{\omega}{k}\right)^2}\sqrt{1 - \frac{r+1}{rc_f^2+1}\left(\frac{\omega}{k}\right)^2}\right\} + \\ + O\left(\sqrt{\omega}\right) = 0. \tag{99}$$

Clearly, two solutions emerge:

1. A Rayleigh wave whose speed is different from zero in the limit $\omega \to 0$ and whose attenuation is of the order $O\left(\sqrt{\omega}\right)$. The relation for the speed reminds the relation (41) with the speeds of bulk waves replaced by the low frequency

limits. Indeed, we have

$$\frac{r+1}{c_s^2} = c_{P1}^2 \frac{\rho_0^S + \rho_0^F}{\mu^S} \equiv \frac{c_{P1}^2}{c_{oS}^2},$$
$$\frac{r+1}{rc_f^2 + 1} = c_{P1}^2 \frac{\rho_0^S + \rho_0^F}{\lambda^S + 2\mu^S + \rho_0^F \kappa} \equiv \frac{c_{P1}^2}{c_{oP1}^2}, \tag{100}$$

and these relations follow from (23), (28). Consequently,

$$\left(2 - \frac{c_{P1}^2}{c_{oS}^2}\left(\frac{\omega}{k}\right)^2\right)^2 - 4\sqrt{1 - \frac{c_{P1}^2}{c_{oS}^2}\left(\frac{\omega}{k}\right)^2}\sqrt{1 - \frac{c_{P1}^2}{c_{oP1}^2}\left(\frac{\omega}{k}\right)^2} = 0. \tag{101}$$

2. A Stoneley wave with speed of propagation of the order $O\left(\sqrt{\omega}\right)$. Hence it goes to zero in the same way as the speed of propagation of P2-wave.

4 Final Remarks

The results for a two-component model of porous solid-fluid mixtures presented in this article should be compared with those obtained by means of Biot's model and with experimental observations. We shall not go into detail of such a comparison in this work. However, it can be easily checked that there is a very good qualitative agreement of both models as far as propagation of acoustic waves is concerned, and these check well with experimental observations. One should mention the following features following from the above considerations.

1. The analysis of acoustic waves based on the model proposed by Wilmański does not contain the flaws arising in Biot's model: the violation of the second law in the case of the simple model (without contributions of higher gradients), and the violation of the material objectivity principle by the contribution of relative acceleration in Biot's model.
2. There exist three bulk modes of propagation: P1-waves, S-waves and P2-waves.
3. Speeds of propagation of these waves agree with experimental observations in both limits of high and low frequencies.
4. There exist two modes of surface waves for impermeable boundaries: Rayleigh waves and Stoneley waves, and again these agree with observations.

In contrast to the high frequency and short wave limits which coincide, the low frequency limit is different from that obtained earlier for long waves. Such a limit exists for $\omega \to 0$ while the long wave limit contains the zone $0 \leq k \leq k_{cr}$ forbidden for the propagation.

Let us finally mention that the motivation for a wave analysis with a real wavenumber k, used in the papers [6–9], rather than with a real frequency ω was primarily based on the observation that this corresponds better to the way in which acoustic waves are initiated in engineering applications. In contrast to the far-field approximation of seismic waves which is usually based on the frequency analysis, engineering applications were primarily concerned with waves initiated by chopping or explosions which led to initial value problems. This is no longer the case. Numerous experiments and measurements are made nowadays by devices producing harmonic vibrations (e.g. [16]) and this leads to boundary value problems in which the real frequency ω is the proper choice of the control variable.

Consequently, the problem of existence or nonexistence of P2-waves and surface Stoneley waves is solely related to the way in which an experiment is performed.

It should be mentioned that similar existence problems appear in all models described by wave equations with damping. In the case of damping in space there exists a critical frequency, and in the case of damping in time there exists a critical wavenumber. The simplest example of such a model is the telegraph equation.

References

1. Aki, K., Richards, P. (1980) Quantitative seismology, theory and methods. W. H. Freeman and Co
2. Ben-Menahem, A., Singh, S. (1981) Seismic waves and sources. Dover
3. Bourbie, T., Coussy, O., Zinszner, B. (1987) Acoustics of porous media. Editions Technip, Paris
4. Brekhovskikh, L. M., Godin, O. A. (1998) Acoustics of layered media. I: Plane and quasi-plane waves. Springer, Berlin
5. Deresiewicz, H., Skalak, R. (1963) On Uniqueness in Dynamic Poroelasticity. Bull. Seismol. Soc. Am. 53:783–788
6. Edelman, I. (2001) Waves on the boundaries of porous media. Physics Doklady, V. 46, N7, 517–521
7. Edelman, I. (2002) On the bifurcation of the Biot slow wave in a porous medium. WIAS-Preprint #738
8. Edelman, I. (2002) Asymptotic analysis of surface waves at vacuum/porous medium interface: Low-frequency range. WIAS-Preprint #745
9. Edelman, I., Wilmański, K. (2002) Asymptotic analysis of surface waves at vacuum/porous medium and liquid/porous medium interfaces. Cont. Mech. Thermodyn., 14,1: 25–44
10. Edelman, I., Wilmański, K., Radkevich, E. (1999) Surfaces waves at a free interface of a saturated porous medium. WIAS-Preprint #513
11. Klimentos, T., McCann, C. (1988) Why is the Biot slow compressional wave not observed in real rocks? Geophysics, 53, 12:1605–1609
12. Lai, C. G. (1998) Simultaneous inversion of Rayleigh phase velocity and attenuation for near-surface site characterization. PhD-Thesis, Georgia Institute of Technology
13. Lai, C. G., Lancellotta, R., Wilmański, K. (2002) Biot's model and nondestructive testing of granular materials by means od acoustic measurements. WIAS–Preprint (in preparation)
14. Plona, T. J. (1980) Observation of a second bulk compressional wave in a porous medium at ultrasonic frequencies. Appl. Phys. Lett., 36:259–261
15. Parker, D. F., Maugin, G. A. (Eds.) (1988) Recent developments in surface acoustic waves. Springer, Berlin
16. Rix, G. J., Lai, C. G., Foti, S. (2001) Simultaneous Measurement of Surface Wave Dispersion and Attenuation Curves. Geotechnical Testing Journal, 24(4):350–358
17. Sato, H., Fehler, M. (1998) Seismic wave propagation and scattering in the heterogeneous earth. Springer, N.Y.

18. Towne, D. H. (1988) Wave Phenomena. Dover Publ., Mineola
19. White, J. E. (1983) Underground sound. Application of seismic waves. Elsevier, Amsterdam
20. Viktorov, I. A. (1981) Zwukowyje powierchnostnyje wolny w twiordych telach [Acoustic surface waves in solids] (in Russian), Moskow
21. Wilmański, K. (1995) On weak discontinuity waves in porous materials. In: Marques, M., Rodrigues, J. (Eds.) Trends in applications of mathematics to mechanics. Longman, N.Y., 71–83
22. Wilmański, K. (1996) Dynamics of Porous Materials under Large Deformations and Changing Porosity. In: Batra, R.C., Beatty, M.F. (Eds.) Contemporary Research in the Mechanics and Mathematics of Materials. 343–358
23. Wilmański, K. (1998) Thermomechanics of Continua, Springer, Berlin
24. Wilmański, K. (1999) Waves in Porous and Granular Materials. In: Hutter, K., Wilmański, K. (Eds.) Kinetic and Continuum Theories of Granular and Porous Media. CISM Courses and Lectures No. 400, Springer Wien New York, 131-186,
25. Wilmański, K. (2001) Thermodynamics of Multicomponent Continua. In: Teisseyre, R., Majewski, E. (Eds.) Earthquake Thermodynamics and Phase Transformations in the Earth's Interior. Academic Press, San Diego, Chapt. 25, 567–671
26. Wilmański, K. (2001) Sound and shock waves in porous and granular materials. In: Ciancio, V., Donato, A., Oliveri, F., Rionero, S. (Eds.) Proceedings "WASCOM 99", 10th Conference on Waves and Stability in Continuous Media, 489–503
27. Wilmański, K. (2002) Propagation of sound and surface waves in porous materials. In: Maruszewski, B. (Ed.) Structured Media, Poznan University of Technology, Poznan, 312–326

Dissipation Influence on Cooling of 2-Dimensional Hard Needle Systems

Agnieszka Chrzanowska and Harald Ehrentraut

Institute of Mechanics, Darmstadt University of Technology, Hochschulstr. 1, 64289 Darmstadt, Germany,
e-mail: {achrzano,ehrentraut}@mechanik.tu-darmstadt.de

received 15 Jul 2002 — accepted 24 Sep 2002

Abstract. We discuss the occurrence of orientational clusters (bundles) which are formed in a freely cooling two-dimensional system of granular hard needles. This phenomenon can be present in the isotropic as well as in the nematic phase. In the nematic phase, bundles are the reason for the global decrease of the order parameter with a simultaneous decrease of the system temperature. Such a feature stands in sharp contrast to genuine liquid crystals where a decrease of temperature is always associated with an increase of the order parameter. Properties of hard needles confined in a pipe are presented for elastic and dissipative walls. It has been observed that even small dissipation at the walls leads to significant density and orientation instabilities. We also detect micro-flows that depend on the distance from the wall.

1 Introduction

Granular materials [16,32,34] differ from molecular systems in many aspects. The main source of these differences is the energy dissipation which, on the one hand, raises the temperature of the constituent particles, and, on the other hand, no longer takes part during the kinetic process. For granular materials, decrease of kinetic energy is dictated by the friction conditions at the collision of two grains. One distinguishes two types of dissipation mechanisms, namely, with respect to normal and tangential friction. These types are based on the notion of the *tangential plane*, which is defined as the plane tangential to both colliding bodies at the point of contact. Normal dissipation takes place when the bodies can slide freely on the surface of the contact and only the momentum perpendicular to the tangential plane is affected by the dissipative interaction. Tangential dissipation, on the other hand, diminishes the part of the momentum that is situated within the tangential plane. Another consequence of the tangential friction is a gain/loss in the particle spin. Including spin, however, entails complications in the kinetic description. Then, it is necessary to introduce the orientational degrees of freedom into the formalism, such as orientation and angular velocity of the particle. Such a description is structurally similar for all the bodies that are anisotropic in shape, but it differs with respect to the type of the collision conditions. For example, the spin of an anisotropic body can change due to the normal momentum exchange and is therefore also affected by the normal dissipation. New complications can occur when the system exhibits long-range orientational order (which takes place when, on the macroscopic scale, orientations of the particles are distributed anisotropically).

This feature, together with the presence of the preferred directions and the orientational order parameters, is symptomatic for liquid crystalline (LC) systems. Depending on the thermodynamic parameters and the type of constituents, one encounters in liquid crystals a rich variety of different phases and symmetries [11]. The simplest of them is the nematic phase. It has the long-range order of uniaxial symmetry and at the same time exhibits disordered distribution of the particles centres of mass. The system of 2D hard needles that is here under consideration exhibits the two-dimensional counterpart of the uniaxial nematic phase. It is worth mentioning that this is the only anisotropic phase that can be formed by 2D hard needles.

Although the name *liquid crystal* is mainly preserved for molecular systems, it often happens that this convention is not followed rigorously. Such a phrase appears also in the context of systems built from particles of larger sizes, like colloids or granulates. There is even mentioning of living liquid crystals as presented in the work of Gruel [13], who studied biological aggregates of granulocytes that exhibit a certain degree of self-steered mobility.

The present paper focuses on granular liquid crystals and is motivated by the question how the orientational properties of liquid crystals are combined with the typical aspects of granular materials. As an example of these systems we have chosen two-dimensional hard needles. The reasons for this choice are the following: Two-dimensional systems have already gained a separate place in studying granular gases [20,21,26,28,29,34]. They offer an easy way of visualisation and help to understand different phenomena as, for instance, formation of clusters. In view of this fact, two-dimensional hard needles (line segments) as the simplest system that exhibits the nematic phase are an ideal material to investigate basic dissipative effects in orientationally ordered media. The conditions of a collision between two needles are easily formulated and can be applied to the future statistical analysis of the dynamical properties. Needles are also a good reference system for more sophisticated and realistic models of anisotropic systems, and they are especially suitable to test theories of transport phenomena in fluids consisting of highly anisometric molecules.

Motivated by an interest in the effects of dissipation, we recently presented a report on the cooling of 2D hard needles [4]. In [4], we showed that the order of the granular system on a long-range scale can decrease with diminishing granular temperature. This unique feature is quite opposite to liquid crystals (with exclusion of a few cases known as re-entrant phases), where lower temperatures entail higher order. We have also given the evidence for the presence of orientational clusters and inelastic collapse. In the present article we present an extended description of these features together with the results of systematic event-driven molecular dynamics simulations and, also, a comparison of the above effects occurring in the isotropic and in the nematic phase.

The paper is organised as follows: Section 2 contains a brief overview of the papers devoted to hard needles. Sections 3 and 4 present the geometry and the rules of the collision between two needles. In Sect. 5, we describe the molecular dynamics technique used and present some typical configurations obtained at different stages of dissipation. Order parameters and collision frequency are discussed in Sect. 6.

Section 7 presents different aspects of the method applied to measure orientational clusters. In Sect. 8 we present features of hard needles confined in a narrow slit. Two different cases are distinguished: the case of purely elastic interactions (Sect. 9) and the case with dissipative walls (Sect. 10). Finally, Sect. 11 gives a summary of the main results.

2 Overview of the Hard Needles "History"

Many properties of hard lines have already been studied both by the density functional method and by the computer simulations. The three-dimensional, non-dissipative case is somewhat trivial since it is structureless. It does not form any anisotropic phase since the excluded volume of two needles is zero. As a consequence, all static properties are those of an ideal gas. Dynamical properties are more interesting. Motivated by the experimental findings on rotational dynamics of long rigid viruses in semi dilute solutions [23,24,35], Frenkel et al. [8,9] performed an extensive molecular dynamics study of 3D hard needles at different densities and, based on the Enskog assumption of uncorrelated successive collisions, they derived theoretical expressions corresponding to the properties measured in simulations. They showed that at moderate densities the Enskog theory is capable to describe correctly the behaviour of the autocorrelation functions for the linear and angular momenta and for the rotational energy. Moreover, they found that the collision frequency formula works almost perfectly at all densities and predicted divergent longitudinal self-diffusion coefficients for the density limit $\rho \to \infty$.

Studies of the behaviour of 3D hard needles at a wall by Poniewierski [30] and by Mao et al. [25] were the first microscopic models of confined liquid crystals. Using the methods of the density functional theory, they could show that the hard lines prefer planar alignment at the wall. This is a consequence of entropy effects. Similar features were found for hard spherocylinders [31] and hard ellipsoids [5], which indicates that this is a general property of the convex hard bodies.

Poniewierski was the first to show that 3D needles, although isotropic in the bulk, exhibit uniaxial nematic symmetry at the wall, and he could calculate the order parameters for them. He also gave arguments on the basis of the Landau theory analysis for the possibility of an uniaxial-biaxial surface transition in which the rotational symmetry in the plane of the wall can be broken. This feature, however, concerns fat particles of finite width. Mao et al., on the other hand, paid more attention to study the surface tension and the depletion potential characteristics.

By means of the Monte Carlo method, Frenkel et al. [7] investigated 2D elastic systems from the point of view of thermodynamic properties. They discussed the equation of state and the type of the phase transition between isotropic and nematic phases. According to their results, it is not certain whether a system made of hard needles has true long-range order (LRO). All the gathered evidence suggests quasi-LRO [7], in which the nematic order parameter vanishes in the thermodynamic limit and all the order parameter correlation functions decay algebraically. The global order parameter S depends on the size of the system (on the total number of the particles N) as $S \sim N^{-kT/2\pi K}$ [7], where K is the two-dimensional Frank constant, T is the thermodynamic temperature and k is the Boltzmann constant. Such type of order, quasi-LRO, is characteristic for 2D nematics, if the potential V acting between two particles is separable, i.e. $V(r_{ij}, \Omega_i, \Omega_j) = \sum_k f_k(r_{ij}) g_k(\Omega_i, \Omega_j)$, $(r_{ij} =$

distance between the centres of the particles i and j, and Ω_i, Ω_j are their orientations), but it is not ruled out for non-separable interactions as those of hard needles. It is also not certain what the rank of the nematic-isotropic transition is. Instead of the common in 3D systems first order transition, the available evidence for 2D hard needles points towards the occurrence of a disclination-unbinding transition of the Kosterlitz-Thouless type [18].

The extremal shape anisotropy of a needle has helped to reveal a two step character of the velocity autocorrelation function (the velocity ACF) of the system composed of uniaxial particles. In [3] the authors of the present paper showed that the velocity ACF of 2D needles is composed of two single exponential-like decays that are associated to the transverse and the longitudinal motion of the particles. Such superposition rule is a general feature of uniaxial molecules and should held in all the systems built from such particles, no matter what phase they form. Similar rules based on the molecules symmetries are expected for more complicated geometries, for instance, for biaxial molecules.

When the particles in the system can interact inelastically, many new phenomena arise. One of the first attempts to investigate properties of an inelastic (granular) system composed of anisotropic constituents is also the study of an ensemble of 3D hard needles by Huthmann, Aspelmeier and Zippelius [17]. The authors of [17] reported on a two-step cooling: (1) a fast exponential decay to a state that is characterised by a time independent ratio of translational to rotational energy; (2) a slow algebraic decay as t^{-2} (t denotes time) of the total kinetic energy. In dense regimes, they observe large-scale structures in the translational velocity field. Neither local ordering nor correlations between velocity and density fields have been encountered.

This overview of the articles devoted to toy models with hard needles we can supplement with the work of Allen and Cunningham [1] about a model of a fluid composed from highly idealised tetrahedral molecules. In this model four hard needles are joined in a rigid unit according to the tetrahedral geometry as those of CBr_4. For such a system the authors of [1] presented the autocorrelation functions obtained from molecular dynamics simulations. They reported on the influence of "chattering" collisions and the well seen deviations from single exponent Enskog-like behaviour even at low densities. In fact, their results for the velocity ACF exhibit a two scale structure (two linear regions on the logarithmic scale), quite similar to the functions for uniaxial molecules from [3]. This is the more surprising, since a tetrahedral molecule is by no means anisotropic. Hopefully, the kinetic theories will explain this strange coincidence. This is also an example that even toy models are not simple enough to give easily all answers.

3 Geometry of Two Colliding Needles

Molecular dynamics simulations are based on monitoring the temporal evolution of the system's microscopic state $\Gamma(t)$. For a N-particle system this state is given in terms of all the coordinates q^N and the conjugate momenta p^N of the constituent particles. If particles are anisotropic, uniaxial and rigid, relevant coordinates of a given particle i are position $\boldsymbol{r}_i$, orientation $\boldsymbol{u}_i$, conjugate momentum $\boldsymbol{p}_i$ and angular momentum $\boldsymbol{J}_i$. The information about all these coordinates at an instant of time t, however, is not sufficient to describe system's kinetics. One needs to specify interactions. If interactions are included, the calculation of appropriate statistical observables can deliver information about macroscopic properties like energy or pressure.

The description of a two-dimensional needle, in which we are here interested in, requires 6 independent variables: $\{x, y, p_x, p_y, \varphi, J_z\}$. Particles are assumed to be hard, and as long as they are not in contact, they do not interact. Interaction takes place at the point of contact and is assumed to be instantaneous. As a result of the collision, the linear and angular momenta are changed. Their new values are easily determined from momentum and energy conservation. In resolving the collision conditions the biggest difficulty lies in finding the point of contact. For a 2D case, a collision takes place whenever a tip of one needle touches another needle at any point (see Fig. 1), and the transfer of the linear momentum takes place within the plane spanned by the two needles. In contrast, 3D collisions [8,9,17] are side-to-side and the momentum transfer is out of plane. As a result, one needs to know trajectories of the needle tips.

Let us take into account a needle of half-length equal to L. The centre of the needle is given by the vector $\boldsymbol{r}_0$, and the orientation by the unit vector $\boldsymbol{u}$. The particle is moving freely both translationally and rotationally, and after time t its centre of mass has the new position

$$\boldsymbol{r} = \boldsymbol{r}_0 + \boldsymbol{V}t, \tag{1}$$

and the new orientation

$$\boldsymbol{u}' = \{\cos(\varphi + \Omega t), \sin(\varphi + \Omega t)\}\,. \tag{2}$$

In the above, $\boldsymbol{V}$ is the vector of the linear velocity, Ω is the angular velocity value and φ is the angle that the orientation vector $\boldsymbol{u}$ makes with the axis X. The coordinates of the two ends of the needle (b - the beginning, e - the end) can be given as (since we tacitly assume L to be positive)

$$\boldsymbol{r}_{b,e} = \boldsymbol{r}_0 \pm L\boldsymbol{u}. \tag{3}$$

The time and the point of contact can be obtained from comparing its X and Y coordinates that are given in terms of the needles i and j properties (see Fig. 1):

$$x^i + V_x^i t \pm L\cos\Phi_i(t) = x^j + V_x^j t + l\cos\Phi_j(t), \tag{4}$$

$$y^i + V_y^i t \pm L\sin\Phi_i(t) = y^j + V_y^j t + l\sin\Phi_j(t), \tag{5}$$

in which $\Phi_i(t)$ stands for $(\varphi_i + \Omega_i t)$ with φ_i being the initial value of the orientation of the i-th particle. These equations are linear with respect to the unknown l (see Fig. 1) and strongly nonlinear with respect to time t. A solution for the point of contact l can be written automatically in terms of time

$$l^{b,e} = \frac{y^{ij} + V_Y^{ij} t \pm L\sin(\varphi_i + \Omega_i t)}{\sin(\varphi_j + \Omega_j t)}, \tag{6}$$

where l can attain values within the interval $(-L < l < L)$. An equation for time t also follows from (4) and (5)

$$\left[x^{ij} + V_x^{ij} t \pm L\cos\Phi_i(t)\right]\sin(\Phi_j(t)) =$$

$$-\left[y^{ij} + V_y^{ij} t \pm L\sin\Phi_i(t)\right]\cos(\Phi_j(t)). \tag{7}$$

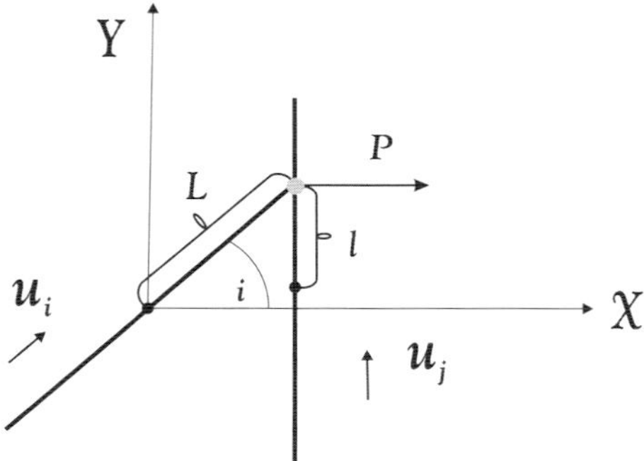

Fig. 1. Geometry of two needles i and j at the point of contact

The double index "ij" indicates here that a given variable is relative with respect to the particle i and j. Because of nonlinearity of the problem, the only way to find solutions is by numerical methods. To obtain a solution of (7), we use the following procedure: We construct a function $\Psi^{b,e}(t) := rhs - lhs$, where rhs and lhs stand for the right- and the left-hand side of (7). Obviously, all the roots of $\Psi^{b,e}(t)$ are the solutions to (7), so solving this equation corresponds to the problem of root finding. In principle, $\Psi^{b,e}(t)$ may have many roots whose existence depends on the parameters in the equation. This is a trivial consequence of the type of the nonlinearity. A similar consequence is a repeated collision between the same pair of particles. To compare, two spherical bodies can collide *only* once. For the collision to reoccur between spheres, an intervention of a third particle or a wall is needed. Anisotropic objects are different. In the case of hard needles there even exist peculiar regimes of kinetic parameters for which the number of elastic collisions of the particles may reach hundreds within a short time [27]. Such a chain collision is called a chattering collision.

As already mentioned, only the smallest root of $\Psi^{b,e}(t)$ is physically interesting. We know that this root will be relatively close to zero if the particles lie in their nearest vicinity. Thus we focus on exploring properties of $\Psi^{b,e}(t)$ only in small time regimes close to zero. The method we apply is "climbing the rope". Starting from the value of $\Psi^{b,e}$ at $t = 0$, we walk along its subsequent values at next times $t = t + \delta$ ($\delta > 0$) until the zero value is crossed and the sign of $\Psi^{b,e}$ is changed. The last pair of values that have opposite signs is used then as the bracketing points for the precise Newton-Raphson method of root finding. To compare, the previous method that has been mainly applied to hard needles in 3D [8,9,17] uses a root searching algorithm that adjusts the step towards the root using the square functions. These functions are based on the assessed maximum values of the first and the second derivatives of $\Psi^{b,e}(t)$. As a result, the method finds *all* the roots in the time interval of interest (for details see [8]). Efficiency of our method depends on the value of δ. For flat parts of the function, which lie far from $y = 0$, there is no need to keep δ small and larger steps are more efficient. For this reason we introduced an adjustment of the step size δ according to the behaviour of the derivative $\Psi^{b,e}$. Namely, $\delta = \delta/C$, where $C = |d\Psi^{b,e}/dt|$. In the vicinities of extrema we also put additional safety conditions on δ, scaling it with certain numbers that have been chosen by testing the routine efficiency. Such a method is not perfect, and it may overlook collisions if the initial value of δ is chosen too large. This can always happen for certain parameters of the needles, especially for those rotating rapidly. If a collision is overlooked, then the needles will overlap. One overlooked root can lead to several particles overlapping. In practice, we encounter a few overlapping pairs (3 to 10) per 5000 collisions (maximum error $\leq 2\%$), a little more in the isotropic phase as in the nematic. If it happens, we employ a clearing procedure which detects overlapping pairs, removes one of their members and replaces it randomly in the system again so that it no longer overlaps. The way of replacement follows: We choose another arbitrary needle from the system and place the removed needle along it at a certain distance. If this leads to a new overlap then the whole procedure is repeated until no overlaps are encountered. This clearing procedure has no meaning for equilibration runs. In the dissipative case, it may slightly add to the cluster formation due to the way a particle is replaced in the system close to another particle.

4 Collision Conditions

During a collision both momenta ($\boldsymbol{P}_i$,$\boldsymbol{P}_j$) and angular momenta ($\boldsymbol{J}_i$,$\boldsymbol{J}_j$) of the actors are changed to new primed values

$$\begin{aligned} \boldsymbol{P}_i' &= \boldsymbol{P}_i + \Delta\boldsymbol{P}, & \boldsymbol{P}_j' &= \boldsymbol{P}_j - \Delta\boldsymbol{P}, \\ \boldsymbol{J}_i' &= \boldsymbol{J}_i + L\boldsymbol{u}_i \times \Delta\boldsymbol{P}, & \boldsymbol{J}_j' &= \boldsymbol{J}_j - l\boldsymbol{u}_j \times \Delta\boldsymbol{P}, \end{aligned} \tag{8}$$

where the exchange of $\Delta\boldsymbol{P}$ (see Fig. 1) has the orientation perpendicular to the needle j, denoted as $\boldsymbol{u}_\perp$, which is in accordance to the geometry of a 2D collision without tangential friction. These conditions (8) are taken in the reference frame whose origin is placed at the point of contact between two colliding particles. The momentum and the angular momentum are automatically conserved:

$$\boldsymbol{J}_i + \boldsymbol{J}_j - L\boldsymbol{u}_i \times \boldsymbol{P}_i - l\boldsymbol{u}_j \times \boldsymbol{P}_j = \boldsymbol{J}_i' + \boldsymbol{J}_j' - L\boldsymbol{u}_i \times \boldsymbol{P}_i' - l\boldsymbol{u}_j \times \boldsymbol{P}_j'. \tag{9}$$

In the elastic case, the value of $\Delta\boldsymbol{P}$ can be obtained from the energy conservation law (m is the mass of a needle and I is its moment of inertia.)

$$\begin{aligned} \frac{P_i^2}{2m} + \frac{P_j^2}{2m} + \frac{J_i^2}{2I} + \frac{J_j^2}{2I} &= \frac{(\boldsymbol{P}_i + \Delta\boldsymbol{P})^2}{2m} + \frac{(\boldsymbol{P}_j - \Delta\boldsymbol{P})^2}{2m} \\ &+ \frac{(\boldsymbol{J}_i + L\boldsymbol{u}_i \times \Delta\boldsymbol{P})^2}{2I} + \frac{(\boldsymbol{J}_j - l\boldsymbol{u}_j \times \Delta\boldsymbol{P})^2}{2I}. \end{aligned} \tag{10}$$

From (10) one gets

$$\Delta P = -\frac{\frac{1}{m}\boldsymbol{u}_\perp \cdot \boldsymbol{P}_{ij} + \frac{L}{I}\mathbf{J}_i \cdot (\boldsymbol{u}_i \times \boldsymbol{u}_\perp) - \frac{l}{I}\boldsymbol{J}_j \cdot (\boldsymbol{u}_j \times \boldsymbol{u}_\perp)}{\frac{L^2}{2I}(\boldsymbol{u}_i \times \boldsymbol{u}_\perp)^2 + \frac{l^2}{2I}(\boldsymbol{u}_j \times \boldsymbol{u}_\perp)^2 + \frac{1}{m}}, \tag{11}$$

where we have used $\Delta\boldsymbol{P} = \Delta P\boldsymbol{u}_\perp$. This expression for $\Delta\boldsymbol{P}$ can be used as a reference for the inelastic case due to the ansatz

$$\boldsymbol{u}_\perp \cdot \Delta\boldsymbol{P}_{inel} = \varepsilon \boldsymbol{u}_\perp \cdot \Delta\boldsymbol{P}_{elas}. \tag{12}$$

The parameter ε is the so called normal restitution parameter. The range of ε spans the interval 0.5 to 1.0 if the same dissipated $\Delta \boldsymbol{P}$ is used in the momentum and angular momentum in (8). Choosing $\varepsilon = 1.0$ corresponds to the elastic case, while $\varepsilon = 0.5$ corresponds to the maximum dissipation during the collision. In the latter case, the particles, after collision, remain in contact and move with the same velocity. Values of ε smaller than 0.5 lead to non-physical overlaps of the particles. In the molecular dynamics simulations described in [4], we used a slightly different collision model in which only momentum was diminished. There, the range of ε without overlaps corresponded to the interval [0.0; 1.0].

Rewriting (11) by use of the velocity of the point of contact V^{imp} as

$$\Delta P = -\frac{V^{imp}}{\frac{L^2}{2I}(\boldsymbol{u}_i \times \boldsymbol{u}_\perp)^2 + \frac{l^2}{2I}(\boldsymbol{u}_j \times \boldsymbol{u}_\perp)^2 + \frac{1}{m}}, \tag{13}$$

one can see that the dissipation model (12) is similar to the ansatz of the contact velocity model (see for instance p.35 in [32]):

$$\boldsymbol{u}_\perp \cdot \boldsymbol{V}^{imp}_{inel} = \varepsilon \boldsymbol{u}_\perp \cdot \boldsymbol{V}^{imp}_{elas}, \tag{14}$$

in the sense that $\mathbf{V}^{imp}_{elas}$ is the only kinetic property on which the dissipation depends.

5 Molecular Dynamics

We have performed Molecular Dynamics (MD) simulations of a 2D system comprising $N = 800$ identical hard needles in the isotropic and in the nematic phase. In simulations, the particles were confined to a rectangular box of unit length and subject to periodic boundary conditions [2,10]. Each member of the system has unit mass and is characterised by its half-length L. To comply with the data from [7] we have used the reduced global density ρ^* as an input parameter for the simulation

$$\rho^* = \frac{N}{A}\,(2L)^2 = \rho\,(2L)^2, \tag{15}$$

where A is the area of the simulation box. With this choice the half-length the needle, L, depends on the input value of ρ^*. Simulations on the freely cooling process have been performed for a range of densities from the interval [3.5; 8.0]. More detailed analysis concerns different stages of clustering in the isotropic phase at $\rho^* = 3.5$ and in the nematic phase at $\rho^* = 8.0$.

As simulation technique we use the MD method of event-driven type. A main objective of this technique is to resolve for the time when two particles are in contact. We assume that this happens always when an end of one needle touches another needle at any point (See Fig. 1). Such assessment of the collision time has to be done for all possible pairs of needles, from which the shortest time t_{col} has to be chosen. Having obtained this value, we update next all particles' coordinates into new values at time $t + t_{col}$. After finishing this procedure the system is exactly at point when two needles are about to collide (interact).

For hard bodies, interactions take place with the help of infinitely strong repulsive forces which, consequently, lead to an infinitely short time of interactions. This assumption is non-physical and serves only for purposes of the simplest theoretical models. As a result of interactions, the momentum and the angular momentum of the particles are changed. For elastic interactions their new values can be obtained from the collisional formulas which are based on the momentum and energy conservation laws. In the inelastic case these values are also changed according to the additional dissipation rules. The update of the momenta finishes the cycle which promotes the system from one collision to another. For statistical purposes this cycle has to be repeated many times; the required number depends on the system size and is limited by the computer capacity.

At first, the MD technique described above has been applied to equilibrate our ensemble. In order to do so we started from a random distribution of the linear and angular velocities adjusted to the temperature $kT = 2$. Within the course of collisions, the system gradually evolves towards the state in which the velocity distributions are of the Maxwellian type, which is due to the equipartition theorem. We found that already after 125 collisions per particle, the Gaussian velocity characteristics is obtained, which is parametrised by a temperature that differs from the imposed initial temperature $kT = 2.0$ by less than 2%. Because of changes in the order parameter S for the system at the density $\rho^* = 8.0$, we performed further equilibration runs for the time corresponding to a total of 200 collisions/particle. The changes of the order parameter were still present indicating that strong fluctuations of the order are really inherent to this system. This feature has already been predicted by the Monte Carlo technique [7].

As a next step, starting from these equilibrium configurations, we undertook investigations on the process of free cooling according to the dissipative collision formulas (12). We performed production runs for the dissipative case with different degrees of dissipation strength. In Fig. 2 we present 2D needle configurations obtained from the simulations. Panels a) and b) show a configuration in the nematic state for $\rho^* = 8$ before (elastic case) and after dissipation (inelastic case). Similarly, panels c) and d) present configurations for the elastic and inelastic cases in the isotropic state at the density $\rho^* = 3.5$. There is a striking change in the particles' organisation in the inelastic case.

Under dissipation the system has developed thick bundles comprised of several needles together with spaces free of particles. In the nematic case the needles are still oriented towards the original preferred orientation, however, the global order is diminished (see Fig. 3). Dynamics of the bundles growth depends on the magnitude of the restitution parameter. For large dissipation (small restitution parameter) two particles remain after collision in their neighbourhood. This fact leads to shorter times until next the collisions and, consequently, to larger collision frequency. For small dissipation the needles preserve much of their energy and can still interact with more distant particles, and the growth of bundles is slower. Another factor that influences dynamics of the cluster growth is the initial density. In panel f) we present an isotropic configuration after dissipation whose mean temperature is comparable to that of the nematic case from b). A quick comparison of these two panels convinces about stronger clustering in more dense systems. Such a statement, however, requires confirmation from the side of quantitative measurements. Finally, panel e) shows a close-up of a bundle that has been taken from the system in b).

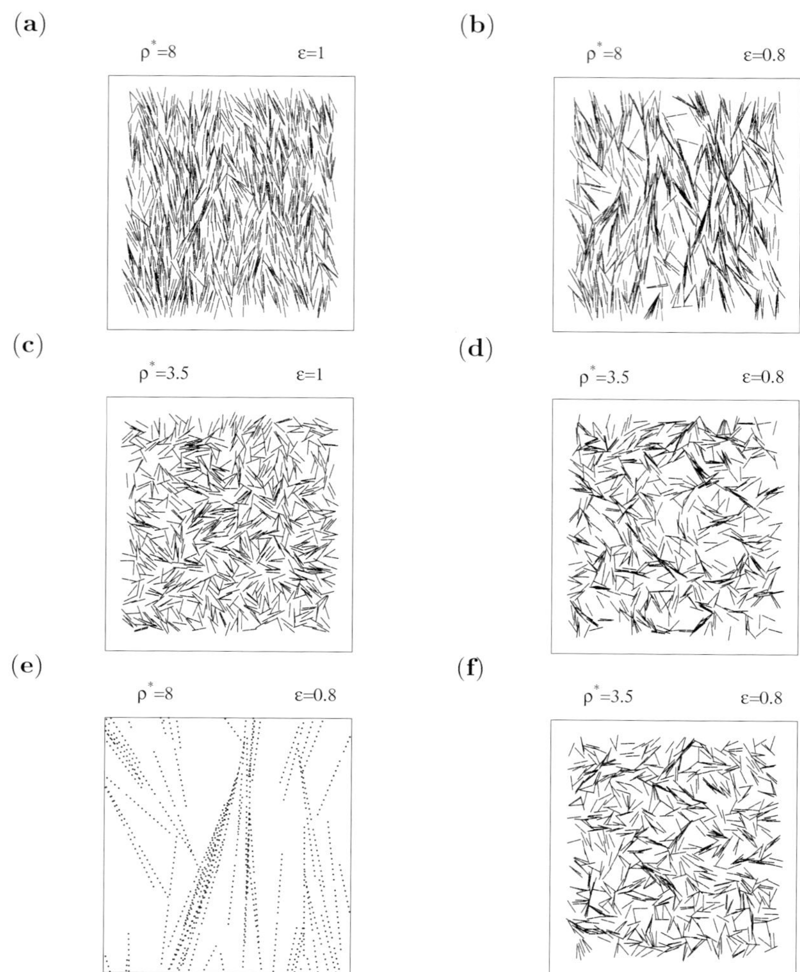

Fig. 2. Typical configurations of the 2D hard needle system. (**a**) and (**c**) are the equilibrium configurations for the nematic ($\rho^* = 8.0$) and for the isotropic phase ($\rho^* = 3.5$) at the temperature $kT = 2.0$. (**b**) and (**d**) are obtained after dissipation: an orientationally ordered state at the density $\rho^* = 8.0$ and the temperature $kT = 0.41$ (after 10000 collisions) and a disordered state at the density $\rho^* = 3.5$ and the temperature $kT = 0.172$ (after 10000 collisions). (**e**) shows a close-up of a bundle from (**a**). (**f**) presents an isotropic configuration obtained at the temperature $kT = 0.418$ (after 5000 collisions), which is comparable to the temperature of the system from (**b**)

6 The Order Parameter Tensor and the Collision Frequency

The two-dimensional order parameter tensor $\boldsymbol{Q}$ of a N-particle system differs from its three dimensional counterpart and is defined by

$$\boldsymbol{Q} = \frac{1}{N} \sum_{i=1}^{N} (2\boldsymbol{u}_i \otimes \boldsymbol{u}_i - \boldsymbol{I}) \tag{16}$$

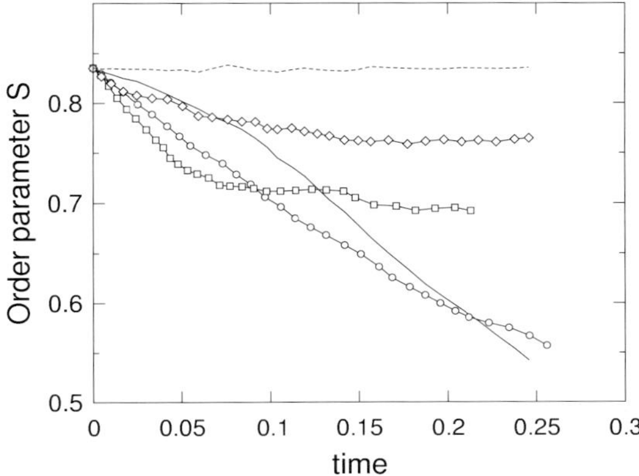

Fig. 3. Order parameter for different restitution parameters ε obtained during 30000 collisions for the nematic phase at the density $\rho^* = 8.0$. The *dashed line* corresponds to $\varepsilon = 1.0$ (no dissipation), the *solid line* to $\varepsilon = 0.9$, *open circles* to $\varepsilon = 0.8$, *squares* to $\varepsilon = 0.7$ and *diamonds* to $\varepsilon = 0.6$. Different lengths of the curves are due to different collision frequencies

where $\boldsymbol{u}_i$ is the orientation of the particle i confined to the XY-plane and $\boldsymbol{I}$ is the two dimensional unit matrix. $\boldsymbol{Q}$ is traceless and has eigenvectors that point towards the director – the direction of the most preferred orientation in the system and the vector perpendicular to the director. Eigenvalues of $\boldsymbol{Q}$ are given by $\{S, -S\}$, where S is the scalar order parameter – a measure of the order degree. In the isotropic phase, $S = 0$, and for perfectly aligned particles, $S = 1$ holds.

Formula (16) is a convenient recipe how to calculate the tensor from the information about the microscopic state of the system. Usually, the result is non-diagonal due to the fact that the director does not perfectly lie along the X (or Y) - axis. Diagonalisation, which, in other terms, is the rotation of our initial frame of reference brings these two directions in coincidence. Hence, after diagonalisation of $\boldsymbol{Q}$ we find the global order parameter S and the director.

In Fig. 3, we present the decay of the order parameter obtained for different restitution parameters. As can be foreseen, newly created voids in the system are also an opportunity for the particles (or whole bundles) to rotate to more "off-the director positions". As a consequence, we observe a diminishing tendency of the order parameter upon cooling of the system, which we refer to as orientational cooling. For the values of ε from the interval $[1.0; 0.75]$ we observe a general tendency that stronger dissipation is followed by stronger decrease of the system order. These results have convinced us of a possibility to obtain a cooling-driven transition to the isotropic phase if the initial thermodynamic conditions are appropriately chosen or the dissipative runs are appropriately long. For values of the restitution parameter ε smaller than 0.75, the particles in the system lose their ability to rotate during the collision and, effectively, strong clustering of the ordered needles occurs. The curves in Fig. 3 have different lengths although the total number of collisions has been kept the same for all the cases studied. The time used here for X axis (and also in some next figures) is a dimensionless number, whose values are bound to the chosen molecular dynamics parameters kT, L and m. This fact indicates different collision rates. Figure 4 presents the cumulative number of collisions over the passing time for the cases with different restitution parameters ε (values of ε are described in the legend of Fig. 4).

In the equilibrium state, the particles collide with one collision frequency which depends on the system density, and the dependence of the collision number upon the time is a linear function. When dissipation is present, this is no longer true. The collisions of the dissipative system with ε from the interval $[1.0; 0.7]$ are less frequent than in the equilibrium state even though at later steps of cooling density instabilities are present. Then, for values of ε below 0.7 the situation changes dramatically. On average, the collision rate (dN_{col}/dt) is growing upon diminishing ε, and at a certain value between 0.6 and 0.7 the profile becomes similar as in the equilibrium case. At the same time, we observe small steps in the curves that are almost parallel to the N_{col} axis, which denotes divergent collision frequency. This is an evidence of inelastic collapses which last over a maximum of a few hundreds of collisions and are dissolved afterwards. Note that inelastic collapse is one of the most characteristic features of freely cooled granular systems [12].

The behaviour of $N_{col}(t)$ depends on the strength of dissipation as well as on the stage of cooling. The fact that particles are bound to clusters is not sufficient to explain the increase of the collision rate. One can see from Fig. 4 that the clusters obtained, for instance, with $\varepsilon = 0.9$ and $\varepsilon = 0.6$ can exhibit very different collision frequencies although their general appearance is very similar. This rather surprising behaviour is the result of the strong relation between the system energy and the density inhomogeneities.

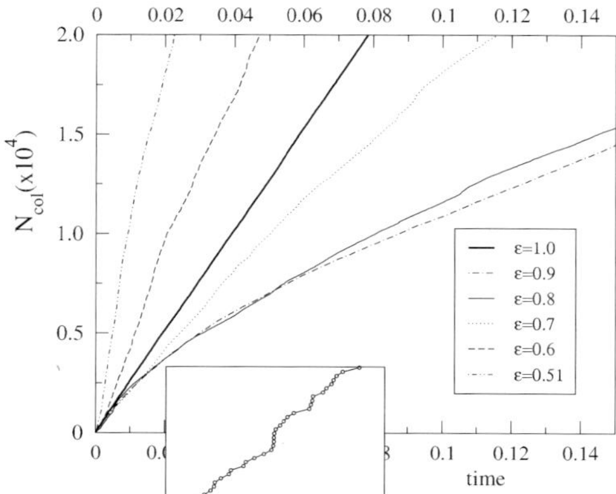

Fig. 4. Number of collisions for different restitution parameters vs. time. The inset profile shows a magnified part of the collision number from a clustering state; vertical steps corresponding to divergent collision frequency are examples of the inelastic collapse

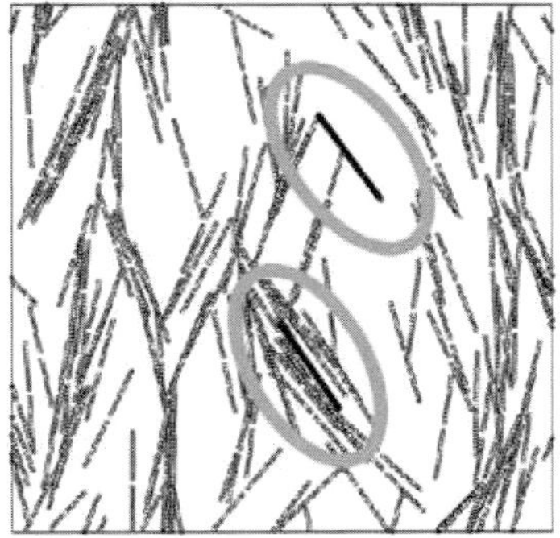

Fig. 5. Representation of the concept of the nearest neighbourhood boxes. The boxes are not fixed in space and their orientation depends on the particle to which they are associated. (The size of the ellipses is not the same as given in the text)

7 Quantitative Description of Orientational Clusters

In simple fluids, short-range structures can be investigated by means of the radial distribution function. This function is of importance for two main reasons: a) it is accessible by use of experimental methods as neutron, X-ray or light scattering and b) it plays a central role in kinetic theories [14]. By definition, the radial distribution function is the ratio of the average density at a distance r from any given particle to the density at a distance r from a particle in the ideal gas under the condition that overall densities are the same. This function, however, is not a good tool for the detection of inhomogeneities. For an inhomogeneous system the radial distribution exhibits strong fluctuations at short distances which depend on the width of the histogram cell. This is, of course, the result of the fact that the clusters are irregular both in size and in orientation, and that also their positions are at random. In two dimensions we can see where these clusters are and how they look, in particular, whether they are strongly developed or not. If we assume much larger systems and a more general, three-dimensional case, this purely optical detection of clusters is no longer possible. Thus, the problem arises to find a method or a measure which is capable of capturing the features of the clusters. In view of this need we came up with a concept of, what we name henceforth the nearest neighbourhood density. Such a function allows us, to a smaller or larger degree, to quantify the process of the bundles creation.

7.1 Density Inhomogeneities

We consider an ellipsoidal box around the centre of each particle (see Fig. 5), whose long axis corresponds to the particle orientation and which has a length equal to $2L$. The length of the short axis is taken as $L/2$. The boxes are not fixed in space and their cumulative area does not sum up to the system volume. We calculate the number of particles x_N whose centres are found within each individual box and assign it to the number of average density within the box $\rho^*_{box} = x_N/A_{elip}$, where

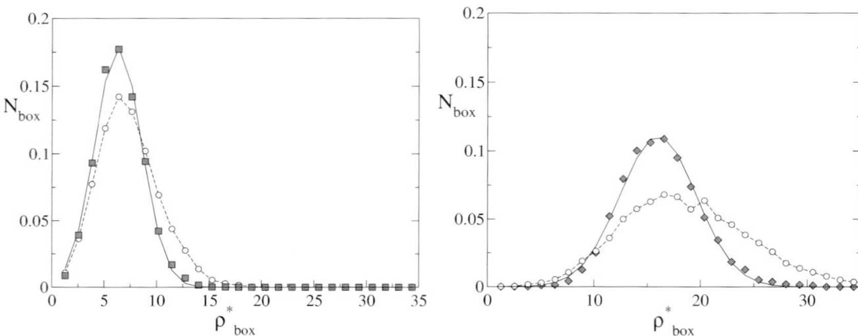

Fig. 6. Nearest neighbourhood density N_{box}; *diamonds* correspond to the non dissipative case and the *solid line* is the fitted Gaussian profile; *circles* are obtained for the system after dissipation; the *dashed line* is used to guide the eye. The left panel depicts the isotropic phase with box mean density $\tilde{\rho}_{box} = 6.32$ and the global density $\rho^*_{box} = 3.5$. The right panel shows the nematic phase with box mean density $\tilde{\rho}_{box} = 16.0$ and the global density $\rho^*_{box} = 8.0$. The temperature for the dissipated system in both cases is $kT = 0.51$.

A_{elip} is the ellipse area. From this data of N neighbourhoods we collect the sets with the same number of particles and build upon them a histogram that shows the number of boxes $N_{box}(\rho^*_{box})$.

The resulting distribution functions (normalised to unity) for the isotropic and the nematic phases are presented in Fig. 6. The profiles obtained for the elastic case fit very well to the Gaussian curve which has its maximum at a certain number larger than the global density. In general, the curves obtained after dissipation are *non symmetrical* with a well pronounced slope towards higher densities. This slope is a direct evidence of new structures – bundles. To obtain the profiles in Fig. 6, we performed averaging over 21 different configurations for the elastic case, and over 10 configurations for the inelastic case. We had to do so since single configurations turned out to be too small with too few clusters to provide enough information for the construction of "smooth" probabilistic functions $N_{box}(\rho^*_{box})$.

The method of nearest neighbourhoods suffers from a certain degree of arbitrariness due to the shape of the neighbourhood box. The density at which one finds the maximum of the distribution profiles is not an universal number. We will name it the *box mean density* $\tilde{\rho}_{box}$. For large and circular neighbourhoods the value of $\tilde{\rho}_{box}$, as it can be expected, lies close to the global system density ρ^*. In this case, however, the information about clusters and their orientations is lost. In the case of anisotropic ellipsoidal boxes, as proposed above, the box mean density is shifted even in the equilibrium case. For the system at $\rho^* = 8.0$, it is almost twice as large, $\tilde{\rho}_{box} = 16$, and for $\rho^* = 3.5$ it is $\tilde{\rho}_{box} = 6.32$. This shift of density, as it can be seen from Fig. 7, depends only on the aspect ratio of the axes of the ellipse assumed as the neighbourhood. It does not depend on the ellipse size. When $\tilde{\rho}_{box}$ is considered in relation to the global density as $\tilde{\rho}_{box}/\rho^*$, then it turns out to be well described by a universal curve which depends only on the box anisotropy and approaches unity for isotropic boxes (see the inset in Fig. 7).

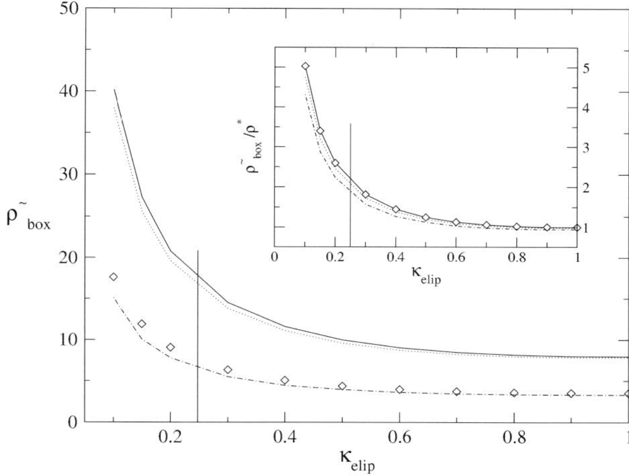

Fig. 7. Neighbourhood mean density $\tilde{\rho}_{box}$ as a function of the box parameters. The curves are obtained from the equilibrium configurations for $\rho^* = 8.0$ where the large ellipsoid axis is $a_{elip} = 10L$ (*solid line*) and $a_{elip} = 2L$ (*dotted line*), and for $\rho^* = 3.5$ where the large ellipsoid axis is $a_{elip} = 10L$ (*diamonds*) and $a_{elip} = 2L$ (*dash-dotted line*). For spherical boxes (the ellipse aspect ratio $\kappa = b_{elip}/a_{elip} = 1$) the mean density $\tilde{\rho}_{box}$ approaches the system's global density ρ^*. The inset shows the ratio of $\tilde{\rho}_{box}$ and ρ^* for the same box parameters as in the main picture (we use also the same symbols). Straight vertical lines indicate the box dimensions which we have used in this paper to analyse profiles of N_{box}. The figure shows that different systems with different sizes of the neighbourhood boxes can scale to the same curve and that the increase of the neighbourhood mean density is only the result of the box anisotropy. $\tilde{\rho}_{box}$ is dimensionless which comes from (15) and the assumption of non-unit values of the area A and the needle half length L

Knowledge about $\tilde{\rho}_{box}$ is crucial for further calculation of the first moments of the function N_{box} that is defined as

$$M_1 := \int_0^\infty (\rho^*_{box} - \tilde{\rho}_{box}) N_{box}(\rho^*_{box}) d\rho^*_{box} . \tag{17}$$

M_1 vanishes for all those functions $N_{box}(\rho^*_{box})$, which are symmetric about $\tilde{\rho}_{box}$. At the same time, any deviation from symmetric profiles leads to a nonzero value of the moment M_1, and the stronger the asymmetry of the profile is, the larger is the absolute value of M_1. Due to this behaviour, M_1 can be regarded as a measure of the profiles asymmetry. In case of the function $N_{box}(\rho^*_{box})$, asymmetry of the profiles is a result of the clustering process. To show this, we present in Fig. 8 gradual changes of the function $N_{box}(\rho^*_{box})$ at different stages of cooling. One observes that the maximum of the function systematically lowers upon cooling with a simultaneous growth of the slope on the side of larger densities. This is the case in the isotropic as well as in the nematic phase and is caused by the fact that the needles gather together in tight bundles. The curves in Fig. 8 are parametrised by

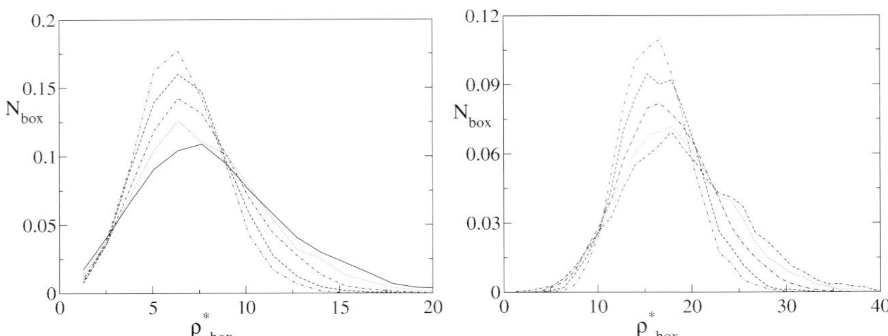

Fig. 8. Time evolution of the nearest neighbourhood density N_{box}. The left panel shows results for an isotropic system at the density $\rho^* = 3.5$ and the temperatures (in units of kT): *dotted line* – 2.0, *dashed line* – 0.885, *dash-dotted line* – 0.505, *double-dot-dashed line* – 0.347, *solid line* – 0.24. The right panel shows N_{box} for the nematic phase at the density $\rho^* = 8.0$ and the temperatures: *dotted line* – 2.0, *dashed line* – 0.759, *dot-dashed line* – 0.615, *double-dot dashed line* – 0.536, *solid line* – 0.486

the mean granular temperature (kT). If needed, the explicit dependence on time can be obtained from the energy dissipation.

By use of (17), one can investigate the time evolution of formation of the bundles. In order to calculate the integral in (17), we use the standard Gauss method. For this method, however, we need to know the values of the integrated function at the Gaussian points. We interpolate them in the following way: Each three successive points in growing order of the data for ρ^*_{box} are uniquely assigned to a parabola, by use of which we can calculate a value of N_{box} at the Gaussian point that lies within the interval built on these three points. Next we drop the smallest point in ρ^*_{box} and choose the next one lying close to the previously largest ρ^*_{box} and calculate N_{box} at the same Gaussian point as before. Usually this value of N_{box} is not the same as the one from the previous step, so we utilise the arithmetic averaging. No such average is taken for the triplets of the points at the beginning and at the end of the data set.

In Fig. 9 we show the moment M_1 as a function of $ln(kT)$ for the nematic and the isotropic phase, respectively. It has been obtained by use of (17), where particular values of $\tilde{\rho}_{box}$ have been found from the least square fit of the Gaussian function to the data of N_{box} obtained for the elastic case.

The obtained profiles exhibit a two-step character: the initial stages of cooling are given by slowly ascending convex curves, whereas at later stages they take on a linear character. The latter type of behaviour indicates that late phases of cooling depend logarithmically on the granular temperature. This statement is only a suggestion based on the results of a few simulations. A more systematic study is needed for its support. The point when the profiles change their character can be associated to the point terminating the homogeneous state and starting density instabilities. At this point bundles of well ordered needles are being gradually built together with the occurrence of very dilute or empty regions.

7.2 Differences Between Clusters of Hard Spheres and Hard Needles

It is constructive now to compare basic features of the cluster formation in anisotropic systems made of hard needles and in isotropic systems made of hard spheres, respectively. In the latter case, a comprehensive study of cluster growth has been presented by Luding and Herrmann in [19]. In order to avoid inelastic collapse effects that stop the computer calculations, the authors of [19] introduced a finite contact duration time together with an elastic contact energy. This allowed them to perform simulations over different stages of the cluster formation for hard spheres. They distinguished three main regimes of the cluster behaviour. The first regime is the homogeneous cooling state, where the clusters are small and grow very slowly. In the second regime the cluster growth intensifies and is characterised by decrease in the number of clusters due to their fusion, growth of the mean cluster size and growth of the largest cluster. Finally, the third regime corresponds to the state when most of the particles are gathered in one huge cluster that reaches the size of the system.

In comparison to the hard needles case, the basic difference concerns the third step. Because of the strong shape anisotropy and possibility to dissipate kinetic energy through rotational motion, the particles have not of much chance to build one big cluster. Most probable is then a transition to the overall isotropic phase caused by the reorientations of whole clusters. This case, however, has not yet been studied.

7.3 Order Parameter of the Neighbourhood Boxes

The order parameter as introduced by (16) forms an objective measure of the system order only if the considered ensembles are large. If, for instance, we take into account only a single particle, the resulting order is $S = 1$, which indicates perfect alignment. On the other hand, for two particles the isotropic order, $S = 0$, is obtained only when the particles are perpendicular. If the system contains several particles, the calculated order is still far from the macroscopic value. This can best be seen in the left panel of Fig. 10, which shows the order parameter of the boxes for the system at density $\rho^* = 3.5$. For this density, the phase is isotropic and the

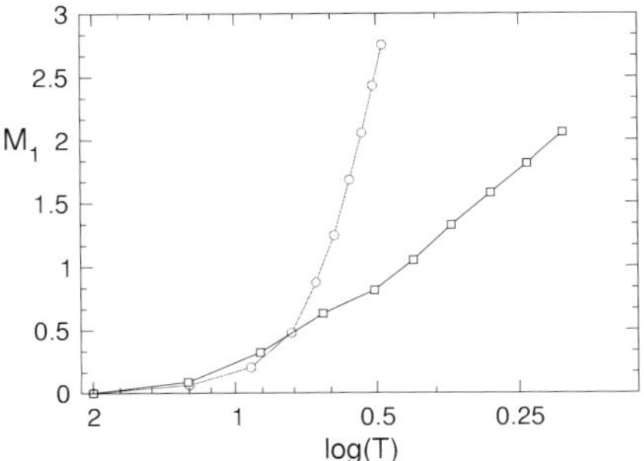

Fig. 9. First moment of the function N_{box}. *Circles* are for the nematic phase at density $\rho^* = 8.0$, and *squares* for the isotropic phase at $\rho^* = 3.5$

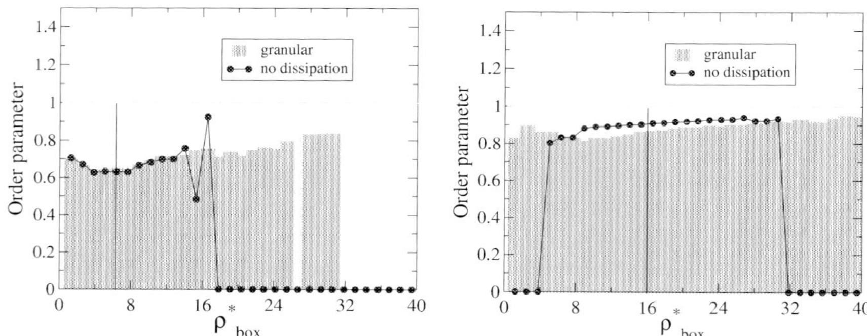

Fig. 10. Order parameter S_{box} of the nearest neighbourhood boxes vs. their density ρ^*_{box}. *Dots* are the data of the order S_{box} for the system in equilibrium. *Shadowed area* is the histogram type presentation of the order parameter S_{box} of the ellipsoidal boxes for the system after dissipation, where the restitution parameter was $\varepsilon = 0.8$. If no box with a given density is present, no order parameter is assigned. The panel on the left side is for the isotropic case at the density $\rho^* = 3.5$, where the temperature of the cooled system is $kT = 0.201$, and the panel on the right side is for the nematic phase with $\rho^* = 8.0$, where the temperature of the cooled system is $kT = 0.486$. Vertical lines show the position of the mean box density, $\tilde{\rho}_{box} = 6.41$ and $\tilde{\rho}_{box} = 16.0$, for the isotropic and the nematic case, respectively

macroscopic order parameter is found to be about zero. At the same time, the box order is between 0.6 and 0.8 and gradually grows with increase of the box density. These higher values of S are bound to the artefact of the boxes containing only few particles as discussed above. A similar effect has already been mentioned Fig. 7.

Despite the fact that S_{box} is a relative measure of the order, we can still compare data of different boxes and interpret the differences among them in relation to the properties of the local structure. Moreover, S_{box} is sensitive to the global phase symmetry. In the equilibrium case (black circles in Fig. 10), the order parameters are different in different phases even if the box density is the same. In the nematic phase they are systematically larger, although the differences are not very big. More features are revealed in the freely cooling system. Looking at the data obtained for this system (shadowed area in Fig. 10), it is clearly seen that in the granular case new boxes have been formed, whose orientational properties differ from the boxes from the system in equilibrium. These new boxes are a direct evidence of the orientational clusters, which are present in the isotropic (Fig. 10 (left)) as well as in the nematic phase (Fig. 10 (right)). The order parameter of newly created boxes in the nematic phase is very high, almost close to perfect alignment, whereas in the isotropic phase its behaviour is uneven. This irregular behaviour in the isotropic phase indicates presence of very well ordered as well as less ordered bundles. Taking into account the granular temperatures associated to the above two cases, $kT = 0.486$ for the nematic and $kT = 0.201$ for the isotropic phase we can conclude that highly ordered bundles are less frequent to encounter if the cooling is imposed on the isotropic phase.

8 Hard Needles in a Pipe

Properties of fluids confined to very small spaces are dramatically different from the bulk media. If the slit size is comparable to the particles size capillary effects come into play which encompass phenomena like density inhomogeneities, capillary and surface phase transitions, novel critical points, "frustrated" states of order or changes in the viscosity properties [6,33]. They are mostly produced by entropic effects but can be significantly moderated by appropriate treatment of the surfaces. These new properties have been found useful in many ways and during the last 30 years, confined fluids have become extremely important in different areas of technological applications. Examples of highest moment can be found in the liquid crystal industry, in the nanotechnology and in biological systems.

8.1 General Remarks

In order to study basic properties induced by the confinement one often turns to the simplified systems. In molecular simulations, the latter are usually composed of perfectly hard bodies such as spheres, ellipsoids, needles or spherocylinders that are placed between two structureless walls [5,15,22,25]. In this section, hence, we present an example of the configurational properties of the simplest system of 2D hard needles that is confined in a pipe. The first part of the section concerns an equilibrated system in which the particles interact elastically. To this system we introduce in the second part an additional effect of dissipative walls. A comparison between two different phases, nematic and isotropic, is presented.

A confined system is difficult to treat in computer simulations, since periodic boundary conditions in the direction normal to the walls cannot be imposed. This fact entails the necessity to use large dimensions of the substrates to ensure sufficiently large statistics. A drawback of using big surfaces is, then, that long simulation times are needed. Alternatively one can average over a number of equivalent ensembles. However, are the systems in the pipe really equivalent to these model substitutes?

In the elastic case of 2D hard needles spontaneous micro-flows can develop along the pipe which may depend on the distance from the wall even if the average momentum of the whole system is put to zero. Once developed they can pertain to longer times. The direction of the relative flow between the surface areas and the middle of the pipe can establish itself in a somewhat stochastic way and, after a sufficiently long time, can be expected to change. However, it is difficult to say how much time would be needed in reality for a given pipe to reverse the micro-flow. One of the factors that can steer the motion of the needles can be the particle's diffusivity. In case of 2D infinitely thin hard lines, the higher order indicates also that the needles can diffuse more easily. As they approach perfect alignment, the probability of a collision goes to zero, because of the needles' zero width, and, consequently, the diffusivity becomes infinite. Such a situation prevails in the vicinity of the walls. If a needle happens to find itself close to the wall, it will certainly be ordered along the wall direction and, thus, will be expected to have higher diffusivity along the pipe. This example indicates that kinematic statistical averages, such as the velocity profiles transverse to the walls, may depend strongly on the length of the pipe and on the time of equilibration. Configurational properties, on the other hand, will rather weekly or not depend on the possible micro-flows.

8.2 Wall-Particle Collision Formulas

A collision between the hard line and the wall can be considered as a collision between two hard lines, when one of them is infinitely long and heavy. The geometry is as in Fig. 1. The particle j represents now the wall W onto which the needle i bumps with one of its ends. We assume (as always) that there is no tangential friction. Thus, the exchange of the momentum $\boldsymbol{\Delta P}$ takes place in the direction normal to the wall. The value of $\boldsymbol{\Delta P}$ is given by the collision rules which must now be generalised as for a binary mixture to suit the assumptions made. For two different needles having the indices, say, i for a particle and W for the wall, a general expression for ΔP yields

$$\Delta P = \frac{A}{B}, \tag{18}$$

where

$$A = 2(\boldsymbol{v}_i - \boldsymbol{v}_W) \cdot \boldsymbol{u}_s + 2\left(L_i \Omega_i \boldsymbol{u}_i - L_W \Omega_W \boldsymbol{u}_W\right) \times \boldsymbol{u}_s \tag{19}$$

and

$$B = -\frac{1}{m_i} - \frac{1}{m_W} - \frac{L_i^2 \left(\boldsymbol{u}_i \times \boldsymbol{u}_s\right)^2}{I_i} + \frac{L_W^2 \left(\boldsymbol{u}_W \times \boldsymbol{u}_s\right)^2}{I_W}. \tag{20}$$

In the above equations, $\boldsymbol{u}_i, \boldsymbol{u}_s$ $(=\boldsymbol{u}_\perp)$ and $\boldsymbol{u}_W$ $(=\boldsymbol{u}_j)$ are as in Fig. 1 and L_i and L_W are the distances between the collision point and the particles centres. I_i and I_W are the moment of inertia of the particle i and the wall, respectively. The wall is at rest, i.e. $\Omega_W = v_W = 0$, and infinitely heavy $m_W = \infty$ (m_W is the mass of the wall), so effectively ΔP depends only on the properties of the colliding particle i:

$$\Delta P = \frac{2\boldsymbol{v}_i \cdot \boldsymbol{u}_s + 2L_i \Omega_i \boldsymbol{u}_i \times \boldsymbol{u}_s}{-\dfrac{1}{m_i} - \dfrac{L_i^2 \left(\boldsymbol{u}_i \times \boldsymbol{u}_s\right)^2}{I_i}}. \tag{21}$$

The formulas (18)-(21) for the particle-wall collision and (11) for the particle-particle collision are the key elements of the molecular dynamics program. Using them, we simulated a system of 160 needles situated between two structureless walls that are separated by a distance of 0.2 (1/5 of the simulating box dimensions). This width corresponds to two lengths of the needles that are used for the nematic phase (the number density $\rho^* = 8.0$) and about three lengths of the needles used for the isotropic phase (the number density $\rho^* = 3.5$). Such a small wall separation makes the features caused by the walls more visible. At first we performed the equilibration runs with 100000 collision which gives an average of 625 collisions per particle. Contrary to the bulk system, this average does not correspond to the real mean number of collisions that a chosen particle undergoes. Much of the simulation time is consumed by the chattering particles close to the surfaces. This phenomenon poses the biggest difficulty in equilibrating the system.

9 Non-dissipative Pipe

The system studied consists of $N = 160$ particles contained in a pore of width $L_{sep} = 0.2$ (the wall separation). The needles undergo a prolonged equilibrating

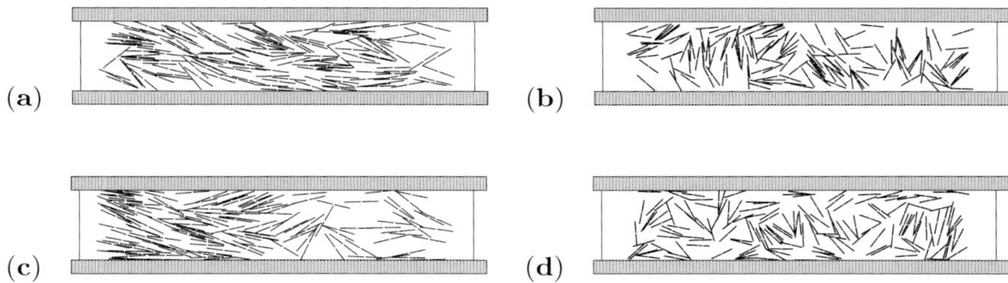

Fig. 11. Hard needles confined between two walls. (**a**) and (**b**) are the equilibrium configurations for the densities $\rho^* = 8.0$ and $\rho^* = 3.5$; (**c**) and (**d**) show the configurations for the similar densities after dissipation of 25 % of the energy caused by the dissipative collisions with the walls

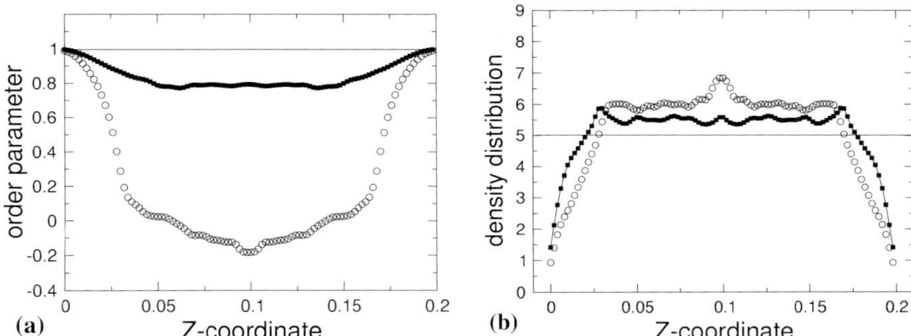

Fig. 12. Order parameter (**a**) and density distribution (**b**) of the needles in the non-dissipative pipe obtained after averaging over 250 configurations. *Open circles* refer to the isotropic configuration at the density $\rho^* = 3.5$ and *filled circles* to the nematic one at the density $\rho^* = 8.0$.

run. Next, a sequence of configurations (about 250) is stored within a production run. This large number was needed to obtain smooth profiles of the mean density distribution and the order parameters. Some snapshots of equilibrium configurations we present in Figs. 11a,b. Figure 11a shows a nematic phase at the density $\rho^* = 8.0$ and Fig. 11b an isotropic phase at the density $\rho^* = 3.5$. The separation of two walls in the first case is twice larger than the length of the needle and in the second, isotropic case three times larger. Such small values of separations are indispensable to investigate influence of the walls. For larger separations this influence rapidly diminishes with a distance from the substrate and the system becomes of a bulk type.

The mean density and orientational order profiles of 3D needles confined between walls are already known from density functional theory [25,30] and Monte Carlo simulations [25]. The 3D case is a simple example: the needles do not form any liquid crystalline phase. Two-dimensional needles in the isotropic phase are very similar to the 3D case results. As given in Fig. 12, isotropic profiles, represented by *open circles*, consist of two main parts. The first one is a wall interfacial area, where the rotations of the particles are hindered by the walls. The other part is a bulk part. For the density distribution, the connection between these two regions is cusp-like, and for the order parameter it resembles a kink. In the wall region, the density (always normalised to unity) is much smaller than in the middle of the sample. This is a general feature of all types of the hard bodies at a wall and is named *depletion effect*. The order parameter directly at the wall is equal to unity, which results from (16) for particles aligned perfectly parallel to the walls. It gradually diminishes, and beyond the surface area establishes its value about zero. In the isotropic phase, the position of the cusp and the kink are the same for the density distribution and for the order parameter and complies with the half length of the needle ($L = 0.033$). This is no longer true for the nematic case (*filled circles* in Fig. 12). Here, the cusp in the density profile is closer to the wall than the half length of the particle ($L = 0.05$), whereas in the order parameter it is still about L. For the nematic phase the value of the order parameter in the middle of the pipe is established about $S = 0.8$ which corresponds to the bulk value of S (see also Fig. 3).

All the density distributions f_{dens} we present are normalised to unity. To obtain them, the following scaling formula has been applied

$$f_{dens} = \frac{f_{hist}}{N_{conf}} \frac{N_{grid}}{L_{sep} N} . \tag{22}$$

In (22), f_{hist} is the number of particles that are found in a given histogram cell for all the configurations considered, N_{conf}. N_{grid} is the number of cells in the histogram for which reasonable values are taken from the interval $[20; 50]$. The wall separation L_{sep} is set to 0.2 and the number of particles is $N = 160$.

Figure 13 shows the component of the linear velocity of the needles, which is along the direction of the pipe V_x. The oscillations about the zero level are quite strong with visible differences between the wall and the middle areas. The magnitude of them is assessed as 1/5 of the width of the Maxwellian distribution that characterises the bulk system for the assumed internal energy ($kT = 2.0$), hence we interpret them as micro-flows. These profiles were obtained after averaging over 250 configurations that were realized for a *single* pipe. In contrast to the density and order parameter curves, we do not expect the velocity profiles to have universal character, and for another pipe they can be completely different. The only regular feature is the fact that the total momentum in the X-direction must be zero.

10 Dissipative Pipe

In Figs. 11c,d we presented snapshots of the configurations of the hard needles that are obtained in the dissipative pipe where the particles among themselves still interact elastically. The restitution parameter during the collision of the needle with the wall is taken to be $\varepsilon = 0.8$. The snapshots are obtained in the system which has lost about 25% of its energy. At this stage of cooling we do not observe large aggregations of the needles close to the walls, although there is a visible increase of

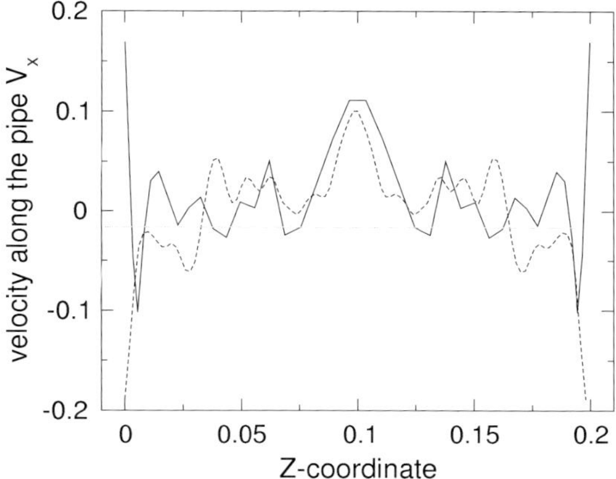

Fig. 13. Averaged velocity of the needles along the pipe plotted versus the distance from the walls. The *solid line* is the profile for the isotropic, while the *dashed line* is for the nematic phase

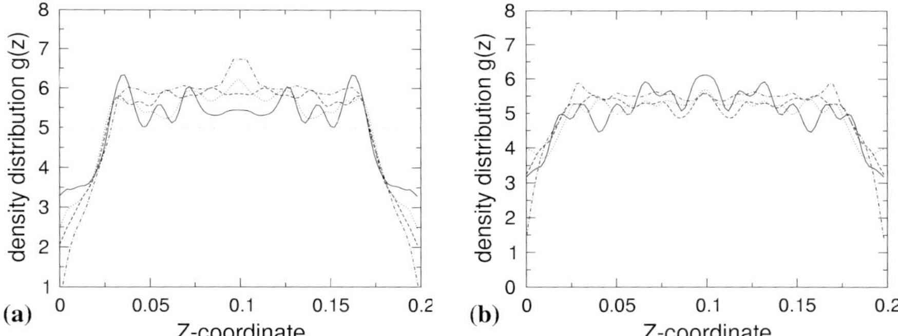

Fig. 14. Density distributions of hard needles at different temperatures for (**a**) the isotropic and (**b**) the nematic case . The *dotted line* corresponds to $kT = 2.0$, the *dashed line* to $kT = 1.8$, the *dash-dotted line* to $kT = 1.7$ and the *solid line* to $kT = 1.6$. The restitution parameter used in the process of dissipation is $\varepsilon = 0.8$

the particles in the surface areas, more pronounced in the isotropic phase. Such an observation means that the depletion effect is weakened, as it was expected.

Looking at the profiles in Fig. 14 one can see that the parts at the walls are getting systematically higher but still do not reach the average level as in the middle of the pipe. Another characteristic feature observed are strong density instabilities which, in spite of quite large statistics used, are still present. The difference in regularity of the profiles between the fully elastic case and the case where the cooling takes place through the dissipative walls is clearly visible. The same concerns the orientational profiles in Fig. 15, which may lead to the conclusion that even small losses of the system energy (like 10%) cause significant density and order parameter instabilities.

11 Summary and Conclusions

Simulations of a freely evolving dissipative gas of two-dimensional hard needles were presented. In particular:

a) Spatial and orientational properties were examined with respect to different values of the restitution parameter.

b) The time-dependence of the observed density and orientation instabilities (bundles) were monitored and compared in the isotropic and in the nematic phase.

c) In order to quantify the growth of the orientational clusters, we introduced the nearest neighbourhood density function N_{box} based on asymmetrical boxes that are associated to each needle. Properties of this function were examined. In an equilibrium state its shape agrees with the Gauss distribution. During the evolution of the system, N_{box} gradually lowers its maximum and builds up a slope on the side of larger densities. This is a direct effect of density instabilities. Asymmetry of the neighbourhood density can be illustrated best by means of its first moments which, after the initial stage, take on a linear dependence vs. the logarithm of the granular temperature. The crossover to this state can be assumed as the end of the homogenous state and the start-up of an intensified process of the cluster formation. Contrary to the system made of hard spheres, hard needles do not tend to gather themselves in one huge cluster. Instead, they remain in several thick blocks which themselves may exhibit particle like behaviour such as, for instance, collective reorientation.

d) When the initial system is prepared in a nematic phase, then its order parameter systematically diminishes upon dissipation. This reduction of the order is connected with the growth of the effective free volume that is accessible to the bundles. Although the needles in individual clusters are getting well ordered with respect to each other, the angular distribution of the bundles becomes more random. A picture of the order within the bundles can be reconstructed from the behaviour of the order function of the nearest neighbourhoods. In the nematic regime the order of the particles within the bundles approaches perfect alignment whereas for the isotropic case the order of the bundles has irregular character. Based on this molecular dynamics results we anticipate the possibility of the nematic-isotropic transition. More detailed thermodynamic parameter studies and much longer runs are required for this purpose.

e) We performed basic molecular dynamics simulations of 2D hard needles that are confined to a slit. In this case the main change in the behaviour comes from the interaction of the system particles with the confining walls. To introduce these interactions we assumed that the walls are modelled by also hard needles but of infinite length and mass. This assumption requires generalisation of the collision formulas for the case of a binary mixture. Then, a limit for the infinite mass is taken for one of the needles. The geometry of the considered case chooses the direction across the slit (pipe). Along this direction the physical properties of the ensemble take on characteristic profiles. In the analysis of the features of confined 2D hard needles we focus on two types of profiles: density and order parameter distributions. Two cases are studied, the nematic and the isotropic case, in a narrow slit, whose

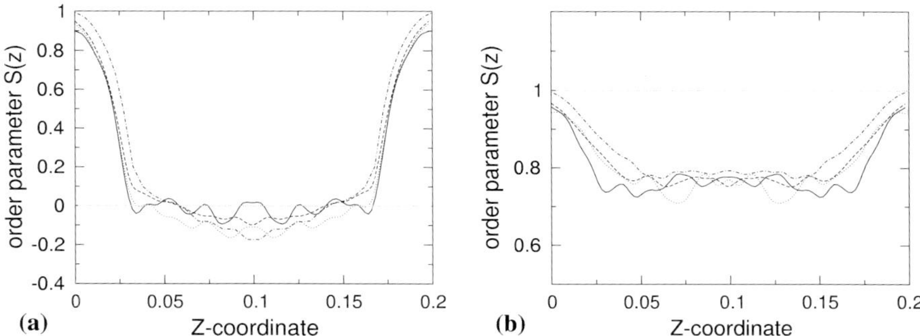

Fig. 15. Order parameter profiles at different temperatures for (**a**) the isotropic and (**b**) the nematic case . The *dotted line* corresponds to $kT = 2.0$, the *dashed line* to $kT = 1.8$, the *dash-dotted line* to $kT = 1.7$ and the *solid line* to $kT = 1.6$. The restitution parameter used in the process of dissipation is $\varepsilon = 0.8$

size is comparable to the length of the needles. The density profiles at the walls are lower than in the middle part of the slit. This is due to the so-called depletion effect which is characteristic for all hard bodies at the walls. Surface area joins the bulk area through a cusp (for the density) or kink (for the order parameter) structure, whose position depends on the type of the phase and the length of the needles. The order parameter at the wall takes on the value of perfect alignment ($S = 1$), and then, while increasing the distance from the wall, slowly descends to the appropriate bulk value.

f) Introducing the effect of dissipation for the needle-wall interaction with the simultaneously preserved at the same time elastic condition for the needle-needle collisions we observe two additional features: softening of the depletion and destabilisation of the bulk area of the profiles.

Future investigations on 2D hard needles are planned to include the kinetic Enskog-like theory of the elastic case and, next, its extension for the dissipative needles. Especially interesting is the question how the restitution parameter influences the behaviour of autocorrelation functions. The so far worked out methods will be applied also to the case without periodic boundary conditions, which corresponds to a highly idealised model of spreading of rodlike polymers on surfaces.

References

1. Allen M. P., Cunningham I. C. H. (1986) A computer simulation study of idealized model tetrahedral molecules. Molec. Phys. **58**, 615–625
2. Allen M. P., Tildesley D. J. (1987) Computer Simulation of Liquids. Clarendon Press, Oxford
3. Chrzanowska A., Ehrentraut H. (2002) Velocity correlations of two-dimensional hard needles from molecular dynamics Systems of Granular Needles. Phys. Rev. E **66**, 012201-1–012201-4
4. Chrzanowska A., Ehrentraut H. (2002) Orientational and Translational Cooling in Two Dimensional Systems of Granular Needles. Tech. Mechanik **22**, 56–65

5. Chrzanowska A., Teixeira P., Ehrentraut H., Cleaver D. (2001) Ordering of hard particles between hard walls. J. Phys. Condens. Matter **13**, 4715–4726
6. Evans R. (1990) Fluids adsorbed in narrow pores: Phase equilibria and structure. J. Phys. Cond. Matter **2**, 8989–9007 and references therein;
7. Frenkel D., Eppenga R. (1985) Evidence for algebraic orientational order in a two-dimensional hard-core nematic. Phys. Rev. A **31**, 1776–1787
8. Frenkel D., Maguire J. F. (1983) Molecular dynamics study of the dynamical properties of an assembly of infinitely thin hard rods. Mol. Phys. **49,** 503–541
9. Frenkel D., Maguire J. F. (1981) Molecular dynamics study of infinitely thin hard rods: scaling behaviour of transport properties. Phys. Rev. Lett. **47**, 1025–1028
10. Frenkel D., Smit B. (1996) Understanding Molecular Simulation. Academic Press Ltd, London
11. de Gennes P. G., Prost J. (1993) The Physics of Liquid Crystals, OUP
12. Goldhirsch I., Zanetti G. (1993) Clustering instability in Dissipative Gases. Phys. Rev. Lett. **70**, 1619–1622
13. Gruel H. (1998) Fluid self-organized machines. Liq. Cryst. **24**, 49–66
14. Hansen J. P., McDonald I. R. (1986) Theory of simple liquids. Academic Press Limited, London
15. Henderson J. R., van Swol F. (1984) On the interface between a fluid and a planar wall. Theory and simulations of a hard sphere fluid at a hard wall. Mol. Phys. **51**, 991–1010
16. Herrmann H. J., Luding S. (1998) Modelling granular media on the computer. Cont. Mech. Therm. **10**, 189–231
17. Huthmann M., Aspelmeier T., Zippelius A. (1999) Granular cooling of hard needles. Phys. Rev. E **60**, 654–659
18. Kosterlitz J. M., Thouless D. (1973) Ordering, metastability and phase transition in two-dimensional systems. J. Phys. C **6**, 1181–1203
19. Luding S., Herrmann H. J. (1999) Cluster growth in Freely Cooling Granular Media. Chaos **9** (3), 673–681
20. Luding S., Huthmann M., McNamara S., Zippelius A. (1998) Homogeneous cooling of rough, dissipative particles: Theory and simulations. Phys. Rev. E **58**, 3416–3425
21. Luding S., McNamara S. (1998) How to handle the inelastic collapse of a dissipative hard-sphere gas with the TC model. Granular Matter **1**, 113–128
22. Lupkowski M., van Swol F. (1990) Computer simulation of fluids interacting with fluctuating walls. J. Chem. Phys. **93**, 737–745
23. Maguire J. F. (1981) Rotational Dynamics of Rigid Rods. J.Chem. Soc. Faraday Trans. II **77**, 513–518
24. Maguire J. F., McTague J. P., Rondelez F. (1980) Rotational diffusion of sterically interacting rodlike macromolecules. Phys. Rev. Lett. **45**, 1891–1894
25. Mao Y., Bladon P., Lekkerkerker H. N. W., Cates M. E. (1997) Density profiles and thermodynamics of rod-like particles between parallel walls. Mol. Phys. **92**, 151–159
26. McNamara S., Young W. R. (1996) Dynamics of a freely evolving, two-dimensional granular medium. Phys. Rev. E **53**, 5089–5100
27. Mukoyama A., Yoshimura Y. (1997) Hundreds of collisions between two hard needles. J. Phys. A **30**, 6667–6670
28. Olafsen J. S., Urbach J. S. (1998) Clustering, Order, and Collapse in a Driven Granular Monolayer. Phys. Rev. Lett. **81**, 4369–4372;

29. Olafsen J. S., Urbach J. S. (1999) Velocity distributions and density fluctuations in a 2D granular gas. Phys. Rev. E **60**, R2468–R2471
30. Poniewierski A. (1993) Ordering of hard needles at hard wall. Phys. Rev. E **47**, 3396–3403
31. Poniewierski A., Holyst R. (1988) Nematic alignment at a solid substrate: The model of hard spherocylinders near a hard wall. Phys. Rev. A **38**, 3721–3727
32. Pöschel, T., Luding, S., (Eds.) (2001) Granular Gases, Springer-Verlag Berlin Heidelberg and references therein
33. Ravikovitch P. I., Vishnyakov A., Neimark A. V. (2001) Density functional theories and molecular simulations of adsorption and phase transitions in nanopores. Phys. Rev. E **64**, 011602-1–01160220
34. http://www.ica1.uni-stuttgart.de/ lui/REFS/references.html
35. Zero K. M., Pecora R. (1982) Rotational and Translational Diffusion in Semidilute Solutions of Rigid-Rod Macromolecules. Macromolecules **15**, 87–93

Induced Anisotropy in Rapid Flows of Nonspherical Granular Materials

Harald Ehrentraut and Agnieszka Chrzanowska

Institute of Mechanics, Darmstadt University of Technology, Hochschulstraße 1, 64289 Darmstadt, Germany, e-mail:{harald, achrzano}@mechanik.tu-darmstadt.de

received 28 Oct 2002 — accepted 11 Jan 2003

Abstract. Flow induced formation of internal order is one of the mechanisms which are responsible for changes in the effective constitutive properties of granular materials. Especially, if the grain shapes differ significantly from a sphere the possibility of flow alignment exists and it takes not much imagination to visualize that a sliding motion of the particles is affected by their mutual alignment. The approach we are taking here is based on a continuum mechanical description of a Cosserat medium with additional internal variables describing the local order of the grain orientations. An analytical description of the problem will be presented, the validity of the SAVAGE–HUTTER theory for anisotropic granular media will be discussed, and experimental findings are compared to the theoretical results.

1 Granular Dynamics

Granular materials are special for many reasons: First of all, they appear in many parts of nature and they play an important role in many technical applications. Secondly, granular materials behave quite differently at rest and under flow conditions; where elastoplastic effects dominate the material response of undisturbed samples, even small perturbations can switch their solid-like response to fluid-like material behavior. Thirdly, granular materials are neither really macroscopic nor microscopic systems – although the overall material properties depend on the interaction of a huge number of particles (thus resembling systems subjected to the methods of statistical mechanics like molecular gases or fluids) the particles themselves are macroscopic objects. Consequently, the interaction of these constituents is dissipative which distinguishes granular media from typical molecular systems.

This hybrid behavior of granular media shows up also in the methods used to model their constitutive properties. Dissipative molecular dynamics is certainly a good way to gain some insight about these materials. However, if the number of model particles is strongly increased, if a three-dimensional model is needed or if the particle shapes are rather irregular the necessary computational power easily exceeds all available resources. Thus, computer simulations based on particle dynamics are very useful to test special cases but they do not provide a universal method to solve engineering problems. On the other hand, a continuum mechanical approach can be used. As it is the case in all continuum theories the amount of microscopical information is so much reduced that the mean behavior of the constituents can be described, but all effects concerning only a small number of particles are not covered by this approach. Normally, this does not pose a problem for typical materials (solids, liquids, or gases); continuum theories are being used, since the constituents are of microscopic size and typical length scales of particle dimensions and macroscopically relevant distances differ by many orders of magnitude. In granular dynamics, however, the difference between both length scales is often much less. In a rock avalanche, e.g., the trajectory of a single boulder might be very important. The deposition of scattered debris at the outrun of an avalanche is also not covered by continuum methods and the spontaneous blocking of granular flows in a constriction (like at the outlet of a silo) is initially being caused by wedging of just a few particles. Thus, there is no universal theory in sight which can cover all aspects of granular behavior but one has to decide which phenomenons should be modeled and which method is suited best for this purpose.

The approach taken in the present paper belongs clearly to the continuum theoretical side. Thus, there are no "particles" entering the description and no microscopic evolution equations are being used. If we speak, however, occasionally of particles it is just to refer to an underlying microscopic description and to motivate ideas about physical properties which influence either the dynamical equations (balances) or the constitutive relations. Details about the transition from a purely microscopic theory to a continuum description through some averaging procedure ("coarse graining" or "time smoothing") can be found in many places in the literature.

1.1 Balance Equations and Constitutive Relations

Let us start from a very basic assumption about the grains of a granular medium: We assume that they are nearly rigid, the grain–grain as well as the grain–wall (or grain–bottom) contact involves friction, and each grain is capable of a gliding and spinning motion (resulting in six degrees of freedom for each grain). The grains may have any shape (although we shall consider mostly special, simple cases), but the size of the grains is assumed to be fairly constant – otherwise we have to resort to a mixture theory which is, of course, possible but complicates the description further and is out of scope for this treatise. We do not consider a matrix fluid filling the gaps between the grains, i.e. we are dealing with dry granular materials.

Since a volume element for the continuum mechanical description must contain a large number of particles and an ensemble of rigid bodies *does not* behave again rigidly, we have to write down the balance equations of a mesomorphic Cosserat medium: the constituents of the medium are deformable, couple stresses exist for the internal interaction and the angular momentum contains not only moment of momentum but also spin. If we denote the mass density by ρ, the velocity field by $\boldsymbol{v}$, the Cauchy stress tensor by $\underline{\underline{\mathbf{t}}}$, the specific external force density by $\boldsymbol{f}$, the specific spin density by $\boldsymbol{s}$, the couple stresses by $\underline{\underline{\pi}}$ and the external specific torque density by $\boldsymbol{m}$, the balance equations are as follows (the symbol $^\top$ denotes transposition):

$$\frac{\partial}{\partial t}\rho + \nabla \cdot (\rho \boldsymbol{v}) = 0\,, \tag{1a}$$

$$\frac{\partial}{\partial t}(\rho \boldsymbol{v}) + \nabla \cdot \Big(\rho \boldsymbol{v}\boldsymbol{v} - \underline{\underline{\mathrm{t}}}^\top\Big) = \rho \boldsymbol{f}\,, \tag{2a}$$

$$\frac{\partial}{\partial t}\Big(\rho(\boldsymbol{x} \times \boldsymbol{v} + \boldsymbol{s})\Big) + \nabla \cdot \Big(\rho \boldsymbol{v}(\boldsymbol{x} \times \boldsymbol{v} + \boldsymbol{s}) - (\boldsymbol{x} \times \underline{\underline{\mathrm{t}}})^\top - \underline{\underline{\pi}}^\top\Big) = \rho\Big(\boldsymbol{x} \times \boldsymbol{f} + \boldsymbol{m}\Big)\,. \tag{3a}$$

Equation (1a) constitutes (of course) conservation of mass, (2a) is the momentum balance, and the balance of angular momentum (3a) does not reduce to a symmetry requirement of the stress tensor. To go further, we assume incompressible, gravity-driven flows – ρ is a constant ρ_0, $\boldsymbol{f}$ is the gravitational acceleration $\boldsymbol{g}$ and no external torques are present, i.e. $\boldsymbol{m} = \boldsymbol{0}$ – and we remove the contributions due to the balances of mass and momentum in the balance of angular momentum. The new set of balances reads now

$$\nabla \cdot \boldsymbol{v} = 0\,, \tag{1b}$$

$$\rho_0\Big(\frac{\partial}{\partial t}\boldsymbol{v} + \nabla \cdot (\boldsymbol{v}\boldsymbol{v})\Big) = \nabla \cdot \underline{\underline{\mathrm{t}}}^\top + \rho_0 \boldsymbol{g}\,, \tag{2b}$$

$$\rho_0\,\frac{d}{dt}\boldsymbol{s} = \boldsymbol{\epsilon} : \underline{\underline{\mathrm{t}}} + \nabla \cdot \boldsymbol{\pi}^\top\,. \tag{3b}$$

Here, $\boldsymbol{\epsilon}$ denotes the LEVI-CIVITA tensor and

$$\frac{d}{dt} = \frac{\partial}{\partial t} + \boldsymbol{v} \cdot \nabla$$

is the material time derivative. In addition, the balance of energy is needed. If one eliminates all contributions due to conservation of mass, momentum and angular momentum from the balance of energy, one obtains the balance equation for the internal energy e

$$\rho\frac{d}{dt}e + \nabla \cdot \boldsymbol{q} - \underline{\underline{\mathrm{t}}} : \nabla \boldsymbol{v} - \underline{\underline{\pi}} : \nabla \boldsymbol{\omega} + \boldsymbol{\omega} \cdot \boldsymbol{\epsilon} : \underline{\underline{\mathrm{t}}} = \rho r\,. \tag{4a}$$

Here, $\boldsymbol{q}$ denotes the heat flux (conductive transport of energy), $\boldsymbol{\omega}$ is the angular velocity – related by $\boldsymbol{s} = \underline{\underline{\boldsymbol{\Theta}}} \cdot \boldsymbol{\omega}$ ($\underline{\underline{\boldsymbol{\Theta}}}$: tensor of inertia) to the spin field $\boldsymbol{s}$ – and r is the supply of energy. For the granular media we want to discuss, the energy supply vanishes – $r = 0$ – and if we do not consider vibrating walls or a vibrating bed the conductive exchange of energy between medium and environment is also negligible, i.e. $\boldsymbol{q} = \boldsymbol{0}$. Then, the balance of internal energy is simply determined by the powers of stresses and couple stresses

$$\rho\frac{d}{dt}e = \underline{\underline{\mathrm{t}}} : \nabla \boldsymbol{v} + \underline{\underline{\pi}} : \nabla \boldsymbol{\omega} - \boldsymbol{\omega} \cdot \boldsymbol{\epsilon} : \underline{\underline{\mathrm{t}}}\,. \tag{4b}$$

Before we continue, a short discussion of the relevant terms in the balances is in order. Tangential friction between particles plays an important role for the stability of a granular material and also, of course, for the torque distribution inside the medium. However, if the granular medium is rapidly moving the main interaction between its constituents is due to collisions where the momentum transfer (for a-spherical particles!) is also connected with an exchange of spin. Thus, tangential friction is of less importance for granular flows and we shall disregard its contributions in the following, meaning that we set the couple stress tensor $\underline{\underline{\pi}}$ to zero,

$$\underline{\underline{\pi}} \approx \underline{\underline{\mathbf{0}}} . \tag{5}$$

Nevertheless, one should be aware of the limits of this approximation: For very small flow velocities and in situations where blocking and formation of arcs inside the material becomes a possibility, we can no longer exclude the couple stresses from our equations. If we decompose the stress tensor $\underline{\underline{\mathbf{t}}}$ into its isotropic part $-p\underline{\underline{\mathbf{1}}}$, the symmetric traceless part $\overline{\underline{\underline{\mathbf{t}}}}$ and a skew-symmetric part $\underline{\underline{\mathbf{t}}}^a$,

$$\underline{\underline{\mathbf{t}}} = -p\underline{\underline{\mathbf{1}}} + \overline{\underline{\underline{\mathbf{t}}}} + \underline{\underline{\mathbf{t}}}^a , \tag{6}$$

then the term $\underline{\underline{\mathbf{t}}} : \nabla \boldsymbol{v}$ reduces to

$$\begin{aligned} \underline{\underline{\mathbf{t}}} : \nabla \boldsymbol{v} &= -p \underbrace{\nabla \cdot \boldsymbol{v}}_{=0} + \overline{\underline{\underline{\mathbf{t}}}} : \overline{\nabla \boldsymbol{v}} + \underline{\underline{\mathbf{t}}}^a : (\nabla \boldsymbol{v})^a \\ &= \underline{\underline{\mathbf{t}}} : \underline{\underline{\mathbf{D}}} + \underline{\underline{\mathbf{t}}} : \underline{\underline{\mathbf{W}}} , \end{aligned} \tag{7}$$

where $\underline{\underline{\mathbf{D}}} := \overline{\nabla \boldsymbol{v}}$ is the stretching or shear rate tensor and $\underline{\underline{\mathbf{W}}} := (\nabla \boldsymbol{v})^a$ is the spin or vorticity tensor. Finally, we can express $\underline{\underline{\mathbf{W}}}$ easily by the vorticity vector $\boldsymbol{\Omega} := 1/2\, \nabla \times \boldsymbol{v}$

$$\underline{\underline{\mathbf{W}}} = \boldsymbol{\epsilon} \cdot \boldsymbol{\Omega} , \tag{8}$$

and arrive at the set of simplified balance equations for granular flows

$$\nabla \cdot \boldsymbol{v} = 0 , \tag{9a}$$

$$\rho_0 \Big(\frac{\partial}{\partial t} \boldsymbol{v} + \nabla \cdot (\boldsymbol{v}\boldsymbol{v}) \Big) = \nabla \cdot \underline{\underline{\mathbf{t}}}^\top + \rho_0 \boldsymbol{g} , \tag{9b}$$

$$\rho_0 \frac{d}{dt} \boldsymbol{s} = \boldsymbol{\epsilon} : \underline{\underline{\mathbf{t}}} , \tag{9c}$$

$$\rho_0 \frac{d}{dt} e = \underline{\underline{\mathbf{t}}} : \underline{\underline{\mathbf{D}}} + (\boldsymbol{\Omega} - \boldsymbol{\omega}) \cdot \boldsymbol{\epsilon} : \underline{\underline{\mathbf{t}}} . \tag{9d}$$

For spherical particles the stresses *are* symmetric and consequently the spin $\boldsymbol{s}$ is a constant of motion. Furthermore, we obtain $\boldsymbol{s} = \mathbf{0}$ in that case since this holds certainly true for the material at rest and conservation of spin implies the vanishing of $\boldsymbol{s}$ always. This, of course, is a direct consequence of our disregard of tangential friction – the particles are sliding, but rolling does not occur.

In addition to the differential equations (9a) – (9d) boundary conditions are needed. Here, we have to distinguish between free boundaries and boundaries where the granular medium is in contact with a wall or the bottom. Let $F^S(\boldsymbol{x}, t) := z - z^S(x, y, t) = 0$ be the implicit definition of the free surface and $F^B(\boldsymbol{x}, t) := z^B(x, y, t) - z = 0$ likewise define the contact between granular material and the bottom (if walls are present, additional functions have to be considered). z^S and

z^B are the z-coordinates of free and bottom surface, respectively, measured in normal direction from the reference surface point (x, y). All surfaces are subjected to kinematic boundary conditions ($\boldsymbol{v}^S$ and $\boldsymbol{v}^B$ are the mapping velocities of the free and basal surfaces, respectively):

$$\frac{\partial F^S}{\partial t} + \boldsymbol{v}^S \cdot \nabla F^S = 0\,, \tag{10}$$

$$\frac{\partial F^B}{\partial t} + \boldsymbol{v}^B \cdot \nabla F^B = 0\,. \tag{11}$$

In many applications, the basal topography will be independent of time, $F^B(\boldsymbol{x}) = 0$, and (11) reduces to the geometric condition $\boldsymbol{v}^B \cdot \nabla F^B = 0$ that bottom velocities are tangential to the basal surface – there is neither in- nor outflux at the base. However, kinematic boundary conditions are not enough to produce a well-posed problem. Here, they have to be supplemented by traction boundary conditions which brings us to the problem of constitutive relations for granular materials. If we would have considered couple stresses, too, torque boundary conditions would also have been necessary.

The most simple (and probably most successful) material law for granular media is of the MOHR-COULOMB type. Following again SAVAGE and HUTTER [21], dry friction with a simple proportionality between normal pressure and shear stress S for the sliding material is assumed. In addition, a yield stress $N \tan\phi$ with a material constant ϕ, called the internal friction angle, shall describe the plastic deformation of the yielding material. Thus, the relation between S and N for the yielding material is simply

$$|S| = N \tan\phi\,. \tag{12}$$

To formulate traction boundary conditions we assume a traction free upper surface (the very small friction between air and the granular material is being neglected) and a COULOMB law for friction between the sliding material and the fixed base,

$$\underline{\underline{\mathbf{t}}}^S \cdot \boldsymbol{n}^S = \mathbf{0}\,, \tag{13}$$

$$-N^B \boldsymbol{n}^B - \underline{\underline{\mathbf{t}}}^B \cdot \boldsymbol{n}^B = \hat{\boldsymbol{v}}^B N^B \tan\delta\,, \tag{14}$$

is being assumed (a hat $\hat{}$ denotes a unit vector, e.g. $\hat{\boldsymbol{v}} := \boldsymbol{v}/\|\boldsymbol{v}\|$). The normal pressure $N^B = -\boldsymbol{n}^B \cdot \underline{\underline{\mathbf{t}}}^B \cdot \boldsymbol{n}^B$ at the bottom is related to the stress by an additional material constant δ, the basal angle of friction. By $\boldsymbol{n}^B$ and $\boldsymbol{n}^S$ we denote the (outward pointing) normal vectors at the base and at the upper free surface, respectively.

1.2 Savage–Hutter Equations for Anisotropic Media

If we want to describe the granular motion along an incline or, more complicated, the evolution of an avalanche following the complex topography of a mountain slope, we have to reformulate the balances (9a) – (9d) for topography-adapted coordinates and apply a proper scaling to simplify the resulting equations. Both has been done first by SAVAGE & HUTTER [21] for the one-dimensional case and has been extended by GRAY, GREVE, HUTTER, KOCH and WIELAND up to the full, three-dimensional equations [10,11]. Let us briefly recall the procedure and see where anisotropic effects may enter the equations.

To begin, we introduce curvilinear coordinates and the appropriate local GAUSSIAN basis to parameterize the bottom topography. Let $\{\boldsymbol{i}, \boldsymbol{j}, \boldsymbol{k}\}$ denote the canonical CARTESIAN basis of three-dimensional space and let

$$\boldsymbol{c}(x) = a(x)\boldsymbol{i} + b(x)\boldsymbol{k}, \quad (x \in [0, L]) \tag{15}$$

be a parameterized curve in the $\boldsymbol{i}, \boldsymbol{k}$-plane where x denotes the arc-length, a and b are the (differentiable) component functions and L is the total length of the curve. Then we can define a reference surface by parallel translation of the curve in the $\boldsymbol{j}$-direction and measure the topography by its elevation in normal direction with respect to the reference surface (compare Fig. 1). Since we denoted the arc-length of

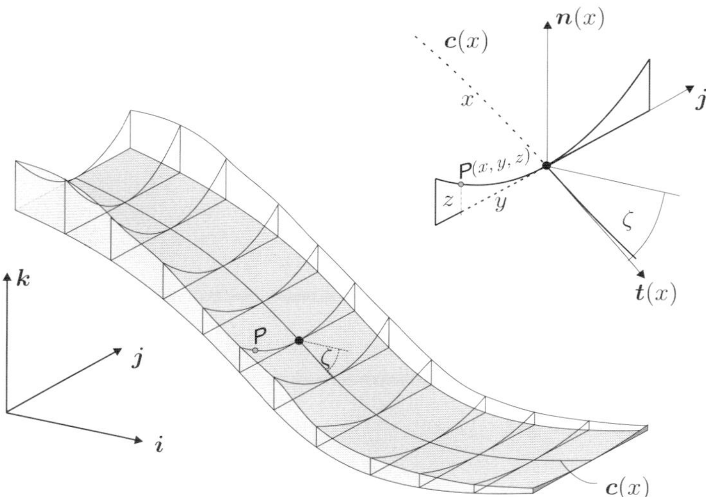

Fig. 1. Definition of the SAVAGE–HUTTER coordinates for a complex bed topography

$\boldsymbol{c}(x)$ by x, it is natural to denote the coordinate in $\boldsymbol{j}$-direction by y (x and y yield a parameterization of the reference surface) and the elevation in normal direction by z. In addition to the CARTESIAN basis we have the local GAUSSIAN basis $\{\boldsymbol{b}_x, \boldsymbol{b}_y, \boldsymbol{b}_z\}$ defined by

$$\boldsymbol{b}_q := \frac{\partial \boldsymbol{x}(x, y, z)}{\partial q} \qquad \text{for } q = x, y, z\,. \tag{16}$$

It is advantageous to use also the basal inclination angle $\zeta(x)$ of the curve $\boldsymbol{c}$ with respect to the horizontal $\boldsymbol{i}$-direction – with the convention $\zeta < 0$ for downhill and $\zeta \geq 0$ for horizontal and uphill directions. If we denote the unit tangential vector $\boldsymbol{c}'(x)$ by $\boldsymbol{t}(x)$ and the upper normal vector orthogonal to the reference plain by

$\boldsymbol{n}(x)$, we obtain finally

$$\boldsymbol{b}_x = \big(1 - z\zeta'\big)\,\boldsymbol{t}(x)\,, \quad \boldsymbol{b}_y = \boldsymbol{j}\,, \quad \boldsymbol{b}_z = \boldsymbol{n}(x)\,, \tag{17}$$

$$\boldsymbol{b}^x = \frac{1}{1 - z\zeta'}\,\boldsymbol{t}(x)\,, \quad \boldsymbol{b}^y = \boldsymbol{j}\,, \quad \boldsymbol{b}^z = \boldsymbol{n}(x)\,, \tag{18}$$

$$\big((g_{ij})\big) = \mathrm{diag}\Big(1 - z\zeta', 1, 1\Big) \ (\text{ i.e. } ds^2 = (1 - z\zeta')^2 dx^2 + dy^2 + dz^2\ )\,, \tag{19}$$

for the basis $\{\boldsymbol{b}_q\}$, the dual basis $\{\boldsymbol{b}^q\}$ and the matrix $((g_{ij}))$ of metric coefficients. The local bases are therefore orthogonal but not normalized. The co-variant derivative ∇ along the surface, expressed in the natural basis and coordinates, is consequently

$$\nabla = \boldsymbol{b}^x \frac{\partial}{\partial x} + \boldsymbol{b}^y \frac{\partial}{\partial y} + \boldsymbol{b}^z \frac{\partial}{\partial z}\,. \tag{20}$$

To abbreviate and to be consistent with literature we use $\Psi := (1 - z\zeta)^{-1}$ and $\kappa := \zeta'$ (κ is the curvature of $\boldsymbol{c}$). For any differentiable vector field $\boldsymbol{v} = u\boldsymbol{t} + v\boldsymbol{j} + w\boldsymbol{n}$, expressed in the local, *normalized* basis by its (x, y, z)-dependent component functions u, v, w, we arrive at the formula

$$\begin{aligned}\nabla \cdot \boldsymbol{v} =& \Psi \frac{\partial u}{\partial x} + \frac{\partial v}{\partial y} + \frac{\partial w}{\partial z} - \Psi\kappa w \\ =& \frac{\partial(\Psi u)}{\partial x} + \frac{\partial v}{\partial y} + \frac{\partial w}{\partial z} - \Psi\kappa w - \Psi^2\kappa' uz\end{aligned} \tag{21}$$

for the divergence of a vector field, and

$$\begin{aligned}\nabla \cdot \underline{\underline{\mathbf{T}}} =& \Big(\frac{\partial}{\partial x}(\Psi T_{xx}) + \frac{\partial}{\partial y}T_{yx} + \frac{\partial}{\partial z}T_{zx} - \Psi\kappa(T_{xz} + T_{zx}) - \Psi^2\kappa' zT_{xx}\Big)\boldsymbol{t} \\ &+ \Big(\frac{\partial}{\partial x}(\Psi T_{xy}) + \frac{\partial}{\partial y}T_{yy} + \frac{\partial}{\partial z}T_{zy} - \Psi\kappa T_{zy} - \Psi^2\kappa' zT_{xy}\Big)\boldsymbol{j} \\ &+ \Big(\frac{\partial}{\partial x}(\Psi T_{xz}) + \frac{\partial}{\partial y}T_{yz} + \frac{\partial}{\partial z}T_{zz} - \Psi\kappa(T_{zz} - T_{xx}) - \Psi^2\kappa' zT_{xz}\Big)\boldsymbol{n}\end{aligned} \tag{22}$$

for the divergence of a tensor field $\underline{\underline{\mathbf{T}}}$ of second order. Please note that no symmetry of $\underline{\underline{\mathbf{T}}}$ was assumed and therefore the formula in (22) differs slightly from the expression used in the extended SAVAGE–HUTTER theory by GRAY, WIELAND & HUTTER [10].

Now we can rewrite (9a) and (9b) using curvilinear coordinates. We obtain for the balance of mass (u, v, w shall denote the velocity components)

$$\frac{\partial}{\partial x}(\Psi u) + \frac{\partial v}{\partial y} + \frac{\partial w}{\partial z} - \Psi\kappa w - \Psi^2\kappa' zu = 0\,. \tag{23}$$

From the balance of momentum (9b) we derive the set of equations

$$\rho_0\Big\{\frac{\partial u}{\partial t}+\frac{\partial}{\partial x}(\Psi u^2)+\frac{\partial}{\partial y}(uv)+\frac{\partial}{\partial z}(uw)-2\Psi\kappa uw-\Psi^2\kappa' zu^2\Big\}$$

$$=-\rho_0 g\sin\zeta+\frac{\partial}{\partial x}(\Psi t_{xx})+\frac{\partial}{\partial y}t_{xy}+\frac{\partial}{\partial z}t_{xz}-\Psi\kappa(t_{zx}+t_{xz})-\Psi^2\kappa' zt_{xx}\,, \tag{24}$$

$$\rho_0\Big\{\frac{\partial v}{\partial t}+\frac{\partial}{\partial x}(\Psi uv)+\frac{\partial}{\partial y}v^2+\frac{\partial}{\partial z}(vw)-\Psi\kappa vw-\Psi^2\kappa' zuv\Big\}$$

$$=\frac{\partial}{\partial x}(\Psi t_{yx})+\frac{\partial}{\partial y}t_{yy}+\frac{\partial}{\partial z}t_{yz}-\Psi\kappa t_{yz}-\Psi^2\kappa' zt_{yx}\,, \tag{25}$$

$$\rho_0\Big\{\frac{\partial w}{\partial t}+\frac{\partial}{\partial x}(\Psi uw)+\frac{\partial}{\partial y}(vw)+\frac{\partial}{\partial z}w^2-\Psi\kappa(w^2-u^2)-\Psi^2\kappa' zuw\Big\}$$

$$=-\rho_0 g\cos\zeta+\frac{\partial}{\partial x}(\Psi t_{zx})+\frac{\partial}{\partial y}t_{zy}+\frac{\partial}{\partial z}t_{zz}-\Psi\kappa(t_{zz}-t_{xx})-\Psi^2\kappa' zt_{zx}\,. \tag{26}$$

We denote the components of the specific spin density $\boldsymbol{s}=\alpha\boldsymbol{t}+\beta\boldsymbol{j}+\gamma\boldsymbol{n}$ by α, β and γ. Then the spin balance (9c) results in

$$\rho_0\Big\{\frac{\partial\alpha}{\partial t}+\Psi u\frac{\partial\alpha}{\partial x}+v\frac{\partial\alpha}{\partial y}+w\frac{\partial\alpha}{\partial z}-\Psi\kappa u\gamma\Big\}=t_{yz}-t_{zy}\,, \tag{27}$$

$$\rho_0\Big\{\frac{\partial\beta}{\partial t}+\Psi u\frac{\partial\beta}{\partial x}+v\frac{\partial\beta}{\partial y}+w\frac{\partial\beta}{\partial z}\Big\}=t_{zx}-t_{xz}\,, \tag{28}$$

$$\rho_0\Big\{\frac{\partial\gamma}{\partial t}+\Psi u\frac{\partial\gamma}{\partial x}+v\frac{\partial\gamma}{\partial y}+w\frac{\partial\gamma}{\partial z}+\Psi\kappa u\alpha\Big\}=t_{xy}-t_{yx}\,. \tag{29}$$

The balance equations (9a) – (9c) provide information about the mechanical behavior of a granular flow. The balance of internal energy (9d), however, becomes important whenever the dissipated energy during the sliding motion of the material becomes so large that a noticeable increase of internal energy (or temperature, the conjugated thermodynamic variable) occurs. Thus, the importance of (9d) depends on the flow situation: Under "normal" conditions the heat generated by bed and internal friction is quite small and we can disregard the energy equation and concentrate on the mechanical equations alone. On the other hand, an increase of temperature may modify friction angles and other constitutive variables. If the dissipation in the flow is therefore high (as, e.g., in alpine rock slides), the energy equation becomes important and we have to treat the full system of equations.

For our purposes here, we shall assume granular flows with small dissipation and consider (9d) only in connection with the second law of thermodynamics, but *not* as a necessary equation for solving the problem of motion.

The thickness of flow avalanches is usually much smaller than their lateral extensions, i.e. an avalanche is a shallow object and we can reduce the dimensionality of the flow problem by depth integration. Indeed, this was done by Savage & Hutter after scaling the balances of mass and momentum. Following the approach by Gray, Hutter, Wieland and Tai [22] we postpone any scaling and integrate the equations (9a) – (9c) directly over the thickness $h=z^S-z^B$ of the granular body. This average will be denoted by a bar,

$$\bar g(x,y,t):=\frac{1}{h(x,y,t)}\int\limits_{z^B(x,y,t)}^{z^S(x,y,t)} g(x,y,z,t)\,dz\,. \tag{30}$$

Since the derivatives with respect to z turn into surface terms of the free and bottom surface, the boundary conditions (10) – (14) must be rewritten, too, and we obtain

$$-\frac{\partial z^S}{\partial t} - \Psi^S u^S \frac{\partial z^S}{\partial x} - v^S \frac{\partial z^S}{\partial y} + w^S = 0\,, \tag{31}$$

$$\frac{\partial z^B}{\partial t} + \Psi^B u^B \frac{\partial z^B}{\partial x} + v^B \frac{\partial z^B}{\partial y} - w^B = 0\,, \tag{32}$$

for the kinematic boundary conditions,

$$t^S_{xx} \Psi^S \frac{\partial z^S}{\partial x} + t^S_{xy} \frac{\partial z^S}{\partial y} - t^S_{xz} = 0\,, \tag{33}$$

$$t^S_{yx} \Psi^S \frac{\partial z^S}{\partial x} + t^S_{yy} \frac{\partial z^S}{\partial y} - t^S_{yz} = 0\,, \tag{34}$$

$$t^S_{zx} \Psi^S \frac{\partial z^S}{\partial x} + t^S_{zy} \frac{\partial z^S}{\partial y} - t^S_{zz} = 0\,, \tag{35}$$

for the stress-free surface and

$$-t^B_{xx} \Psi^B \frac{\partial z^B}{\partial x} - t^B_{xy} \frac{\partial z^B}{\partial y} + t^B_{xz} = \Big(\|\nabla F^B\| \hat{u}^B \tan\delta + \Psi^B \frac{\partial z^B}{\partial x} \Big) N^B\,, \tag{36}$$

$$-t^B_{yx} \Psi^B \frac{\partial z^B}{\partial x} - t^B_{yy} \frac{\partial z^B}{\partial y} + t^B_{yz} = \Big(\|\nabla F^B\| \hat{v}^B \tan\delta + \Psi^B \frac{\partial z^B}{\partial y} \Big) N^B\,, \tag{37}$$

$$-t^B_{zx} \Psi^B \frac{\partial z^B}{\partial x} - t^B_{zy} \frac{\partial z^B}{\partial y} + t^B_{zz} = \Big(\|\nabla F^B\| \hat{w}^B \tan\delta - 1 \Big) N^B \tag{38}$$

for the sliding boundary condition at the base.

The depth integrated balance of mass has, of course, been already obtained by Gray, Wieland & Hutter and reads [10]

$$\frac{\partial h}{\partial t} + \frac{\partial}{\partial x}(h\overline{\Psi u}) + \frac{\partial}{\partial y}(h\bar{v}) - \kappa' h\overline{\Psi^2 z u} - 2\kappa h \overline{\Psi w} = 0\,. \tag{39}$$

Integration of the balances of momentum yields

$$\begin{aligned}\rho_0\Big\{\frac{\partial}{\partial t}(h\bar{u}) + \frac{\partial}{\partial x}(h\overline{\Psi u^2}) + \frac{\partial}{\partial y}(h\overline{uv}) - 2h\kappa\overline{\Psi uw} - h\kappa'\overline{\Psi^2 zu^2}\Big\} \\ = -\rho_0 hg\sin\zeta + \frac{\partial}{\partial x}(h\overline{\Psi t_{xx}}) + \frac{\partial}{\partial y}(h\overline{t_{xy}}) - h\kappa\overline{\Psi(t_{zx}+t_{xz})} \\ -h\kappa'\overline{\Psi^2 zt_{xx}} + N^B(\Psi^B\frac{\partial z^B}{\partial x} + \|\nabla F^B\|\hat{u}^B\tan\delta)\,,\end{aligned} \tag{40}$$

$$\begin{aligned}\rho_0\Big\{\frac{\partial}{\partial t}(h\bar{v}) + \frac{\partial}{\partial x}(h\overline{\Psi vu}) + \frac{\partial}{\partial y}(h\overline{v^2}) - h\kappa\overline{\Psi vw} - h\kappa'\overline{\Psi^2 zuv}\Big\} \\ = \frac{\partial}{\partial x}(h\overline{\Psi t_{yx}}) + \frac{\partial}{\partial y}(h\overline{t_{yy}}) - h\kappa\overline{\Psi t_{zy}} \\ -h\kappa'\overline{\Psi^2 zt_{xy}} + N^B(\Psi^B\frac{\partial z^B}{\partial x} + \|\nabla F^B\|\hat{v}^B\tan\delta)\,,\end{aligned} \tag{41}$$

$$\begin{aligned}\rho_0\Big\{\frac{\partial}{\partial t}(h\bar{w}) + \frac{\partial}{\partial x}(h\overline{\Psi uw}) + \frac{\partial}{\partial y}(h\overline{wv}) - h\kappa\overline{\Psi(w^2-u^2)} - h\kappa'\overline{\Psi^2 zuw}\Big\} \\ = -\rho_0 hg\cos\zeta + \frac{\partial}{\partial x}(h\overline{\Psi t_{zx}}) + \frac{\partial}{\partial y}(h\overline{t_{zy}}) - h\kappa\overline{\Psi(t_{zz}-t_{xx})} \\ -h\kappa'\overline{\Psi^2 zt_{zx}} + N^B(1 - \|\nabla F^B\|\hat{w}^B\tan\delta)\,.\end{aligned} \tag{42}$$

The only difference to the equations obtained by Gray, Wieland & Hutter is again the symmetry of the stresses which we do not presuppose at this time. Thus, we have additional three equations from the spin balance ($\boldsymbol{s} = \alpha\boldsymbol{t} + \beta\boldsymbol{j} + \gamma\boldsymbol{n}$)

$$\rho_0\Big\{\frac{\partial}{\partial t}(h\bar{\alpha}) + \frac{\partial}{\partial x}(h\overline{\Psi u\alpha}) + \frac{\partial}{\partial y}(h\overline{v\alpha}) - \kappa(\overline{\Psi\alpha w} + \overline{\Psi\gamma u})\Big\} = \overline{t_{yz}} - \overline{t_{zy}}\,, \tag{43}$$

$$\rho_0\Big\{\frac{\partial}{\partial t}(h\bar{\beta}) + \frac{\partial}{\partial x}(h\overline{\Psi u\beta}) + \frac{\partial}{\partial y}(h\overline{v\beta}) - \kappa\overline{\Psi\beta w}\Big\} = \overline{t_{zx}} - \overline{t_{xz}}\,, \tag{44}$$

$$\rho_0\Big\{\frac{\partial}{\partial t}(h\bar{\gamma}) + \frac{\partial}{\partial x}(h\overline{\Psi u\gamma}) + \frac{\partial}{\partial y}(h\overline{v\gamma}) - \kappa(\overline{\Psi\gamma w} - \overline{\Psi\alpha u})\Big\} = \overline{t_{xy}} - \overline{t_{yx}}\,, \tag{45}$$

which supplement the "normal" set of equations of the Savage–Hutter theory.

The next step in the derivation of the final set of equations is the scaling which determines the physically relevant terms. We assume:

- The aspect ratio $\epsilon = H/L$ (H being the maximum thickness of the avalanche and L denoting the typical length of the flowing body) of the granular avalanche is in most flow situations rather small (i.e. $\epsilon \ll 1$),
- the curvature $\kappa = \tilde{\kappa}/R$ with R/L is at most of order ϵ,
- the shear/normal stress ratios t_{ij}/t_{kk} ($i \neq j$ and i, j, k taking values in (x, y, z)) are of order ϵ [11], and
- typical "vertical" velocities w are also of ϵ-order of magnitude.

Then the Savage–Hutter equations are obtained from the balances of mass and momentum. The interested reader can find a detailed presentation in [10] which we shall not repeat here.

The agreement between experiment and the predictions of the Savage–Hutter theory as reported in [10] is very convincing, and the question might be asked if it is really necessary to amend the set of equations in the case of non-spherical grains or if the *same* set of equations is capable to describe anisotropic flows.

Indeed, the crucial terms here are the shear stress differences $t_{ij} - t_{ji}$ which act as torque density in the spin balance. One of the assumptions of the SAVAGE–HUTTER theory is that all shear stresses are of the same order of magnitude. In this approximation, their difference would be much smaller than a typical shear stress and the asymmetry of the stress tensor turns out to be (at least) a second order effect. However, we can tackle the problem of the spin balance also from a dynamical point of view. Since granular flows are rather dense, each grain is confined in a narrow region of space bounded by its nearest neighbors which form some kind of cage. Especially when the grain shape differs significantly from a sphere the rotational mobility of single grains is therefore strongly hindered and it is reasonable to assume that typical values of the spin $h\alpha$ (or any other spin component) are at least as small as the vertical velocity w, i.e. we would have to assume a scaling law

$$h(\alpha, \beta, \gamma) = (gL)^{1/2} \epsilon^{\sigma} (\tilde{\alpha}, \tilde{\beta}, \tilde{\gamma}) \quad \text{with } \sigma \geq 1. \tag{46}$$

Then the equations (43) – (45) tell us that the shear stress differences are of the same orders, too, and the asymmetry of the stress tensor in the momentum balance is a correction term of higher order than those contributing to the SAVAGE–HUTTER equations.

Thus we can conclude: The SAVAGE–HUTTER equations in the form of [10] provide an adequate description even for anisotropic granular materials. For a granular system without any tangential friction of the grains the asymmetry of the stress tensor becomes only important when higher order effects are investigated. The additional equations for anisotropic granular media in the limit of the SAVAGE–HUTTER theory reduce consequently to a symmetry statement:

$$\overline{t_{yz}} - \overline{t_{zy}} = 0\,, \tag{47}$$

$$\overline{t_{zx}} - \overline{t_{xz}} = 0\,, \tag{48}$$

$$\overline{t_{xy}} - \overline{t_{yx}} = 0 \tag{49}$$

However, we can *not* start from symmetric a priori expressions for the stresses. The symmetry conditions (47) – (49) provide additional information which determine the internal structure of the granular medium.

This fact becomes of particular importance when we consider processes which are capable of orienting the particles in the granular material. Such processes have to break the isotropic symmetry of space which, however, occurs in many situations. Probably the most important mechanisms of symmetry breaking for granular materials occurs in shear bands. Flow and gradient direction in a simple shear distinguish certain spatial directions and the vorticity of the flow field yields an effective torque acting on the particles in the flow. Therefore, we expect some influence of the shear on the internal structure of the granular material. The exploitation of the symmetry conditions (47) – (49) to derive an analytical description of the shear induced anisotropy is subject of the next section.

2 Constitutive Relations and Order Parameters

In an anisotropic granular material, which is slowly flowing down a flat incline, nothing spectacular happens: A stationary plug flow develops, the grains are randomly oriented (see Fig. 2) and virtually no shear layer occurs at the bottom if the basal friction is not too high. The situation changes when we increase the mean

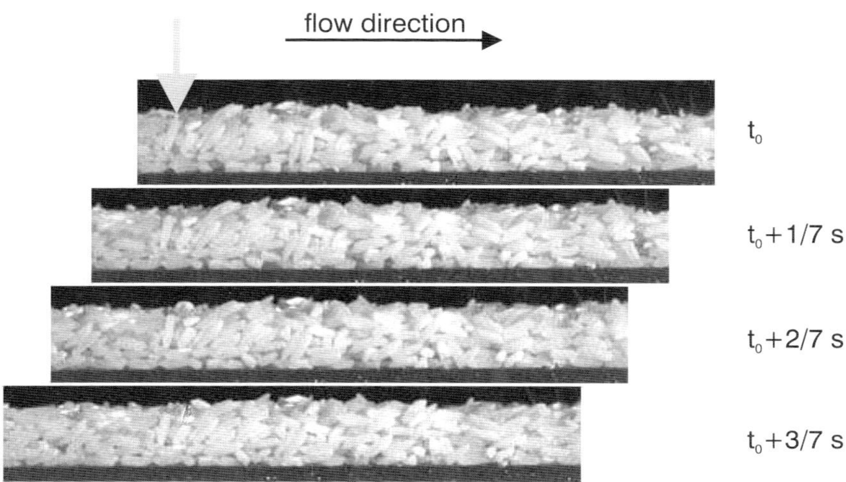

Fig. 2. Plug flow of rice. The whole body glides along the incline without any structural changes in the interior (see, e.g.,the vicinity of the grain marked by the gray arrow). The photographs are vertically aligned such that the same grains occur in columns

velocity of the flow (compare Fig. 3). Now, the velocities increase from the base to the free surface and the grains tend to orient in the shear flow. Although the examples shown in Figs. 2 and 3 pertain to rather shallow flows, the general phenomena can clearly be seen. In a shear layer the internal order is increased and the main orientation of the grains is determined by the flow direction (among other variables, as we shall see), whereas in a plug flow the uniform velocity field has no ordering capacity.

If we want to gain a better understanding of the flow induced order we have to introduce order parameters and we have to develop a constitutive theory which involves these order parameters as internal variables.

2.1 Definition of Order Parameters

Since a flowing anisotropic granular material can be regarded as a general anisotropic fluid we can follow classical ideas developed for molecular gases and liquid crystals to define suitable order parameters. Since our particles are (mostly) gliding we can neglect one rotational degree of freedom, i.e., the orientation of a

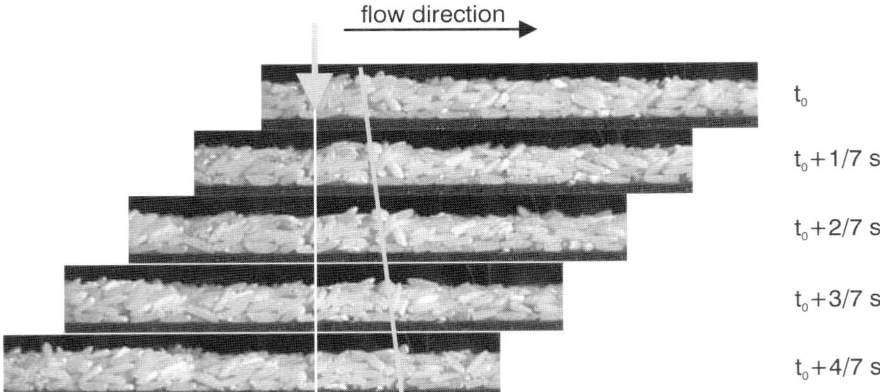

Fig. 3. Shear flow of rice. The photographs are vertically aligned with respect to grains at the base (white line). The top layer moves with a faster velocity, thus producing a simple shear (tilted gray line)

single grain is sufficiently prescribed by *one* unit vector which can be aligned along a characteristic axis of the grain (e.g. its symmetry axis). We adopt the notations of WALDMANN & HESS [12] and Hess [13,14] and start with an *orientation distribution function* (ODF) for the granular axes $\boldsymbol{n}$,

$$f: \mathbb{R}^3 \times \mathrm{S}^2 \times \mathbb{R} \longrightarrow \mathbb{R}_0^+ , \tag{50}$$

which is normalized

$$\oint_{\mathrm{S}^2} f(\boldsymbol{x}, \boldsymbol{n}, t)\, d^2 n = 1 \tag{51}$$

and symmetric

$$f(\boldsymbol{x}, \boldsymbol{n}, t) = f(\boldsymbol{x}, -\boldsymbol{n}, t) , \tag{52}$$

since for the grains we are considering here, orientations $\boldsymbol{n}$ and $-\boldsymbol{n}$ are indistinguishable. Then $f(\boldsymbol{x}, \boldsymbol{n}, t)\, d^2 n\, d^3 x$ denotes the probability to find grains with orientations in the solid angle $d^2 n$ inside the volume element $d^3 x$. Hence-force, we shall denote the *ensemble average* with respect to the ODF by angular brackets

$$< G > := \oint_{\mathrm{S}^2} f(\boldsymbol{x}, \boldsymbol{n}, t) G(\boldsymbol{x}, \boldsymbol{n}, t)\, d^2 n . \tag{53}$$

Since f itself is a rather complicated object we take some of its moments to characterize the local order in the material. For doing so, we choose a set of tensor functions and write the orientation distribution function as an infinite tensor series with respect to this basis. Our base functions are the Cartesian components of symmetric irreducible tensors, restricted to the unit sphere (for more information about this tensor family see the small booklet of HESS & KÖHLER [17] or

Ehrentraut & Muschik [9]), which are closely related to spherical harmonics and form a $L^2(\mathrm{S}^2, \mathbb{R})$-basis for the square-integrable functions on the unit sphere. Here, the short comment shall be sufficient that symmetric irreducible tensors can be obtained from the ℓ-fold tensorial products of the unit vector $\boldsymbol{n}$ after removing all sorts of internal contractions such that the resulting tensor vanishes when a summation over an arbitrary index pair is performed. For second order tensors, the symmetric irreducible part of the tensor is identical to the symmetric traceless part! The crucial proposition for functions on the sphere reads now:

Let f be a square-integrable function on S^2. Then f can be expressed as a series of symmetric irreducible tensors (indicated by the bracket $\overline{\cdots}$)

$$f(\boldsymbol{n}) = \frac{1}{4\pi}\left\{ f_0 + \sum_{\ell=1}^{\infty} \frac{(2\ell+1)!!}{\ell!} a_{\mu_1\cdots\mu_\ell}\ \overline{n_{\mu_1}\cdots n_{\mu_\ell}} \right\} \tag{54}$$

with Greek indices denoting Cartesian components, Einsteins *summation convention used for them,*

$$a_{\mu_1\cdots\mu_\ell} = \oint_{S^2} f(\boldsymbol{n})\ \overline{n_{\mu_1}\cdots n_{\mu_\ell}}\ d^2n \quad (\ell \in \mathbb{N}), \tag{55}$$

$$f_0 = \oint_{S^2} f(\boldsymbol{n})\, d^2n, \tag{56}$$

and $(2\ell+1)!!$ denoting the product of all odd integers smaller or equal to $2\ell+1$.

For the ODF the absolute term fulfills $f_0 = 1$ due to normalization and all odd moments vanish due to symmetry. Thus we have

$$f(\cdot) = \frac{1}{4\pi}\left\{ 1 + \sum_{\ell\ \mathrm{even}}^{\infty} \frac{(2\ell+1)!!}{\ell!} a(\boldsymbol{x},t)_{\mu_1\cdots\mu_\ell}\ \overline{n_{\mu_1}\cdots n_{\mu_\ell}} \right\} \tag{57}$$

with the *alignment tensors*

$$a_{\mu_1\cdots\mu_\ell} = \oint_{\mathrm{S}^2} f(\boldsymbol{n})\ \overline{n_{\mu_1}\cdots n_{\mu_\ell}}\ d^2n \quad (\ell \in 2\mathbb{N}). \tag{58}$$

Of special interest is the first non-vanishing anisotropy moment, the alignment tensor of second order $\underline{\underline{\mathbf{a}}}$, the fourth alignment tensor influences the constitutive functions in special cases, but higher order alignment tensors are usually considered to be of minor importance.

2.2 Transversal Isotropy

The alignment tensor of ℓ-th order has in general $2\ell+1$ independent components. In many situations, however, the orientation distribution function possesses a transversally isotropic (uniaxial) symmetry which must not be confused with a symmetry of the particles. This symmetry can be expressed by the presence of of a spatial direction, specified by a unit vector $\boldsymbol{d}$, such that the dependence of f on $\boldsymbol{n}$ reduces to a dependence on the projection of $\boldsymbol{n}$ on $\boldsymbol{d}$:

$$f(\boldsymbol{x},\boldsymbol{n},t) = g(\boldsymbol{x},\boldsymbol{n}\cdot\boldsymbol{d},t) \text{ with a suitable function } g. \tag{59}$$

In this case the alignment tensors take a very simple form. Indeed, they are determined by the symmetry axis $\boldsymbol{d}$, called *director*, and *one* scalar parameter which is the ensemble average of the ℓ-th Legendre polynomial P_ℓ, i.e.

$$a_{\mu_1\cdots\mu_\ell} = < P_\ell(\boldsymbol{n}\cdot\boldsymbol{d}) > \overline{d_{\mu_1}\cdots d_{\mu_\ell}} \ . \tag{60}$$

Traditionally, the second order parameter $< P_2 > \equiv S$ is called MAIER–SAUPE order parameter; if more than one scalar order parameter is used, they are denoted as $S_\ell := < P_\ell >$, i.e. S and S_2 refer to the same quantity.

To close this small investigation on order parameters we remark that S can vary in the interval $[-0.5, 1]$, that perfect alignment (the ODF collapses to the Dirac δ-function) forces all order parameters to be one ($S_\ell = 1$), and that a random distribution ensures $S_2 = S_4 = \cdots = S_\ell = \cdots = 0$.

2.3 A Remark on Evolution Equations for Alignment Tensors

The balance equations contain no information about the degree of order in the medium, but the macroscopic order parameters (alignment tensors) can be defined and therefore evolution equations are needed. To obtain these equations one can start from a mesoscopic level where the orientation variable $\boldsymbol{n}$ is treated at the same footing as the spatial variable $\boldsymbol{x}$. All physical quantities are then considered to be fields depending on time, position and orientation [2]. The newly formulated, mesoscopic balance equations contain additional flux terms due to orientation changes and can be found in different places in the literature [2–4,6,20]. For the mesoscopic mass density $\rho^{\text{meso}}(\boldsymbol{x},\boldsymbol{n},t) := \rho f$, conservation of mass results in

$$\frac{\partial\rho^{\text{meso}}}{\partial t} + \nabla_x\cdot\left(\boldsymbol{v}\rho^{\text{meso}}\right) + \nabla_n\cdot\left(\boldsymbol{\omega}\times\boldsymbol{n}\rho^{\text{meso}}\right) = 0\,, \tag{61}$$

which can be easily reformulated to yield an evolution equation for the ODF

$$\frac{\partial}{\partial t}f + \nabla_x\cdot(\boldsymbol{v}f) + \nabla_n\cdot(\boldsymbol{\omega}\times\boldsymbol{n}f) + f\left(\frac{\partial}{\partial t} + \boldsymbol{v}\cdot\nabla_x\right)\ln\rho = 0\,. \tag{62}$$

Since the alignment tensors are moments of f, the wanted equations can be derived by obtaining the moments of (62). However, this proves to be a rather complicated task and we shall not perform the calculations here, but discuss some of the problems which will arise.

Calculating the moments of f means multiplication by $\overline{n_{\mu_1}\cdots n_{\mu_\ell}}$ and integration over the unit sphere. This is no problem for the first term containing the time derivative and also the second term provides no difficulties if we assume that the velocity fields $\boldsymbol{v}$ does not depend on $\boldsymbol{n}$, which can be experimentally justified for many interesting cases. However, we cannot assume the same for $\boldsymbol{\omega}\times\boldsymbol{n}$. Since a uniform tangential vector field on the sphere can only be the null vector field (the fact that any continuous tangential vector field on the sphere S^2 vanishes at least at one point is a well-known theorem of differential geometry and it is often referred to as the proposition of the "combed hedgehog"). Consequently, we have to deal with nonlinear terms and this leads to non-linear coupling of *all* evolution equations for the alignment tensors. Thus closure relations are needed to obtain a finite set of equations, but even then one has to tackle a system on nonlinear, coupled, partial differential equations. Since we do not need these relations here, this particular line of thought will not be followed here any longer. The interested reader can find a general form of the alignment tensor balances in BLENK et al. [2] and informations on how to close the system of equations and to extract physical properties in different papers by HESS [13–16].

2.4 Viscosity Coefficients

We already saw that constitutive modeling for anisotropic granular flows can greatly benefit from liquid crystal theories [7]. Since the rheological behavior of a liquid crystal is, in addition to its optical properties, responsible for its performance in a liquid crystal display, many experiments have been performed to measure the viscosity coefficients and to examine the complex interaction between flow and local order in the fluid.

Normally, shear rates in liquid crystalline flows are low and the fluid behaves NEWTONian. However, the internal structure of the fluid increases the number of viscosity coefficients compared to an isotropic liquid. One of the first attempts to study the rheology of nematics is due to LESLIE [18]. He used a state space based on the equilibrium fields ρ and T (temperature), together with the simplest non-equilibrium variable for flows, the shear rate tensor $\underline{\underline{\gamma}}\ \overline{\nabla_x \boldsymbol{v}}$, and as structural variables, director $\boldsymbol{d}$ and its co-rotational ("JAUMANN") derivative $\boldsymbol{D} := \dot{\boldsymbol{d}} - \boldsymbol{\Omega} \times \boldsymbol{d}$, where $\boldsymbol{\Omega}$ is the vorticity of the flow field. If we want a *linear* relation between spatial derivatives of the velocity field (note that $\boldsymbol{D}$ contains $\boldsymbol{\Omega} = \frac{1}{2}\nabla_x \times \boldsymbol{v}!$) and the non-equilibrium part of the stress tensor, the most general ansatz based on this state space is [5]

$$\begin{aligned} t_{\nu\mu}^{\text{non-eq.}} &= \alpha_1 d_\nu d_\mu d_\lambda d_\kappa \gamma_{\lambda\kappa} + \alpha_2 d_\nu D_\mu + \alpha_3 d_\mu D_\nu \\ &\quad + \alpha_4 \gamma_{\nu\mu} + \alpha_5 d_\nu d_\lambda \gamma_{\lambda\mu} + \alpha_6 d_\mu d_\lambda \gamma_{\lambda\nu} \\ &\quad + \zeta_1 d_\lambda d_\kappa \gamma_{\lambda\kappa} \delta_{\mu\nu} + \zeta_2 d_\nu d_\mu \nabla_\lambda v_\lambda + \zeta_3 \nabla_\lambda v_\lambda \delta_{\mu\nu} \,. \end{aligned} \tag{63}$$

The first six coefficients (α's) are called LESLIE coefficients, the last two do not appear when incompressibility of the fluid is assumed. If we decompose (63) into the symmetric traceless, skew-symmetric and isotropic parts, we derive new coefficients, which are simple linear combinations of the old ones, but which are more adapted to the physical meaning of the terms, namely

$$\begin{aligned} \overline{t_{\nu\mu}^{\text{non-eq.}}} &= 2\eta\gamma_{\nu\mu} + 2\tilde{\eta}_1\ \overline{\overline{d_\nu d_\lambda \gamma_{\lambda\mu}}} + 2\tilde{\eta}_2\ \overline{d_\nu D_\mu} \\ &\quad + \tilde{\eta}_3\ \overline{d_\nu d_\mu d_\lambda d_\kappa \gamma_{\lambda\kappa}} + \zeta_2\ \overline{d_\nu d_\mu}\ \nabla_\lambda v_\lambda \end{aligned} \tag{64}$$

$$(t_{\nu\mu}^{\text{non-eq.}})^a = -\gamma_1 (d_\nu D_\mu)^a - \gamma_2 \left(\overline{d_\nu d_\lambda}\ \gamma_{\lambda\mu}\right)^a \,, \tag{65}$$

$$\frac{1}{3} t_{\lambda\lambda}^{\text{non-eq.}} = \eta_V \nabla_\lambda v_\lambda + \kappa d_\lambda d_\kappa \gamma_{\lambda\kappa} \,, \tag{66}$$

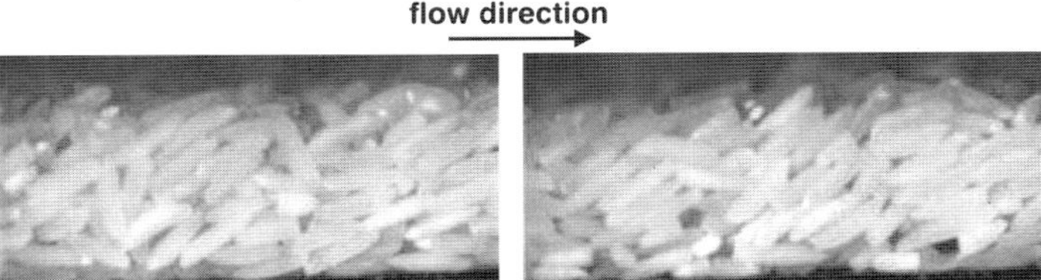

Fig. 4. Bottom layer of flowing rice. The grains are aligned and tilted along the flow direction

with

$$\eta = \frac{1}{2}\left(\alpha_4 + \frac{1}{3}(\alpha_5 + \alpha_6)\right) , \tag{67}$$

$$\tilde{\eta}_1 = \frac{1}{2}(\alpha_5 + \alpha_6), \quad \tilde{\eta}_2 = \frac{1}{2}(\alpha_2 + \alpha_3), \quad \tilde{\eta}_3 = \frac{1}{2}\alpha_1, \tag{68}$$

$$\gamma_1 = \alpha_3 - \alpha_2, \quad \gamma_2 = \alpha_6 - \alpha_5 , \tag{69}$$

$$\eta_V = \frac{1}{3}\zeta_2 + \zeta_3, \quad \kappa = \zeta_1 + \frac{1}{3}(\alpha_1 + \alpha_5 + \alpha_6) . \tag{70}$$

Disappointingly enough, (64)–(66) are not very helpful if we want to study the influence of the order parameters on the viscosities. Since these equations are macroscopic relations the (scalar) order parameters must be buried in the coefficients and the functional dependence is unclear. However, if we take (64)–(66) as *mesoscopic* equations and replace the director field $\boldsymbol{d}$ by the variable $\boldsymbol{n}$ and $\boldsymbol{D}$ by $\boldsymbol{N}(\cdot) := (\boldsymbol{\omega} - \boldsymbol{\Omega}) \times \boldsymbol{n}$,

$$\begin{aligned} \overline{t^{\text{non-eq.}}_{\nu\mu}} &= 2\eta\gamma_{\nu\mu} 2\tilde{\eta}_1 \overline{\overline{n_\nu n_\lambda \gamma_{\lambda\mu}}} \; 2\tilde{\eta}_2 \overline{n_\nu N_\mu} \\ &\quad + \tilde{\eta}_3 \overline{n_\nu n_\mu n_\lambda n_\kappa \gamma_{\lambda\kappa}} + \zeta_2 \overline{n_\nu n_\mu} \nabla_\lambda v_\lambda, \end{aligned} \tag{71}$$

$$(t^{\text{non-eq.}}_{\nu\mu})^a = -\gamma_1 (n_\nu N_\mu)^a - \gamma_2 \left(\overline{n_\nu n_\lambda} \; \gamma_{\lambda\mu}\right)^a , \tag{72}$$

$$\frac{1}{3} t^{\text{non-eq.}}_{\lambda\lambda} = \eta_V \nabla_\lambda v_\lambda + \kappa n_\lambda n_\kappa \gamma_{\lambda\kappa} , \tag{73}$$

we can calculate the macroscopic stresses as ensemble averages [8] if we assume that velocity fluctuations are negligible. For dense granular flows this approximation is experimentally well justified, and we can proceed further.

2.5 Flow Alignment

In order to examine flow alignment experimentally, rice was sent down an incline and the shear layer at the bottom was observed. Figure 4 shows two pictures taken in consecutive experiments. One sees clearly that the single grains are neither oriented randomly – the close confinement of the grains enforces their mutual alignment – nor that the alignment of the grains is independent on the flow. Since the friction between chute (made of plexiglass with a thin coating to prevent electrostatic attraction) and grains was rather low the shear rate is also not very high and a NEWTONian relation between shear rate and stress should apply. In addition, the geometry of the grains effectively hinders a rolling motion – the rice grains glide along the incline and the angular velocity $\boldsymbol{\omega}$ can be neglected. This experimental result confirms the assumptions of the extended SAVAGE–HUTTER theory we discussed in Sect. 1.2

Let us assume for simplicity that the orientation distribution of the grains is transversally isotropic. If only limited information about the functional form of the ODF is available, this assumption can always be justified by a maximum entropy argument as the most likely form agreeing with the data [6]. Figure 5 shows the histogram of orientations of the grains together with an transversally isotropic ODF fitted to the data.

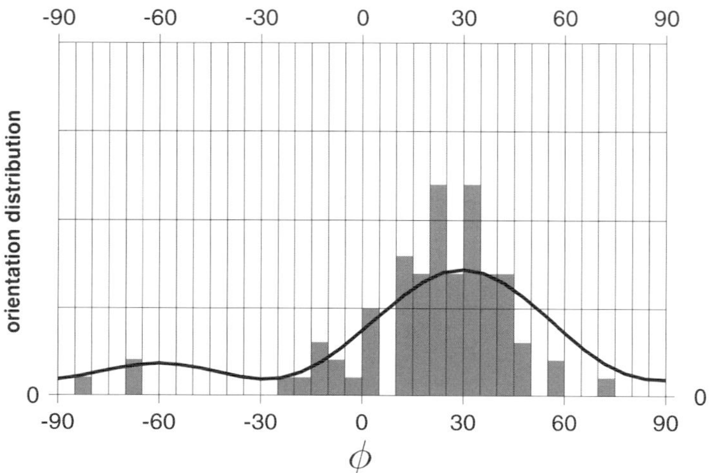

Fig. 5. Orientation distribution of rice grains in the shear zone. The histogram is obtained from the pictures in Fig. 4, the curve represents the ODF up to alignment tensors of fourth order which were obtained from the histogram

Then we can use scalar order parameters and a director $\boldsymbol{d}$ to express the alignment tensors, and we make the ansatz for $\boldsymbol{d}$

$$\boldsymbol{d} = \cos\phi\, \boldsymbol{t} + \sin\phi\, \boldsymbol{n} \tag{74}$$

with a flow alignment angle ϕ and x- and z-coordinate measured in flow (downhill) and gradient direction (perpendicular to the chute), respectively. It can be shown that another equilibrium solution of the symmetry relation (75) exists, where $\boldsymbol{d}$ points along the third, horizontal direction, but this equilibrium is unstable.

We already argued that in granular flows individual rotations of the grains are greatly constrained and that the contribution of the spin balance results in a simple symmetry statement for the stresses. Since the constitutive equations (71) – (73) do not obey this symmetry restriction by construction, we can extract additional information from the averaged relation

$$< (t^{\text{non-eq.}}_{\nu\mu})^a >= -\gamma_1 < n_\nu N_\mu >^a -\gamma_2 < \overline{n_\nu n_\lambda}\ \gamma_{\lambda\mu} >^a \ . \tag{75}$$

We can perform the calculations to evaluate the averaged expression for the stress tensor and insert the special ansatz (74) for the director to obtain equations which still contain the viscosity coefficients of an ordered reference fluid. Now, BAALSS & HESS [1] related these "ordered" coefficients to a reference viscosity η_{ref} and the axes ratio Q of ellipsoids resembling the shape of a single particle

$$\eta^{\mathrm{ord}} = \left(1 + \frac{1}{6}\left(Q - Q^{-1}\right)^2\right)\eta^{\mathrm{ref}} , \tag{76}$$

$$\tilde{\eta}_1^{\mathrm{ord}} = \frac{1}{2}\left(Q - Q^{-1}\right)^2 \eta^{\mathrm{ref}} , \tag{77}$$

$$\tilde{\eta}_2^{\mathrm{ord}} = \frac{1}{2}\left(Q^{-2} - Q^2\right)\eta^{\mathrm{ref}} , \tag{78}$$

$$\tilde{\eta}_3^{\mathrm{ord}} = -\tilde{\eta}_1^{\mathrm{ord}} , \tag{79}$$

$$\gamma_1^{\mathrm{ord}} = \left(Q - Q^{-1}\right)^2 \eta^{\mathrm{ref}} = 2\tilde{\eta}_1^{\mathrm{ord}} , \tag{80}$$

$$\gamma_2^{\mathrm{ord}} = \left(Q^{-2} - Q^2\right)\eta^{\mathrm{ref}} = 2\tilde{\eta}_2^{\mathrm{ord}} . \tag{81}$$

Thus, the symmetry condition (75) results in an equation containing the wanted flow alignment angle, order parameters from the averaging procedure and the aspect ratio Q of the grains – the reference viscosity cancels from (75). Thus we can conclude that material properties of the grains like roughness, which would appear in the reference viscosity, do not affect the flow alignment as long as grains do not stick together or the linear approximation used for the stress tensor breaks down.

After a lengthy, but simple calculation we arrive at an expression for the alignment angle

$$\cos(2\phi) = -\frac{2 - 5S}{3S}\frac{Q - Q^{-1}}{Q + Q^{-1}} . \tag{82}$$

The aspect ratio Q of a rice grain is easily measured. For the variety used it was $Q \approx 3.5$. The order parameter S can be directly measured, too, since we know the distribution function from Fig. 5. But in principle the situation so far is unsatisfactory. What conditions determine the local order in the rice? Certainly, initial conditions are not responsible for the order observed, since the grain distribution at the very beginning of the experiment was nearly random. Thus, the order must be established by the flow.

2.6 Coupling of Flow and Order Parameters

What mechanism can be responsible for the ordering of the grains? Since the particles interact only by collisions, the answer is simple: Energetic considerations based on interaction potentials can be ruled out, and the remaining thermodynamic force is entropy, which appears here in terms of excluded volumes.

A flowing granular material is in non-equilibrium since friction is present and the system dissipates energy. As long as the friction coefficient between bottom and grains is not very high, most of the energy is dissipated internally during the inelastic grain-grain collisions. Consequently, high dissipation inhibits the flow and smaller dissipation allows for quicker motion of the particles. If the local order parameters influence the dissipation – and we shall see that they do – the lowest possible dissipation would result in the best motion of the granules and therefore determine the order parameters.

Thus we claim a *minimum dissipation principle* for granular flows: The internal dissipation of the granular medium has to be assumed minimum to specify the local order parameters.

A short look at the balance for internal energy (9d) shows that the dissipation is mainly due to the dissipation potential $\Psi := \underline{\underline{t}} : \nabla \boldsymbol{v}$ when we neglect couple stresses. Thus we have to calculate the dissipation potential Ψ which is rather simple since we already know the stresses and $\nabla \boldsymbol{v}$ can be approximated by the assumption of a simple shear flow. However, it should be noted, that Ψ contains order parameters, the aspect ratio of the particles, the flow alignment angle, which can be expressed by (82), *and* the reference viscosity of (76)–(81). Thus Ψ depends in a complicated way on the order parameters and, of course, on material properties which were of no importance for the calculation of the alignment angle! However, η_{ref} affects the absolute value of the dissipation, but *not* the minimum of Ψ plotted against the order parameter. Figure 6 depicts the functional dependence of the dissipation in

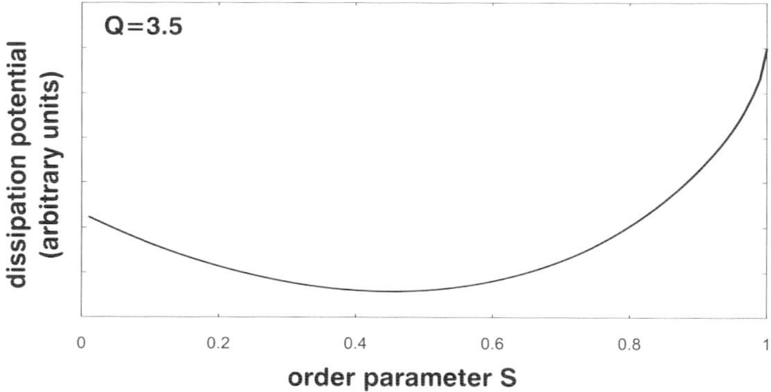

Fig. 6. Dissipation potential for axes ratio Q=3.5. The minimum is assumed for $S \approx 0.5$

the flow on the order parameter S. The minimum is assumed roughly for $S \approx 0.5$, the flow alignment angle which results is slightly less than $40°$, which is in good agreement with the mean alignment angle of Fig. 5. Thus we have been able to solve the problem completely and the theoretical solution agrees nicely with the experimental observations.

3 Conclusions

Anisotropic fluids provide a wide field for investigations for anyone who is interested in continuum theories and structured media. On the previous pages a continuum thermo-mechanical framework was outlined which was inspired by the works of Ericksen, Leslie, Hess and many others on liquid crystals and has been adopted for granular materials. Certainly, it will have its applications also in other fields of material sciences where anisotropy is present.

It turned out that the SAVAGE–HUTTER theory applies as well to anisotropic granular media as to "normal" ones; the SAVAGE–HUTTER equations yield a suitable framework to calculate shallow granular flows in general. Nevertheless, the basic assumption of symmetric shear stresses cannot be upheld for anisotropic grains. Since, on the other hand, the spin balance reduces just to the same symmetry statement, the distinction between these two situations seems to be very artificial.

However, one should be careful not to confuse these two concepts: Any stress tensor which is a priori symmetric requires constitutive relations obeying this very symmetry *under all circumstances*! If we allow, however, for skew-symmetric parts in general and conclude, that these parts must vanish in certain situations, we can extract useful physical information like the flow alignment angle. In addition, one should keep im mind, that the symmetry statement is an approximation and that we have disregarded all couple stresses (tangential friction) to derive the statement. Hence, there are flow situations where neglection of the spin contributions leads to misjudgment. Simple examples are very thin flows when rolling of the particles becomes important.

For a simple but very important flow type (simple shear) we could determine the material response of the granular medium. The grains align within the flow field and the local order changes – resulting in a decreased effective viscosity of the yielding material. Since failure of granular structures often starts in shear bands, this phenomenon of "shear thinning" can significantly influence the dynamics of granular failure with induced anisotropy. Here, much future work will be needed to obtain a general model which can describe the transition from rest to fully developed flow.

Acknowledgments: Finally, we like to thank our colleagues for their support and many valuable discussions: Prof. K. HUTTER as the leader of the group "Fluid Dynamics", Dr. W. ECKART, Dr. J.M.N.T. GRAY, Dr. N. KIRCHNER, Dr. Y.-S. TAI, MSc. S. PUDASAINI and, of course, Dr. M. WEIS and Dipl.-Phys. B. MÜGGE, who kept our computers alive and kicking.

References

1. Baalss D.,Hess S. (1988) The viscosity coefficients of oriented nematic and nematic discotic liquid crystals; affine transformation model. Z Naturforsch 43a:662–670
2. Blenk S., Ehrentraut H., Muschik W. (1991) Statistical foundation of macroscopic balances for liquid crystals in alignment tensor formulation. Physica A 174:119–138
3. Blenk S., Ehrentraut H., Muschik W. (1992) Macroscopic constitutive equations for liquid crystals induced by their mesoscopic orientation distribution. Int J Engng Sci 30(9):1127–1143
4. Blenk S., Ehrentraut H., Muschik W. (1993) A continuum theory for liquid crystals describing different degrees of orientational order. Liquid Crystals 14(4):1221–1226
5. de Gennes, P.G. (1974) The Physics of Liquid Crystals. Clarendon Press, Oxford
6. Ehrentraut, H. 81996) A Unified Mesoscopic Continuum Theory of Uniaxial and Biaxial Liquid Crystals. Wissenschaft- und Technik Verlag, Berlin

7. Ehrentraut, H. (2001) Anisotropic Fluids: From Liquid Crystals to Granular Materials. In: Straughan B., Greve R., Ehrentraut H., Wang Y. (Eds.) Continuum Mechanics and Applications in Geophysics and the Environment, Springer, Berlin Heidelberg, 18 – 43
8. Ehrentraut H., Hess S. (1995) Viscosity coefficients of partially aligned nematic and nematic discotic liquid crystals. Phys Rev E 51(3):2203 – 2212
9. Ehrentraut H., Muschik W. (1998) On symmetric irreducible tensors in d-dimensions. ARI 51:149–159
10. Gray J.M.N.T., Wieland M., Hutter K. (1999) Gravity-driven free surface flow of granular avalanches over complex basal topography. Proc R Soc Lond A 455:1841–1874
11. Greve R., Koch T., Hutter K. (1994) Unconfined flow of granular avalanches along a partly curved surface. I. Theory. Proc R Soc Lond A 445:399–413
12. Hess S., Waldmann L. (1966) Kinetic theorie for a dilute gas of particles with spin. Z Naturforsch 21a:1529–1546
13. Hess S. (1975) Irreversible thermodynamics of nonequilibrium alignment phenomena in molecular liquids and in liquid crystals. Z Naturforsch 30a:728–733
14. Hess S. (1975) Irreversible thermodynamics of nonequilibrium alignment phenomena in molecular liquids and in liquid crystals II. Z Naturforsch 30a:1224–1232
15. Hess S. (1976) Fokker-Planck-equation approach to flow alignment in liquid crystals. Z Naturforsch 31a:1034–1037
16. Hess S. (1976) Pre- and post-transitional behaviour of the flow alignment and flow-induced phase transition in liquid crystals. Z Naturforsch 31a:1507–1513
17. Hess S., W. Köhler W. (1980) Formeln zur Tensor-Rechnung. Palm & Enke, Erlangen
18. Leslie F.M. (1968) Some constitutive equations for liquid crystals. Arch Rat Mech Anal 28:265–283
19. Muschik W., Ehrentraut H., Papenfuss C., Blenk S. (1995) Mesoscopic theory of liquid crystals. In: Brey J.J., Marro J., Rubi J.M., San Miguel M. (Eds.) 25 Years of Non-Equilibrium Statistical Mechanics, Lecture Notes in Physics 445, Springer, Berlin Heidelberg, 303–311
20. Muschik W., Ehrentraut H., Papenfuss C., Blenk S. (1996) Mesoscopic theory of liquid crystals. In: Shiner J.S. (Ed.) Entropy and Entropy Generation, Kluwer, Dordrecht, 101 – 109
21. Savage S.B., Hutter K. (1989) The motion of a finite mass of granular material down a rough incline. J Fluid Mech 199, 177–215
22. Tai Y.-C. (2000) Dynamics of Granular Avalanches and their Simulations with Shock-Capturing and Front-Tracking Numerical Schemes. Shaker Verlag, Aachen

Part III

Porous and Granular Materials. Subscale- and Micromechanical Effects.

Modelling Particle Size Segregation in Granular Mixtures

Nina P. Kirchner and Kolumban Hutter

Institute of Mechanics III, Darmstadt University of Technology, Hochschulstr. 1, 64289 Darmstadt, Germany,
e-mail: {kirchner,hutter}@mechanik.tu-darmstadt.de

received 19 Aug 2002 — accepted 12 Sep 2002

Abstract. A striking phenomenon to be observed e.g. in (natural, industrial and laboratory) avalanches is the effect of "inverse grading": during the avalanching motion, the individual particles forming the body of the avalanche are re-distributed in such a way that large particles gather on top and at the front, while small particles concentrate close to the bottom and the rear. This segregation according to particle size has been our motivation to develop a thermodynamically consistent continuum-mechanical model which is applicable to granular mixtures consisting of a finite number of constituents. The thermo-mechanical model is characterised by N additional evolution equations of "configurational force-type" for N scalar-valued, internal variables: the volume fractions of the constituents. Moreover, a partial abandonment of the principle of phase separation is suggested to account for interactions between various constituents on the constitutive level. The main results obtained from a rigorous exploitation of the entropy principle according to Müller & Liu are presented and discussed.

1 Introduction and Motivation

Granular materials are encountered in many different forms and various contexts in everyday life – in the form of rice, muesli, washing powder or sand. As such, they are defined as a collection of a large number of discrete, identifiable solid particles with interstices filled with a fluid or a gas. A granular material can either be composed of particles belonging to a single constituent, or it may consist of particles belonging to vastly distinct constituents. In the latter case, one speaks of *granular mixtures* or *multiphase mixtures*, although the system is immiscible in the sense that a homogeneous distribution of particles cannot be achieved. The granular mixture thus displays an inherent geometrical structure.

In fields such as production technology and process engineering, feeding and discharging of particulate materials into and from storage systems of any kind (e.g. silos and hoppers) are typical operations of bulk solid handling giving rise to *granular flows.* Once deposited, it is observed that the granular material piles up in a heap, displaying solid-like rather than the earlier observed fluid-like behaviour: Granular materials are hybrids in the sense that a characterisation of the granular body as "solid" or "fluid" can only be accomplished depending on the conditions the body is exposed to. Typical examples of such particulate materials in industrial

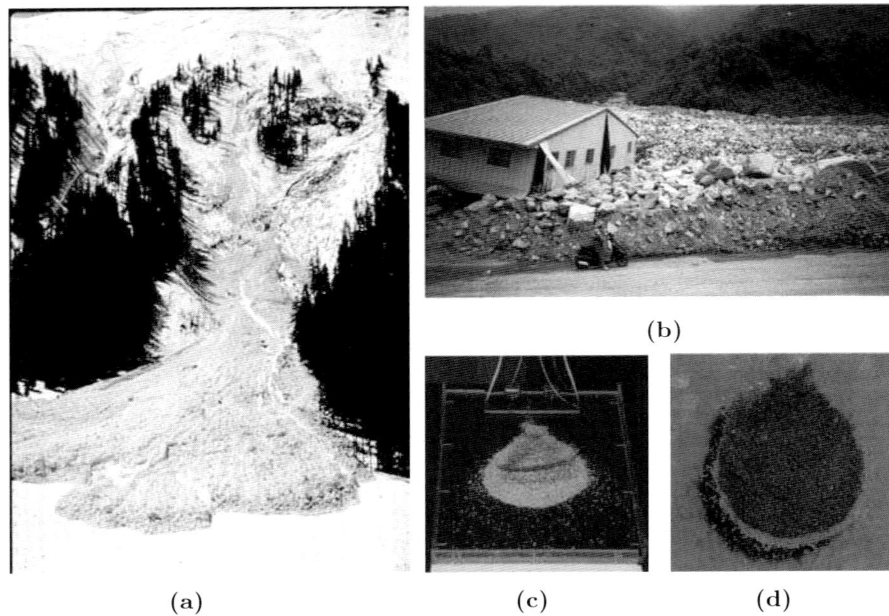

Fig. 1. **(a)** Deposit of a wet snow avalanche (Courtesy of the Swiss Federal Institute of Snow and Avalanche Research, Davos, Switzerland) **(b)** Debris flow deposit from a disastrous flow event on 31. July ∼01. Aug. 1996 in Taiwan. In the picture the road in the front has been cleared. It demonstrates particle size separation. The free surface of the deposit is covered by large boulders, whilst the lower part consists of the fine material (Courtesy of Prof. Hung, Ju-Jiang) **(c)** Deposit of a laboratory avalanche viewed from the top. The large white particles are located at the top and at the front **(d)** Deposit of the same laboratory avalanche viewed from the bottom. The small black particles are located at the bottom and in rearward regions. The semi-annulus of white particles is a manifestation of the larger particles at the front

applications range from cereals or rice in the food processing industries to pharmaceuticals and construction materials such as gravel, sand and cement in the chemical and civil engineering sector, respectively. In a geomechanical context, snow, rock or powder avalanches, debris or pyroclastic flows or the formation of dunes are typical examples of granular flows.

A characteristic feature observed in the above mentioned granular materials (which are in fact granular *mixtures*) is the process of *segregation*, by which we mean that grains with like properties tend to accumulate in certain regions of the granular body. In general, the segregation takes place while the granular body is in motion and the overall material behaviour is thus fluid-like or gaseous. *Segregation according to particle size* is the most common separation mechanism and can be observed in deposits of debris and snow avalanches, where big chunks of snow are mostly found at the front and top, whereas smaller ones are more likely to be found at the bottom and in rearward regions of the deposit, see Fig. 1a,b. Laboratory experiments, in which a granular mixture of sugar (large white particles) and poppy seeds (small black particles) is moving along an inclined chute and comes to rest in a horizontal run-out zone reproduce these observations very well, see Fig. 1c,d. The same can be observed when emptying a box of cereals: at the bottom, small particles are left – larger particles, which have the tendency to rise to the surface, have been present in the first servings but are no longer present in the last ones, turning the remainder of the cereals into a serving of crumbles. This phenomenon is known as the "brazil nut" effect. In such (and similar) cases, it is (at least from a process engineering point of view) essential to control the spatial distribution of the constituents in the storage container to reduce the risk of being left – at some time – with a low quality mixture which does no longer contain individual constituents in sufficient concentrations. Processes have thus to be started which counteract the segregational process.

Particle segregation in granular mixtures may not only or exclusively be triggered by differences in particle size: it is also possible that particles segregate due to differences in shape, surface properties, density, or resilience. Segregation according to particle size seems however to be dominant, see [34,43]. In the remainder of this article, we will thus focus on *segregation according to particle size*, and we do not explicitly distinguish between surface segregation (treated e.g. by [6,32,33]) and segregation taking place in the bulk material. Moreover, we will consider only *dry* granular mixtures, for which the interstices between the particles are filled by air. In general, this need not be the case, but granular mixtures with one or several fluids in its interstices, known as *debris flows* or *mud flows*, shall not be addressed here.

It is hoped that with the above description, enough motivation for considering dry granular mixtures has been given. To be precise, we will from now on work with a dry granular mixture of $2 \leq N < \infty$ constituents. The grains of the mixture vary in size, of which N typical values are used as representatives giving rise to the formation of N classes of constituents according to which the grains can be classified. Since interaction between grains of different constituents triggers the redistribution of grains as e.g. observed during segregation processes, a model has to be formulated which properly accounts for this interaction if we eventually want to be able to understand, model and predict the effects of particle size segregation. The framework within which this analysis will be carried out is that of continuum thermodynamics.

To be fair, mention should be made about alternative methods with which segregation processes have been or are being described. In doing so, we restrict attention to continuum formulations, leaving event driven dynamics as e.g. treated in [15] unmentioned. The *concept of information entropy* has been developed in [36] to derive a first model explaining inverse grading. This approach relies on gravity as the only cause of the segregation mechanism and is difficult to generalise. A *mixture model of essentially diffusive nature* has been proposed in [11]. This model is extended to a full thermodynamic theory in [10]. It should be mentioned that a "discrete" mixture model, in which the particle diameters are present only in discrete steps, is a restriction. A continuous distribution of particle size would have to be accounted for by a "mixture theory with continuous diversity", see [8,9], but has so far not been pursued in the context of particle size segregation.

The model proposed here is presented and discussed in Sect. 2, and the constitutive principles applied are outlined in Sect. 3. Section 4 deals with the exploitation

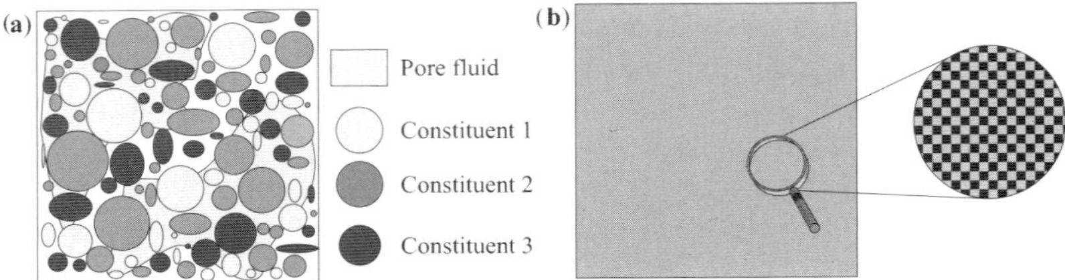

Fig. 2. A granular mixture consisting of discrete, identifiable solid particles which are dispersed in a gaseous and a fluid phase. **(a)** The internal structure of the granular mixture **(b)** The uniform and idealised mixture material after the homogenisation process

of the entropy principle; individual subsections are devoted to the method of exploitation as suggested by MÜLLER and LIU, and selected implications (valid in non-equilibrium) that can be deduced. The specialisation to the case of thermodynamic equilibrium is addressed in Sect. 5 and serves to highlight selected peculiarities of the proposed model. Finally, the article is drawn to a close with concluding remarks in Sect. 6.

2 Mixture Model and Modelling of Internal Variables

To deal with granular mixtures, we will focus in this work on models that are based on continuum mechanical theories. The vast field of modelling methods stemming from statistical mechanics, molecular dynamics or cellular automata has to be excluded from our considerations in order not to exceed the scope of this article. We refer, however, to [1,15,19] and the many references therein.

The first systematic continuum mechanical mixture models have been derived in the late 1950's [39,41] as an extension of the concepts of "classical" continuum mechanics, by which we mean continuum mechanics dealing with a single constituent continuum. These early mixture models have been proposed to describe the phenomena of diffusion and chemical reaction in mixtures consisting of N fluid constituents. To account for the fact that the constituents are *miscible* at the molecular (microscopical) level, it is commonly assumed that different constituents are simultaneously present in a so-called representative volume element (RVE). This simultaneous presence of constituents is achieved by a (real or imagined) homogenisation process which causes the individual constituents to blend into an uniform idealised mixture material. On the one hand, the homogenisation process is the key factor for the application for continuum mechanical field theories, on the other hand, however, certain information about the mixture is lost during such a process: Granular mixtures are essentially *immiscible* at the microscopic level and display hence an inherent geometrical structure, which is not reflected in the homogenised mixture material, see Fig. 2.

Remark 1. To model multi-fluid flow in a porous medium, GRAY & HASSINAZADEH [14] have proposed an averaging procedure for microscale equations, leading to explicit relations between constituent quantities at the micro- and the macroscale, respectively. Their approach accounts for interfaces at the microscopic level at various degrees of complexity; in the simplest case, these interfaces reduce to singular surfaces across which certain quantities (e.g. the density) may suffer from jumps. The "elements of macroscale conservation equations for multiphase flow in porous media" are described in [13]. ◇

Consequently, granular mixtures require a different treatment if their internal structure shall be respected and reflected in the continuum mechanical model. Indeed, we are challenged by the task to provide suitable amendments to those early models in order to make them applicable to the granular systems which we wish to investigate. Such extensions concern the postulation of additional balance equations for internal variables accounting for the inherent geometrical structure as well as an appropriate modification of constitutive principles.

Remark 2. As far as the modelling of internal variables is concerned, different approaches have been pursued, and a few of them shall now be briefly addressed. A more detailed overview and additional references can e.g. be found in [31].

- GOODMAN & COWIN recognised the *volume distribution function* as a kinematical quantity and suggested that it be governed by an equation of balance (with non-vanishing flux term) which they called the *balance of equilibrated forces*, see [12]. In their model, "kinetic energy" and "powers of working" associated with the volume distribution function are postulated to exist and which give rise to additional terms in the balance of internal energy.
- A different approach has been proposed by BOWEN [3], who treats the *volume fraction* as an internal state variable as do more recently SVENDSEN & HUTTER, [37]. They postulate an evolution equation for the volume fraction that has the same structure as the balance of mass (and thus does not contain a flux term).
- WILMANSKI has in [44] proposed to model *porosity* by a balance-like evolution equation containing a flux term. To derive this equation, a multicomponent system consisting of a solid constituent forming the matrix and N fluid constituents filling the interstitial pore space is considered. The $N+1$ volume fractions are replaced by the single scalar field of *porosity* n. The reason for this reduction is the assumption that the volume contribution of the pores is sufficient to describe the internal structure of the granular material. ◇

Before addressing the above mentioned amendments in greater detail, let us introduce the following definitions.

Definition 1. Let K_α, $\alpha = 1, \dots, N$ denote the constituents of a granular mixture. Each K_α is characterised by a *partial mass density* $\rho_\alpha(\boldsymbol{x}, t)$, which is its mass density with respect to the entire mixture volume. The *true mass density* of K_α (which refers to the volume of K_α only) is denoted by $\rho_\alpha^t(\boldsymbol{x}, t)$. We define

$$\rho = \sum_{\alpha=1}^{N} \rho_\alpha$$

to be the *mass density of the granular mixture.* The *volume fraction* $\nu_\alpha(\boldsymbol{x}, t)$ is defined as the ratio of the volume of K_α to the total mixture volume and can hence be written as

$$\nu_\alpha := \rho_\alpha / \rho_\alpha^t;$$

obviously, $\nu_\alpha \in [0,1]$ holds. The mixture is called *saturated* or *unsaturated*, respectively, if

$$\sum_{\alpha=1}^{N} \nu_\alpha = 1 \qquad \text{or} \qquad \sum_{\alpha=1}^{N} \nu_\alpha < 1$$

is fulfilled, respectively. Finally, if $\boldsymbol{v}$ is the *barycentric velocity* given by

$$\boldsymbol{v} := \sum_{\alpha=1}^{N} \frac{\rho_\alpha}{\rho} \boldsymbol{v}_\alpha := \xi_\alpha \boldsymbol{v}_\alpha,$$

where $\boldsymbol{v}_\alpha$ is the *velocity* and ξ_α is the *mass fraction* of K_α, we introduce for each K_α its *diffusion velocity relative to the mixture (barycentric) velocity*, $\boldsymbol{u}_\alpha := \boldsymbol{v}_\alpha - \boldsymbol{v}$. ⊠

Remark 3. It is worth mentioning that "saturation" is not uniquely defined: Soil mechanicians call a granular mixture "saturated" if the pore space is entirely filled by a fluid, and "unsaturated" if the pore space is partly filled by a fluid and partly by air or even voids. ◇

With these definitions at hand, we can now proceed to introduce an appropriate mixture model for granular materials. Since the interstitial constituent may – in particular when being a fluid – play an important dynamical role, we will work with a mixture model involving N balances of mass and N balances of linear and angular momentum. However, only the mixture balance of energy and the mixture balance of entropy are considered. The former is justified whenever no phase changes arise and the individual constituents response to changes in thermal conditions takes place infinitely fast, so that in fact a single temperature field is sufficient to describe the entire mixture. The latter is physically more reasonable, cf. the discussion in [40]. The (localised) thermo-mechanical balance relations for each constituent K_α read

$$\begin{aligned} \mathfrak{c}_\alpha &= \acute{\rho_\alpha} + \rho_\alpha \operatorname{div} \boldsymbol{v}_\alpha, \\ \mathfrak{m}_\alpha &= \boldsymbol{v}_\alpha \mathfrak{c}_\alpha + \rho_\alpha \acute{\boldsymbol{v}}_\alpha - \operatorname{div} \boldsymbol{T}_\alpha - \rho_\alpha \boldsymbol{b}_\alpha, \\ \mathfrak{M}_\alpha &= \boldsymbol{r} \times \mathfrak{m}_\alpha + \operatorname{ax} \boldsymbol{T}_\alpha, \qquad\qquad \alpha = 1, \dots, N \\ 0 &= \rho \dot{\varepsilon} - \boldsymbol{T} \cdot \boldsymbol{D} + \operatorname{div} \boldsymbol{q} - \rho r, \\ \pi &= \rho \dot{\eta} + \operatorname{div} \boldsymbol{\phi} - \rho \sigma, \end{aligned} \tag{1}$$

where $\mathfrak{c}_\alpha$, $\mathfrak{m}_\alpha$ and $\mathfrak{M}_\alpha$ are the constituent growth terms of mass, linear and angular momentum, respectively. Stating equations (1), we have assumed sufficient differentiability of all involved fields to avoid having to deal with jump conditions. The quantities $\boldsymbol{T}_\alpha$ and $\boldsymbol{b}_\alpha$ denote the constituent Cauchy stress tensor and a constituent volume force such as e.g. gravity, respectively, $\boldsymbol{r}$ is the position vector of the particle under consideration (from the origin relative to which the angular momentum is calculated), and $\operatorname{ax} \boldsymbol{T}_\alpha$ denotes the axial vector of $\boldsymbol{T}_\alpha$. Further, $\acute{(\cdot)} := \partial(\cdot)/\partial t + \nabla(\cdot)\boldsymbol{v}_\alpha$ is the material derivative calculated with respect to the constituent velocity $\boldsymbol{v}_\alpha$ of K_α, whereas $(\cdot)^{\cdot}$ is the material derivative calculated with respect to the mixture (barycentric) velocity. Note that $(1)_4$ is the *mixture* balance of energy – as a conservation equation, the growth term on its left-hand side is identical to zero. In general, the growth terms arising in the *constituent* balances do *not* vanish, however, they are restricted by TRUESDELL's third metaphysical principle, see [40]. Further, ε is the mixture specific internal energy, $\boldsymbol{T}$ and $\boldsymbol{D}$ denote the mixture Cauchy stress and the stretching tensor, respectively, $\boldsymbol{q}$ is the mixture heat flux and r an external mixture energy supply such as e.g. radiation. The mixture specific entropy is denoted by η, and $\boldsymbol{\phi}$, σ and π are its flux, supply, and production, respectively.

Remark 4. It is in principle possible to work with constituent balances of constituent internal energy, ε_α and entropy, η_α. If this is done, TRUESDELL'S third metaphysical principle requires that the mixture balance of internal energy and entropy, $(1)_{4,5}$, are obtained by a summation of their constituent counterparts. The application of this principle provides us – among others – with explicit formulas governing the relation between the mixture quantities and their constituent analoga. In particular, we find that

$$\varepsilon = \sum_{\alpha=1}^{N} \xi_\alpha \varepsilon_\alpha + \tfrac{1}{2} \sum_{\alpha=1}^{N} \xi_\alpha \boldsymbol{u}_\alpha \cdot \boldsymbol{u}_\alpha, \ \eta = \sum_{\alpha=1}^{N} \xi_\alpha \eta_\alpha, \ \boldsymbol{\phi} = \sum_{\alpha=1}^{N} (\boldsymbol{\phi}_\alpha + \rho_\alpha \eta_\alpha \boldsymbol{u}_\alpha),$$

$$\boldsymbol{q} = \sum_{\alpha=1}^{N} (\boldsymbol{q}_\alpha - \boldsymbol{T}_\alpha \boldsymbol{u}_\alpha + \rho_\alpha \varepsilon_\alpha \boldsymbol{u}_\alpha + + \tfrac{1}{2} \rho_\alpha (\boldsymbol{u}_\alpha \cdot \boldsymbol{u}_\alpha) \boldsymbol{u}_\alpha)$$

hold, where $\boldsymbol{\phi}_\alpha$ and $\boldsymbol{q}_\alpha$ are constituent fluxes of entropy and energy, respectively. All other quantities have been defined above. Obviously, the mixture internal energy, its flux and the mixture entropy flux are not just the (possibly weighted) sum of their constituent counterparts but contain "diffusive" contributions caused by the diffusive motion of the individual constituents in the mixture. The constituent quantities ε_α, $\boldsymbol{q}_\alpha$, η_α and $\boldsymbol{\phi}_\alpha$ are *constitutive* quantities playing an important role at a later stage of our investigations, see Def. 2 in Subsect. 4.1. ◇

To account for the inherent geometrical structure of the granular material, we treat the volume fractions ν_α of each constituent K_α as *internal* (or *structural*) scalar-valued variables, for each of which an additional balance law will be postulated.

Remark 5. Structural variables can be found in many theories which aim at relating the *macroscopic* response of a system to the structure which its constituents exhibit at the *microscale*. Examples range from the (scalar valued) *isotropic damage variable*, d, (introduced in [20]) and the vector valued *director* (introduced in [7] and today used e.g. in liquid crystal theories, see [2]) to tensor valued *structure-* or *fabric tensors* entering e.g. the modelling of induced anisotropy in ice, see [28]. ◇

To describe segregation in granular mixtures, we pursue a phenomenologically motivated approach proposed by PASSMAN et al. [31]: they extended GOODMAN & COWIN's concept, see Remark 2, to be applicable to a granular mixture of $N < \infty$ constituents. They suggested that the evolution of each of the N volume fractions ν_α, which play the role of the internal variables accounting for the immiscibility of the granular mixture, be governed by a balance of *equilibrated* or *configurational* force. We use the attribute "configurational" because ν_α are variables describing some properties of configuration at the microscale at the macroscale. It has nothing in common with ESHELBY's configurational force balance. The balance equation for the volume fractions as proposed in [31] reads

$$0 = \rho_\alpha k_\alpha \grave{\grave{\nu}}_\alpha + \mathfrak{c}_\alpha k_\alpha \grave{\nu}_\alpha - \operatorname{div} \boldsymbol{h}_\alpha - \rho_\alpha (f_\alpha + l_\alpha) - \mathfrak{g}_\alpha, \quad \alpha = 1, \ldots, N. \tag{2}$$

Here, k_α is a constant coefficient, $\boldsymbol{h}_\alpha$ is the constituent configurational flux, f_α and l_α are a constituent production and supply term, respectively, and $\mathfrak{g}_\alpha$ denotes the constituent growth term. To reduce the complexity of $(1)_{1-3}$ and (2), we will henceforth assume that the constituent growth terms $\mathfrak{c}_\alpha$ and $\mathfrak{g}_\alpha$ vanish identically, and that, moreover, the exchange of angular momentum happens only through the exchange of linear momentum, $\mathfrak{M}_\alpha = \boldsymbol{r} \times \mathfrak{m}_\alpha$. Despite these simplifications, additional terms occur when following the approach of [12]: the volume fractions ν_α, which are regarded as kinematic quantities, give rise to "kinetic energy" and "powers of working" which are postulated as

$$E_\alpha := \int\limits_V \rho_\alpha k_\alpha \frac{\grave{\nu}_\alpha^2}{2} \mathrm{d}V, \qquad P_\alpha := \oint\limits_{\partial V} \grave{\nu}_\alpha \boldsymbol{h}_\alpha \cdot \boldsymbol{n} \mathrm{d}a, \quad \alpha = 1, \ldots, N.$$

Here, $\boldsymbol{n}$ is the unit normal vector on the surface ∂V of the volume V occupied by the granular mixture. These terms affect the (localised) form of the energy balance, see $(3)_5$ below, and we will hence work with the following set of equations:

$$\begin{aligned}
0 &= \grave{\rho}_\alpha + \rho_\alpha \operatorname{div} \boldsymbol{v}_\alpha, \\
\mathfrak{m}_\alpha &= \rho_\alpha \grave{\boldsymbol{v}}_\alpha - \operatorname{div} \boldsymbol{T}_\alpha - \rho_\alpha \boldsymbol{b}_\alpha, \\
\boldsymbol{0} &= \operatorname{ax} \boldsymbol{T}_\alpha \quad \Longleftrightarrow \quad \boldsymbol{T}_\alpha = \boldsymbol{T}_\alpha^{\mathrm{T}}, \qquad\qquad \alpha = 1, \ldots, N \\
0 &= \rho_\alpha k_\alpha \grave{\grave{\nu}}_\alpha - \operatorname{div} \boldsymbol{h}_\alpha - \rho_\alpha (f_\alpha + l_\alpha), \\
0 &= \rho \dot{\varepsilon} - \boldsymbol{T} \cdot \boldsymbol{D} + \operatorname{div} \boldsymbol{q} - \rho r - \sum_{\alpha=1}^N \boldsymbol{h}_\alpha \cdot \nabla \grave{\nu}_\alpha - \sum_{\alpha=1}^N \rho_\alpha f_\alpha \grave{\nu}_\alpha, \\
\pi &= \rho \dot{\eta} + \operatorname{div} \boldsymbol{\phi} - \rho \sigma.
\end{aligned} \tag{3}$$

Complementing (3) by constitutive relations will yield a continuum mechanical model for the material under consideration.

Remark 6. Let us briefly comment upon the assumptions imposed on the constituent growth terms. As we are dealing with a granular mixture whose constituents differ e.g. in particle size, it is likely that (during a deformation) large grains split up into pieces of smaller size (this process is sometimes referred to as the *fragmentation* of grains). In such a case, we would have to account for a negative exchange term in the mass balance equation referring to the "large" constituent, while positive exchange terms would arise in one or several mass balance equations referring to "smaller" constituents. Likewise, the inverse process, namely that several small grain fragments become one large grain e.g due to the action of some (mechanical, thermal or other) agent, can occur. However, since the focus in this work is on the phenomenon of particle segregation, processes such as fragmentation and grain-growth are assumed to be negligible, resulting in $\mathfrak{c}_\alpha = 0$ for all $\alpha = 1, \ldots, N$.

Turning to the constituent balances of angular momentum, it is noted that the assumption $\mathfrak{M}_\alpha = \boldsymbol{r} \times \mathfrak{m}_\alpha$ implies that the balance of angular momentum reduces to a symmetry condition imposed on the partial Cauchy stress tensors, see $(3)_3$. In other words, we are neglecting an intrinsic spin of the particles which has to be accounted for in the case of so-called *polar* continua. Here, we deal only with *non-polar* media.

The reason for assuming $\mathfrak{g}_\alpha = 0$ for all α is very simple: whereas one can relatively easy assign a meaning to the term $\rho_\alpha k_\alpha \grave{\grave{\nu}}_\alpha$ in (2), it is difficult to come up with (at least partly justified) physically motivated expressions for the remaining configurational quantities. Whereas the flux $\boldsymbol{h}_\alpha$ and the production f_α will be shown to be of relevance in the thermodynamic analysis, l_α does not affect the latter. Since at most few is known about the exchange terms $\mathfrak{g}_\alpha$, and since – as we strongly suspect – they also do not have a significant effect on the thermodynamic analysis, we assumed them to be vanishing identically, following thus [42]. However, if desired, there would be no particular difficulty in relaxing the assumption $\mathfrak{g}_\alpha = 0$. ◇

Summarising what has been obtained so far, we state that the usual thermo-mechanical balance laws have been complemented by N additional evolution equations for the volume fractions ν_α which act as internal structural variables. The first amendment to adapt early mixture models to granular mixtures has thus been made. However, this is not sufficient. The second amendment will concern the postulates adopted in formulating the constitutive equations, to which we turn in Sect. 3.

3 Constitutive Principles

Let us start with a few general remarks on constitutive principles. We have already noted that (3) has to be complemented by constitutive equations to close the system of (functional differential) equations. To do so, we have to specify the *state space* $\mathbb{S}$, which comprises the independent variables of the theory and which is the domain of the constitutive equations. If we denote by $\mathcal{C}$ a constitutive quantity, we can express this in an abstract notation as

$$\mathcal{C} = \hat{\mathcal{C}}(\mathbb{S}) \quad \forall\, \mathcal{C} \in \mathbb{C}, \tag{4}$$

where $\mathbb{C}$ is the set of all constitutive quantities. Upon inserting the constitutive equations into the balance equations, the latter become the so-called *field equations*, and any solution to those is called a *thermodynamic process*. Note that as soon as we deal with mixtures rather than with single constituent materials, the quantities arising in (4) need careful specification.

To illustrate this with an example, consider a simple heat-conducting, single-constituent fluid. The state space for this material is given by $\mathbb{S} := \{\rho, \theta, \nabla\theta\}$, where θ denotes the (empirical) temperature, while the set of constitutive quantities is $\mathbb{C} := \{\boldsymbol{T}, \boldsymbol{q}, \varepsilon\}$. The well-known constitutive principles put forth in [41] request that the constitutive quantities are given by an expression of the form (4) so that e.g. $\boldsymbol{T} = \hat{\boldsymbol{T}}(\rho, \theta, \nabla\theta)$.

Let us now turn to mixtures with $2 \leq N < \infty$ constituents, where each constituent has its "own" state space $\mathbb{S}_\alpha$ and its "own" set of constitutive quantities, $\mathcal{C}_\alpha \in \mathbb{C}_\alpha$. The question on which state space a constitutive quantity $\mathcal{C}_\alpha$ should depend has been answered differently, see [31,41], and has given rise to the so-called *principle of equipresence* on the one hand, and the *principle of phase separation* on the other hand. To highlight the differences between these two principles, we consider again a simple heat conduction fluid, which is now assumed to consist of N constituents.

The *principle of equipresence* states the following: Each (constituent) constitutive quantity should depend on *all* independent variables used in the material modelling, that is,

$$\mathcal{C} = \hat{\mathcal{C}}(\mathbb{S}) \;\forall\; \mathcal{C} \in \mathbb{C}, \text{ where } \mathbb{C} := \cup_{\alpha=1}^{N} \mathbb{C}_\alpha\,, \quad \mathbb{S} := \cup_{\alpha=1}^{N} \mathbb{S}_\alpha, \;\; \alpha = 1, \dots, N.$$

Applied to our simple example, this means that e.g. $\boldsymbol{T}_\alpha = \hat{\boldsymbol{T}}_\alpha(\rho_1, \dots, \rho_N, \theta_1, \dots, \theta_N, \nabla\theta_1, \dots, \nabla\theta_N)$, and it is obvious that the principle of equipresence gives rise to extremely complex and complicated equations as soon as larger state spaces and more than a few constituents are considered.

To cure this shortcoming, the *principle of phase separation* has been formulated which states that constitutive quantities of constituent K_α should only depend on variables that belong to K_α:

$$\mathcal{C}_\alpha = \hat{\mathcal{C}}_\alpha(\mathbb{S}_\alpha) \;\forall\; \mathcal{C}_\alpha \in \mathbb{C}_\alpha\,, \qquad \alpha = 1, \dots, N.$$

Applied to our simple example, this means that e.g. $\boldsymbol{T}_\alpha = \hat{\boldsymbol{T}}_\alpha(\rho_\alpha, \theta_\alpha, \nabla\theta_\alpha)$, and it is obvious that *the principle of phase separation is too restrictive if we wish to account for pronounced interaction between various constituents on a constitutive level.*

Remark 7. Dealing with an N-constituent material, the thermo-mechanical balance equations may contain non-vanishing growth (or *exchange*) terms, cf. (1). Irrespective of whether the principle of phase separation or the principle of equipresence is applied, those growth terms depend "by construction" on all independent variables, that is, on $\mathbb{S} := \cup_{\alpha=1}^{N} \mathbb{S}_\alpha, \;\; \alpha = 1, \dots, N.$ ◇

Turning back to segregation in granular mixtures, we *conjecture* that the application of the principle of phase separation is hardly able to capture the segregational processes: the only terms capturing interaction between the constituents are the momentum growth terms $\mathfrak{m}_\alpha = \hat{\mathfrak{m}}_\alpha(\cup\mathbb{S}_{\alpha=1}^{N})$, cf. Remark 7. Moreover, the evolution of the internal variable ν_α – introduced to account for the internal structure of the granular material – is completely unaffected by (and in fact independent of) the evolution of the corresponding quantity ν_β if the principle of phase separation is applied, see $(3)_4$. To allow for more "coupling effects" between the constituents of the mixture, we thus suggest a *partial abandonment of the principle of phase separation.* The former has to be such that a sufficient amount of interaction between constituents is captured while at the same time, the sophisticated equations that would arise when applying the principle of equipresence (which accounts by definition for a maximum of constituent interaction) should be avoided.

We thus suggest the following constitutive modelling: The state space $\mathbb{S}_\alpha$ of each constituent K_α, $\alpha = 1, \dots, N$, of the mixture is given by

$$\mathbb{S}_\alpha := \{\overset{t}{\rho}_\alpha, \nabla\overset{t}{\rho}_\alpha, \nu_1, ..., \nu_N, \nabla\nu_1, ..., \nabla\nu_N, \dot{\nu}_1, ...\dot{\nu}_N, \theta, \nabla\theta, \boldsymbol{D}_\alpha\}, \tag{5}$$

where $\boldsymbol{D}_\alpha$ is the constituent stretching tensor. All other quantities have been defined above. We treat the true mass densities as independent variables to allow for fluid or gaseous (compressible) constituents. A pronounced interaction between the constituents is with this definition of the state space $\mathbb{S}_\alpha$ introduced on a constitutive level by letting a constitutive quantity of K_α depend on *all structural variables* $\nu_1, \dots, \nu_N$ and their temporal and spatial derivatives. The constituent constitutive quantities are gathered in the set

$$\mathbb{C}_\alpha := \{\boldsymbol{T}_\alpha, \boldsymbol{q}_\alpha, \varepsilon_\alpha, \boldsymbol{h}_\alpha, \eta_\alpha, \boldsymbol{\phi}_\alpha, f_\alpha\}, \qquad \alpha = 1, \dots, N, \tag{6}$$

so that we work with

$$\mathcal{C}_\alpha = \hat{\mathcal{C}}_\alpha(\mathbb{S}_\alpha) \qquad \forall \quad \alpha = 1, \ldots, N,$$

where $\mathbb{S}_\alpha$ is as in (5) and $\mathcal{C}_\alpha$ as in (6). Note that the mixture constitutive quantities ε, η and $\boldsymbol{\phi}$ do not appear in (6); their domain is the set $\mathbb{S} := \cup_{\alpha=1}^{N}\mathbb{S}_\alpha$. We have included ε_α, $\boldsymbol{q}_\alpha$, η_α and $\boldsymbol{\phi}_\alpha$ in (6) despite the fact that they are not arising explicity in the balance equations (1), cf. Remark 4. The supply term related to the equilibrated forces, l_α, see $(3)_5$, is *not* included as a constitutive quantity in (6) – the reason for this will be given in Sect. 4.

4 Exploitation of the Entropy Principle

The exploitation of the entropy principle is – especially for the mixture model with complex material behaviour proposed here – accompanied by tedious and long calculations. In order to avoid being distracted by technical details, we thus omit the latter (they can be found in [21]) and present the results finally obtained. The general procedure of exploiting the entropy principle is outlined in Subsect. 4.1; equipped with the ideas *how* the results are derived, the latter are presented and discussed without digression in Subsects. 4.2–4.4.

4.1 The Approach of Müller & Liu

Before the exploitation of the entropy principle is now carried out according to [24,25,29], let us comment briefly upon assumptions that have tacitly been made in the two preceding sections.

As far as entropy η is concerned, it is required to be a scalar-valued, objective and additive quantity such that it gives rise to the balance law $(1)_5$. Moreover, it is determined by a general constitutive equation, $\eta = \hat{\eta}(\mathbb{S})$. Likewise, the entropy flux $\boldsymbol{\phi}$ is an objective vector-valued function given by $\boldsymbol{\phi} = \hat{\boldsymbol{\phi}}(\mathbb{S})$, implying in particular that no a-priori assumptions on the relation between heat flux and entropy flux have been made. The entropy supply σ is given by a linear combination of all other supplies.

Following the rational thermodynamics interpretation of the second law of thermodynamics, it is required that all solutions to the field equations must satisfy the second law, or, in other words, that the entropy production π must be non-negative for all thermodynamic processes. The second law is thus interpreted as a restriction to the material response, since it implies that the constitutive equations have to be chosen such that the second law of thermodynamics is fulfilled irrespective of the processes which a body under consideration may experience.

Within the field of rational thermodynamics, approaches of various degrees of complexity have been undertaken to fulfil the entropy principle; among them, one finds classical irreversible thermodynamics [27], the large group of Clausius-Duhem Coleman-Noll approaches (see e.g. [5]), and the approach proposed by Müller [29] complemented by its method of exploitation as suggested by Liu [24,25] (for review articles covering such approaches, cf. e.g. [16,18,30]). We will here invoke the entropy principle of Müller & Liu to reduce the complexity of the constitutive functions and to derive the response of constitutive quantities such as e.g. the Cauchy stress in thermodynamic equilibrium.

The exploitation of the entropy principle starts from a modified entropy inequality in which all balance equations are considered as constraints and are thus subtracted from the entropy inequality $(1)_5$. Dealing with *saturated* mixtures, for which the subsequent analysis is carried out, the saturation constraint $\sum_{\alpha=1}^{N} \nu_\alpha = 1$, cf. Def. 1, must likewise be accounted for (note that the saturation constraint can be re-written as $0 = \sum_{\alpha=1}^{N} \grave{\nu}_\alpha - \nabla \nu_\alpha \cdot \boldsymbol{u}_\alpha$). We will thus work with the following form of the entropy inequality:

$$
\begin{aligned}
\theta\pi = \theta\rho\dot{\eta} + \theta \operatorname{div}\boldsymbol{\phi} - \rho\sigma - \sum_{\alpha=1}^{N} \theta\lambda_\alpha^{\text{mass}} (\grave{\nu}_\alpha + \rho_\alpha \operatorname{div} \boldsymbol{v}_\alpha) \\
- \sum_{\alpha=1}^{N} \theta \boldsymbol{\lambda}_\alpha^{\text{mom}} (\rho_\alpha \grave{\boldsymbol{v}}_\alpha - \operatorname{div} \boldsymbol{T}_\alpha - \rho_\alpha \boldsymbol{b}_\alpha - \mathfrak{m}_\alpha) \\
- \sum_{\alpha=1}^{N} \theta \lambda_\alpha^{\text{equi}} (\rho_\alpha k_\alpha \grave{\grave{\nu}}_\alpha - \operatorname{div} \boldsymbol{h}_\alpha - \rho_\alpha (f_\alpha + l_\alpha)) \\
- \theta\lambda^{\text{sat}} \sum_{\alpha=1}^{N} (\grave{\nu}_\alpha - \nabla\nu_\alpha \cdot \boldsymbol{u}_\alpha) \\
- \Big(\rho\dot{\varepsilon} - \boldsymbol{T}\cdot\boldsymbol{D} + \operatorname{div}\boldsymbol{q} - \sum_{\alpha=1}^{N} \boldsymbol{h}_\alpha \cdot \nabla\grave{\nu}_\alpha + \sum_{\alpha=1}^{N} \rho_\alpha f_\alpha \grave{\nu}_\alpha - \rho r\Big) \geq 0. \qquad (7)
\end{aligned}
$$

Here, $\lambda_\alpha^{\text{mass}}$, $\lambda_\alpha^{\text{equi}}$ and λ^{sat} are scalar-valued Lagrange multipliers, $\boldsymbol{\lambda}_\alpha^{\text{mom}}$ is vector-valued and the Lagrangian multiplier associated with the balance of energy has been defined as $\lambda^{\text{ener}} := 1/\theta$, where θ is the reciprocal value of the absolute temperature and a positive number. As has been shown in [26] for single fluids, this identification is justifiable since $\dot{\theta}$ is not contained in the state space $\mathbb{S}_\alpha$, see (6), and the result carries over to mixtures as well.

Remark 8. Strictly speaking, the explicit form of the Lagrangian multiplier λ^{ener} should be *determined* in the course of the investigations, just as it is the case for $\boldsymbol{\lambda}_\alpha^{\text{mom}}$, $\lambda_\alpha^{\text{equi}}$ and $\lambda_\alpha^{\text{mass}}$, see (11). However, we have chosen to assign to λ^{ener} the value $1/\theta$ for several reasons: We decided to combine the results known for mixtures of fluids [26] and the ones derived for granular, single-constituent materials in [21,23]; with a modified ideal-wall postulate (inspired by [26] but adapted to granular mixtures) it is in principle possible to reduce the functional dependence of λ^{ener}. The more complicated this postulate is, the smaller will the constitutive dependence of λ^{ener} be, until it is finally reduced to $\lambda^{\text{ener}} = \hat{\lambda}^{\text{ener}}(\theta)$. However, since such a modified ideal-wall postulate is physically questionable (readers may find the modified postulate somewhat artifical for granular materials, because it is difficult to imagine ideal walls to exist in reality in granular media) we decided to assign to λ^{ener} the value $1/\theta$ right away, knowing that this is a result which should actually be *derived* rather than *postulated.* Moreover, anticipating the occurrence of extremely complex equations on the constituent level, we decided to simplify the theory by letting $\lambda^{\text{ener}} = 1/\theta$, which has several computational advantages.

As soon as more insight is gained into the mechanisms of segregation in granular mixtures (e.g. by numerical experiments for binary or ternary mixtures), one must think about letting λ^{ener} unconstrained and aim at *deriving* its proper expression. ◇

Following [25], we require that the entropy source σ is known in terms of all other sources, that is,

$$\rho\sigma = \sum_{\alpha=1}^{N} \rho_\alpha \boldsymbol{\lambda}_\alpha^{\text{mom}} \cdot \boldsymbol{b}_\alpha + \sum_{\alpha=1}^{N} \lambda_\alpha^{\text{equi}} \rho_\alpha l_\alpha + r\rho/\theta. \tag{8}$$

Remark 9. Inequality (7) and relation (8) disclose the difference exhibited by the two different exploitations of the entropy inequality according to COLEMAN & NOLL and MÜLLER & LIU. In the latter approach, supply terms may vanish, if so, (8) is trivially satisfied but (7) remains unchanged. In the former, it is assumed that supplies are arbitrary, implying that corresponding Lagrange multipliers must vanish. So, for non-vanishing l_α the configurational force balance will likely affect constitutive behaviour when the MÜLLER-LIU approach is used but not in COLEMAN-NOLL's approach. ◇

In general, the Lagrange multipliers may be independent variables or constitutive quantities. The following analysis will lead us to the conclusion that $\lambda_\alpha^{\text{mass}}$, $\boldsymbol{\lambda}_\alpha^{\text{mom}}$, $\lambda_\alpha^{\text{equi}}$ can be given by some constitutive relations, while no restriction on λ^{sat} is exerted; λ^{sat} hence represents an independent variable. In Subsect. 4.2 and Sect. 5, this issue is addressed in more detail. If the differentiations in inequality (7) are carried out with regard to the state spaces $\mathbb{S}_\alpha$ and $\mathbb{S}$, respectively, one arrives after lengthy but straightforward calculations at an expression for π which reads in principle as follows:

$$\theta\pi = \sum_{\alpha=1}^{N} \Big(\sum_i f_i^{(\alpha)}(\mathbb{S}) s_i^{(\alpha)} + \sum_j g_j^{(\alpha)}(\mathbb{S}) n_j^{(\alpha)} \Big) + f_r \geq 0. \tag{9}$$

Here, $f_i^{(\alpha)}$ and $g_j^{(\alpha)}$ are coefficient functions preceding variables $s_i^{(\alpha)}$ which are *contained* in $\mathbb{S}_\alpha$ and variables $n_j^{(\alpha)}$ *exceeding* the state space $\mathbb{S}_\alpha$, respectively. The indices i and j "count" the elements of $\mathbb{S}_\alpha$, while α counts – as before – the constituents of the granular mixture. The quantities $n_j^{(\alpha)}$ are often simply called the *higher derivatives*. The function f_r collects terms associated with variables that belong to none of the just described classes. Note that $s_i^{(\alpha)}$ and $n_j^{(\alpha)}$ may be scalar-, vector- or tensor-valued (and that therefore, the same holds true for the coefficient functions), but for simplicity of notation, this is not reflected in the symbols chosen in (9). In order to satisfy the entropy principle, the coefficient functions preceding the higher derivatives $n_j^{(\alpha)}$ in (9) are each required to vanish identically, $g_j^{(\alpha)} \equiv 0$, see [24].

Inequality (9) simplifies considerably due to the vanishing of the coefficient functions and becomes the so-called *residual inequality* which is in essence of the form

$$\theta\pi = \sum_{\alpha=1}^{N} \sum_i f_i^{(\alpha)}(\mathbb{S}) s_i^{(\alpha)} + \tilde{f}_r \geq 0, \tag{10}$$

where a modified function $\tilde{f}_r$ accounts for possible changes in f_r due to the vanishing coefficient functions $g_j^{(\alpha)}$. This residual inequality will in Sect.5 be evaluated in equilibrium, but before this is done it is sketched how additional information, valid also in non-equilibrium, can be deduced from the results obtained so far.

The vanishing of coefficient functions $g_j^{(\alpha)}$ gives rise to the so-called *Liu-equations* of which we – for the present modelling approach – obtain a total number of $15\,N$. To present the wealth of information deducible from these equations would exceed the scope of this article, the interested reader is instead referred to [21] where the Liu-equations are exploited rigorously. We will here restrict ourselves to highlighting selected results that are obtained in the course of exploiting the Liu-equations. These results will be addressed in the following Subsects. 4.2–4.4, in which a central role is played by the *constituent* and *mixture Helmholtz free energy*:

Definition 2. The *constituent Helmholtz free energy* ψ_α is defined as

$$\psi_\alpha := \varepsilon_\alpha - \theta\eta_\alpha\,, \qquad \alpha = 1, \ldots, N.$$

With this, the *mixture Helmholtz free energy* ψ is defined as

$$\psi := \sum_{\alpha=1}^{N} \xi_\alpha \psi_\alpha = \sum_{\alpha=1}^{N} \xi_\alpha \varepsilon_\alpha - \theta\eta = \varepsilon - \tfrac{1}{2}\sum_{\alpha=1}^{N} \xi_\alpha \boldsymbol{u}_\alpha \cdot \boldsymbol{u}_\alpha - \theta\eta,$$

where the last two equalities are obtained from employing the relations between mixture quantities and their respective counterparts as given in Remark 4. ⊠

4.2 Exploitation of the Liu-equations: Consequences for the Lagrange Multipliers

Exploiting the entropy principle according to Müller & Liu, see e.g. [24,29], necessitates the introduction of additional unknowns – the Lagrangian multipliers – which are associated with the balance equations and possibly other constraining conditions. For the present model, we have introduced a total number of $3\,N + 2$ such multipliers in (7). To one of them, λ^{ener}, the value $1/\theta$ has been assigned on the basis of [26], so that we are left with $3\,N + 1$ unknown Lagrangian multipliers.

The Liu-equations determine $3\,N$ of those remaining $3\,N + 1$ multipliers: for each $\alpha = 1, \ldots, N$, we have

$$\begin{aligned} \lambda_\alpha^{\text{mass}} &= -\frac{\psi_\alpha - \psi}{\theta} - \frac{\rho_\alpha^t}{\theta}\frac{\partial \psi_\alpha}{\partial \rho_\alpha^t}, \\ \lambda_\alpha^{\text{equi}} &= -\frac{1}{\theta \rho_\alpha k_\alpha}\sum_{\beta=1}^{N} \rho_\beta \frac{\partial \psi_\alpha}{\partial \dot{\nu}_\beta}, \\ \boldsymbol{\lambda}_\alpha^{\text{mom}} &= -\frac{\boldsymbol{u}_\alpha}{\theta}. \end{aligned} \tag{11}$$

We observe that

- the Lagrange multipliers $\lambda_\alpha^{\text{mass}}$ and $\lambda_\alpha^{\text{equi}}$ explicitly account for interaction between the constituents. Since the *mixture* Helmholtz free energy ψ contributes to each $\lambda_\alpha^{\text{mass}}$, a broad spectrum of possible interactions is captured. However, information on how the constituent interaction takes place is in this case left rather unspecified. This is different for $\lambda_\alpha^{\text{equi}}$, where interactions between the constituents can easily be detected and identified as changes in the constituent

Helmholtz free energy ψ_α caused by variations in temporal and spatial changes of the volume fractions of *all* constituents. Had the principle of phase separation not been partly abandoned, see (5), the sum over all β in $(11)_2$ would have given rise to only one non-zero term, namely for $\beta = \alpha$ to $\rho_\alpha \partial\psi_\alpha / \partial\dot{\nu}_\alpha$. Indeed,

- the Helmholtz energies ψ_α play an important role: once a constitutive relation for ψ_α is postulated (and thus, ψ is determined as well, see Def. 2), the $2N$ Lagrange multipliers related to the balances of mass and equilibrated force are explicitly determined. In this sense, the constituent Helmholtz free energies ψ_α serve as "potentials", a property which we will encounter (and discuss) again in Subsect. 4.4 and Sect. 5. Further,
- the Lagrange multipliers associated with the constituent balances of linear momentum do not vanish (as is usually the case for single constituent materials) – each of them is in fact proportional to the diffusion velocity of the respective constituent and has dimension [m/(K s)]. Finally,
- the Lagrangian multiplier related to the saturation constraint, λ^{sat}, can *not* be determined by the Liu-equations. In fact, λ^{sat}, is an independent quantity with dimension $[\text{kg}^2/(\text{m s}^2)]$ and thus called *saturation pressure*[1]. It is shown in Sect. 5 that the saturation pressure is one of three pressures naturally arising in this model.

4.3 Exploitation of the Liu-equations: Consequences for the Heat Flux–Entropy Flux Relation

The central result is the statement that the constituent entropy fluxes are *not* collinear to the constituent heat fluxes. The non-collinearity of these fluxes might be surprising for those usually working with single constituent materials and the COLEMAN–NOLL approach of exploiting the entropy principle: in the latter, heat flux and entropy flux are *by definition* collinear to each other, see [5]. For mixtures, a collinearity of heat flux and entropy flux on the constituent level is often postulated, however, the resulting overall mixture flux of entropy is then *not* collinear to the mixture heat flux. To avoid these a-priori assignments, MÜLLER [29] suggested to treat the entropy flux as an independent constitutive quantity. To investigate the heat flux–entropy flux relations on the constituent level under such circumstances, we now define the vector-valued quantity

$$\boldsymbol{l}_\alpha := -\boldsymbol{\phi}_\alpha + \frac{1}{\theta}\boldsymbol{q}_\alpha - \lambda_\alpha^{\text{equi}}\boldsymbol{h}_\alpha, \quad \alpha = 1, \dots, N.$$

Starting from this expression, we explore symmetry group properties to deduce additional constraining conditions. If in particular the conditions of isotropy for the vector valued quantity $\boldsymbol{l}_\alpha$ defined above are considered, one finds in combination with the Liu-equations that the constituent entropy flux $\boldsymbol{\phi}_\alpha$ is *not* collinear to the constituent heat flux $\boldsymbol{q}_\alpha$. To be precise, we find $\boldsymbol{l}_\alpha = \boldsymbol{0}$, or, equivalently,

$$\boldsymbol{\phi}_\alpha = \frac{1}{\theta}\boldsymbol{q}_\alpha - \lambda_\alpha^{\text{equi}}\boldsymbol{h}_\alpha, \quad \alpha = 1, \dots, N. \tag{12}$$

[1] In [17], the role of the saturation pressure in a soil mechanical context is highlighted.

It is worth emphasising that the constitutive complexity of the Lagrange multipliers $\lambda_\alpha^{\text{equi}} = \hat{\lambda}_\alpha^{\text{equi}}(\mathbb{S})$ is reduced considerably if some of the Liu-equations are combined with the conditions of isotropy applied to $\boldsymbol{l}_\alpha$: we find that

$$\hat{\lambda}_\alpha^{\text{equi}}(\mathbb{S}) = \tilde{\lambda}_\alpha^{\text{equi}}(\overset{t}{\rho}_\alpha, \nu_1, \ldots, \nu_N, \grave{\nu}_1, \ldots, \grave{\nu}_N, \theta), \quad \alpha = 1, \ldots, N \tag{13}$$

holds. A major "tool" to derive these far-reaching simplifications are the so-called *generalised Gibbs equations* which can be deduced from the Liu-equations but shall not be discussed here. We refer the interested reader instead to [21,22] and Sect. 6, where a few comments on the generalised Gibbs equations are made.

4.4 Exploitation of the Liu-equations: A Representation of the Constituent Configurational Fluxes h_α

We now turn to the constituent configurational fluxes $\boldsymbol{h}_\alpha$ arising in the additional balance law for the internal structural variable ν_α, $\alpha = 1, \ldots, N$, see (2). The explicit representations of these fluxes that will now be presented reveal clearly and amply the effects of a partial abandonment of the principle of phase separation. Recall that the former have already been discussed in Subsect. 4.2 when focusing on the Lagrange multiplier $\lambda_\alpha^{\text{equi}}$.

The constituent configurational fluxes are for each $\beta = 1, \ldots, N$ given by

$$\begin{aligned} \boldsymbol{h}_\beta = \frac{1}{\left(1 - \theta \dfrac{\partial \lambda_\beta^{\text{equi}}}{\partial \grave{\nu}_\beta}\right)} \Bigg[& \underbrace{2 \sum_{\alpha=1}^{N} \rho_\alpha \frac{\partial \psi_\alpha}{\partial (\nabla \nu_\beta \cdot \nabla \nu_\beta)} \nabla \nu_\beta}_{=:\mathcal{A}_{\beta\beta}^{(\alpha)}} + \underbrace{\theta \sum_{\alpha=1, \alpha \neq \beta}^{N} \frac{\partial \lambda_\alpha^{\text{equi}}}{\partial \grave{\nu}_\beta} \boldsymbol{h}_\alpha}_{\{1\}} \\ & + \underbrace{\sum_{\gamma=1, \gamma \neq \beta}^{N} \sum_{\alpha=1}^{N} \rho_\alpha \frac{\partial \psi_\alpha}{\partial (\nabla \nu_\gamma \cdot \nabla \nu_\beta)} \nabla \nu_\gamma}_{=:\mathcal{A}_{\gamma\beta}^{(\alpha)}} + \underbrace{\sum_{\alpha=1}^{N} \rho_\alpha \frac{\partial \psi_\alpha}{\partial \grave{\nu}_\beta} (\boldsymbol{u}_\alpha - \boldsymbol{u}_\beta)}_{\{2\}} \Bigg] . \end{aligned} \tag{14}$$

We will now comment upon the terms emerging from the abandonment of the principle of phase separation:

- The term abbreviated by $\mathcal{A}_{\beta\beta}^{(\alpha)}$ clearly results from an abandonment of the principle of phase separation in the gradients of the volume fractions: for fixed β, each constituent Helmholtz free energy ψ_α, $\alpha = 1, \ldots, N$ captures and contributes interaction effects triggered by spatial changes in the volume fraction ν_β.
- What is however suspected to be even more important for segregational processes is the interaction expressed in the term abbreviated by $\mathcal{A}_{\gamma\beta}^{(\alpha)}$. Here, variations in ψ_α, $\alpha = 1, \ldots, N$, which are due to changes in the geometrical location of constituent K_β with respect to all other constituents $K_\gamma, \gamma = 1, \ldots, N, \gamma \neq \beta$ are collected. We will encounter the expressions $\mathcal{A}_{\beta\beta}^{(\alpha)}$, $\mathcal{A}_{\gamma\beta}^{(\alpha)}$ again in Sect. 5.
- Terms $\{1\}$ and $\{2\}$ exist in this form as long as the constituent Helmholtz energy ψ_α (and with this, likewise the Lagrange multiplier $\lambda_\alpha^{\text{equi}}$) is allowed to depend on $\grave{\nu}_\beta$ for all $\beta = 1, \ldots, N$. If this is not the case[2], the terms in $\{1\}$

[2] Previous works as e.g. [42] apply the principle of phase separation and postulate that $\partial \psi_\alpha / \partial \grave{\nu}_\alpha := 0$ holds, hence, terms corresponding to $\{1\}$ and $\{2\}$ do not exist in that particular setting.

and {2} are by definition equal to zero. Note that term {2} is an expression of Darcy-type.

We observe that (14) constitutes N algebraic equations for N configurational fluxes. Further, since $\lambda_\alpha^{\rm equi}$ is determined once constitutive equations for the Helmholtz energies are prescribed, see Subsect. 4.2, the same applies now to the configurational fluxes. It is important to realize that although each $\boldsymbol{h}_\beta$ has been introduced as an independent constitutive quantity, its constitutive response is entirely governed by the Helmholtz free energies.

Let us briefly address the issue of existence of solutions to the system (14). If we – for simplicity of notation – introduce "generalised" vectors $\boldsymbol{h} := (\boldsymbol{h}_1, \dots, \boldsymbol{h}_N)$ and $\boldsymbol{r} := (\boldsymbol{r}_1, \dots, \boldsymbol{r}_N)$, whose "components" are in the former case the constituent configurational fluxes and are in the latter case not specified, and introduce also an $N \times N-$matrix $\boldsymbol{A}$ containing the coefficients of the configurational fluxes, (14) can be re-written as $\boldsymbol{A}\boldsymbol{h} = \boldsymbol{r}$, where $\boldsymbol{h}$ is the vector of unknowns. Clearly, if $\det \boldsymbol{A} \neq 0$, (14) has a unique solution, that is, the configurational fluxes $\boldsymbol{h}_1, \dots, \boldsymbol{h}_N$ are uniquely determined. However, if $\det \boldsymbol{A} \neq 0$ fails to hold, (14) may either have infinitely many or no solutions at all. Since the entries of $\boldsymbol{A}$ are given in essence in terms of (partial derivatives of) the Helmholtz energies, the existence of solutions to (14) will crucially depend on the constitutive forms postulated for ψ_α, $\alpha = 1, \dots, N$.

Remark 10. To illustrate the above comments, we specify (14) to a binary mixture, that is, we let $N = 2$. Then, two alternative representations of (14) read

$$\begin{cases} \boldsymbol{h}_1 = a\boldsymbol{h}_2 + \boldsymbol{r}_1, \\ \boldsymbol{h}_2 = b\boldsymbol{h}_1 + \boldsymbol{r}_2, \end{cases} \qquad \text{and} \qquad \underbrace{\begin{pmatrix} 1 & -a \\ -b & 1 \end{pmatrix}}_{=\boldsymbol{A}} \underbrace{\begin{pmatrix} \boldsymbol{h}_1 \\ \boldsymbol{h}_2 \end{pmatrix}}_{=\boldsymbol{h}} = \underbrace{\begin{pmatrix} \boldsymbol{r}_1 \\ \boldsymbol{r}_2 \end{pmatrix}}_{=\boldsymbol{r}}, \tag{15}$$

respectively, where (with $\tilde{\boldsymbol{u}} := \boldsymbol{u}_2 - \boldsymbol{u}_1$)

$$a = \frac{\theta}{1 - \theta(\partial\lambda_1^{\rm equi}/\partial\grave{\nu}_1)} \cdot \frac{\partial\lambda_2^{\rm equi}}{\partial\grave{\nu}_1}, \qquad b = \frac{\theta}{1 - \theta(\partial\lambda_2^{\rm equi}/\partial\grave{\nu}_2)} \cdot \frac{\partial\lambda_1^{\rm equi}}{\partial\grave{\nu}_2},$$

$$\boldsymbol{r}_1 = \frac{1}{1 - \theta\frac{\partial\lambda_1^{\rm equi}}{\partial\grave{\nu}_1}} \left[2(\mathcal{A}_{11}^{(1)} + \mathcal{A}_{11}^{(2)})\nabla\nu_1 + (\mathcal{A}_{21}^{(1)} + \mathcal{A}_{21}^{(2)})\nabla\nu_2 + \rho_2\frac{\partial\psi_2}{\partial\grave{\nu}_1}\tilde{\boldsymbol{u}}\right],$$

$$\boldsymbol{r}_2 = \frac{1}{1 - \theta\frac{\partial\lambda_2^{\rm equi}}{\partial\grave{\nu}_2}} \left[2(\mathcal{A}_{22}^{(1)} + \mathcal{A}_{22}^{(2)})\nabla\nu_2 + (\mathcal{A}_{12}^{(1)} + \mathcal{A}_{12}^{(2)})\nabla\nu_1 + \rho_1\frac{\partial\psi_1}{\partial\grave{\nu}_2}\tilde{\boldsymbol{u}}\right].$$

Provided that $\det \boldsymbol{A} \neq 0$, that is, that $1 - ab \neq 0$, see $(15)_2$, the system $\boldsymbol{A}\boldsymbol{h} = \boldsymbol{r}$ has a unique solution, namely

$$\boldsymbol{h}_1 = \frac{a\boldsymbol{r}_2 + \boldsymbol{r}_1}{1 - ab}, \qquad \boldsymbol{h}_2 = \frac{ab\boldsymbol{r}_2 - b\boldsymbol{r}_1}{1 - ab} + \boldsymbol{r}_2. \tag{16}$$

For the binary system under consideration, $\det \boldsymbol{A} \neq 0$ reads as follows:

$$\frac{\partial\lambda_2^{\rm equi}}{\partial\grave{\nu}_1} \cdot \frac{\partial\lambda_1^{\rm equi}}{\partial\grave{\nu}_2} \cdot \frac{\theta^2}{1 - \theta\dfrac{\partial\lambda_1^{\rm equi}}{\partial\grave{\nu}_1}} \cdot \frac{1}{1 - \theta\dfrac{\partial\lambda_2^{\rm equi}}{\partial\grave{\nu}_2}} \neq 1. \tag{17}$$

As mentioned above, the existence of solutions to (14) depends crucially on the forms postulated for ψ_α, $\alpha = 1, \ldots, N$, and hence, see $(11)_2$, on those for $\lambda_\alpha^{\text{equi}}$. In fact, it is conjectured that (17) constrains in particular the choice of coefficients entering constitutive postulates for ψ_α and $\lambda_\alpha^{\text{equi}}$, respectively. Our conjecture is nourished from the following considerations: first, let us assume isothermal conditions, so that $\theta(\boldsymbol{x}, t)$ takes on a constant value, say $\theta(\boldsymbol{x}, t) := \tilde{c}$. Second, let us within a Gedanken-experiment postulate the following forms for $\lambda_{1,2}^{\text{equi}}$:

$$\lambda_1^{\text{equi}} := c_1^{(1)} \dot{\nu}_1 \theta + c_2^{(1)} \dot{\nu}_2 \theta + c_3^{(1)} \dot{\nu}_1^2 + c_4^{(1)} \dot{\nu}_2^2 + c_5^{(1)} \dot{\nu}_1 \dot{\nu}_2, \tag{18a}$$

$$\lambda_2^{\text{equi}} := c_1^{(2)} \dot{\nu}_1 \theta + c_2^{(2)} \dot{\nu}_2 \theta + c_3^{(2)} \dot{\nu}_1^2 + c_4^{(2)} \dot{\nu}_2^2 + c_5^{(2)} \dot{\nu}_1 \dot{\nu}_2. \tag{18b}$$

These forms are motivated[3] by their "counterpart" arising in a single-constituent theory, where the simplest representation of λ^{equi} is given by (cf. [22])

$$\lambda^{\text{equi}} = c_1 \dot{\nu} \theta + c_2 \dot{\nu}^2.$$

Whereas insertion of $\theta = \tilde{c}$ into (17) and substitution of a linearised form of (18a), (18b) into (17) yields

$$\frac{c_1^{(2)} c_2^{(1)} \tilde{c}^4}{(1 - \tilde{c}^2 c_1^{(1)})(1 - \tilde{c}^2 c_2^{(2)})} \neq 1,$$

and leads hence to a purely algebraic problem constraining the four unknowns $c_1^{(1)}$, $c_2^{(1)}$, $c_1^{(2)}$, $c_2^{(2)}$, insertion of the "full" forms given in (18a),(18b) leads to

$$\frac{\left[c_1^{(2)} \tilde{c} + 2 c_3^{(2)} \dot{\nu}_1 + c_5^{(2)} \dot{\nu}_2\right]\left[c_2^{(1)} \tilde{c} + 2 c_4^{(1)} \dot{\nu}_2 + c_5^{(1)} \dot{\nu}_1\right] \tilde{c}^2}{\left[1 - \tilde{c}(c_1^{(1)} \tilde{c} + 2 c_3^{(1)} \dot{\nu}_1 + c_5^{(1)} \dot{\nu}_2)\right]\left[1 - \tilde{c}(c_2^{(2)} \tilde{c} + 2 c_4^{(2)} \dot{\nu}_2 + c_5^{(2)} \dot{\nu}_1\right]} \neq 1. \tag{19}$$

Obviously, (19) couples the algebraic problem of determining the 10 unknown parameters $c_i^{(j)}$, $i = 1, \ldots, 5, j = 1, 2$ to the solution of the field equations given e.g. in (3), since derivatives of the primary unknowns ν_1, ν_2 enter (19). It is seen that even for simple problems (isothermal binary mixtures, simple constitutive postulates), we are confronted with the necessity of solving complex systems of (partial and/or ordinary as well as possibly algebraic) equations. Obviously, solutions will be sought for by employing numerical techniques, but so far, their application to the theory presented here has not yet been tackled. It is, however, the next step to be undertaken. ◇

5 Selected Results in Thermodynamic Equilibrium

We will now evaluate inequality (10) in thermodynamic equilibrium.

[3] Note that we have not proved that these representations are in fact admissible, and, in particular, do not contradict the Poincaré conditions, see [22]. These simple forms have been chosen just for the sake of illustrating purposes, and the thermodynamic analysis performed in the sequel in the state of thermodynamic equilibrium, see Sect. 5, is in fact based on the assumption that the Lagrange multipliers $\lambda_\alpha^{\text{equi}}$ depend quadratically (and not linearly) on the time variations of the volume fractions.

Definition 3. A granular mixture is in *thermodynamic equilibrium*, if the thermodynamic processes taking place on the constituent level are time-independent and characterised by a homogeneous temperature field and vanishing[4] velocity field: $\nabla\theta = \mathbf{0}$, $\boldsymbol{v}_\alpha = \mathbf{0}$. As a consequence, the constituent stretching tensor as well as the mixture velocity field vanish: $\boldsymbol{D}_\alpha = \mathbf{0}, \boldsymbol{v} = \mathbf{0}$. Furthermore, $\grave{\nu}_\alpha = \partial\nu_\alpha/\partial t + \nabla\nu_\alpha \cdot \boldsymbol{v}_\alpha = 0$ holds. The evaluation of functionals $\mathcal{C}$ in thermodynamic equilibrium is denoted by a lower index $|_\mathrm{E}$ as in $\mathcal{C}|_\mathrm{E}$. ⊠

Let us now focus on the residual inequality (10), which is for convenience repeated here:

$$\theta\pi = \sum_{\alpha=1}^{N}\sum_i f_i^{(\alpha)}(\mathbb{S})s_i^{(\alpha)} + \tilde{f}_r \geq 0.$$

In the sum on the right hand side, $s_i^{(\alpha)}$ assumes all "values"[5] contained in the set $\mathbb{E} := \{\grave{\nu}_1, \ldots, \grave{\nu}_N, \nabla\theta, \boldsymbol{v}_1, \ldots, \boldsymbol{v}_N, \boldsymbol{D}_1, \ldots, \boldsymbol{D}_N\}$; these quantities vanish in thermodynamic equilibrium. Moreover, $s_i^{(\alpha)}$ assumes the values $\nabla\rho_1^t, \ldots, \nabla\rho_N^t$, $\nabla\nu_1, \ldots, \nabla\nu_N$, whereas $\tilde{f}_r$ is a sum of N terms, in which the α-th summand is proportional to $\lambda_\alpha^{\mathrm{equi}}$ for $\alpha = 1, \ldots, N$.

Postulate 1 To guarantee that the entropy production assumes its minimum in thermodynamic equilibrium, $\pi|_\mathrm{E} = 0$, it is postulated that $\lambda_\alpha^{\mathrm{equi}}|_\mathrm{E} := 0$ holds for each $\alpha = 1, \ldots, N$. ♠

By construction, Postulate 1 implies the identical vanishing of $\tilde{f}_r$ in thermodynamic equilibrium. Moreover, it can be shown that due to Postulate 1, the coefficient functions preceding the variables $\nabla\rho_1^t, \ldots, \nabla\rho_N^t$ and $\nabla\nu_1, \ldots, \nabla\nu_N$ vanish also identically, so that indeed $\pi|_\mathrm{E} = 0$ holds. Viewing π as a function of $\mathbb{E}$ with parameter set $\cup_{\alpha=1}^N \mathbb{S}_\alpha \setminus \mathbb{E}$ implies that the following conditions must necessarily be satisfied:

$$\left.\frac{\partial\pi}{\partial\grave{\nu}_\gamma}\right|_\mathrm{E} = 0, \quad \left.\frac{\partial\pi}{\partial\nabla\theta}\right|_\mathrm{E} = \mathbf{0}, \quad \left.\frac{\partial\pi}{\partial\boldsymbol{v}_\gamma}\right|_\mathrm{E} = \mathbf{0}, \quad \left.\frac{\partial\pi}{\partial\boldsymbol{D}_\gamma}\right|_\mathrm{E} = \mathbf{0}, \;\; \gamma = 1, \ldots, N. \tag{20}$$

The evaluation of these conditions is simplified considerably if we assume that the Lagrange multipliers $\lambda_\alpha^{\mathrm{equi}}$, which have in [21] been shown to be of the form $\lambda_\alpha^{\mathrm{equi}} = \hat{\lambda}_\alpha^{\mathrm{equi}}(\rho_\alpha^t, \nu_1, \ldots, \nu_N, \grave{\nu}_1, \ldots, \grave{\nu}_N, \theta)$, see also (13), depend *quadratically* on the time variations of the volume fractions:

$$\lambda_\alpha^{\mathrm{equi}} = \tilde{\lambda}_\alpha^{\mathrm{equi}}(\rho_\alpha^t, \nu_1, \ldots, \nu_N, (\grave{\nu}_1)^2, \ldots, (\grave{\nu}_N)^2, \theta).$$

Further, the following quantities will be needed:

[4] It should be mentioned that it is actually sufficient to postulate $\boldsymbol{v}_\alpha = \boldsymbol{v} = const.$, which implies in particular vanishing diffusion velocities: $\boldsymbol{u}_\alpha = \mathbf{0}$ holds for all $\alpha = 1, \ldots, N$ and is an objective statement. However, to establish a formal "conformity" with the definition of thermodynamic equilibrium as used in single-constituent continuum mechanics (where $\boldsymbol{v}|_\mathrm{E} = \mathbf{0}$ is postulated), we require $\boldsymbol{v}_\alpha|_\mathrm{E} = \mathbf{0}$ which implies the vanishing of the barycentric velocity, see Def. 1.

[5] This terminology is a little bit sloppy, since the independent variables are in fact field quantities. It is hoped that nonetheless, no confusion arises.

Definition 4. The *constituent configuration pressure* $\wr_\beta$ and the *constituent thermodynamic pressure* $\wp_\beta$ are defined as

$$\wr_\beta := \sum_{\alpha=1}^{N} \rho_\alpha \frac{\partial \psi_\alpha}{\partial \nu_\beta}, \qquad \wp_\beta := (\rho_\beta^t)^2 \frac{\partial \psi_\beta}{\partial \rho_\beta^t}. \qquad \boxtimes$$

Let us now present and discuss what is obtained from $(20)_{1,4}$, namely[6]

$$f_\gamma|_{\mathrm{E}} = \frac{1}{\rho_\gamma}\Big(-\sum_{\beta,\alpha=1}^{N} \frac{\partial}{\partial \nu_\beta}\Big(\frac{\partial \boldsymbol{q}_\alpha}{\partial \dot{\nu}_\beta}\Big|_{\mathrm{E}} - \theta \frac{\partial \phi_\alpha}{\partial \dot{\nu}_\beta}\Big|_{\mathrm{E}}\Big)\cdot \nabla \nu_\beta - (\wr_\gamma - \wp_\gamma + \lambda^{\mathrm{sat}})|_{\mathrm{E}}\Big), \tag{21}$$

$$\begin{aligned} \boldsymbol{T}_\gamma|_{\mathrm{E}} = &- \underbrace{\nu_\gamma \wp_\gamma|_{\mathrm{E}} \boldsymbol{I}}_{\{1\}} - \underbrace{2\mathcal{A}_{\gamma\gamma}^{(\alpha)}|_{\mathrm{E}} \nabla \nu_\gamma \otimes \nabla \nu_\gamma}_{\{2\}} \\ &- \underbrace{\sum_{\beta \neq \gamma, \beta=1}^{N} \mathcal{A}_{\beta\gamma}^{(\alpha)} \mathrm{sym}(\nabla \nu_\beta \otimes \nabla \nu_\gamma)|_{\mathrm{E}}}_{\{3\}}. \end{aligned} \tag{22}$$

First, take a closer look at the production of equilibrated force of constituent K_γ: $f_\gamma|_{\mathrm{E}}$ is balanced by a sum of terms of dimension $[\mathrm{kg}^2/(\mathrm{ms}^2)]$; in particular, $\lambda^{\mathrm{sat}}|_{\mathrm{E}}$ is hence interpreted as a pressure. Since it is the Lagrange multiplier associated with the saturation constraint, λ^{sat} is termed *saturation pressure*, cf. Subsect.4.2. Note that λ^{sat} has in the previous sections not been determined by other constitutive quantities (such as e.g. the Helmholtz free energy) – λ^{sat} is in fact an independent quantity.

Equation (21) shows clearly that the theory gives rise to *three* pressure terms, namely the thermodynamical pressure $\wp_\gamma$, the configurational pressure $\wr_\gamma$ and the saturation pressure λ^{sat} discussed above. The first captures a particular constituent compressibility and is hence unaffected by the remaining constituents, see Def. 4. The configurational pressure, which collects the (density weighted) response of *all* constituent Helmholtz free energies to changes in a given volume fraction, is influencing the evolution of the volume fraction of a particular constituent K_γ since f_γ is the production term in (2). In thermodynamic equilibrium, the configurational pressure asserts that the production of equilibrated force of K_γ, $f_\gamma|_{\mathrm{E}}$, is influenced by all mixture constituents (even in cases when entropy flux and heat flux are collinear to each other).

In (22), we have used the abbreviations $\mathcal{A}_{\gamma\gamma}^{(\alpha)}$ and $\mathcal{A}_{\beta\gamma}^{(\alpha)}$ introduced in (14). We observe that in equilibrium, the constituent Cauchy stress $\boldsymbol{T}_\gamma$ contains an *isotropic* part related exclusively to quantities belonging to phase γ: this part is given by the under-braced term labelled $\{1\}$. It is complemented by an *anisotropic*[7] contribution caused by dyadic products of combinations of the gradient of all volume fractions, cf. the under-braced terms labelled by $\{2\}$ and $\{3\}$, respectively. Terms $\{2\}$ and $\{3\}$

[6] The implications of $(20)_{2,3}$ are for brevity not given here, the reader is referred to [21] where details can be found.

[7] The term "anisotropic" is used here in the sense that $\{2\}$ and $\{3\}$ are contributions to $\boldsymbol{T}|_{\mathrm{E}}$ which are not proportional to the identity tensor but have entries on the off-diagonals, which implies the existence of shear stresses in equilibrium.

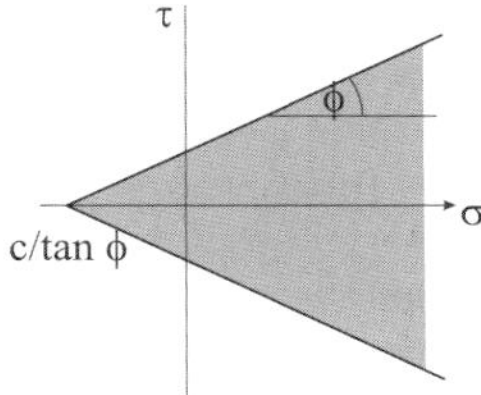

Fig. 3. The Mohr–Coulomb yield criterion. Yielding occurs when shear and normal stresses are related to each other by (3), i.e. when the point (τ, σ) lies on one of the lines bounding the cone. Non-vanishing shear stresses can be supported in equilibrium if the point (τ, σ) lies within the shaded area

take all constituents of the mixture into account. In particular, {2} is for a fixed constituent γ given by the sum of (density weighted) changes of all constituent Helmholtz free energies due to variations in the (norm of) the gradient of the volume fraction ν_γ, since $\mathcal{A}^{(\alpha)}_{\gamma\gamma} = \sum_\alpha \rho_\alpha \partial\psi_\alpha / \partial(\nabla\nu_\gamma \cdot \nabla\nu_\gamma)$. Recalling that a scalar product of two vectors provides information about the angle between these vectors, {3} incorporates the (density weighted) sum of the variations of all Helmholtz free energies induced by changes in the "geometrical" location of constituent K_γ with respect to all other constituents, since $\mathcal{A}^{(\alpha)}_{\beta\gamma} = \sum_\alpha \rho_\alpha \partial\psi_\alpha / \partial(\nabla\nu_\beta \cdot \nabla\nu_\gamma)$. Interestingly, $\boldsymbol{T}_\gamma|_{\mathrm{E}}$ does not have a contribution from the saturation pressure.

The fact that non-vanishing shear stresses can exist in equilibrium is a typical feature of granular materials and is in its simplest form described by the *Mohr–Coulomb yield criterion.* For a cohesive material, it reads

$$\|\tau\| = c + (\tan\phi)\,\sigma. \tag{23}$$

Here, τ and σ are the shear and the normal stress, c is the cohesion and ϕ the angle of internal friction, see Fig. 3. The Mohr–Coulomb criterion expresses that in equilibrium, non-vanishing shear stresses are possible as long as the point (τ, σ) lies in the shaded area depicted in Fig. 3, and that yielding will occur when shear and normal stresses are related to each other by (23), i.e. when the point (τ, σ) lies on one of the lines bounding the cone. Equation (22) tells us that {2} and {3} are a sum of Mohr–Coulomb terms. Since the anisotropic contributions {2}, {3} to $\boldsymbol{T}_\gamma|_{\mathrm{E}}$, see (22), are related to the equilibrated stress vector $\boldsymbol{h}_\gamma$ via the coefficient functions $\mathcal{A}^{(\alpha)}_{\gamma\gamma}$ and $\mathcal{A}^{(\alpha)}_{\beta\gamma}$, the above discussion can be summarised by the statement that the existence of the equilibrated stress vectors arising in the balance equations for the structural variables gives rise to Mohr-Coulomb friction in equilibrium (this has already been observed in [35]). Equation (22) highlights, moreover, the important role played by the Helmholtz free energies: the representation of $\boldsymbol{T}_\gamma$ shows that the constituent Cauchy stresses (in equilibrium) are already known if constitutive relations for all constituent Helmholtz free energies are provided. The same applies to the equilibrated stress $\boldsymbol{h}_\gamma$, which is even in non-equilibrium determined as soon as constitutive relations for ψ_α are postulated, see (14). Since $\boldsymbol{\phi}_\gamma$ and $\boldsymbol{q}_\gamma$ are related to $\boldsymbol{h}_\gamma$ via (12), they are also known if the Helmholtz free energies are given, implying further that f_γ is also determined by the latter. Instead of having to postulate constitutive relations for all quantities collected in $\mathbb{C}_\alpha$, see Sect. 3, one only has to formulate adequate postulates for the Helmholtz energies, and the non-equilibrium parts of the Cauchy stresses and equilibrated force production terms, respectively.

With these remarks, we draw the discussion of selected features of our model capturing segregational effects in granular mixtures to a close. Section 6 addresses a few specific issues, which could for reasons of brevity not be dealt with in this article.

6 Additional Remarks and Conclusion

Three aspects which have so far not or only very shortly been addressed shall now be briefly commented upon.

First, we turn to the *generalised Gibbs equations* which can be deduced from the Liu-equations, see Subsect. 4.1. For a single-constituent material, the generalised Gibbs equations relate – among others – the differentials of entropy and (internal) energy in the following form: $\mathrm{d}\eta = \lambda^{\mathrm{ener}}\mathrm{d}\varepsilon + \mathrm{d}\mathcal{P}$. Here, λ^{ener} is the Lagrange multiplier associated with the balance of internal energy and $\mathcal{P}$ is a sum of differentials of variables belonging to the state space $\mathbb{S}$. It is a-prior unknown whether $\mathrm{d}\eta$ is an *exact* differential – a property which (if fulfilled) guarantees that the entropy η is in fact a well defined scalar potential. In general, the conditions of Poincaré and Frobenius, see e.g. [4], provide the answer to the question whether $\mathrm{d}\eta$ is exact or not. Moreover, the Poincaré conditions are "constructive" in the sense that should $\mathrm{d}\eta$ not be exact, explicit conditions *making* $\mathrm{d}\eta$ exact can be deduced. As a rule of thumb, additional constraints on the Lagrange multipliers as well as the constitutive functions ("hidden" in $\mathcal{P}$) are then obtained. A detailed investigation of the Poincaré conditions for a single constituent material with an additional balance law accounting for an internal, structural variable can be found in [22]. For mixtures, the situation is more complicated, and a detailed investigation of the Poincaré conditions has for the present model not yet been carried out. It is, however, needed to derive additional constraining conditions on e.g. the Lagrange multipliers $\lambda_\alpha^{\mathrm{equi}}$, $\alpha = 1, \ldots, N$, as has already been mentioned in the footnote in the context of (18a), (18b).

Second, it is worthwhile to check whether the partial abandonment of the principle of phase separation is a "proper" extension of the earlier works of Passman et al. and Wang & Hutter, see [31,42]. Dealing with a mixture theory for fluid suspensions, Passman et al. strictly employed the principle of phase separation and exploited the entropy principle using the Coleman–Noll approach. Their results have been re-investigated by Wang & Hutter, who used the Müller–Liu approach instead. An essential assumption made in [31,42] is that ψ_α is *independent* of $\grave{\nu}_\alpha$, which in turn implies that $\lambda_\alpha^{\mathrm{equi}}$ vanishes identically. Moreover, certain contributions to the equilibrium Cauchy stress are then not-existent, as has been pointed out already in the footnote in the context of the discussion of (14). Speaking of a "proper" extension, we must clarify what is meant by this terminology. "Proper" is to be understood in the following sense: if we – for the present model in which the principle of phase separation has been *partly* abandoned – impose the same restrictions and constraints as imposed in [31,42], our results should "collapse" to those obtained in these articles. Let us check this for two selected examples, namely the expressions for the constituent configurational fluxes and the constituent Cauchy stresses as given in (14) and (22). Imposing the principle of phase separation and using $\partial\psi_\alpha/\partial\grave{\nu}_\alpha := 0$, so that $\lambda_\alpha^{\mathrm{equi}} = 0$ follows as well as the collinearity of heat flux and entropy flux, see (12), (14) and (22) reduce to

$$\boldsymbol{h}_\beta = 2\rho_\beta \frac{\partial\psi_\beta}{\partial(\nabla\nu_\beta\cdot\nabla\nu_\beta)}, \tag{24a}$$

$$\begin{aligned}\boldsymbol{T}_\beta|_{\mathrm{E}} &= -\nu_\beta\wp_\beta|_{\mathrm{E}}\boldsymbol{I} - 2\mathcal{A}_{\beta\beta}^{(\alpha)}|_{\mathrm{E}}\nabla\nu_\beta\otimes\nabla\nu_\beta \\ &= -\nu_\beta\wp_\beta|_{\mathrm{E}}\boldsymbol{I} - 2\rho_\beta\frac{\partial\psi_\beta}{\partial(\nabla\nu_\beta\cdot\nabla\nu_\beta)}|_{\mathrm{E}}\nabla\nu_\beta\otimes\nabla\nu_\beta .\end{aligned} \tag{24b}$$

These relations are identical with those obtained in [31,42], so that the extension proposed here is consistent.

Lastly, the issue of postulating constitutive relations for the non-equilibrium parts of the Cauchy stresses and the equilibrated force productions is briefly addressed. Recall that the latter two expressions are – besides the Helmholtz free energies – the only quantities for which constitutive relations have to be postulated, see Sect. 5. For the case of a single-constituent granular material with an additional internal variable, specific constitutive relations for a rough granular material, *roughness* being here the internal variable, have been proposed in [38,23]. In these works, applications of this particular model to simple shear flows as well as first numerical investigations are presented. They may serve as a guideline when an increasing number of constituents (as encountered for mixtures) is considered, explicit non-equilibrium contributions to $\boldsymbol{T}_\alpha$ and f_α, $\alpha = 1, \ldots, N$ have however in the present setting not yet been formulated.

Let us summarise what has been obtained: the results presented in this article have shown that a partial abandonment of the principle of phase separation provides a means to account for the redistribution of grains in a granular mixture. The system of forces emerging in such a case is, however, very complex and should in a first approach be investigated for simple (i.e. binary or ternary) mixtures. We point out that only due to the abandonment of the principle of phase separation in the gradients of the volume fractions, thermodynamic driving forces separating the constituents can be captured. These driving forces are the Helmholtz energies, whose dependence on *all* gradients of volume fractions enters the expression for the configurational fluxes, see (14). In models where phase separation is strictly employed, the configurational fluxes act only within a particular constituent, see (24a), (24b), and can thus not capture forces driving the constituents apart.

References

1. Baxter, G. W., Behringer, R. P. (1991) Cellular automata models for the flow of granular materials. Phys. Rev. **D**, **51**, 465–471
2. Blenk, S., Ehrentraut, H., Muschik, W. (1991) Statistical foundation of macroscopic balances for liquid crystals in alignment tensor formulation. Physica A **174**, 119–138
3. Bowen, R. M. (1984) Diffusion models implied by the theory of mixtures. In: Rational thermodynamics (Ed. Truesdell, C.) Springer, Berlin–New York
4. Bowen, R. M., Wang, C.-C. (1976) Introduction to vectors and tensors. 1st edn. Plenum Press, New York–London

5. Coleman, B. D., Noll, W. (1963) The thermodynamics of elastic materials with heat conduction and viscosity. Arch. Rat. Mech. Anal. **13**, 167–178
6. Drahun, J. A., Bridgewater, J. (1983) The mechanism of free surface size segregation. Powder Technology **36**, 39–53
7. Duhem, P. (1893) Le potentiel thermodynamique at la pression hydrostatique. Ann. Sci. École Norm. Sup. **10** (Series 3), 187–230
8. Faria, S. H. (2001) Mixtures with continuous diversity: general theory and application to polymer solutions. Continuum Mech. Thermodyn. **13** (2), 91–120
9. Faria, S. H., Hutter, K. (2002) A systematic approach to the thermodynamics of single and mixed flowing media with microstructure. Part I: balance equations and jump conditions. Continuum Mech. Thermodyn., published online, see http://dx.doi.org/10.1007/s001610200084
10. Fried, E., Roy, B. C. (2003) Gravity-induced segregation of cohesionless granular mixtures. This volume
11. Fried, E., Gurtin, M. E. , Hutter, K. (Submitted for publication to J. Non-Newtonian Fluid Mech.) A void-based description of compaction and segregation in flowing granular materials
12. Goodman M. A., Cowin S. C. (1972) A continuum theory for granular materials. Arch. Ration. Mech. Anal. **44**, 249–266
13. Gray, W. G. (1999) Elements of a systematic procedure for the derivation of macroscale conservation equations for multiphase flow in porous media. In: Hutter, K., Wilmanski, K. (Eds.) Kinetic and continuum theories of granular and porous media. CISM courses and lectures, No 400, Springer Wien–New York
14. Gray, W. G., Hassanizadeh, S. M. (1996) Macroscale continuum mechanics for multiphase porous-media flow including phases, interfaces, common lines and common points. Advances in Water Resources **21**, 261–281
15. Herrmann, H. J., Luding, S. (1998) Modeling granular media on the computer. Contin. Mech. Thermodyn. **10**, 189–231
16. Hutter, K. (1977) The foundations of thermodynamics, its basic postulates and implications. A review of modern thermodynamics. Acta Mechanica **27**, 1–54
17. Hutter K., Laloui, L., Vulliet, L. (1998) Thermodynamically based mixture models of saturated and unsaturated soils. Mech. Cohes.-Frict. Mater. **4**, 295–338
18. Hutter, K. Wang Y. (to appear) Phenomenological thermodynamics and entropy principles. In: Greven, A., Keller, G., Warnecke, G. (Eds.) Entropy
19. Jenkins, J. T., Richmann, M. W. (1985) Kinetic theory for plane flows of a dense gas of identical, rough, inelastic, circular disks. Phys. Fluids **28**, 3485–3494
20. Kachanov, L. M. (1986) Introduction to Continuum Damage Mechanics. Martinus Nijhoff Publishers, Dordrecht
21. Kirchner, N. P. (2001) Thermodynamics of Structured Granular Media. Ph. D. Thesis, Institute of Mechanics, Darmstadt University of Technology
22. Kirchner, N., Hutter, K. (2002) Elasto-plastic behaviour of a granular material with an additional scalar degree of freedom. In: Ehlers, W., Bluhm, J. (Eds.) Porous Media. Theory, Experiments and Numerical Applications. Springer Berlin–New York
23. Kirchner, N. P., Teufel, A. (2002) Thermodynamically consistent modelling of abrasive granular materials. I. Non-equilibrium theory. Proc. R. Soc. Lond. A **458**, 1–24
24. Liu, I.-S. (1972) Method of lagrange multipliers for exploitation of the entropy principle. Arch. Rat. Mech. Anal. **46**, 131–148

25. Liu, I.-S. (1973) On the entropy supply in a classical and a relativistic fluid. Arch. Rat. Mech. Anal. **50**, 111–117
26. Liu, I.-S., Müller, I. (1984) Thermodynamics of mixtures of fluids. In: Truesdell, C. (Ed.) Rational thermodynamics. Springer, Berlin–New York
27. Meixner, J. (1943) Zur Thermodynamik irreversible Prozesse. [On the thermodynamics of irreversible processes.] Z. physik. Chemie **538**, 235–263
28. Morland, L. W., Staroszczyk, R. (1998) Viscous response of polar ice with evolving fabric. Continuum Mech. Thermodyn. **10**, (3), 135–152
29. Müller, I. (1971) Die Kältefunktion, eine universelle Funktion in der Thermodynamik viskoser wärmeleitender Flüssigkeiten. [The coldness function, a universal function in the thermodynamics of viscous, heat-conducting fluids.] Arch. Rat. Mech. Anal. **40**, 1–36
30. Muschik, W., Papenfuss, C., Ehrentraut, H. (2001) A sketch of continuum thermodynamics. J. Non-Newtonian Fluid Mech. **96**, 255–290
31. Passman S. L., Nunziato J. W., Walsh E. K. (1984) A theory of multiphase mixtures. In: Truesdell, C. (Ed.) Rational Thermodynamics. Springer Berlin–New York
32. Prigozhin, L. (1993) A variational problem of bulk solid mechanics and free surface segregation. Chemical Engineering Sciences **48**, No. 21, 3647–3656
33. Pudasaini, S.P., Mohring, J. (2002) Mathematical model and experimental validation of free surface size segregation for polydisperse granular materials. Granular Matter **4**, 45–56
34. Ristow, G. H. (1998) Flow properties of granular materials in three-dimensional geometries. Habilitation, Philipps-Universität Marburg, pp. 1–111
35. Savage, S. B. (1979) Gravity flow of cohesionless granular materials in chutes and channels. J. Fluid Mech. **92**, 53–96
36. Savage, S. B., Lun, C. K. K. (1988) Particle size segregation in inclined chute flow of dry cohesionless granular solids. J. Fluid Mech. **189**, 311–335
37. Svendsen, B., Hutter, K. (1995) On the thermodynamics of a mixture of isotropic materials with constraints. Int. J. Engng Sci. **33**, 2021–2054
38. Teufel, A. (2001) Simple flow configurations in hypoplastic abrasive materials. Diploma Thesis, Institute of Mechanics, Darmstadt University of Technology
39. Truesdell, C. (1957) Sulle basi della termomeccanica (italian) [On the foundations of thermomechanics]. Academia Nazionale dei Lincei, rendiconti della Classe di Scienze Fisiche, Matematiche e Naturali **8**, No. 22, 33–38, 158–166
40. Truesdell, C. (1969) Rational Thermodynamics. 1st edition, McGraw-Hill, NY
41. Truesdell, C., Toupin, R. (1965) The classical field theories. In *Handbuch der Physik III/1(Encyclopedia of Physics III/1)*, Ed. Flügge, S., Springer Berlin–New York
42. Wang, Y., Hutter, K. (1999) A constitutive model for multi-phase mixtures and its application in shearing flows of saturated soil-fluid mixtures. Granular Matter **1**(4), 163–181
43. Williams, J. C. (1976) The segregation of particulate materials. A review. Powder Technology **15**, 245–251
44. Wilmanski, K. (1999) Mechanics of continuous media. Springer, Berlin–New York

Gravity-Induced Segregation of Cohesionless Granular Mixtures

Eliot Fried and Bidhan C. Roy

Department of Theoretical and Applied Mechanics, University of Illinois, Urbana, IL 61801, USA, e-mail: {e-fried,bcroy}@uiuc.edu

received 22 Jul 2002 — accepted 23 Sep 2002

Abstract. Working with the context of a theory proposed recently by Fried, Gurtin and Hutter [6], we consider a one-dimensional problem involving a granular mixture of $K > 2$ discrete sizes bounded below by an impermeable base, above by an evolving free surface, and subject to gravity. We demonstrate the existence of a solution in which the medium segregates by particle size. For a mixture of small and large particles ($K = 2$), we use methods of Smoller [30] to show that the segregated solution is unique. Further, for a mixture of small, medium, and large particles ($K = 3$), we use LeVeque's [18] CLAWPACK to construct numerical solutions and find that these compare favourably with analytical predictions.

1 Introduction

A granular material is a collection of solid particles together with an interstitial fluid such as air or water. Generally, granular materials consist not of identical particles, but, rather, of various particle types that may differ in size, shape, density, resilience, and roughness. Of interest in this article is the process of size-based particle segregation. Although it is relatively easy to achieve homogenous mixing with miscible fluids, it is very difficult to do so in a granular mixture involving particles of different sizes. For example, stirring a container of freeze-dried coffee causes large grains to rise to the surface.

Applications in which size-based segregation occur and are of importance are abundant. While mineral processing technologies exploit the tendency for granular materials to segregate, industrial mixing technologies must counteract this tendency. Though vital to the chemical, pharmaceutical, powder metallurgy, glass, ceramic, paint, food, and construction industries, separation and mixing technologies are presently limited by a reliance upon empirically-based heuristics. It therefore seems reasonable to expect that an enhanced understanding of the mechanisms underlying segregation will lead to advances in a broad spectrum of industrial enterprises.

Theoretical studies of granular flows have been undertaken using molecular dynamics, statistical physics, and continuum mechanics. The approach based on molecular dynamics involves modelling the granular material as an assembly of rigid bodies, taking into account translational and rotational degrees of freedom, and allowing for hysteretic interactions. Euler's equations of motion are formulated and solved numerically. Numerical results obtained using molecular dynamics exhibit mixed agreement with observations. For example, in a simulation of plane Couette flow, Thompson and Grest [31] employed a Hookean-type elastic force model with 750 soft particles of equal radii; a plug-like motion of the core and a boundary shear layer of 6–12 particle diameters in thickness was found, showing good agreement with experimental results of Hanes and Inman [8]; however, the shear stress did not show a quadratic dependence on the mean shear rate as observed in the experiments of Bagnold [4]. Agreement with experiment aside, even with the most advanced computing resources currently available, the inherent memory demands of molecular dynamics make it infeasible for systems of more than 10^6 particles. Applications of molecular dynamics to size-based segregation can be found in Dolgunin and Ukolov [5], Gallas, Herrmann, Pöschel and Sokolowski [7], Herrmann and Luding [9], Hirshfeld and Rapport [10], Hong, Quinn and Luding [11], Moaker, Shinbrot and Muzzio [21], Ohtsuki, Takemoto, Hata, Kawai and Hayashi [22], Pöschel, Schwager and Salreena [23], Ristow [24], Shinbrot, Zeggio and Muzzio [27], Shoichi [28], Smith, Baxter, Tuzun and Hayes [29].

In the limit as the number of particles becomes infinitely large, molecular dynamics is replaced by a statistical approach in which moments of a Boltzmann type equation are used. Interactions between the individual bodies are expressed by the collision operator, taking into account energy losses during collisions. The number of moments taken determines the complexity of the theory, which is continuous for fields that are statistical averages of quantities that exhibit large fluctuations on the microscale. The important results are the evolution equations for the density, the velocity, and the granular temperature (Jenkins and Savage [15], Lun, Savage, Jeffrey and Chepurniy [19], Makse and Kurchan [20]). Basic to these kinetic models are the assumptions that momentum transfer occurs via collisions and that only binary collisions occur. This implies that the granular phase must be sufficiently dispersed, which for dry granular flows subject to gravity, seems plausible for only very rapid motions. However, many flows of practical interest fall into the intermediate regime where both frictional contacts and particle-particle collisions are significant (Ancey, Evesque and Coussot [1], Johnson, Nott and Jackson [16]). Applications of this statistical approach in the study of segregation problem can be found in Arnarson and Jenkins [2], Arnarson and Willits [3], Jenkins [13], Jenkins and Mancini [14], among others.

Continuum mechanical models are purely phenomenological descriptions and are restricted to macroscopic length scales that extend over many particle diameters. Closure conditions are based on standard invariance requirements and thermodynamical restrictions and may account for microstructural effects. Insofar as monodisperse granular materials are concerned, continuum-level theories have been exploited to considerable advantage (Hutter and Rajagopal [12], Savage [26], Wang and Hutter [32]).

Here, we use a recent continuum model developed by Fried, Gurtin and Hutter [6] to study gravity-driven segregation and compaction of polydisperse granular material. Based on a kinematical treatment of voids, the basic physical laws allow for segregation via diffusion of different particle types and voids. This theory ignores inertial effects and is thus restricted to situations involving relatively slow flow. For a cohesionless granular material consisting of particles that differ only by size, this theory yields evolution equations for the volume-weighted mixture-velocity $\boldsymbol{v}$, the pressure p, and $K \geq 1$ particulate volume fractions φ_k constrained to obey

$$0 \leq \varphi_k \leq 1 \qquad \text{and} \qquad \varphi^{\mathrm{P}} = \sum_{k=1}^{K} \varphi_k \leq \mathring{\varphi}, \tag{1}$$

with φ^{P} the total particulate volume fraction and $\mathring{\varphi}$ the volume fraction at random close-packing. Writing $\boldsymbol{f}$ for the external body force per unit mass, these equations consist of the constraint

$$\operatorname{div} \boldsymbol{v} = 0, \tag{2}$$

which enforces the requirement that all volume be accounted for (locally) by particles and voids, the force balance

$$\operatorname{div} \boldsymbol{S} + \sum_{k=1}^{K} m_k \varphi_k \boldsymbol{f} = \operatorname{grad} p, \tag{3}$$

with $\boldsymbol{S}$ the extra stress (given constitutively as an isotropic function of the volume fractions $(\varphi_1, \varphi_2, \ldots, \varphi_K)$ and the strain-rate $\boldsymbol{D} = \frac{1}{2}(\operatorname{grad} \boldsymbol{v} + (\operatorname{grad} \boldsymbol{v})^{\top})$, which, by (2), must obey $\operatorname{tr} \boldsymbol{D} = 0$) and m_k the density of a particle of type k, and the particulate volume balances

$$\frac{\mathrm{D}\varphi_k}{\mathrm{D}t} = -\operatorname{div}\big(\varphi_k \alpha_k(\imath_{\boldsymbol{D}}) h(\varphi^{\mathrm{P}}) \boldsymbol{f}\big), \qquad k = 1, 2, \ldots, K, \tag{4}$$

in which $\alpha_k > 0$ is the effective mobility of particles of type k, depending on the list, $\imath_{\boldsymbol{D}} = (|\boldsymbol{D}|, \det \boldsymbol{D})$, of invariants of the strain–rate $\boldsymbol{D}$ and h is the compaction function. The mobilities are stipulated to be positive (even when $\boldsymbol{D} = \boldsymbol{0}$). Thus, if we assume a direct correspondence between particle size and particle mobility, with smaller particles being less mobile than large particles, no generality is lost by taking

$$\alpha_1(\imath_{\boldsymbol{D}}) > \alpha_2(\imath_{\boldsymbol{D}}) > \cdots > \alpha_K(\imath_{\boldsymbol{D}}) > 0 \tag{5}$$

for all $\boldsymbol{D}$. Further, it is assumed that

$$h(\varphi^{\mathrm{P}}) > 0 \;\text{ for }\; 0 < \varphi^{\mathrm{P}} < \mathring{\varphi} \qquad \text{and} \qquad h(\varphi^{\mathrm{P}}) = 0 \;\text{ for }\; \varphi^{\mathrm{P}} \geq \mathring{\varphi}. \tag{6}$$

Due to the hyperbolicity of (4), the evolution equations (2)–(4) must be supplemented by relations which hold across shock surfaces across which the particulate volume fraction, the strain-rate, and the extra stress suffer jump discontinuities. Considering such a shock and writing $\boldsymbol{n}_{\text{shock}}$ for its unit normal field and V_{shock} for the associated (scalar normal) velocity relative to the mixture, the relations associated with the particulate volume balances are

$$[\![\varphi_k]\!] V_{\text{shock}} = [\![\varphi_k \alpha_k(\imath_{\boldsymbol{D}}) h(\varphi^{\mathrm{P}})]\!] \boldsymbol{f} \cdot \boldsymbol{n}_{\text{shock}}, \qquad k = 1, 2, \ldots, K, \tag{7}$$

where, given a field g, $[\![g]\!] = g^{+} - g^{-}$, with g^{+} the limit of g, on the shock surface, taken from the region into which $\boldsymbol{n}_{\text{shock}}$ points and g^{-} the corresponding limit from the other side of the shock surface.

In addition to the evolution equations and shock relations, the theory delivers boundary conditions at free surfaces and impermeable solid boundaries. Focusing only on the conditions associated with the particulate volume balances, the conditions that hold at a free surface with unit normal $\boldsymbol{n}_{\text{free}}$ directing into the region of pure voids and associated (scalar normal) velocity V_{free} relative to the mixture are

$$\alpha_k(\imath_{\boldsymbol{D}})h(\varphi^{\mathrm{P}}) = \frac{V_{\text{free}}}{\boldsymbol{f} \cdot \boldsymbol{n}_{\text{free}}}, \tag{8}$$

for any particle type k present at the surface, while the conditions

$$\varphi_k = 0 \qquad \text{or} \qquad \alpha_k(\imath_{\boldsymbol{D}})h(\varphi^{\mathrm{P}}) = 0, \qquad k = 1, 2, \ldots, K, \tag{9}$$

must hold at an impermeable boundary.

In what follows, we consider a one-dimensional specialisation of the equations listed above. Under this specialisation, which we discuss in Sect. 2, the velocity field is constant and the governing equations reduce to a hyperbolic system. Our analysis is based on a particular packing function, which we introduce in Sect. 3. A non-dimensionalisation, which renders the problem in terms of packing fractions as opposed to volume fractions, is performed in Sect. 4. In Sect. 5, we pose a problem for a granular mixture bounded above by the free surface and below by an impermeable base. For a system with K particle sizes, we obtain a solution to this problem which, at steady state, involves K different layers—with the presence of particles of a given size in a given layer determined by their size. For the case $K = 3$, we use LeVeque's [18] CLAWPACK to construct numerical solutions. This robust code requires a user-supplied Riemann solver. Our Riemann solver is based on the flux splitting scheme of Roe [25] and incorporates proper entropy corrections and flux limiters. In Sect. 6, we apply tools developed by Smoller [30] to establish the uniqueness of the solution presented in Sect. 5 in the case $K = 2$.

2 A Class of One-Dimensional Problems

We suppose that the body force is purely gravitational, viz.,

$$\boldsymbol{f} = \boldsymbol{g}, \tag{10}$$

with $\boldsymbol{g}$ the gravitational acceleration. We let x denote the Cartesian coordinate in the direction $\boldsymbol{e} = -\boldsymbol{g}/|\boldsymbol{g}|$ and seek solutions of the evolution equations for which $\boldsymbol{v}$ and p are functions of x and t, with

$$\boldsymbol{v} = v\boldsymbol{e}. \tag{11}$$

Then, since $\operatorname{div}\boldsymbol{v} = 0$, v must be constant and we may, without loss in generality, assume that $v \equiv 0$. Thus, $\boldsymbol{D} = \boldsymbol{0}$ and the extra stress $\boldsymbol{S}$ either vanishes or is indeterminate (Fried, Gurtin and Hutter [6]). For the indeterminate case, we take $\boldsymbol{S} = \sigma\boldsymbol{e} \otimes \boldsymbol{e}$, with σ a scalar. Force balance requires that

$$\frac{\partial(p - \sigma)}{\partial x} = -\sum_{k=1}^{K} m_k \varphi_k |\boldsymbol{g}|, \tag{12}$$

which, given the volume fractions φ_k, determines the difference $p - \sigma$ between the pressure and the extra stress. Equation (12) holds with $S = 0$ when the constitutive relation for $\boldsymbol{S}$ is well-defined at $\boldsymbol{D} = \boldsymbol{0}$.

In view of the foregoing, the particulate volume balances (4) take the form

$$\frac{\partial\varphi_k}{\partial t}=\alpha_k g\frac{\partial\big(\varphi_k h(\varphi^{\mathrm{P}})\big)}{\partial x},\qquad k=1,2,\ldots,K,\tag{13}$$

where, for simplicity, we use α_k to denote the values of the effective mobilities for $\boldsymbol{D}=\boldsymbol{0}$.

Regarding the shock relations, we choose $\boldsymbol{n}_{\mathrm{shock}}=\boldsymbol{e}$, so that $\boldsymbol{f}\cdot\boldsymbol{n}_{\mathrm{shock}}=-|\boldsymbol{g}|$. Thus, (7) becomes

$$[\![\varphi_k]\!]V_{\mathrm{shock}}=-\alpha_k|\boldsymbol{g}|[\![\varphi_k h(\varphi^{\mathrm{P}})]\!],\qquad k=1,2,\ldots,K.\tag{14}$$

Similarly, the condition (8) that hold for any particle of type k at a free surface becomes

$$\alpha_k h(\varphi^{\mathrm{P}})=-\frac{V_{\mathrm{free}}}{|\boldsymbol{g}|}.\tag{15}$$

3 Constitutive Specialisation of the Packing Function

For simplicity, we choose the packing function h to be of the particular form

$$h(\varphi^{\mathrm{P}})=\varphi^*-\sum_{k=1}^{K}\varphi_k,\tag{16}$$

which obeys the requirements expressed in (6).

4 Non-dimensionalisation

On introducing a characteristic length L and a characteristic time T, we define dimensionless quantities

$$\tilde{x}=\frac{x}{L},\quad \tilde{t}=\frac{t}{T},\quad \tilde{\varphi}_k=\frac{\varphi_k}{\varphi^*},\quad \tilde{\varphi}^{\mathrm{P}}=\frac{\varphi^{\mathrm{P}}}{\varphi^*},\quad \tilde{V}=\frac{VT}{L},\tag{17}$$

and

$$\tilde{\alpha}_k=\frac{\alpha_k|\boldsymbol{g}|T\varphi^*}{L},\tag{18}$$

where V is used to denote the velocity of a generic shock or a generic free surface. We refer to $\tilde{\varphi}_k$ as the *packing fraction*.

Bearing in mind the constitutive specialisation (16) and dropping tildes, the hyperbolic system (13) becomes

$$\frac{\partial\varphi_k}{\partial t}=\alpha_k\frac{\partial}{\partial x}\Big(\varphi_k(1-\varphi^{\mathrm{P}})\Big),\qquad k=1,2,\ldots,K,\tag{19}$$

and the shock relations (14) become

$$[\![\varphi_k]\!]V_{\mathrm{shock}}=-\alpha_k[\![\varphi_k(1-\varphi^{\mathrm{P}})]\!],\qquad k=1,2,\ldots,K.\tag{20}$$

Further, the boundary condition (15) for a particle of type k at a free surface becomes

$$\alpha_k(1-\varphi^{\mathrm{p}}) = -V_{\mathrm{free}}, \tag{21}$$

while the boundary condition (9) at an impermeable solid surface becomes

$$\varphi_k = 0 \qquad \text{or} \qquad \alpha_k(1-\varphi^{\mathrm{p}}) = 0, \qquad k = 1, 2, \ldots, K. \tag{22}$$

By virtue of (1) and $(17)_{3,4}$, the packing fractions must obey

$$0 \le \varphi_k \le 1 \qquad \text{and} \qquad \varphi^{\mathrm{p}} = \sum_{k=1}^{K} \varphi_k \le 1. \tag{23}$$

5 Solution to a Particular Problem

We now consider an open container that occupies the interval $0 \le x \le 1$, with $x = 0$ an impermeable base. We assume that, initially, each particulate packing fraction φ_k has a prescribed constant value, viz.,

$$\varphi_k(x, 0) = \mathring{\varphi}_k, \qquad 0 \le x \le 1. \tag{24}$$

Further, we require that

$$0 < \mathring{\varphi}^{\mathrm{p}} = \sum_{k=1}^{K} \mathring{\varphi}_k < 1, \tag{25}$$

so that, initially, the mixture is loosely packed and has a free surface at the top $x = 1$ of the container.

Granted the initial conditions (24), the boundary conditions (9) at the base $x = 0$, and the condition (21) for any particle type k at the free surface, we seek solutions of the hyperbolic system (19) that consist of uniform states separated by shocks across which the jump conditions (20) hold.

As a candidate for such a solution, we consider a generalisation of the solution obtained for a mixture of small and large particles ($K = 2$) by Fried, Gurtin and Hutter [6]. This candidate can be depicted (Fig. 1) as the union of $2K$ regions, $\mathcal{R}_1, \mathcal{R}_2, \ldots, \mathcal{R}_{2K}$, containing particles of various types and in various proportions. For $k = 1, 2, \ldots, K$, regions $\mathcal{R}_k$ and $\mathcal{R}_{2K-k+1}$ contain particles of types $k, k+1, \ldots, K$. In particular, region $\mathcal{R}_1$ corresponds to the initial state, so that the mixture is given by (24). While the mixtures in regions $\mathcal{R}_1, \mathcal{R}_2, \ldots, \mathcal{R}_K$ are loosely packed, those in regions $\mathcal{R}_K, \mathcal{R}_{K+1}, \ldots, \mathcal{R}_{2K}$ are closely packed. Region $\mathcal{R}_1$ is bounded by a segregation shock $\mathcal{S}^1_{\mathrm{seg}}$ emanating from the free surface and a compaction shock $\mathcal{S}_{\mathrm{com}}$ emanating from the base. These shocks meet at time T_1, at which point they are both deflected. Subsequent to T_1, $\mathcal{S}^1_{\mathrm{seg}}$ becomes horizontal while $\mathcal{S}_{\mathrm{com}}$ continues to travel upwards. In the region $\mathcal{R}_{2K}$ lying between the base, the portion of $\mathcal{S}_{\mathrm{com}}$ adjacent to $\mathcal{R}_1$, and the horizontal portion of $\mathcal{S}^1_{\mathrm{seg}}$, the packing fraction φ_i^{2K} of particles of type i is given by

$$\varphi_i^{2K} = \mathring{\varphi}_i + \frac{\alpha_i(1 - \sum_{l=1}^{K} \mathring{\varphi}_l)\mathring{\varphi}_i}{\sum_{l=1}^{K} \alpha_l \mathring{\varphi}_l}, \tag{26}$$

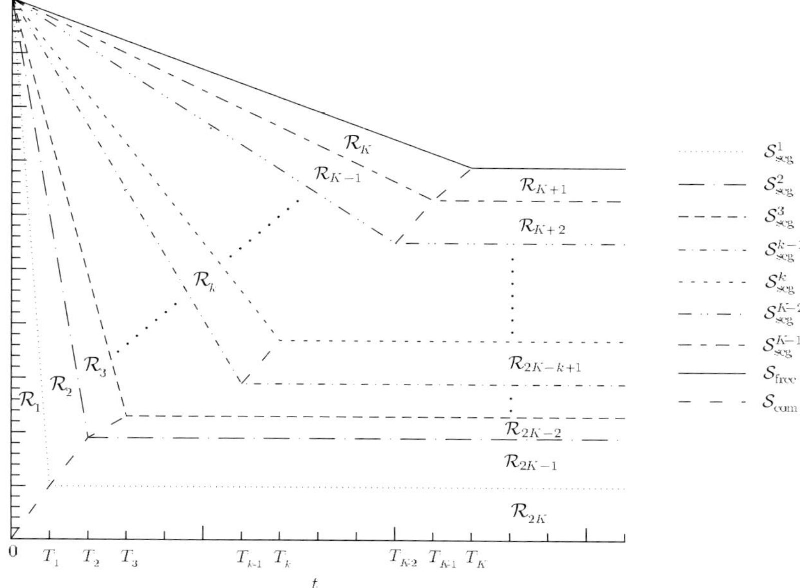

Fig. 1. Segregation, by gravity, of a granular aggregate involving particles of K sizes in a fixed container. At any time t, the solution consists of uniform states separated by shocks

for $i = 1, 2, \ldots, K$. In addition, until time T_1, the velocities of $\mathcal{S}^1_{\mathrm{seg}}$ and $\mathcal{S}_{\mathrm{com}}$ are given by

$$V_{\mathcal{S}^1_{\mathrm{seg}}} = \alpha_1 (1 - \sum_{l=1}^{K} \mathring{\varphi}_l) \qquad \text{and} \qquad V_{\mathcal{S}_{\mathrm{com}}} = - \sum_{l=1}^{K} \alpha_l \mathring{\varphi}_l. \tag{27}$$

At time T_2, $\mathcal{S}_{\mathrm{com}}$ meets another segregation shock $\mathcal{S}^2_{\mathrm{seg}}$ emanating from the free surface. At this time, $\mathcal{S}_{\mathrm{com}}$ and $\mathcal{S}^2_{\mathrm{seg}}$ are deflected. Subsequent to T_2, $\mathcal{S}_{\mathrm{com}}$ continues to travel upward and $\mathcal{S}^2_{\mathrm{seg}}$ becomes horizontal. This process continues until time T_{K-1}, when $\mathcal{S}_{\mathrm{com}}$ is deflected upward and $\mathcal{S}^{K-1}_{\mathrm{seg}}$ becomes horizontal. For $k = 2, \ldots, K$, the packing fraction φ^k_i of particles of type i in region $\mathcal{R}_k$ is given by the positive square root of the quadratic equation

$$\alpha_i (\varphi^k_i)^2 - \Big(\alpha_i (1 - \sum_{j=k, \neq i}^{K} \varphi^k_j) - \alpha_{k-1} (1 - \sum_{j=k-1}^{K} \varphi^{k-1}_j) \Big) \varphi^k_i$$
$$- \varphi^{k-1}_i \Big(1 - \sum_{j=k-1}^{K} \varphi^{k-1}_j \Big) (\alpha_{k-1} - \alpha_i) = 0, \tag{28}$$

for $i = 1, 2, \ldots, K$. In addition, the velocity of $\mathcal{S}_{\rm seg}^k$ until T_k and the velocity of $\mathcal{S}_{\rm com}$ between T_{k-1} and T_k are given by

$$V_{\mathcal{S}_{\rm seg}^k} = \alpha_k(1 - \sum_{l=k}^{K} \varphi_l^k) \qquad \text{and} \qquad V_{\mathcal{S}_{\rm com}} = -\sum_{l=k}^{K} \alpha_l \varphi_l^k. \tag{29}$$

At time T_K, $\mathcal{S}_{\rm com}$ and $\mathcal{S}_{\rm free}$ meet and $\mathcal{S}_{\rm free}$ becomes horizontal. As mentioned above, the region $\mathcal{R}_{K+1}$ contains only the largest particles — those of type K. For $k = 2, \ldots, K$, the particulate packing fraction φ_i^{2K-k+1} in region $\mathcal{R}_{2K-k+1}$ is given by

$$\varphi_i^{2K-k+1} = \varphi_i^k + \frac{\alpha_i(1 - \sum_{l=k}^{K} \varphi_l^k)\varphi_i^k}{\sum_{l=1}^{K} \alpha_l \varphi_l^k}, \tag{30}$$

for $i = k, \ldots, K$.

To verify that the candidate described above is a solution, we have only to show that the following conditions are satisfied:

(i) the solid-boundary condition (cf. $(22)_2$)

$$\alpha_l\Big(1 - \sum_{i=1}^{K} \varphi_i^{2K}\Big) = 0, \qquad l = 1, 2, \ldots, K, \tag{31}$$

for $t > 0$;

(ii) for each $k = 2, 3, \ldots, K$, the jump conditions (cf. (20))

$$(\varphi_l^k - \varphi_l^{2K-k+1})V_{\mathcal{S}_{\rm com}} = \alpha_l \varphi_l^k\Big(1 - \sum_{i=k}^{K} \varphi_i^k\Big) \qquad \text{on} \qquad \mathcal{S}_{\rm com}, \tag{32}$$

for $l = k, k+1, \ldots, K$ and $T_{k-1} < t < T_k$, and, for $0 < t < T_k$,

$$[\![\varphi_l^k]\!] V_{\mathcal{S}_{\rm seg}^{k-1}} = \alpha_l [\![\varphi_l^k(1 - \sum_{i=k}^{K} \varphi_i^k)]\!] \qquad \text{on} \qquad \mathcal{S}_{\rm seg}^{k-1}, \tag{33}$$

for $l = k, k+1, \ldots, K$;

(iii) the interface condition (cf. (9))

$$V_{\rm seg}^k = 0 \qquad \text{on} \qquad \mathcal{S}_{\rm seg}^k, \tag{34}$$

for $t \geq T_k$;

(v) the free-surface conditions (cf. (21))

$$V_{\rm free} = \alpha_K(1 - \varphi_K^K) \qquad \text{on} \qquad \mathcal{S}_{\rm free}, \tag{35}$$

for $0 < t < T_K$, and

$$V_{\rm free} = 0 \qquad \text{on} \qquad \mathcal{S}_{\rm free}, \tag{36}$$

for $t \geq T_K$.

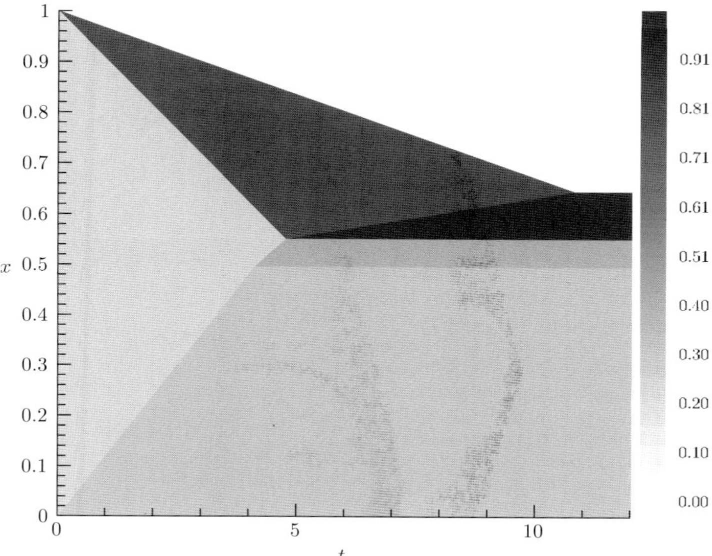

Fig. 2. Characteristic plot of species type $k = 3$. Here, $(\alpha_1, \alpha_2, \alpha_3) = (0.3, 0.2, 0.1)$ and $(\varphi_1, \varphi_2, \varphi_3) = (0.2, 0.2, 0.2)$

To establish that (31) holds for $t > 0$, we substitute for φ_i^{2K} from (26) and sum over the index i. Similarly, (32) follows directly if we substitute for φ_l^{2K-k+1} from (30) and for $V_{\mathcal{S}_{\text{com}}}$ from (29). Also, on substituting for $V_{\mathcal{S}_{\text{seg}}^{k-1}}$ from (29), it can be shown that the packing fraction φ_l^k satisfying (33) is given by (28). Hence, (33) follows if we can show that the packing fractions in question are determined uniquely by the quadratic equation (28). Since $\alpha_{k-1} > \alpha_k$ for $k = 2, \ldots, K$ and the packing in the region is not close, it follows that (28) has only one real positive root, φ_l^k, and that $0 < \varphi_l^k < 1$. Hence, the only positive root of (28) is indeed the packing fraction. For $t \geq T_k$, the shock $\mathcal{S}_{\text{seg}}^k$ separates two compacted zones. From (9), it follows directly that $\alpha_i(1 - \sum_{j=k}^{K} \varphi_j^{2K-k+1}) = 0$. Hence, (34) holds for $t \geq T_k$. Since the shock $\mathcal{S}_{\text{free}}$ separates the region $\mathcal{R}_K$ (which contains only particles of type K, with a packing fraction φ_K^K) from the free surface (where $\varphi_k = 0$, $k = 1, \ldots, K$), it follows directly as a consequence of (21) that $V_{\text{free}} = \alpha_K(1 - \varphi_K^K)$ for $0 < t < T_K$. Hence, (35) holds for $0 < t < T_K$. Finally, since, for $t \geq T_K$, the shock $\mathcal{S}_{\text{free}}$ separates the compacted zone $\mathcal{R}_{K+1}$ (where $\varphi_K^{K+1} = 1$) from the free surface, it follows that $1 - \varphi_K^{K+1} = 0$ and, hence, by (9), we have $V_{\text{free}} = 0$ for $t \geq T_k$. Thus, (36) holds for $t \geq T_k$.

Consistent with intuitive expectations, the solution exhibits both segregation and compaction. Segregation is by particle size, with the region closest to the free surface consisting only of the largest particles. As an illustration, we consider the particular case of a mixture containing particles of three sizes, with effective mobilities

$$\alpha_1 = 0.3, \qquad \alpha_2 = 0.2, \qquad \text{and} \qquad \alpha_3 = 0.1, \tag{37}$$

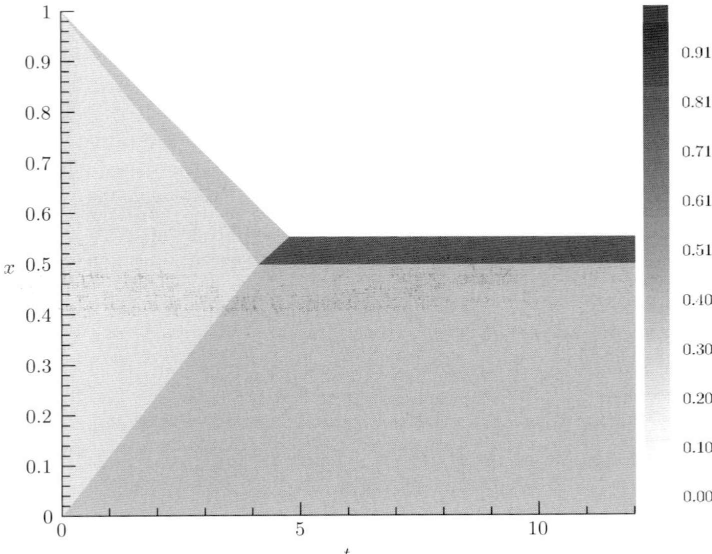

Fig. 3. Characteristic plot of species type $k = 2$. Here, $(\alpha_1, \alpha_2, \alpha_3) = (0.3, 0.2, 0.1)$ and $(\varphi_1, \varphi_2, \varphi_3) = (0.2, 0.2, 0.2)$

and use CLAWPACK to generate a solution to the problem described above, with

$$\mathring{\varphi}_k = 0.2, \qquad k = 1, 2, 3. \tag{38}$$

Figures 2–4 show characteristic plots for each of the particle types and Fig. 5 shows profiles of the packing fractions as functions of position at various times. These plots are a result of the numerical solutions obtained using CLAWPACK. In keeping with the general results presented above, these plots exhibit shocks that separate states in which the particulate packing fractions are uniform. At each time, the domain consists of four distinct regions. Prior to the instant $t = 4.16$ $(\equiv T_1)$, these regions are demarcated by the free surface, which moves with velocity $V_{\text{free}} = 0.03$, segregation shocks $\mathcal{S}^1_{\text{seg}}$ and $\mathcal{S}^2_{\text{seg}}$ with velocities $V_{\mathcal{S}^1_{\text{seg}}} = 0.16$ and $V_{\mathcal{S}^2_{\text{seg}}} = 0.09$, and a compaction shock $\mathcal{S}_{\text{com}}$ with velocity $V_{\mathcal{S}_{\text{com}}} = -0.12$. The region $(\mathcal{R}_3)$ closest to the free surface consists only of the largest particles with $\varphi_3 = 0.672680$. In the region $(\mathcal{R}_2)$ between $\mathcal{S}^2_{\text{seg}}$ and $\mathcal{S}^1_{\text{seg}}$, $\varphi_1 = 0$, $\varphi_2 = 0.309717$, and $\varphi_3 = 0.219433$. In the region $(\mathcal{R}_1)$ between $\mathcal{S}^1_{\text{seg}}$ and $\mathcal{S}^2_{\text{com}}$, the mixture is in its initial state. Initially at the base, the medium is closely packed with $\varphi_1 = 0.40$, $\varphi_2 = 0.35$, and $\varphi_3 = 0.25$. Between time $t = 4.16$ $(\equiv T_1)$ and time $t = 4.77$ $(\equiv T_2)$, the velocity of $\mathcal{S}_{\text{com}}$ is $V_{\mathcal{S}_{\text{com}}} = -0.08$. At $t = 4.16$ $(\equiv T_1)$, a horizontal layer $(\mathcal{R}_6)$ develops adjacent to the basal region and the packing fractions within this layer are given by $\varphi_1 = 0.399996$, $\varphi_2 = 0.333329$, and $\varphi_3 = 0.266674$. Subsequent to $t = 4.77$ $(\equiv T_2)$, the velocity of $\mathcal{S}_{\text{com}}$ becomes $V_{\mathcal{S}_{\text{com}}} = -0.06$. Time $t = 4.77$ $(\equiv T_2)$ signals the development of a layer $(\mathcal{R}_5)$ of closely packed large particles, where $\varphi_1 = 0$, $\varphi_2 = 0.657399$, and $\varphi_3 = 0.342599$. At time $t = 10.91$ $(\equiv T_3)$, $\mathcal{S}_{\text{com}}$ intersects the free surface and a steady state is achieved.

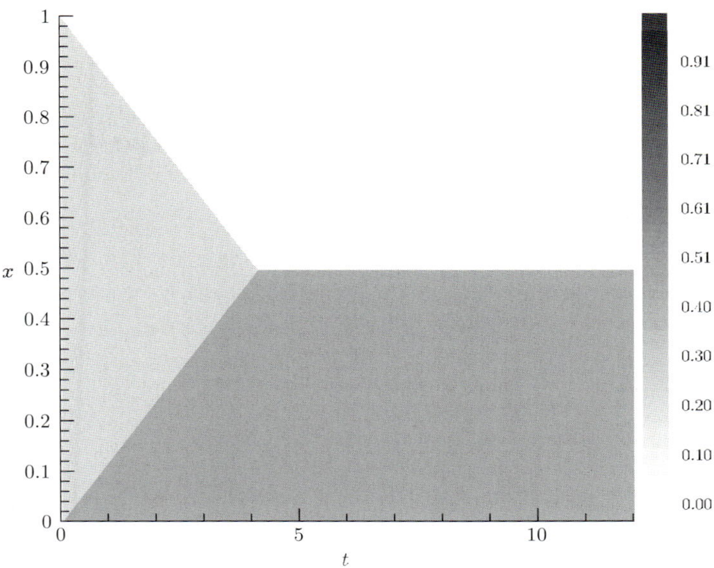

Fig. 4. Characteristic plot of species type $k = 1$. Here, $(\alpha_1, \alpha_2, \alpha_3) = (0.3, 0.2, 0.1)$ and $(\varphi_1, \varphi_2, \varphi_3) = (0.2, 0.2, 0.2)$

Table 1. Comparison of the analytical and numerical values of the distribution of particles of all three types in various compaction zones

Compaction zone	Packing fraction	Analytical value	Numerical value	Relative error
	φ_1	0.400000	0.399996	0.000010
$\mathcal{R}_6$	φ_2	0.333333	0.333329	0.000012
	φ_3	0.266667	0.266674	0.000026
$\mathcal{R}_5$	φ_2	0.657400	0.657399	0.000002
	φ_3	0.342600	0.342599	0.000002
$\mathcal{R}_4$	φ_3	1.000000	0.999998	0.000002

Table 1 compares the analytical and the numerical values of the distribution of all the three types of particles at the steady state in the zones where close packing is achieved. The maximum relative error is less than 2.6×10^{-5}.

6 Uniqueness of the Segregated Solution for a Mixture of Small and Large Particles

In general, it would be desirable to understand whether the solution presented in Sect. 5 is unique. Here, as a step toward this goal, we use methods developed by Smoller [30] to resolve this issue for the case of a mixture of small and large particles.

6.1 Riemann Problem

Toward addressing the uniqueness of the segregated solution for a mixture of small and large particles, we consider the system

$$\left.\begin{aligned}\frac{\partial\varphi_1}{\partial t} &= \alpha_1\frac{\partial}{\partial x}\Big(\varphi_1(1-\varphi_1-\varphi_2)\Big),\\ \frac{\partial\varphi_2}{\partial t} &= \alpha_2\frac{\partial}{\partial x}\Big(\varphi_2(1-\varphi_1-\varphi_2)\Big),\end{aligned}\right\} \tag{39}$$

which arises upon specialising (19) to the case $K = 2$, for (x,t) belonging to $(-\infty,\infty)\times(0,\infty)$ and subject to the initial conditions

$$\varphi_k(x,0) = \begin{cases}\varphi_k^< & \text{if } x<0,\\ \\ \varphi_k^> & \text{if } x>0,\end{cases} \tag{40}$$

with $\varphi_k^{\lessgtr}$ constant. For the moment, we interpret (39) in the weak sense, so that the discontinuity conditions that must hold across shocks, free surfaces, and solid boundaries are implicitly satisfied. In view of (5),

$$\alpha_1 > \alpha_2 > 0. \tag{41}$$

Further, in view of (23), (φ_1,φ_2) must belong to

$$\mathcal{A} = \{\varphi_1,\varphi_2) : 0\le\varphi_1\le 1, 0\le\varphi_2\le 1, \varphi_1+\varphi_2\le 1\}, \tag{42}$$

which we refer to as the set of admissible packing fractions.

Generally, we require only that $(\varphi_1^{\lessgtr},\varphi_2^{\lessgtr})$ belong to $\mathcal{A}$. This includes special problems corresponding to less than closely packed mixtures of semi-infinite extent that are bounded above by free surfaces ($\varphi_1^< = \varphi_2^< = 0$ and $\varphi_1^> + \varphi_2^> < 1$) or below by solid bases ($\varphi_1^< + \varphi_2^< < 1$ and $\varphi_1^> + \varphi_2^> = 1$). However, our primary interest is in problems where $\varphi_1^{\lessgtr}$ and $\varphi_2^{\lessgtr}$ obey $0 < \varphi_1^{\lessgtr} + \varphi_2^{\lessgtr} < 1$.

The Jacobian of the system (39) is simply

$$\begin{bmatrix}\alpha_1(1-2\varphi_1-\varphi_2) & -\alpha_1\varphi_1\\ -\alpha_2\varphi_2 & \alpha_2(1-\varphi_1-2\varphi_2)\end{bmatrix}, \tag{43}$$

which has (real) eigenvalues

$$\begin{aligned}\lambda_\pm(\varphi_1,\varphi_2) &= \frac{1}{2}\Big(\alpha_1(1-2\varphi_1-\varphi_2)+\alpha_2(1-2\varphi_2-\varphi_1)\Big)\\ &\quad \pm\frac{1}{2}\sqrt{\Big(\alpha_1(1-2\varphi_1-\varphi_2)-\alpha_2(1-2\varphi_2-\varphi_1)\Big)^2+4\alpha_1\alpha_2\varphi_1\varphi_2}\,.\end{aligned} \tag{44}$$

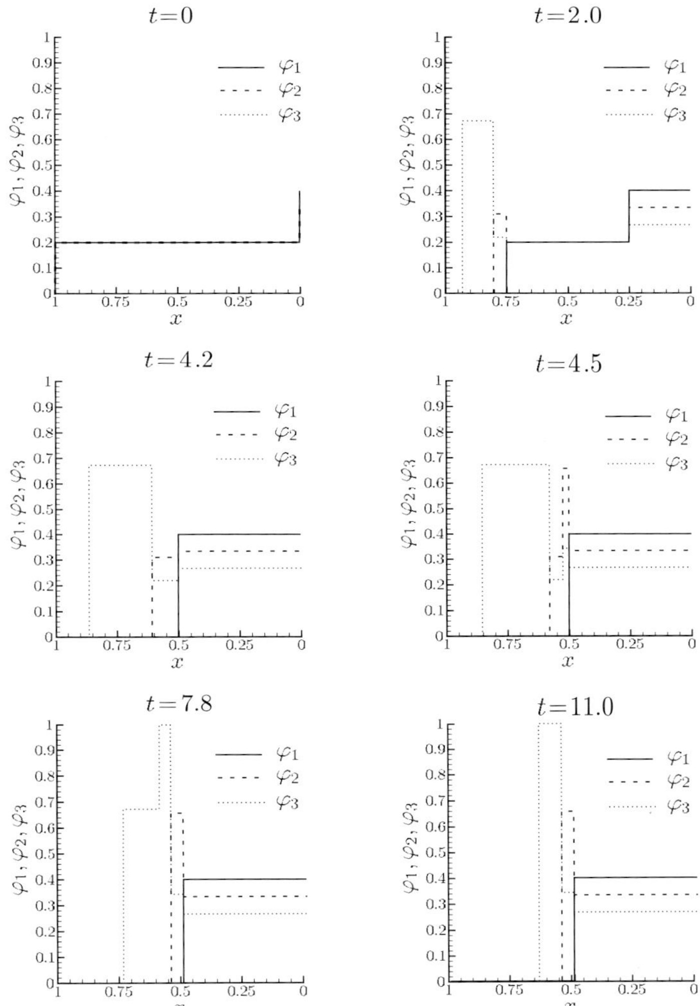

Fig. 5. The distribution of particles in space at a given time. Here, $(\alpha_1, \alpha_2, \alpha_3) = (0.3, 0.2, 0.1)$ and $(\varphi_1, \varphi_2, \varphi_3) = (0.2, 0.2, 0.2)$

By inspection,

$$0 \leq \lambda_-(\varphi_1, \varphi_2) < \lambda_+(\varphi_1, \varphi_2). \tag{45}$$

Hence, (39) is hyperbolic and we may conclude that solutions to the Riemann problem (39)–(40) must be uniform states separated by either (i) discontinuities (which may be compaction or segregation shocks, free surfaces, or solid boundaries) or (ii) rarefactions. Since one of the eigenvalues defined in (44) may vanish, (39) admits solutions involving static discontinuities.

Preliminary Results Concerning Discontinuities Here, we consider the problem of determining all uniform states (φ_1, φ_2) in $\mathcal{A}$ that may be connected to $(\varphi_1^<, \varphi_2^<)$ by a discontinuity emanating from the origin of the (x,t)-plane and moving with velocity V. Such a state must satisfy not only the jump conditions

$$\left.\begin{aligned} \alpha_1\varphi_1(1-\varphi_1-\varphi_2) - \alpha_1\varphi_1^<(1-\varphi_1^<-\varphi_2^<) &= V(\varphi_1-\varphi_1^<), \\ \alpha_2\varphi_2(1-\varphi_1-\varphi_2) - \alpha_2\varphi_2^<(1-\varphi_1^<-\varphi_2^<) &= V(\varphi_2-\varphi_2^<), \end{aligned}\right\} \tag{46}$$

but also the Lax [17] entropy condition, which in the present context requires that either

$$\lambda_-(\varphi_1,\varphi_2) < V < \lambda_+(\varphi_1,\varphi_2) \qquad \text{and} \qquad V < \lambda_-(\varphi_1^<,\varphi_2^<) \tag{47}$$

or

$$\lambda_-(\varphi_1^<,\varphi_2^<) < V < \lambda_+(\varphi_1^<,\varphi_2^<) \qquad \text{and} \qquad \lambda_+(\varphi_1,\varphi_2) < V. \tag{48}$$

Eliminating V between $(46)_1$ and $(46)_2$ yields a quadratic equation

$$A(\varphi_1,\varphi_1^<,\varphi_2^<)\varphi_2^2 + B(\varphi_1,\varphi_1^<,\varphi_2^<)\varphi_2 + C(\varphi_1,\varphi_1^<,\varphi_2^<) = 0 \tag{49}$$

for φ_2 in terms of φ_1, $\varphi_1^<$, and $\varphi_2^<$. Here,

$$\left.\begin{aligned} A(\varphi_1,\varphi_1^<,\varphi_2^<) &= \alpha_2(\varphi_1-\varphi_1^<) - \alpha_1\varphi_1, \\ B(\varphi_1,\varphi_1^<,\varphi_2^<) &= \alpha_1\varphi_1\varphi_2^< + \alpha_1\varphi_1 \\ &\quad - \alpha_1\varphi_1^<(1-\varphi_1^<-\varphi_2^<) - \alpha_1\varphi_1^2 \\ &\qquad - \alpha_2\varphi_1 + \alpha_2\varphi_1^2 - \alpha_2\varphi_1\varphi_1^<, \\ C(\varphi_1,\varphi_1^<,\varphi_2^<) &= \alpha_1\varphi_1^<\varphi_2^<(1-\varphi_1^<-\varphi_2^<) \\ &\quad - \alpha_1\varphi_1\varphi_2^< + \alpha_1\varphi_1^2\varphi_2^< \\ &\qquad - \alpha_2\varphi_1^<\varphi_2^<(1-\varphi_1^<-\varphi_2^<) \\ &\qquad\quad + \alpha_1\varphi_1\varphi_2^<(1-\varphi_1^<-\varphi_2^<). \end{aligned}\right\} \tag{50}$$

The roots

$$d_\pm(\varphi_1,\varphi_1^<,\varphi_2^<) = \frac{-B(\varphi_1,\varphi_1^<,\varphi_2^<)}{2} \mp \frac{\sqrt{B^2(\varphi_1,\varphi_1^<,\varphi_2^<) - 4A(\varphi_1,\varphi_1^<,\varphi_2^<)C(\varphi_1,\varphi_1^<,\varphi_2^<)}}{2A(\varphi_1,\varphi_1^<,\varphi_2^<)} \tag{51}$$

of the quadratic (49) describe curves within $\mathcal{A}$. A straightforward calculation, too lengthy to display here, shows that the curve

$$\mathcal{D}_-(\varphi_1,\varphi_1^<,\varphi_2^<) = \{(\varphi_1,\varphi_2) : \varphi_1^< < \varphi_1 < 1, \varphi_2 = d_-(\varphi_1,\varphi_1^<,\varphi_2^<)\} \tag{52}$$

describes all pairs (φ_1,φ_2) consistent with the jump conditions (46) and the inequalities (47). Similarly, the curve

$$\mathcal{D}_+(\varphi_1,\varphi_1^<,\varphi_2^<) = \{(\varphi_1,\varphi_2) : \varphi_1^< < \varphi_1 < 1, \varphi_2 = d_+(\varphi_1,\varphi_1^<,\varphi_2^<)\} \tag{53}$$

describes all pairs (φ_1,φ_2) consistent with the jump conditions (46) and the inequalities (48). A direct argument shows that, for given $(\varphi_1^<,\varphi_2^<))$, the curve determined by $d_+(\cdot,\varphi_1^<,\varphi_2^<)$ decreases monotonically over its domain. Thus, given $\varphi_2 = d_+(\varphi_1,\varphi_1^<,\varphi_2^<)$, we may write $\varphi_1 = \mathcal{D}_+^{-1}(\varphi_2,\varphi_1^<,\varphi_2^<)$. Figure 6 depicts $\mathcal{D}_+$ and $\mathcal{D}_-$ for a generic choice of $(\varphi_1^<,\varphi_2^<)$.

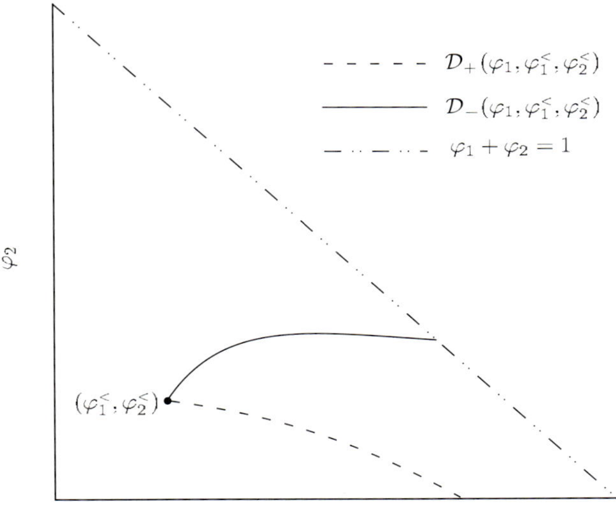

Fig. 6. The phase portrait of discontinuities (φ_1, φ_2) connecting to the right of a generic point $(\varphi_1^<, \varphi_2^<)$

Preliminary Results Concerning Rarefactions Here, we consider the problem of determining all rarefactions (φ_1, φ_2) taking values in $\mathcal{A}$, with

$$\varphi_k(x,t) = \phi_k(\xi), \qquad \xi = \frac{x}{t}, \tag{54}$$

that may be continuously connected to $(\varphi_1^<, \varphi_2^<)$ across a line emanating from the origin of the (x,t)-plane. Such a state must satisfy not only the system

$$\left.\begin{aligned} \xi\phi_1'(\xi) + \alpha_1\big(\phi_1(\xi)(1-\phi_1(\xi)-\phi_2(\xi))\big)' = 0,\\ \xi\phi_2'(\xi) + \alpha_2\big(\phi_2(\xi)(1-\phi_1(\xi)-\phi_2(\xi))\big)' = 0, \end{aligned}\right\} \tag{55}$$

which arises on inserting (54) in (39), but also, since the (non-negative) eigenvalues $\lambda_\pm(\phi_1(\xi), \phi_2(\xi))$ must increase with ξ, either

$$\lambda_-(\varphi_1^<, \varphi_2^<) < \lambda_-(\phi_1, \phi_2) \tag{56}$$

or

$$\lambda_+(\varphi_1^<, \varphi_2^<) < \lambda_+(\phi_1, \phi_2). \tag{57}$$

On writing (55) in the form

$$\begin{bmatrix} \alpha_1(1-2\phi_1(\xi)-\phi_2(\xi))+\xi & -\alpha_1\phi_1(\xi) \\ -\alpha_2\phi_2(\xi) & \alpha_2(1-\phi_1(\xi)-2\phi_2(\xi))+\xi \end{bmatrix} \begin{bmatrix} \phi_1'(\xi) \\ \phi_2'(\xi) \end{bmatrix} = \begin{bmatrix} 0 \\ 0 \end{bmatrix}, \tag{58}$$

and recalling that the eigenvalues of the Jacobian (43) are unique, it follows that $\xi = -\lambda_\pm(\phi_1, \phi_2)$.

We ignore the case where ϕ_1 and ϕ_2 are constant, which yields the trivial solution $\phi_k = \varphi_k^<$ and $\phi_2 = \varphi_2^<$, and assume that neither ϕ_1' nor ϕ_2' vanishes. Then, from (58) we obtain a pair,

$$\frac{\mathrm{d}\phi_2}{\mathrm{d}\phi_1} = \frac{\alpha_1(1 - 2\phi_1 - \phi_2) - \lambda_\pm(\phi_1, \phi_2)}{\alpha_1 \phi_1}, \tag{59}$$

of first order differential equations which must be solved subject to the inequalities (56) and (57) and the initial condition

$$\phi_2 = \varphi_2^< \qquad \text{when} \qquad \phi_1 = \varphi_1^<. \tag{60}$$

As solutions we find

$$\phi_2 = r_\pm(\phi_1, \varphi_1^<, \varphi_2^<) \tag{61}$$

with

$$r_\pm(\phi_1, \varphi_1^<, \varphi_2^<) = \frac{-B(\phi_1, \varphi_1^<, \varphi_2^<)}{2} \pm \frac{\sqrt{B^2(\phi_1, \varphi_1^<, \varphi_2^<) - 4A(\phi_1, \varphi_1^<, \varphi_2^<)C(\phi_1, \varphi_1^<, \varphi_2^<)}}{2A(\phi_1, \varphi_1^<, \varphi_2^<)} \tag{62}$$

and A, B, and C as defined in (50). A direct calculation, too lengthy to display here, shows that the curve

$$\mathcal{R}_-(\phi_1, \varphi_1^<, \varphi_2^<) = \{(\phi_1, \phi_2) : \varphi_1 < \varphi_1^<, \phi_2 = r_-(\phi_1, \varphi_1^<, \varphi_2^<)\} \tag{63}$$

describes all rarefactions (ϕ_1, ϕ_2) satisfying the constraint (56). Similarly, the curve

$$\mathcal{R}_+(\phi_1, \varphi_1^<, \varphi_2^<) = \{(\phi_1, \phi_2) : \varphi_1 < \varphi_1^<, \phi_2 = r_+(\phi_1, \varphi_1^<, \varphi_2^<)\} \tag{64}$$

describes all rarefactions (ϕ_1, ϕ_2) satisfying the constraint (57). A straightforward argument shows that, for given $(\varphi_1^<, \varphi_2^<)$, the curve determined by $r_+(\cdot, \varphi_1^<, \varphi_2^<)$ decreases monotonically over its domain. Thus, given $\phi_2 = r_+(\phi_1, \varphi_1^<, \varphi_2^<)$, we may write $\phi_1 = \mathcal{R}_+^{-1}(\phi_2, \varphi_1^<, \varphi_2^<)$. Figure 7 depicts $\mathcal{R}_+$ and $\mathcal{R}_-$ for a generic choice of $(\varphi_1^<, \varphi_2^<)$.

Solutions of the Riemann Problem Given packing fractions $\varphi_1^<$ and $\varphi_2^<$ consistent with $0 < \varphi_1^< + \varphi_2^< < 1$, the curves

$$\mathcal{C}_-(\varphi_1, \varphi_1^<, \varphi_2^<) = \mathcal{D}_-(\varphi_1, \varphi_1^<, \varphi_2^<) \cup \mathcal{R}_-(\varphi_1, \varphi_1^<, \varphi_2^<) \tag{65}$$

and

$$\mathcal{C}_+(\varphi_1, \varphi_1^<, \varphi_2^<) = \mathcal{D}_+(\varphi_1, \varphi_1^<, \varphi_2^<) \cup \mathcal{R}_+(\varphi_1, \varphi_1^<, \varphi_2^<) \tag{66}$$

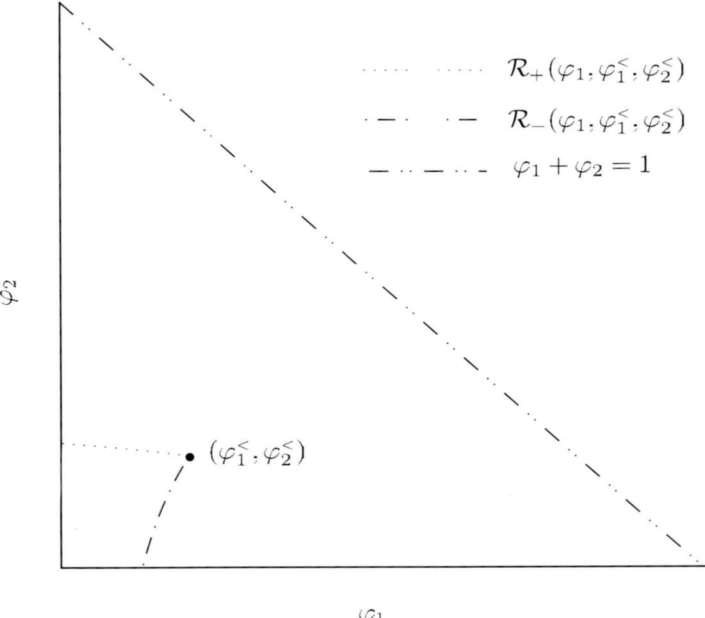

Fig. 7. The phase portrait of rarefactions (φ_1, φ_2) connecting to the right of generic point $(\varphi_1^<, \varphi_2^<)$

divide the region $\mathcal{A}$ into four subregions $\mathcal{A}_1(\varphi_1^<, \varphi_2^<)$, $\mathcal{A}_2(\varphi_1^<, \varphi_2^<)$, $\mathcal{A}_3(\varphi_1^<, \varphi_2^<)$, and $\mathcal{A}_4(\varphi_1^<, \varphi_2^<)$ (see Figure 8), with

$$\left.\begin{aligned}
\mathcal{A}_1(\varphi_1^<, \varphi_2^<) &= \{(\varphi_1, \varphi_2) \in \mathcal{A} : \varphi_1^< < \varphi_1, \\
&\qquad \mathcal{D}_-(\varphi_1, \varphi_1^<, \varphi_2^<) \le \varphi_2 \le \mathcal{D}_+(\varphi_1, \varphi_1^<, \varphi_2^<)\} \\
\mathcal{A}_2(\varphi_1^<, \varphi_2^<) &= \{(\varphi_1, \varphi_2) \in \mathcal{A} : \varphi_2^< > \varphi_2, \\
&\qquad \mathcal{R}_-^{-1}(\varphi_2, \varphi_1^<, \varphi_2^<) \le \varphi_1 \le \mathcal{D}_+^{-1}(\varphi_2, \varphi_1^<, \varphi_2^<)\}, \\
\mathcal{A}_3(\varphi_1^<, \varphi_2^<) &= \{(\varphi_1, \varphi_2) \in \mathcal{A} : \varphi_1^< > \varphi_1, \\
&\qquad \mathcal{R}_+(\varphi_1, \varphi_1^<, \varphi_2^<) \le \varphi_2 \le \mathcal{R}_-(\varphi_1, \varphi_1^<, \varphi_2^<)\}, \\
\mathcal{A}_4(\varphi_1^<, \varphi_2^<) &= \mathcal{A} \setminus (\mathcal{A}_1(\varphi_1^<, \varphi_2^<) \cup \mathcal{A}_2(\varphi_1^<, \varphi_2^<) \cup \mathcal{A}_3(\varphi_1^<, \varphi_2^<)).
\end{aligned}\right\} \quad (67)$$

Hence, if $(\varphi_1^>, \varphi_2^>)$ lies in any of the four regions $\mathcal{A}_i(\varphi_1^<, \varphi_2^<)$, $i = 1, 2, 3, 4$, we can find an admissible state (φ_1, φ_2) lying in $\mathcal{C}_-(\varphi_1, \varphi_1^<, \varphi_2^<)$ and also determine the nature of the solution (i.e., the presence of discontinuities or rarefactions) as follows: Consider the family $\Im = \{\mathcal{C}_+(\varphi_1, \bar{\varphi}_1, \bar{\varphi}_2) : (\bar{\varphi}_1, \bar{\varphi}_2) \in \mathcal{C}_-(\varphi_1, \varphi_1^<, \varphi_2^<)\}$ of curves. Since the region $\mathcal{A}$ is closed and bounded, the (φ_1, φ_2)-plane is covered univalently by $\Im$; i.e. through each point $(\varphi_1^>, \varphi_2^>)$, there passes exactly one curve $\mathcal{C}_+(\varphi_1, \bar{\varphi}_1, \bar{\varphi}_2)$ belonging to $\Im$.

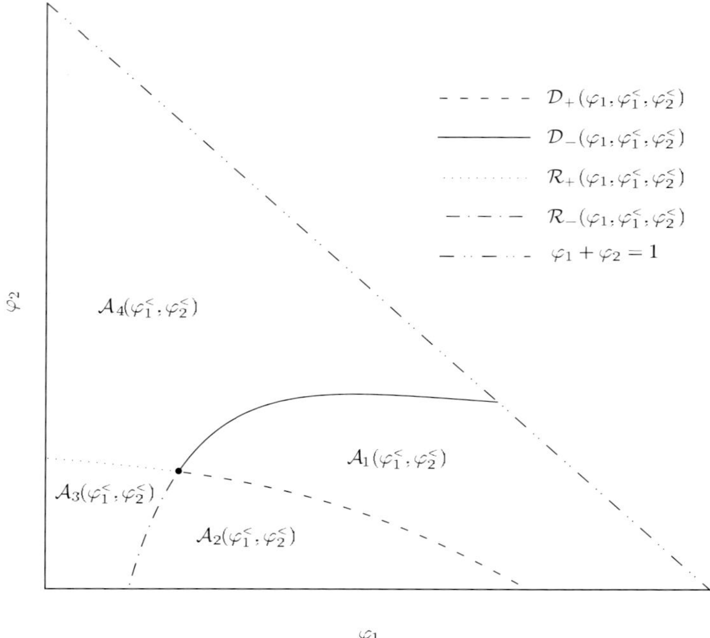

Fig. 8. The regions $\mathcal{A}_1(\varphi_1^<,\varphi_2^<)$, $\mathcal{A}_2(\varphi_1^<,\varphi_2^<)$, $\mathcal{A}_3(\varphi_1^<,\varphi_2^<)$, and $\mathcal{A}_4(\varphi_1^<,\varphi_2^<)$ corresponding to a generic point $(\varphi_1^<,\varphi_2^<)$

- Let $(\varphi_1^>,\varphi_2^>)$ lie in $\mathcal{A}_1(\varphi_1^<,\varphi_2^<)$. As shown in Fig. 9a, for each $(\varphi_1^>,\varphi_2^>)$, there is a unique point $(\bar\varphi_1,\bar\varphi_2)\in\mathcal{C}_-(\varphi_1,\varphi_1^<,\varphi_2^<)$ for which the curve $\mathcal{C}_+(\varphi_1,\bar\varphi_1,\bar\varphi_2)$ belongs to $\Im$ and passes through $(\varphi_1^>,\varphi_2^>)$. However, in $\mathcal{A}_1(\varphi_1^<,\varphi_2^<)$,
$$\mathcal{C}_-(\varphi_1,\varphi_1^<,\varphi_2^<)=\mathcal{D}_-(\varphi_1,\varphi_1^<,\varphi_2^<)$$
and
$$\mathcal{C}_+(\varphi_1,\bar\varphi_1,\bar\varphi_2)=\mathcal{D}_+(\varphi_1,\bar\varphi_1,\bar\varphi_2).$$
Hence, $(\bar\varphi_1,\bar\varphi_2)$ is connected to $(\varphi_1^<,\varphi_2^<)$ on the right by a discontinuity that satisfies the constraint (47). Similarly, $(\varphi_1^>,\varphi_2^>)$ is connected to $(\bar\varphi_1,\bar\varphi_2)$ on the right by a discontinuity that satisfies the constraint (48). The problem is completely solved once $(\bar\varphi_1,\bar\varphi_2)$ is determined. To this purpose, we solve $\varphi_2^> = d_+(\varphi_1^>,\bar\varphi_1,d_-(\bar\varphi_1,\varphi_1^<,\varphi_2^<))$ for $\bar\varphi_1$. Given $\bar\varphi_1$, we have $\bar\varphi_2=d_-(\bar\varphi_1,\varphi_1^<,\varphi_2^<)$.
- Let $(\varphi_1^>,\varphi_2^>)$ lie in $\mathcal{A}_2(\varphi_1^<,\varphi_2^<)$. As shown in Fig. 9b, for each $(\varphi_1^>,\varphi_2^>)$, there is a unique point $(\bar\varphi_1,\bar\varphi_2)\in\mathcal{C}_-(\varphi_1,\varphi_1^<,\varphi_2^<)$ for which the curve $\mathcal{C}_+(\varphi_1,\bar\varphi_1,\bar\varphi_2)$ belongs to $\Im$ and passes through $(\varphi_1^>,\varphi_2^>)$. However, in $\mathcal{A}_2(\varphi_1^<,\varphi_2^<)$,
$$\mathcal{C}_-(\varphi_1,\varphi_1^<,\varphi_2^<)=\mathcal{R}_-(\varphi_1,\varphi_1^<,\varphi_2^<)$$
and
$$\mathcal{C}_+(\varphi_1,\bar\varphi_1,\bar\varphi_2)=\mathcal{D}_+(\varphi_1,\bar\varphi_1,\bar\varphi_2).$$

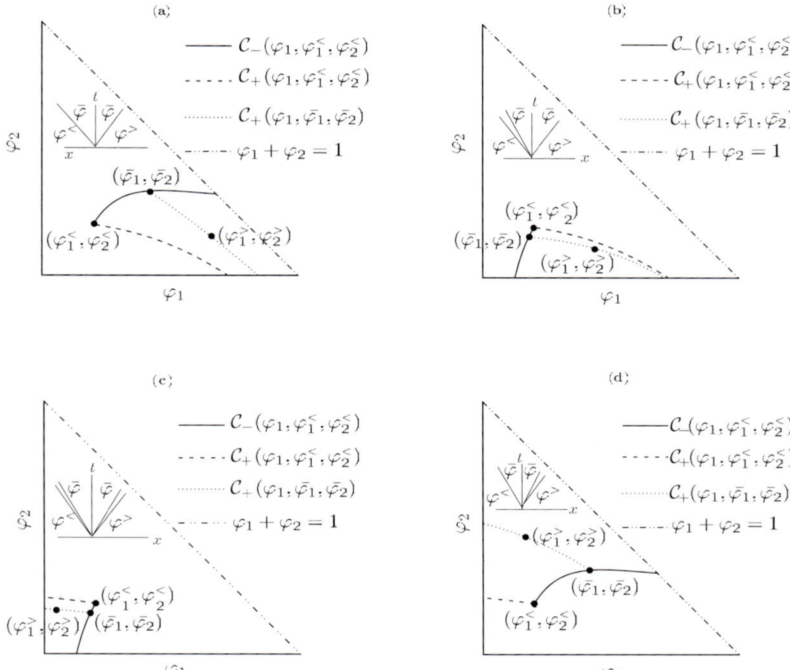

Fig. 9. Solution of the Riemann problem when $(\varphi_1^>,\varphi_2^>)$ lies in **(a)** $\mathcal{A}_1(\varphi_1^<,\varphi_2^<)$, **(b)** $\mathcal{A}_2(\varphi_1^<,\varphi_2^<)$, **(c)** $\mathcal{A}_3(\varphi_1^<,\varphi_2^<)$, and **(d)** $\mathcal{A}_4(\varphi_1^<,\varphi_2^<)$

Hence, $(\bar\varphi_1,\bar\varphi_2)$ is connected to $(\varphi_1^<,\varphi_2^<)$ on the right by a rarefaction that satisfies the constraint (56). Similarly, $(\varphi_1^>,\varphi_2^>)$ is connected to $(\bar\varphi_1,\bar\varphi_2)$ on the right by a discontinuity that satisfies the constraint (48). The problem is completely solved once $(\bar\varphi_1,\bar\varphi_2)$ is determined. To this purpose, we solve $\varphi_2^> = d_+(\varphi_1^>,\bar\varphi_1,r_-(\bar\varphi_1,\varphi_1^<,\varphi_2^<))$ for $\bar\varphi_1$. Given $\bar\varphi_1$, we have $\bar\varphi_2 = r_-(\bar\varphi_1,\varphi_1^<,\varphi_2^<)$.

- Let $(\varphi_1^>,\varphi_2^>)$ lie in $\mathcal{A}_3(\varphi_1^<,\varphi_2^<)$. As shown in Fig. 9c, for each $(\varphi_1^>,\varphi_2^>)$, there is a unique point $(\bar\varphi_1,\bar\varphi_2)\in\mathcal{C}_-(\varphi_1,\varphi_1^<,\varphi_2^<)$ for which the curve $\mathcal{C}_+(\varphi_1,\bar\varphi_1,\bar\varphi_2)$ belongs to $\Im$ and passes through $(\varphi_1^>,\varphi_2^>)$. However, in $\mathcal{A}_3(\varphi_1^<,\varphi_2^<)$,

$$\mathcal{C}_-(\varphi_1,\varphi_1^<,\varphi_2^<) = \mathcal{R}_-(\varphi_1,\varphi_1^<,\varphi_2^<)$$

and

$$\mathcal{C}_+(\varphi_1,\bar\varphi_1,\bar\varphi_2) = \mathcal{R}_+(\varphi_1,\bar\varphi_1,\bar\varphi_2).$$

Hence, $(\bar\varphi_1,\bar\varphi_2)$ is connected to $(\varphi_1^<,\varphi_2^<)$ on the right by a rarefaction that satisfies the constraint (56). Similarly, $(\varphi_1^>,\varphi_2^>)$ is connected to $(\bar\varphi_1,\bar\varphi_2)$ on the right by a rarefaction that satisfies the constraint (57). The problem is completely solved once $(\bar\varphi_1,\bar\varphi_2)$ is determined. To this purpose, we solve $\varphi_2^> = r_+(\varphi_1^>,\bar\varphi_1,r_-(\bar\varphi_1,\varphi_1^<,\varphi_2^<))$ for $\bar\varphi_1$. Given $\bar\varphi_1$, we have $\bar\varphi_2 = r_-(\bar\varphi_1,\varphi_1^<,\varphi_2^<)$.
- Let $(\varphi_1^>,\varphi_2^>)$ lie in $\mathcal{A}_4(\varphi_1^<,\varphi_2^<)$. As shown in Fig. 9d , for each $(\varphi_1^>,\varphi_2^>)$, there is a unique point $(\bar\varphi_1,\bar\varphi_2)\in\mathcal{C}_-(\varphi_1,\varphi_1^<,\varphi_2^<)$ for which the curve $\mathcal{C}_+(\varphi_1,\bar\varphi_1,\bar\varphi_2)$

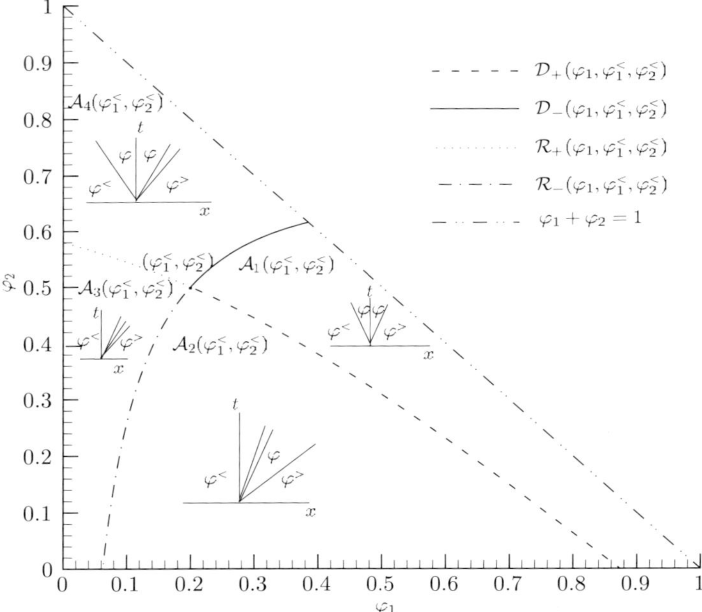

Fig. 10. The states (φ_1, φ_2) connecting to the right of $(\varphi_1^<, \varphi_2^<) = (0.2, 0.5)$. Here, $(\alpha_1, \alpha_2) = (0.4, 0.1)$

belongs to $\mathfrak{I}$ and passes through $(\varphi_1^>, \varphi_2^>)$. However, in $\mathcal{A}_4(\varphi_1^<, \varphi_2^<)$,

$$\mathcal{C}_-(\varphi_1, \varphi_1^<, \varphi_2^<) = \mathcal{D}_-(\varphi_1, \varphi_1^<, \varphi_2^<)$$

and

$$\mathcal{C}_+(\varphi_1, \bar{\varphi}_1, \bar{\varphi}_2) = \mathcal{R}_+(\varphi_1, \bar{\varphi}_1, \bar{\varphi}_2).$$

Hence, $(\bar{\varphi}_1, \bar{\varphi}_2)$ is connected to $(\varphi_1^<, \varphi_2^<)$ on the right by a discontinuity that satisfies the constraint (47). Similarly, $(\varphi_1^>, \varphi_2^>)$ is connected to $(\bar{\varphi}_1, \bar{\varphi}_2)$ on the right by a rarefaction that satisfies the constraint (57). The problem is completely solved once $(\bar{\varphi}_1, \bar{\varphi}_2)$ is determined. To this purpose, we solve $\varphi_2^> = r_+(\varphi_1^>, \bar{\varphi}_1, d_-(\bar{\varphi}_1, \varphi_1^<, \varphi_2^<))$ for $\bar{\varphi}_1$. Once $\bar{\varphi}_1$ is determined, we use $\bar{\varphi}_2 = d_-(\bar{\varphi}_1, \varphi_1^<, \varphi_2^<)$ to find $\bar{\varphi}_2$.

The presence of discontinuities or rarefactions depends on the rate in which voids are generated and the mobilities of particles of type 1 and type 2. In the region $\mathcal{A}_1(\varphi_1^<, \varphi_2^<)$, the rates at which particles of type 1 and type 2 diffuse to fill voids exceeds the rate in which voids are generated, resulting in a solution involving discontinuities. However, in $\mathcal{A}_2(\varphi_1^<, \varphi_2^<)$, $\mathcal{A}_3(\varphi_1^<, \varphi_2^<)$ or $\mathcal{A}_4(\varphi_1^<, \varphi_2^<)$, the rates at which voids are generated exceeds the rates at which particles of either type diffuse. This results in solutions involving rarefactions only (as found in $\mathcal{A}_3(\varphi_1^<, \varphi_2^<)$) or combinations of rarefactions and discontinuities (as found in $\mathcal{A}_2(\varphi_1^<, \varphi_2^<)$ and $\mathcal{A}_4(\varphi_1^<, \varphi_2^<)$).

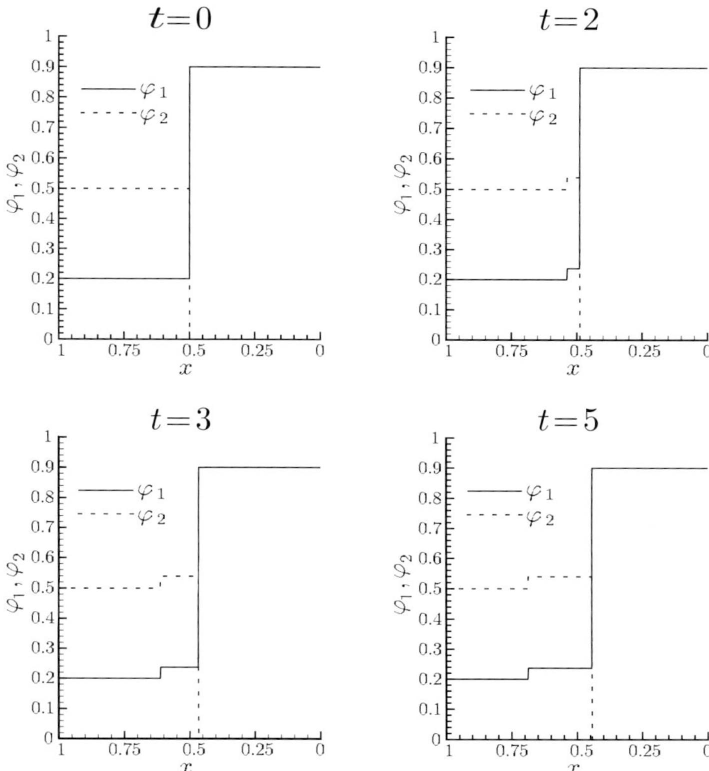

Fig. 11. Solution of the Riemann problem when $(\varphi_1^>, \varphi_2^>) = (0.9, 0.0)$ lies in $\mathcal{A}_1(\varphi_1^<, \varphi_2^<)$

As an illustration, we present numerical results generated by CLAWPACK for the case $(\alpha_1, \alpha_2) = (0.4, 0.1)$ and $(\varphi_1^<, \varphi_2^<) = (0.2, 0.5)$. Figure 10 describes all possible states connecting to the right of $(0.2, 0.5)$. The solution of the Riemann problem (39) for the cases when $(\varphi_1^>, \varphi_2^>)$ lies in $\mathcal{A}_1(\varphi_1^<, \varphi_2^<)$, $\mathcal{A}_2(\varphi_1^<, \varphi_2^<)$, $\mathcal{A}_3(\varphi_1^<, \varphi_2^<)$ or $\mathcal{A}_4(\varphi_1^<, \varphi_2^<)$ is shown in Figs. 11, 12, 13, and 14, respectively.

Figure 15 depicts the influence of the dimensionless effective mobility α_1, in the case $(\varphi_1^<, \varphi_2^<) = (0.2, 0.2)$. As we increase α_1, the area of the region $\mathcal{A}_1(\varphi_1^<, \varphi_2^<)$, where the solution is described only by discontinuities (or where the rate at which the particles diffuse exceeds the rate at which the volume fraction of the voids increases), diminishes. Thus, as the size of particles of type 1 decreases, solutions involving rarefactions become more likely.

6.2 Application

Here, we use the results concerning the Riemann problem to establish the uniqueness of the solution given in Sect. 5 for the particular case $K = 2$ — where the

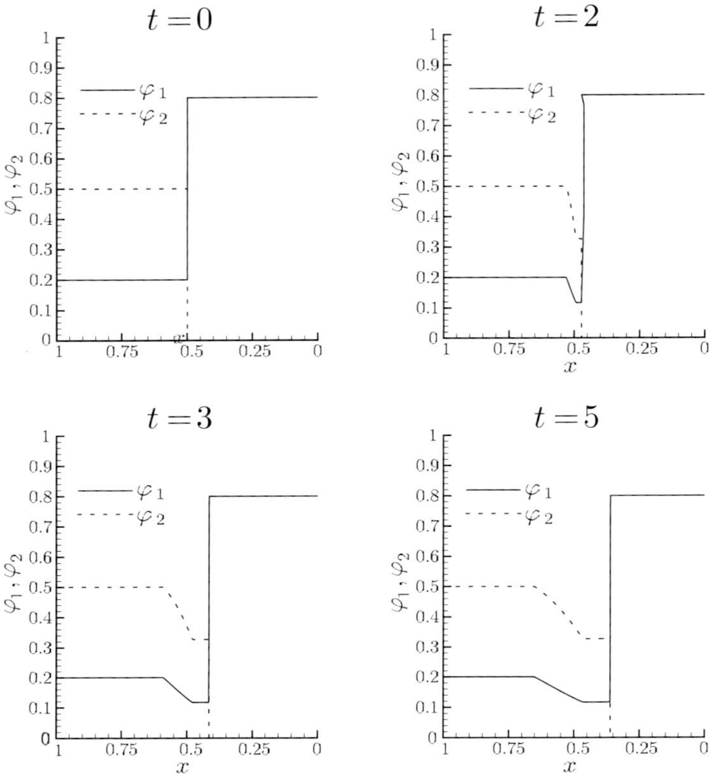

Fig. 12. Solution of the Riemann problem when $(\varphi_1^>, \varphi_2^>) = (0.8, 0.0)$ lies in $\mathcal{A}_2(\varphi_1^<, \varphi_2^<)$

mixture contains particles of only two different sizes. In this case, the solution is the union of:

- a uniform state bounded by a free surface $\mathcal{S}_{\text{free}}$ and a segregation shock $\mathcal{S}_{\text{seg}}$ and involving only large particles with packing fraction

$$\varphi_2 = -\frac{1}{2}\left(\frac{\alpha_1(1-\mathring{\varphi}_1-\mathring{\varphi}_2)}{\alpha_2} - 1\right) + \frac{1}{2}\sqrt{\left(\frac{\alpha_1(1-\mathring{\varphi}_1-\mathring{\varphi}_2)}{\alpha_2} - 1\right)^2 + 4\left(\frac{\alpha_1}{\alpha_2} - 1\right)\mathring{\varphi}_2(1-\mathring{\varphi}_1-\mathring{\varphi}_2)}; \quad (68)$$

- a mixed uniform state bounded by the segregation shock $\mathcal{S}_{\text{seg}}$ and a compaction shock $\mathcal{S}_{\text{com}}$ and involving small and large particles with packing fractions $\varphi_1 = \mathring{\varphi}_1$ and $\varphi_2 = \mathring{\varphi}_2$;

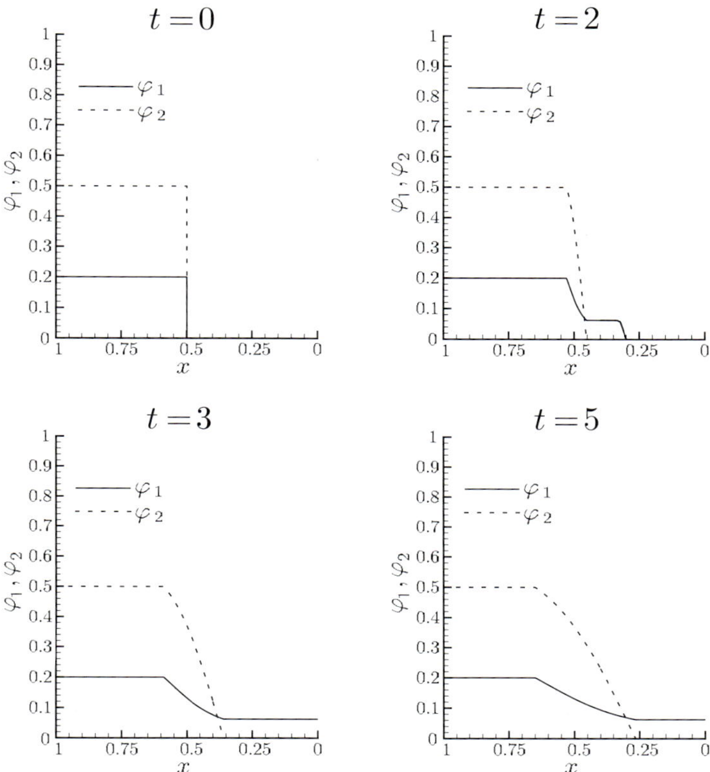

Fig. 13. Solution of the Riemann problem when $(\varphi_1^>, \varphi_2^>) = (0.0, 0.0)$ lies in $\mathcal{A}_3(\varphi_1^<, \varphi_2^<)$

- a mixed uniform state closely-packed bounded by the container base and the compaction shock $\mathcal{S}_{\rm com}$ until the time T_1, by the container base and $\mathcal{S}_{\rm seg}$ after T_1, and involving small and large particles with packing fractions

$$\varphi_1^* = \mathring{\varphi}_1 + \frac{(1 - \mathring{\varphi}_1 - \mathring{\varphi}_2)\alpha_1 \mathring{\varphi}_1}{\alpha_1 \mathring{\varphi}_1 + \alpha_2 \mathring{\varphi}_2}, \tag{69}$$

and

$$\varphi_2^* = \mathring{\varphi}_2 + \frac{(1 - \mathring{\varphi}_1 - \mathring{\varphi}_2)\alpha_2 \mathring{\varphi}_2}{\alpha_1 \mathring{\varphi}_1 + \alpha_2 \mathring{\varphi}_2}; \tag{70}$$

- a uniform closely-packed state bounded by a compaction shock $\mathcal{S}_{\rm com}$ and the segregation shock $\mathcal{S}_{\rm seg}$ until the time T_2, by the free surface and $\mathcal{S}_{\rm seg}$ after T_2 and involving only large particles with packing fraction $\varphi_1 = 1$.

Prior to the instant $T_1 = 1/(\alpha_1(1 - \mathring{\varphi}_2) + \alpha_2 \mathring{\varphi}_2)$, $V_{\rm seg} = \alpha_1(1 - \mathring{\varphi}_1 - \mathring{\varphi}_2)$, $V_{\rm com} = -(\alpha_1 \mathring{\varphi}_1 + \alpha_2 \mathring{\varphi}_2)$ and $V_{\rm free} = \alpha_1(1 - \varphi_2)$, with φ_2 given by (68). Between the instants

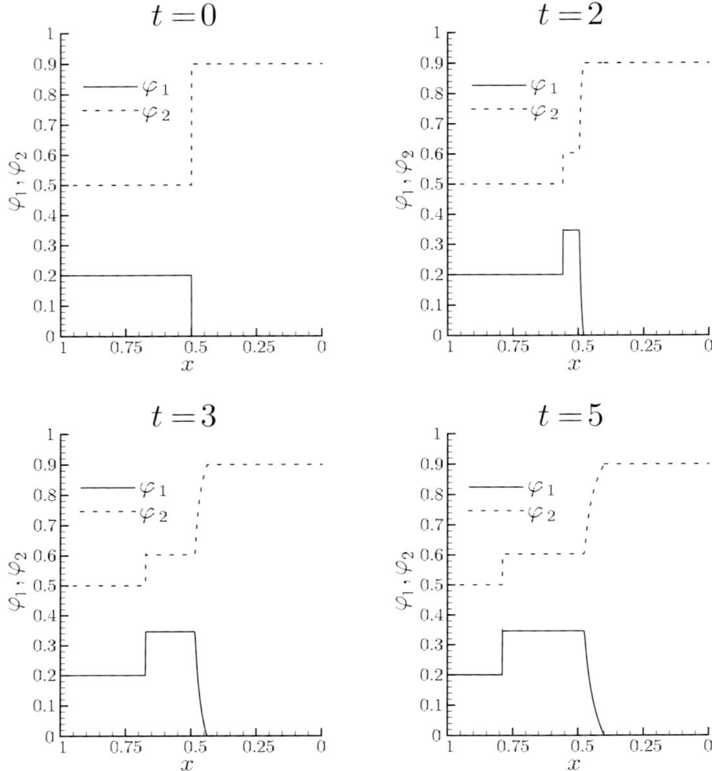

Fig. 14. Solution of the Riemann problem when $(\varphi_1^>, \varphi_2^>) = (0.0, 0.9)$ lies in $\mathcal{A}_4(\varphi_1^<, \varphi_2^<)$

T_1 and $T_2 = [\frac{\alpha_1}{\alpha_2}(1 - \mathring{\varphi}_1 - \mathring{\varphi}_2)]/[\alpha_1(1 - \mathring{\varphi}_2) + \alpha_2\mathring{\varphi}_2]$, $V_{\rm com} = -\alpha_2\varphi_2$, with φ_2 given once again by (68). Subsequent to T_2, $V_{\rm seg} = 0$.

Concerning this solution, we have the following

Theorem 1 *The solution delineated above solves uniquely the problem*

$$\left.\begin{aligned}\frac{\partial\varphi_1}{\partial t} &= \alpha_1\frac{\partial}{\partial x}\Big(\varphi_1(1-\varphi_1-\varphi_2)\Big),\\ \frac{\partial\varphi_2}{\partial t} &= \alpha_2\frac{\partial}{\partial x}\Big(\varphi_2(1-\varphi_1-\varphi_2)\Big),\end{aligned}\right\} \tag{71}$$

subject to the initial conditions

$$\varphi_k(x,0) = \begin{cases} 0 & \textit{if } x > 1,\\ \mathring{\varphi}_k & \textit{if } 0 < x \leq 1,\end{cases} \tag{72}$$

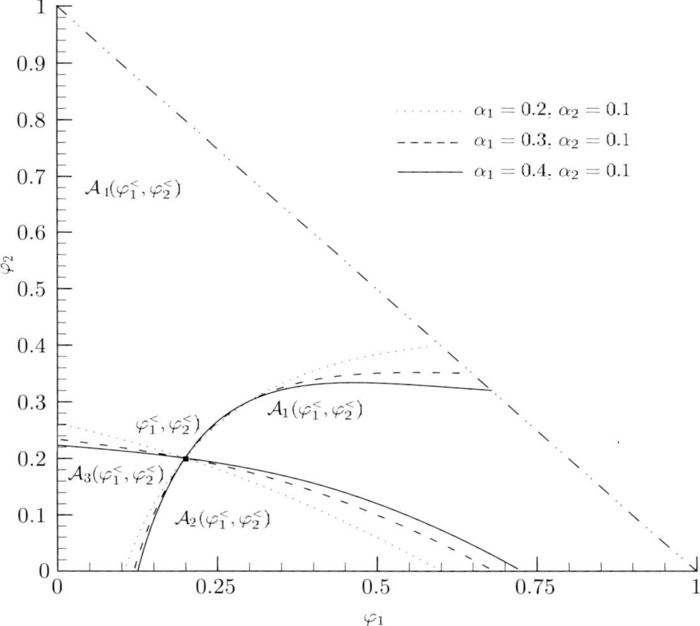

Fig. 15. The effect of varying α_1 on the states (φ_1, φ_2) connecting to the right of $(\varphi_1^<, \varphi_2^<) = (0.2, 0.2)$

the jump conditions

$$\left.\begin{aligned} [\![\varphi_1]\!] V_{\text{shock}} &= -\alpha_1 [\![\varphi_1(1 - \varphi_1 - \varphi_2)]\!], \\ [\![\varphi_2]\!] V_{\text{shock}} &= -\alpha_2 [\![\varphi_2(1 - \varphi_1 - \varphi_2)]\!], \end{aligned}\right\} \tag{73}$$

across any shock (moving with velocity V_{shock}), the condition

$$\alpha_k(1 - \varphi_1 - \varphi_2) = V_{\text{free}} \tag{74}$$

for particles of type k at a free surface (moving with velocity V_{free}), and condition

$$\varphi_1 + \varphi_2 = 1 \tag{75}$$

at the base.

Proof. We first show that rarefactions may not issue from the points $(x, t) = (1, 0)$ and $(x, t) = (0, 0)$ corresponding to the initial free surface and to the base. Suppose that, at $t = 0$, a rarefaction occurs at $x = 1$. Then

$$\lambda_{\pm}(0, 0) < \lambda_{\pm}(\phi_1(\xi), \phi_2(\xi)), \qquad \xi = \frac{x}{t}. \tag{76}$$

However, from (44), $\lambda_-(0, 0) = \alpha_2$ and $\lambda_-(\phi_1, \phi_2) < \alpha_2(1 - \phi_1 - 2\phi_2)$. Thus, by (76), $\alpha_2 < \alpha_2(1 - \phi_1 - 2\phi_2)$ or $\phi_1 + 2\phi_2 < 0$, which cannot occur since ϕ_1 and ϕ_2 must be in

$\mathcal{A}$. Thus, the first characteristic does not give rise to a rarefaction. Similarly, from (44), $\lambda_+(0,0) = \alpha_1$ and $\lambda_+(\phi_1,\phi_2) < \alpha_1$. Thus, by (76), $\alpha_1 = \lambda_2(\phi_1,\phi_2) < \alpha_1$, which cannot hold. Hence, by contradiction, no rarefactions may emanate from $(x,t) = (1,0)$. At the base ($x = 1$), the state (φ_1,φ_2) connecting to the right of the state $(\mathring{\varphi}_1,\mathring{\varphi}_2)$ satisfies the relation $1 - \varphi_1 - \varphi_2 = 0$. Such a state must lie on the intersection of $\mathcal{D}_-(\varphi_1,\mathring{\varphi}_1,\mathring{\varphi}_2)$ with the line $\varphi_1 + \varphi_2 = 1$. Hence, the characteristic at $x = 1$ satisfies only (47) and is not a rarefaction. Thus, by contradiction, no rarefactions may emanate from $(x,t) = (0,0)$.

At the point $(x,t) = (X_1,T_1)$, with $X_1 = (\alpha_1(1-\mathring{\varphi}_1-\mathring{\varphi}_2))/[\alpha_1(1-\mathring{\varphi}_2)+\alpha_2\mathring{\varphi}_2]$, the segregation shock S_{seg} and the compaction shock S_{com} meet. To understand the nature of the solution at $t = T_1$, we solve for (71) subject to the conditions $(\varphi_1(x < X_1, t = T_1), \varphi_2(x < X_1, t = T_1)) = (0,\varphi_2)$ and $(\varphi_1(x \geq X_1, t = T_1), \varphi_2(x \geq X_1, t = T_1)) = (\varphi_1^*,\varphi_2^*)$ where φ_2 is given as in (68) and $(\varphi_1^*,\varphi_2^*)$ as in (69)–(70). If $(\bar{\varphi}_1,\bar{\varphi}_2)$ denotes an admissible state connecting to the right of $(0,\varphi_2)$ by a rarefaction, then (62) together with constraints (56) and (57) imply that $\varphi_1 < 0$. This falls outside the domain $\mathcal{A}$. Hence, a rarefaction cannot emanate from $(x,t) = (X_1,T_1)$.

At the point $(x,t) = (X_2,T_2)$, with $X_2 = (\alpha_2(1-\varphi_2)X_1)/\alpha_1$ and φ_2 given by (68), the free surface S_{free} and the compaction shock S_{com} meet. To understand the nature of the solution at $t = T_2$, we solve for (71) subject to the conditions $(\varphi_1(x < X_2, t = T_2), \varphi_2(x < X_2, t = T_2)) = (0,0)$ and $(\varphi_1(x \geq X_2, t = T_2), \varphi_2(x \geq X_2, t = T_2)) = (0,1)$. Following a proof identical to that used in the case of the point $(x,t) = (1,0)$, we conclude that a rarefaction cannot emanate from (X_2,T_2).

Hence the solution cannot involve rarefactions. To complete the proof, we rely on the analysis of the Riemann problem detailed above.

- At the point $(x,t) = (1,0)$, we can regard $(\varphi_1^<,\varphi_2^<) = (0,0)$ and $(\varphi_1^>,\varphi_2^>) = (\mathring{\varphi}_1,\mathring{\varphi}_2)$. Since the admissible state (φ_1,φ_2) lies in $D_-(\varphi_1,\varphi_1^< = 0,\varphi_2^< = 0)$, it follows that this state is connected to $(\varphi_1^<,\varphi_2^<) = (0,0)$ by a discontinuity (the free surface) that satisfies the constraint (47) together with the jump conditions

$$\left.\begin{aligned} \alpha_1\varphi_1(1-\varphi_1-\varphi_2) &= V_{\text{free}}\varphi_1, \\ \alpha_2\varphi_2(1-\varphi_1-\varphi_2) &= V_{\text{free}}\varphi_2. \end{aligned}\right\} \tag{77}$$

  This implies that $\varphi_1 = 0$ and that $V_{\text{free}} = \alpha_2(1-\varphi_2)$. To determine φ_2, we note that $(\varphi_1^>,\varphi_2^>) = (\mathring{\varphi}_1,\mathring{\varphi}_2)$ lies in $D_+(\varphi_1^> = \mathring{\varphi}_1, 0, \varphi_2)$. Hence, the state $(0,\varphi_2)$ is connected to $(\varphi_1^>,\varphi_2^>) = (\mathring{\varphi}_1,\mathring{\varphi}_2)$ by a discontinuity (the segregation shock) satisfying the constraint (48) along with the jump conditions

$$\left.\begin{aligned} \alpha_1\varphi_1(1-\varphi_1-\varphi_2) - \alpha_1\mathring{\varphi}_1(1-\mathring{\varphi}_1-\mathring{\varphi}_2) &= V_{\text{seg}}(\varphi_1-\mathring{\varphi}_1), \\ \alpha_2\varphi_2(1-\varphi_1-\varphi_2) - \alpha_2\mathring{\varphi}_2(1-\mathring{\varphi}_1-\mathring{\varphi}_2) &= V_{\text{seg}}(\varphi_2-\mathring{\varphi}_2). \end{aligned}\right\} \tag{78}$$

  The above conditions imply that $V_{\text{seg}} = \alpha_1(1-\mathring{\varphi}_1-\mathring{\varphi}_2)$ and that φ_2 is given by (68). Since $\alpha_1 > \alpha_2$ and $1 - \mathring{\varphi}_1 - \mathring{\varphi}_2 > 0$, it follows that (68) has only one positive real root φ_2 and that $\varphi_2 < 1$. Hence, the root of (68) yields a packing fraction.
- At the point $(x,t) = (0,0)$, the state $(\varphi_1^*,\varphi_2^*)$ connecting to the right of $(\mathring{\varphi}_1,\mathring{\varphi}_2)$ satisfies the relation $\varphi_1^* + \varphi_2^* = 1$. However, from (44) it follows that $\lambda_-(\varphi_1^*,\varphi_2^*) = 0$ and $\lambda_+(\varphi_1^*,\varphi_2^*) = -(\alpha_1\varphi_1^*+\alpha_2\varphi_2^*)$. Since one of the eigenvalues

is constant, a contact discontinuity (the compaction shock) emanates from the base. From the jump conditions

$$\left.\begin{aligned} \alpha_1\mathring{\varphi}_1(1-\mathring{\varphi}_1-\mathring{\varphi}_2) &= -V_{\rm com}(\varphi_1^*-\mathring{\varphi}_1), \\ \alpha_2\mathring{\varphi}_2(1-\mathring{\varphi}_1-\mathring{\varphi}_2) &= -V_{\rm com}(\varphi_2^*-\mathring{\varphi}_2), \end{aligned}\right\} \tag{79}$$

and the condition $\varphi_1^* + \varphi_2^* = 1$, it follows that $V_{\rm com} = -(\alpha_1\mathring{\varphi}_1 + \alpha_2\mathring{\varphi}_2)$ and $(\varphi_1^*, \varphi_2^*)$ is given as in (69)–(70).

- At the point $(x,t) = (X_1, T_1)$, we can regard $(\varphi_1^<, \varphi_2^<) = (0, \varphi_2)$ and $(\varphi_1^>, \varphi_2^>) = (\varphi_1^*, \varphi_2^*)$, where $\varphi_1^* + \varphi_2^* = 1$. Since the admissible state $(\bar\varphi_1, \bar\varphi_2)$ lies in $\mathcal{D}_-(\bar\varphi_1, 0, \varphi_2^< = \varphi_2)$, this state must be connected to $(\varphi_1^<, \varphi_2^<) = (0, \varphi_2)$ by a discontinuity (a compaction shock) that satisfies the constraint (47) along with the jump conditions

$$\left.\begin{aligned} \alpha_1\bar\varphi_1(1-\bar\varphi_1-\bar\varphi_2) &= V_{\rm com}\bar\varphi_1, \\ \alpha_2\bar\varphi_2(1-\bar\varphi_1-\bar\varphi_2) - \alpha_2\varphi_2(1-\varphi_2) &= V_{\rm com}(\bar\varphi_2-\varphi_2). \end{aligned}\right\} \tag{80}$$

  This implies that $\bar\varphi_1 = 0$ and that $V_{\rm com} = -\alpha_2\varphi_2$. To determine $\bar\varphi_2$, we note that $(\varphi_1^>, \varphi_2^>) = (\varphi_1^*, \varphi_2^*)$ lies in $\mathcal{D}_+(\varphi_1^*, 0, \bar\varphi_2)$. Hence, the state $(0, \varphi_2)$ is connected to $(\varphi_1^>, \varphi_2^>) = (\varphi_1^*, \varphi_2^*)$ by a discontinuity (the segregation shock) that satisfies the constraint (48) along with the jump conditions

$$\left.\begin{aligned} \alpha_1\varphi_1^*(1-\varphi_1^*-\varphi_2^*) &= -V_{\rm seg}\varphi_1^*, \\ \alpha_2\bar\varphi_2(1-\bar\varphi_1-\bar\varphi_2) - \alpha_2\varphi_2^*(1-\varphi_1^*-\varphi_2^*) &= V_{\rm seg}(\bar\varphi_2-\varphi_2^*). \end{aligned}\right\} \tag{81}$$

  The above conditions imply that $V_{\rm seg} = 0$ and $\varphi_2 = 1$.
- At the point $(x,t) = (X_2, T_2)$, we can regard $(\varphi_1^<, \varphi_2^<) = (0, 0)$ and $(\varphi_1^>, \varphi_2^>) = (0, 1)$. Proceeding exactly as in the case of the point $(x,t) = (0,0)$, it follows that the solution can be described by a contact discontinuity (the free surface) separating the free surface from the compacted layer where $(\varphi_1, \varphi_2) = (0, 1)$.

7 Discussion

Using a model proposed by Fried, Gurtin and Hutter [6], we have studied some simple aspects of size-based segregation occurring under the action of gravity. For a flow with constant velocity, the model reduces to a system of one-dimensional conservation laws. We have presented a solution for a particular initial-value problem involving a mixture of particles of $K \geq 2$ sizes. This solution shows segregation and compaction by particle size. At steady state, this solution consists of layers of closely packed particles, with the upper-most layer consisting only of particles of the largest size. Numerical solutions for $K = 3$ particles were computed using CLAWPACK. Relying on methods developed by Smoller [30], we established the uniqueness of the solution for the case $K = 2$ of a mixture of small and large particles. The issue of uniqueness in the case $K > 2$ remains open.

The problem considered here is idealised in the sense that the flow field is trivial. Furthermore, we have ignored variations of the particulate mobilities with the strain-rate. As discussed by Fried, Gurtin and Hutter [6], such variations should rule out particle diffusion in the absence of sufficient agitation. Under flow conditions more general than those considered here, strain-rate dependence of the mobilities would allow for the existence of regions in which particles would simply move with the mixture. Compaction and segregation would thus be confined to regions of sufficiently high agitation.

References

1. Ancey, C., Evesque, P. and Coussot, P. (1996) Motion of a single bead on a bead row: theoretical investigations. J. Phys. I **6**, 725–751
2. Arnarson, B.Ö., Jenkins J. T. (2000) Particle segregation in the context of species momentum balances. In *Traffic and Granular Flow '99* (eds. D. Helberg, H. J. Herrmann, M. Schreckenberg and D. E. Wolf), Springer, 481–487
3. Arnarson, B.Ö., Willits, J.T. (1998) Thermal diffusion in binary mixtures of smooth, nearly elastic spheres with and without gravity. Phys. Fluids **10**, 1324–1328
4. Bagnold, R.A. (1954) Experiments on a gravity-free dispersion of large solid sphere in a newtonian fluid under shear. Proc. R. Soc. Lond. A **225**, 49–63
5. Dolgunin, V.N., Ukolov, A.A. (1995) Segregation modeling of particle rapid flow. Powder Tech. **83**, 95–103
6. Fried, E., Gurtin, M.E., and Hutter, K. (Submitted for publication) A void based description of Compaction and Segregation in flowing granular materials
7. Gallas, J.A.C., Herrmann, H.J., Pöschel, T. and Sokolowski, S. (1996) Molecular dynamics simulation of size segregation in three dimensions. J. Stat. Phys. **82**, 443–450
8. Hanes, D.M., Inman, D.L. (1985) Observations of rapidly flowing granular-fluid materials. J. Fluid Mech. **150**, 357–380
9. Herrmann, H.J., Luding, S. (1998) Modeling granular material on the computer. Continuum Mech. Thermodyn. **10**, 189–231
10. Hirshfeld, D., Rapaport, D.C. (1997) Molecular dynamics of grain segregation in sheared flow. Phys. Rev. E **56**, 2012–2018
11. Hong, D.C., Quinn, P.V., and Luding, S. (2001) Reverse Brazil nut problem: Competition between percolation and Condensation. Phys. Rev. Lett. **86**, 3423–3426
12. Hutter, K., Rajagopal, K.R. (1994) On flows of granular materials. Continuum Mech. Thermodyn. **6**, 81–139
13. Jenkins, J.T. (1998) Particle segregation in collisional flows of inelastic spheres. In *Physics of Dry Granular Media* (eds. H.J. Herrmann, J.P. Hovig and S. Luding), Kluwer, 645–658
14. Jenkins, J.T. , Mancini, F. (1989) Kinetic theory for mixtures of smooth, nearly elastic spheres. Phys. Fluids A **1**, 2050–2057
15. Jenkins, J.T., Savage, S.B. (1983) Theory for the rapid flow of identical, smooth, nearly elastic spherical particles. J. Fluid Mech. **130**, 187–202
16. Johnson, P.C., Nott, P. and Jackson, R. (1990) Frictional-Collisional equations of motion for particulate flows and their application to chute flows. J. Fluid Mech. **210**, 501–535
17. Lax, P. (1957) Hyperbolic systems of conservation laws, II. Comm. Pure Appl. Math. **10**, 537–566

18. LeVeque, R. (1994) http://www. amath. washington. edu/~claw
19. Lun, C.K.K., Savage, S.B., Jeffrey, D.J. and Chepurniy, N. (1984) Kinetic theory for granular flow: Inelastic particles in couette flow and slightly inelastic particles in a general flow field. J. Fluid Mech. **146**, 223–256
20. Makse, H.A., Kurchan, J. (2002) Testing the thermodynamic approach to granular matter with a numerical model of a decisive experiment. Nature **415**, 614–617
21. Moaker, M., Shinbrot, T. and Muzzio, F.J. (2000) Experimentally validated computations of flow, mixing and segregation of non–cohesive grains in 3D tumbling blenders. Powder Tech. **109**, 58–71
22. Ohtsuki, T., Takemoto, Y., Hata, T., Kawai, S. and Hayashi, A. (1993) Molecular–Dynamics study of cohesionless granular materials— Size segregation by shaking. Int. J. Mod. Phys. B **7**, 1865–1872
23. Pöschel, T., Schwager, T. and Salreena, C. (2000) Onset of fluidization in vertically shaken granular material. Phys. Rev. E **62**, 1361–1367
24. Ristow, G.H. 2000 *Pattern Formation in Granular Materials*, Springer
25. Roe, P.L. (1981) Approximate Riemann solvers, parameter vectors, and difference schemes. J. Comp. Phys. **43**, 357–372
26. Savage, S.B. (1984) Mechanics of rapid granular flow. Adv. Appl. Mech. **24**, 289–366
27. Shinbrot, T., Zeggio, M., Muzzio, F.J. (2001) Computational approaches to granular segregation in tumbling blenders. Powder Tech. **116**, 224–231
28. Shoichi, S. (1998) Molecular-Dynamics simulations of granular axial segregation in a rotating cylinder. Mod. Phys. Lett. B **12**, 115–122
29. Smith, L., Baxter, J., Tuzun, U. and Heyes, D.M. (2001) Granular dynamics simulations of heap formation: Effects of feed rate on segregation patterns in binary granular heap. J. Eng. Mech. **127**, 1000–1006
30. Smoller, J. (1994) *Shock Waves and Reaction-Diffusion Equations*, 2nd edition, Springer
31. Thompson, P.A., Grest, G.S. (1991) Granular flow: Friction and the dilatancy transition. Phys. Rev. Lett. **67**, 1751–1754
32. Wang, Y., Hutter, K. (2001) Granular material theories revisited. In *Geomorphological Fluid Mechanics* (eds. N.J. Balmforth, A. Provenzale), *Lecture Notes in Physics* **582**, Springer

Index

Printing (Computer to Plate): Saladruck Berlin
Binding: Stürtz AG, Würzburg